SO-AZK-427

THE AAAS
SCIENCE BOOK LIST

THIRD EDITION

A Selected and Annotated List of Science and Mathematics Books
for Secondary School Students, College Undergraduates and
Nonspecialists

Compiled by HILARY J. DEASON

WITHDRAWN

AMERICAN ASSOCIATION FOR THE ADVANCEMENT OF SCIENCE

Washington, D.C., 1970

LORETTE WILMOT LIBRARY
NAZARETH COLLEGE

AAAS Miscellaneous Publication 70–8

Copyright © 1970
American Association for the Advancement of Science
1515 Massachusetts Avenue, N.W., Washington, D. C. 20005

All rights reserved

166224

Library of Congress Catalog Card Number 74-105531
International Standard Book Number 0-87168-201-X

Printed in the U.S.A.
by
Geo. W. King Printing Co.
Baltimore, Maryland

CONTENTS

(Subjects listed according to Dewey Decimal Classification)

INTRODUCTION

The AAAS Science Book List is a guide to recreational and collateral reading and to basic reference works in the sciences and mathematics for junior and senior high school students, college undergraduates, and nonspecialist adults. The first edition (1959) and the second edition (1964), both out of print, have been widely used throughout the English-speaking world as acquisition guides for high school, preparatory school, public, and college undergraduate libraries to develop collections in the pure and applied sciences including mathematics, that will serve their readers effectively.

Libraries and Science Education—James B. Conant defined science as "that portion of accumulative knowledge in which new concepts are continuously developing from experiment and observation and lead to further experimentation and observation." (*On Understanding Science,* Yale, 1947, p. 98.) How well he forecast the spirit of revolution in the teaching of science and mathematics which is resulting in an overhauling of elementary, secondary, and college science course content, and in revisions in the training and certification of science and mathematics teachers!

No longer is it good practice to limit science instruction to a single textbook which the student accepts as immutable fact, or to a single laboratory manual in which the student repeats classic exercises yielding known results, labels outline diagrams, etc. Instead, the sciences and mathematics are being presented as dynamic, outgoing, open-ended processes. The instructional materials communicate to and involve the student in the excitement of bibliographic research, experimentation, testing, discovery, and rational evaluation—processes that have characterized the march of science and technology during the entire course of human progress.

New curricula for mathematics, biology, chemistry, earth sciences, space science, general science, oceanography, physics, and miscellaneous interdisciplinary subjects are being perfected and adopted at an accelerating rate in secondary schools and liberal arts colleges.

Collateral reading and reference work forms an integral part of these new curricula. Such collateral reading is considered so important that, where there are gaps in the published literature, special materials are being produced under the auspices of some curriculum study groups.

Evolution of AAAS Book Lists—The American Association for the Advancement of Science, founded in 1848 is the oldest general national scientific organization in North America. It has given rise to many of the specialized scientific, technical, and professional organizations in the United States. The organization presently has 130,000 individual members and 300 affiliated organizations and societies. Its objects stated at the time of its establishment and equally in effect today "are to further the work of scientists, to facilitate cooperation among them, to improve the effectiveness of science in the promotion of human welfare, and to increase public understanding and appreciation of the importance and promise of the methods of science in human progress." In 1941 the AAAS joined with several other scientific societies by establishing a Cooperative Committee on the Teaching of Science and Mathematics to work on science and mathematics educational problems. Thus it became involved in the general problems of science education and the shortages of competent scientific manpower.

The National Science Foundation, established in 1950, has provided financial support for the improvement of science education through the development of new teaching methods and curricula, improvement of facilities for science education, and for the training of science teachers. Recognizing that good school library collections are essential for science education, the American Association for the Advancement of Science at the request of and with the financial support of the National Science Foundation initiated a Traveling High School Library in 1955 which was conducted until 1962. The program consisted of a selected collection of 200 books in the sciences and mathematics available in multiple sets with a catalog for loan to a large number of schools on a rotating basis. A similar collection of 160 titles suitable for elementary schools was established in 1959 and loaned to elementary schools with central library collections until 1964. The general objectives of these traveling library programs were to encourage and stimulate the improvement of the quality and quantity of science and mathematics collections in school libraries; and thus provide opportunities for students to read science and mathematics books to broaden their science background; to provide materials for reference and to supplement and update material in classroom textbooks; to make available better resources for students working on special assignments and projects; and to assist students with appropriate interests and aptitudes in career choices.

In 1955, when the traveling library programs began, the various published bibliographies and catalogs used by school and public libraries for book selection were deficient both in the quality and quantity of the selections in the pure and applied sciences and mathematics. Hence the annotated catalogs of the traveling collections came into widespread use as supplements. In response to many requests for a more comprehensive list, *The AAAS Science Book List* was published in 1959 as a guide to recreational and collateral reading and to basic reference works in the sciences and mathematics for junior and senior high school students, college undergraduates, and nonspecialist adults. The annotated list of some 900 titles came into immediate use as an acquisition guide in both school and public libraries. In 1964 a second edition called *The AAAS Science Book List for Young Adults,* including annotations on 1,376 titles was published to meet the continuing demand. In less than two years the entire edition of 17,000 copies was exhausted.

Upon termination in 1964 of the financial support of the National Science Foundation for the traveling library programs and for subsidizing the publication of annotated book lists, the AAAS Board of Directors decided to continue certain phases of the program under its own auspices.

It had become apparent that book review media and miscellaneous professional periodicals were reviewing and evaluating an inadequate percentage of the current production of trade books in the sciences and mathematics, and most of them were not reviewing textbooks at all. It was also evident that many of the published reviews of science books were concerned chiefly with readability, style, and other qualities and often gave no reliable professional evaluation of the scientific and technical content. With the advice of the AAAS Committee on Publications and a selected group of specialist consultants plans were made for a new book review periodical, *Science Books: A Quarterly Review* which began publication in the spring of 1966. It reviews more than 200 new books in each issue. Several hundred professional reviewers representing the many scientific and technical disciplines prepare personal evaluations of the new books which provide the basis for the published annotations which often involve consultation of more than one specialist. In addition to indicating the level of difficulty reviews in *Science Books* provide a quality evaluation as a guide to book selection and acquisition, "highly recommended," "recommended," "acceptable," or "not recommended." During its first five years of publication *Science Books* has proved its usefulness as indicated by its circulation which is comparable to that of other widely used review media, and by its acceptance as a source of evaluation by publishers of book catalogs, bibliographies, and other educational materials.

Evaluation and selection—The process of compiling the second edition (1964) was reported as follows:

"The selection of this list of scientifically accurate, well-written, and useful books for collateral reading and reference began with a circular letter to book publishers inviting them to send trade books, handbooks, field manuals, biographies of scientists, reference works, and college textbooks considered suitable for secondary school students, college undergraduates, and nonprofessional adult readers. *Publishers' Weekly* (R. R. Bowker Co.), and book announcements and reviews in *Science, Scientific American, Natural History, Sky and Telescope, Geo-Times* (A.G.I.), *The Science Teacher* (N.S.T.A.), and other periodicals were examined regularly. Titles that appeared to be appropriate and that had not been sent voluntarily by the publishers were requested. Books in the 1959 list and the books that had been supplied by the publishers since that year were examined as a staff activity. Preliminary lists, one for each major discipline, were then prepared and sent out to the scientific and technical collaborators for review, recommendation as to inclusion or omission, for voluntary comments, for designation of comparative difficulty, and for suggested additional titles.

"The tabulation of the responses indicated clearly, in almost every instance, whether each title should be retained on the final list, or omitted (the editor made the decisions when the vote was close). The supplementary titles recommended by the respondents were obtained from the publishers and those that were added to the list were usually recommended by more than one advisor and supported by published favorable reviews."

To prepare this third edition of *The AAAS Science Book List** the titles in the second edition were re-evaluated, out-of-print titles with a few exceptions were omitted, out-of-date items were replaced by revised editions or other selections, and prices and other information on titles retained were corrected as needed. Substantial additions to the list are primarily selections from the "highly recommended" and "recommended" books that have been reviewed in *Science Books: A Quarterly Review* (April 1965–May 1970) and are of appropriate levels of difficulty. Additional selections from various sources were included as recommended by various reviewers and consultants.

In the first and second editions single and double asterisks were used to designate titles of greatest usefulness, and hence priority acquisitions, for libraries. By now we believe that librarians are better informed concerning the essential elements of a good science collection than they were in 1959 when we published the first edition. The larger third edition is intended as a guide to building a library collection that will serve the needs of students and others, and offers suggestions for collateral reading and book acquisition that will meet special needs in various localities and accommodate specifics of different programs and curricula in the schools and communities.

---

* In the third edition the words "for young adults" have been dropped from the title because many users felt that the connotation was too restrictive and did not indicate the actual scope of the list. Since the counterpart children's list (grades K-8) is so designated, there should be no confusion.

Scope of the third edition: The 2,441 titles in the third edition are distributed among various subject categories as follows:

030 Encyclopedias, etc., 11	572 Anthropology, 44
150 Psychology, 70	574 Biology, 260
333 Conservation, 17	580 Botany, 98
390 Cultural anthropology, 27	590 Zoology, 314
410 Linguistics, 8	610 Medical sciences, 145
500 Science in general, 157	620 Engineering sciences, 185
510 Mathematics, 229	630 Agricultural sciences, 105
520 Astronomy, 142	660 Industrial technology, 48
530 Physics, 147	720 Urban planning and
540 Chemistry, 90	architecture, 42
549 Mineralogy, 17	770 Photography, 14
550 Geological sciences, 151	910 Geography and exploration, 44
560 Paleontology, 22	913 Archeology, 54

The seventeenth edition of the Dewey Decimal Classification has been followed in organizing this annotated bibliography with one major exception. Biographies are included in the various subject categories instead of in a single place under "920 Collective Biography." This procedure encourages browsing in biographies at the same time the reader is reading for detail or depth in the specifics of the disciplines.

The broader scope of the third edition reflects the increasingly interdisciplinary nature of science education and popular interest in the sciences. Desirable as it might be to group together books that are related to the environmental science, which are of so much current interest and concern, to do so would vitiate the organization of the bibliography for other purposes. The environmental sciences involve many subjects including, but not limited to conservation, sociology, oceanography, ecology, natural history, physiology, public health, and engineering. Oceanography, another popular topic, has geological, geographical, meteorological, biological, and engineering aspects.

Because of the ever-expanding frontiers of knowledge, a continuous acquisition and replacement program is necessary to maintain an adequate and serviceable collection in any library. At the same time it is necessary to recognize and retain those "classics" in the literature of science that never will be out of date.

Paperbound editions—Books in this annotated bibliography are primarily trade, text or library editions in hard covers. In a few instances softcover titles have been listed because they are considered important acquisitions and are available in no other edition. However, if softcover editions of hardback editions are also available the citation and price have been included. The tremendous number of paperbounds, the difficulty of maintaining complete stocks in retail establishments, and the short life of many editions have made the compilation and publication of annotated lists of science paperbounds a somewhat frustrating experience. The latest such bibliography prepared under AAAS auspices is *A Guide to Science Reading,* edited by Hilary J. Deason (New American Library, Signet T3003, 1966, 75¢) which is still in print, but out of date. There is no plan to revise and update it at present. The best source of information on paperbounds currently available is the composite guide, *Paperbound Books in Print* (R. R. Bowker Co.), published monthly with three cumulative issues a year. The cumulative issues contain listings by author, by subject, and by title.

Periodicals—Reference work in the sciences frequently involves research in current and back issues of periodicals. Since the needs of libraries for periodicals and the interests of their users vary so widely, no attempt has been made to include a list of periodicals in this bibliography. An excellent annotated guide to periodicals is now available and it is highly recommended for all libraries: *Magazines for Libraries: For the general reader and public school, junior college, and college libraries,* compiled by Bill Katz and Berry Gargal (R. R. Bowker Co., 1969, 409 pp. $16.95. 69-19208).

Acknowledgments—The second edition's acknowledgments listed a large number of individuals and organizations who had collaborated and offered suggestions. Our thanks to them again is conveyed inasmuch as their work on the second edition provided the foundation for the third. Many of those who assisted became regular reviewers for the periodical *Science Books* and their number has been increased greatly as this publication grew in size and scope. Without naming them individually, which would occupy considerable space, we extend our sincere thanks for their advice and continuing collaboration. (The names of reviewers are listed in each issue of *Science Books.*)

We again acknowledge our indebtedness to the book publishers and their representatives who have cooperated most generously by providing books for review and evaluation, and for their response to our many inquiries. The books annotated for this book list are, for the most part, incorporated in the AAAS library where they serve the entire staff, as well as persons from other organizations and educational institutions interested in scientific and technical literature.

Special thanks are due to Charles D. Maurer, Kent State University, who served as the principal editorial assistant during the long and tedious job of revision and compilation, to Mrs. Patsy Walters who typed most of the manuscript and to Mrs. Joan C. Taylor who completed the manuscript and was responsible for the proofreading and indexing. Miss Sarah Jane Macauley and Miss Ann Thomas, college students, assisted with the indexes.

030 ENCYCLOPEDIAS AND GENERAL WORKS (See also 503 SCIENCE DICTIONARIES, ENCYCLOPEDIAS)

The major encyclopedias and standard dictionaries provide the foundation for a good reference library in the pure and applied sciences. Listed in this section are major works which the editor and his consultants consider to be of superior quality and maximum usefulness in school, college, and public libraries. (Those who desire a complete evaluation of all encyclopedias should consult *General Encyclopedias in Print, 1969; a comparative analysis,* by S. Padraig Walsh; R. R. Bowker Co., 1969, 96 pp. $3.00.)

The American Heritage Dictionary of the English Language. William Morris (Editor). American Heritage and Houghton Mifflin, 1969. 1550 pp. illus. $12.50 (discount to schools and libraries). 76–86995. (JH–SH–C)

A completely new dictionary that became a "non-fiction best-seller" shortly after publication and won immediate acceptance by thousands as a "personal dictionary." It has copious marginal illustrations, up-to-date coverage of scientific and technical terms, notes on synonyms, and word histories. Excellent for students to use at home.

Collier's Encyclopedia. Louis Shores (Editor-in-chief). Crowell-Collier, 1969. 24 vols. illus. ports. maps. $329.50 (discount to schools and libraries). 69–10453. (JH–SH–C)

One of the major general encyclopedias with good coverage in science and technology. There are substantial annual revisions, and a very good yearbook (available to set owners at a reduced price) provides a good resume of new developments and noteworthy events. It is popular and extensively used in high school libraries.

The Encyclopedia Americana. Bernard S. Cayne (Editor-in-chief). Americana, 1969. 30 vols. illus. ports. maps. $375 (discount to schools and libraries). 72–97500. (JH–SH–C)

One of the major general encyclopedias with excellent science and technology. It has six more volumes than the other two major sets. Each article has cross-references and Vol. 30 is a detailed analytical index. Major annual revisions keep the set current and *The American Annual,* an alphabetically arranged supplement, is available to set owners at a discount.

Encyclopedia International. Stanley Schindler (Editor-in-chief). Grolier, 1970. 20 vols. illus. ports. maps. $275 (discount to schools and libraries). 77–88108. (JH–SH)

A new general encyclopedia (first published 1963–64) which is at an intermediate level of difficulty. It has good science and technology coverage in concise articles. There are copious internal cross-references, and reading and study guides are provided with major articles. Good colored maps throughout. It is updated by *Encyclopedia Yearbook,* sold to set owners at a reduced price.

Compton's Encyclopedia and Fact-index. Donald E. Lawson (Editor-in-chief). Compton, 1969. 24 vols. illus. ports. maps. $134 to schools and libraries. 69–10041. (JH)

A survey of knowledge designed for students from the middle grades onward. It is very useful for junior high and some senior high students who need an "easy reference set." Each volume has a text with a fact-index in the back that also provides references to material in other volumes; alphabetically arranged. The recent complete revision and expansion from 15 to 24 volumes has greatly improved the coverage. There is a continuous revision program, and a yearbook is available to set owners at a reduced price.

Encyclopedia Britannica. Warren E. Preece (Editor-in-chief). Britannica, 1969. 24 vols. illus. ports. maps. $459.50 (discount to schools and libraries). 69–10039. (SH–C)

This is the classic and oldest general encyclopedia which is preferred by many scholars. Although now thoroughly American, it retains much of its original British character. An accelerated revision policy during recent years has resulted in improved currency and adequacy of scientific and technical material. Two important features are the index volume which is very complete and contains liberal cross references, and a special atlas of 222 colored maps prepared by Rand McNally. *Britannica Books of the Year* and the new *Britannica Yearbook of Science and the Future* (first published in 1969) are valuable supplements.

New Book of Knowledge: The Children's Encyclopedia. Martha Glauber Shapp (Editor-in-chief). Grolier, 1969, 20 vols. 9442 pp. illus. $200 (discount to schools and libraries). 70–88103. (JH)

Primarily intended for students in grades 3 and up, this set was entirely new when first published in 1966, and is revised annually. At its inception it was strongly oriented toward the sciences and mathematics and the new developments in curricula. It has real value for junior high school students, particularly for those who find major encyclopedias too difficult or lacking in appeal. These are colorfully illustrated books for study and browsing in depth, not for "quick" definitions. Suggestions for science projects and activities are included.

The Random House Dictionary of the English Language (Unabridged). Jess Stein (Editor-in-chief). Random House, 1969. 2059 pp. illus. $25 (discount to schools and libraries). (JH–SH–C)

Completely new when first published in 1966, and revised and reprinted annually, this is the first major work of this kind to be produced with the aid of electronic data processing. In consequence the total time of compilation was greatly shortened and the published work was absolutely current in all respects. An atlas and gazeteer are included. It is especially good for scientific and technical definitions, and has references to various other works in the sciences. It is a thoroughly practical student dictionary, and ideal for non-specialist adults. (A teacher's guide for giving instruction in reference work is available.)

Webster's Third New International Dictionary of the English Language (Unabridged). Philip G. Gove (Editor-in-chief). Merriam, 1966 [c. 1961, 1966.] 2770 pp. illus. $47.50 (discount to schools and libraries). (SH–C)

The third edition (1961) of the most widely used and the oldest unabridged dictionary has restored, in the 1966 printing, some of the omitted but desirable features of the second edition. It has many appendices not found in other unabridged dictionaries, and exceeds all others in total pagination and number of entries.

The World Book Dictionary. Clarence L. Barnhart (Editor-in-chief). Field
Enterprises, 1970. 2 vols. 2415 pp. illus. $50.20 (discount to schools and
libraries). (JH–SH–C)
A first-class dictionary developed and published especially for schoolrooms
and libraries which is revised annually and has many unique features. These
include a graduated vocabulary development inventory up through college
undergraduate level, vocabulary exercises, a section on language and spelling
rules, and a handbook of style. It omits biographical and geographical entries
for which students should consult encyclopedias. It is an ideal teaching
dictionary and has many valuable features for self-study.

The World Book Encyclopedia. Robert O. Zeleny (Editor). Field Enterprises,
1970. 20 vols. illus. ports. maps. $179.80, $189.80, and $199.80 for various
bindings (discount to schools and libraries). (JH–SH)
No other multivolume reference set is so widely sold and used in schools,
libraries, and homes. It is strictly alphabetical, word-by-word, and hence
with the cross-references listed in proper alphabetical sequence, it is self
indexed. Also included are many special features such as transparencies,
maps, four-color art reproductions, suggestions for science projects, and
career guidance information. The percentage of annual revisions is very
large. Many professional people like *World Book* as a quick first reference
even though they own one of the three major sets. *World Book Annual* and
also *World Book Science Annual* are available at reduced cost to set owners
and are valuable sources of new information.

150 PSYCHOLOGY

Berelson, Bernard, and Gary A. Steiner. *Human Behavior: An Inventory of
Scientific Findings.* Harcourt, Brace, 1964. xxiii+712 pp. illus. $13.75 (text
ed. $10.50); abridged ed., 1967, $4.75; paper $2.40; simplified ed., $1.80.
64–11621. (C)
Summarizes what social scientists believe they have substantiated about the
way human beings behave. There are 1045 findings listed in almost every
field of human activity—among them, politics, sex and marriage, race, religion, class and status, communication and public opinion—which are supported by accounts of research by scholars in psychology, anthropology,
sociology and related fields.

Brome, Vincent. *Freud and His Early Circle.* Morrow, 1968. xii+275 pp. $5.00.
68–14813. (C)
A somewhat unique approach to the life of Sigmund Freud, this account is
remarkably dispassionate, well written, and engaging. The fascination with
the curious mixture of bizarre, fanatical, and sometimes tragic personalities
that were woven into Freud's professional life is portrayed. Little attention is
given to psychoanalytic detail.

Brown, J. Marshall, F. K. Berrien, and David L. Russell, with W. D. Wells.
Applied Psychology. Macmillan, 1966. xiii+639 pp. illus. $8.95. 66–19710.
(SH–C)
A novel textbook for an introductory and terminal course at the junior college
level. Some chapters are good reference materials for high school students,

such as those dealing with mental health, crime, gerontology, and consumer psychology. The chapters are relatively independent and can be presented in almost any order desired. Four major divisions deal with the applications of psychology: (1) to adjustment; (2) to industry; (3) to consumer and political issues; and (4) to crime. Good bibliography.

Candland, Douglas K., and James F. Campbell. *Exploring Behavior: An Introduction to Psychology.* Basic, 1961. xii+179 pp. diagrs. $5.95; Fawcett (Premier R268), paper 60¢. 61–16956. (SH)
A simple and clear introduction to the basic principles of modern experimental psychology. The application of the scientific method in psychology is discussed with numerous examples of how it may be applied in the reader's everyday experience.

Carmichael, Leonard. *Basic Psychology.* Random, 1957. 340 pp. $5.95. 56–8883. (SH–C)
The psychology of the "normal, adult, civilized person of our time" written for the general reader.

Comfort, Alex. *The Nature of Human Nature.* Harper & Row, 1966. 222 pp. illus. $5.00. 67–15973. (SH–C)
This lyric essay aims at a description of man not only as a biological and psychological being, but also as a social organism. Encompasses geographic, biologic and social ecology. Man is compared and contrasted with the various species of animal behavior to emphasize the uniqueness of human behavior. Drawing upon psychoanalytic theory, the author presents the irrational and unconscious attributes of man as the ignored but ultimately redeeming dimension of humanity. Provocative and controversial collateral reading for advanced high school and college undergraduates.

Deese, James Earle. *Principles of Psychology.* Allyn and Bacon, 1964. xi+489 pp. illus. $10.60 (text ed. $7.95); student guide, paper $3.25. 64–11565. (SH)
A thorough and outstanding introduction to psychology. The first section covers basic mental processes, including an intriguing chapter on language and thought; part two includes human development, personality, adjustment, intelligence, and statistical methods; the last part deals with the relation of man to his society. The style is direct and perceptive, the book a well-written, soundly organized, up-to-date statement of the field. Extensive glossary, reference and index sections.

Engle, T. L., and L. Snellgrove. *Psychology: Its Principles and Applications.* (5th ed.). Harcourt, Brace, 1969. xi+612 pp. illus. $4.32 (school price) (SH–C)
The book has seven units including "The Science of psychology," "Mental health," "The Family and other small groups," and "You and society," which contains some excellent information on career guidance in general and on psychology in particular. Activities suggested at the end of each chapter are mainly open-ended inquiries that will provide the student with some elementary facility in the collection, tabulation, and analysis of experimental data. References and suggestions for further reading are listed at the end of each chapter. Intended for secondary schools, but useful in a college course in lieu of too-technical texts.

Galanter, Eugene. *Textbook of Elementary Psychology.* Holden-Day, 1966. xiii+419 pp. illus. $8.50. 66–17895. (C)
Covers those areas of psychology (learning, language structure and acquisition, and perception) particularly amenable to the application of mathematical models. Since the book adheres so strictly to the "mathematical models" point of view, many important areas of psychology are ignored. But the material is an excellent collateral source for a general introductory course in experimental psychology.

Haas, Kurt. *Understanding Ourselves and Others.* Prentice-Hall, 1965. xi+370 pp. illus. $8.95. 65–12483. (SH–C)
A survey of the major psychological aspects of human development, personality, abnormal processes, techniques of treatment, and mental health. It is not intended for the detailed introductory psychology courses, but is an excellent supplementary text for first-year psychology courses in personal adjustment and mental health. High school teachers can use it as reference in courses on human relations or "problems in maturing." Thought-provoking questions conclude each chapter. A list of references and an excellent index are appended.

Hilgard, Ernest R., and Richard C. Atkinson. *Introduction to Psychology* (4th ed.). Harcourt, Brace, 1967. xv+686 pp. illus. $8.95. 67–12526. (C)
An excellent introductory psychology textbook intended for the student in a first-year course, this includes many features not found in contemporary works. The treatment is complete and comprehensive, yet not too involved as to discourage the introductory student. Attractive features include a multitude of tables, photographs, and line drawings. Within each of the 24 chapters there are a series of critical discussions dealing with contemporary topics in psychology. This text will stimulate the motivated student.

Hyde, Margaret O., and Edward S. Marks. *Psychology in Action.* McGraw-Hill, 1967. 153 pp. illus. $4.50. 67–10365. (JH–SH)
A balanced and intelligent survey intended to introduce the diversity and application of psychology. Emphasis is on the practical aspects of work in various fields in psychology: clinical, social, counseling, human engineering, and learning theory.

Hyman, Ray. *The Nature of Psychological Inquiry.* (Foundations of Modern Psychology Series.) Prentice-Hall, 1964. xi+116 pp. $4.50, paper $1.95. 64–21828. (C)
Discusses psychological research and how it is carried out, with frequent references to important experiments from the recent history of psychology. All phases of research work are covered, from getting ideas for research projects, through collecting, processing, and interpreting the data, to communicating the results. Stresses many problems in research, among them the use of statistical methods, the use of theories in experiments, and the concept of "control" in scientific observation.

Jones, Ernest. *The Life and Work of Sigmund Freud.* (Abridged ed. by Lionel Trilling and Steven Marcus.) Basic, 1961. xxv+541 pp. illus. $10.00; Doubleday, paper $2.95. 61–15950. (C)
A classic in biography as well as a valuable source for understanding the

development of psychoanalytic psychology, this warm, readable book covers Freud's childhood and adolescence, his long engagement and marriage, early experiments with hypnotism and cocaine, his own self-analysis, battles against distortion of his theories and personal slander, disputes and splits with his colleagues, and the long, painful illness that led to his death in 1939.

Jung, C. G. (ed.). *Man and his Symbols.* Doubleday, 1964. 320 pp. illus. $14.95; Dell, 1968, paper $1.25. 64–18631. (SH–C)

A popular illustrated portrayal of Jung's greatest contribution to psychology—symbolism, particularly as revealed through dreams—in five essays planned by Jung, who wrote the first: "Approaching the unconscious." The others, written by his colleagues are "Ancient myths and modern man," "The process of individuation," "Symbolism in the visual arts," and "Symbolism in an individual analysis." Explanatory notes and an index are included. Absorbing reading for the nonprofessional.

Kalish, Richard A. *The Psychology of Human Behavior.* Wadsworth, 1966. viii+529 pp. illus. $7.50. 66–25626. (SH–C)

A conceptual rather than empirical textbook approach to basic psychology for the non-psychology major and the interested high school student. Emphasis is on mental hygiene, personality, and social behavior; reported through case histories, literature excerpts, and author experiences.

Klagsburn, Francine. *Sigmund Freud.* (Immortals of Science Series.) Watts, 1967. ix+150 pp. $3.95; LB $2.96. 67–10229. (JH–SH)

An interesting, though somewhat idealized, introduction to Freud: his life, his works, and his humanity. Good collateral reading for all students. A complete listing of works by and about Freud is included for those interested.

Lindgren, Henry Clay, Donn Byrne, and L. Petrinovich. *Psychology: An Introduction to the Study of Human Behavior.* Wiley, 1966. ix+560 pp. illus. $8.50. (SH)

A well-written, cleverly illustrated, provocative text. Deals with human development, principles of behavior, the problem of individual differences, emotional and social factors in human behavior, and concludes with a section on the use of psychology in business and in world affairs.

Marx, Melvin H., and William A. Hillix. *Systems and Theories in Psychology.* McGraw-Hill, 1963. xvii+489 pp. $9.50. 62–21803. (C)

Surveys systems and schools in psychology, including Structuralism, Functionalism, Associationism, Behaviorism, Gestalt psychology, and Psychoanalysis. Also reviews contemporary varieties of stimulus-response, field, and personality theories. For the advanced student with a solid background in general psychology.

Miller, George A. *Psychology: The Science of Mental Life.* Harper & Row, 1962. $6.25. 62–9894. (C)

With William James' definition of psychology as his subtitle and theme, Miller alternates chapters on noted figures from the history of psychology with accounts of contemporary problems related to their work. Covers Wilhelm Wundt's introspective psychology; the pragmatic, philosophical psychology of James; Pavlov and conditioning; Freud's psychoanalysis; Binet and intelligence testing; and a discussion of two large modern problems: clinical versus statistical procedures, and the psychology of communication and persuasion.

Murphy, Gardner, and Lois B. Murhpy(eds.). *Western Psychology from the Greeks to William James.* Basic, 1969. x+269 pp. $8.50. 72–78454. (C)
This is a collection of some of the major contributions to the history of philosophical psychology. Gives the reader a first hand panorama of the history of philosophical influences on modern psychology.

Ruch, Floyd L. *Psychology in Life* (7th ed.). Scott, 1967. 490 pp. illus. $6.75, paper $6.50. 63–21139. (C)
Clear, comprehensive, and creatively designed, this is an excellent introduction to psychology. (An abridged version of a more comprehensive work of the same title.) Deals with many areas and points of view, from the early schools of modern psychology to recent advances in learning theory, drugs, mass communication and brainwashing, perception, personality problems and group behavior.

Skinner, B. F. *Science and Human Behavior.* MacMillan, 1963 x+461 pp. $5.95, paper $2.95. 53–7045. (SH)
Convincing and lucid arguments for a behavioristic approach to the study of man by the controversial psychologist who raised his second child in an air-conditioned mechanical box. For Skinner, the way an individual behaves is not the result of internal will, but of certain specifiable conditions that can be discovered and which will lead to our ability largely to predict his actions.

Watson, Robert I. *Great Psychologists: From Aristotle to Freud* (2nd ed.). Lippincott, 1968. xiv+613 pp. $10.00. 68–15731. (SH–C)
A history of psychology based on the lives and personalities of its major contributors. The first half covers the Greeks, medieval theologians, and later philosophers; the second deals with the development of modern psychology as a science, including contributions of James, Freud, and Jung, as well as more functional approaches, Behaviorism, and Gestalt psychology. Detailed bibliography.

152 PSYCHOLOGY—PHYSIOLOGICAL AND EXPERIMENTAL

Crafts, L. W., T. C. Schneirla, E. E. Robinson, and R. W. Gilbert. *Recent Experiments in Psychology* (2nd ed.). McGraw-Hill, 1952. xvii+503 pp. $6.50, paper $4.95. 50–9141. (C)
Each of the 24 chapters deals with experiments in a particular area of investigation. Some of the topics are: "A study of attitudes toward mysticism among college students," "The origin of the cat's responses to rats and mice," "A comparison of the intelligence of 'racial' and national groups in Europe." The investigations are diversified and all involve a rigorous application of the scientific method. A valuable book for students with some psychological background.

Cronbach, Lee J. *Essentials of Psychological Testing* (2nd ed.). Harper & Row, 1960. xxi+650 pp. illus. $9.75. 59–13919. (C)
The first part of this comprehensive text deals with basic concepts, purposes, administration, scoring, and validity of tests. The rest is devoted to the

description and theory of tests in particular areas and their functions—general
ability tests, aptitude, special ability, proficiency and performance, interest
inventories, personality measurements, and personality dynamics. Parts of the
text are technical and it assumes a general knowledge of psychology. A valuable reference work for every college library.

Garrett, Henry E. *Great Experiments in Psychology* (3rd ed.). Appleton-Century-Crofts, 1951. xvii+358 pp. illus. $5.75. 51–11732. (SH)
Accounts of famous experiments carried out by Pavlov, Thorndike, Kohler,
Binet, and other historical figures in psychology. Although old, it is a classic
for supplementary reading. The techniques and drawings may give students
ideas for their own experiments.

Tyler, Leona E. *Tests and Measurements.* (Foundations of Modern Psychology
Series.) Prentice-Hall, 1963. xi+116 pp. illus. $4.50, paper $1.95. 63–7441.
(C)
Based on the concept that dealing with quantities and not qualities alone
differentiates modern scientific psychology from other attempts to understand
human behavior. Describes basic statistical methods, various types of psychological tests, including those of intelligence, personality, and special ability,
and such current topics as factor analysis and projective tests.

152.5 PSYCHOLOGY—MOTIVATION

Bindra, Dalbir, and Jane Stewart (eds.). *Motivation.* Penquin, 1966. 352 pp.
illus. $1.95 (paper). (SH-C)
A collection that provides balance between speculative discussion and experimental results to introduce general background of and empirical research in
motivation. Editorial comment is minimal and excess verbiage has been
deleted. Those dissatisfied with the excerpts are directed to the originals.
Emphasis is somewhat selective.

Haber, Ralph Norman (ed.). *Current Research in Motivation.* Holt, Rinehart
and Winston, 1966. xii+800 pp. illus. $13.95. 66–10205. (C)
The format follows the historical development of motivation research and
presents an overview of the diversity of problems and approaches in motivation. This work will serve as a good reference manual and as collateral
reading.

153 PSYCHOLOGY—INTELLIGENCE

Barnett, S. A. *Instinct and Intelligence: Behavior of Animals and Man.* (Prentice-Hall Series in Nature and Natural History.) (Illus. by Stanley Wyatt.)
Prentice-Hall, 1967. 209 pp. $7.95. 67–15163. (C)
A good introduction to some of the facts and points of view of current ethological approaches to the study of behavior. Emphasis is on innate response
patterns, but includes intelligence and learning. Useful collateral reading.

Festinger, Leon, Henry W. Riecken, and Stanley Schachter. *When Prophecy
Fails.* U. of Minnesota, 1956. vii+256 pp. $4.00; Harper & Row (Torch-
books TB/1132), paper $1.60. 56–11611. (C)
A group that predicted a monstrous flood that would end the world was
observed for several months, unknown to them, by a team of behavioral

scientists who took part in their activities. The result is a research project that gives fascinating insight into the problems of studying and interpreting human behavior, and into the difficult and sometimes humorous trials of the participant-observation method.

Gagne, Robert M. (ed.). *Learning and Individual Differences.* Merrill, 1967. xv+265 pp. illus. $8.95. 67–10538. (SH–C)
A collection of papers dealing with sophisticated issues, new and old, by well-known theoreticians and researchers which require close and careful attention from the reader.

Ginsburg, Herbert, and Sylvia Opper. *Piaget's Theory of Intellectual Development: An Introduction.* (The Prentice-Hall Series in Developmental Psychology.) Prentice-Hall, 1969. xi+237 pp. illus. $6.95; $2.95 (paper). 76–76312.
This is an excellent integration of Piaget's theory and supportive example geared to the advanced college undergraduate. Although purposefully selective the material covered is basic to an understanding of Piaget and is explained in depth. Can be used as either a text or a source book on Piaget.

Mednick, Sarnoff A. *Learning.* (Foundations of Modern Psychology Series.) Prentice-Hall, 1964. x+118 pp. illus. $4.50, paper $1.95. 64–10869. (C)
The theory and facts of learning are presented in terms of supporting research. Tells how learning is studied and introduces the basic vocabulary of the field, then describes classical and operant conditioning, the development of complex habits, training transfer, motivation, and remembering and forgetting. A clear and logical development of the topic.

153.7 PSYCHOLOGY—PERCEPTION

Droscher, Vitus B. *The Magic of the Senses: New Discoveries in Animal Perception.* (Tr. by Ursula Lehrburger and Oliver Coburn.) Dutton, 1969. 298 pp. illus. $8.95. 69–13340. (SH–C)
Covers a vast array of sensory mechanisms and is supplemented with numerous drawings and diagrams. This semipopular introduction to sensory physiology treats both American and European literature.

Hansel, C. E. M. *ESP: A Scientific Evaluation.* Scribner's, 1966. xxi+263 pp. illus. $6.95; paper, $2.45. 66–15979. (SH–C)
In lucid and simple language the author describes ESP, how it is examined, and the obvious experimental error in research designs concerning it. He evaluates what are usually presented as the crucial experiments in extrasensory perception, showing the many possibilities for conscious or unconscious bias and, in some instances, actual cheating. The book serves as a good illustration of the need for rigid scientific control and tight experimentation.

154.63 PSYCHOLOGY—DREAMS

Diamond, Edwin. *The Science of Dreams.* Doubleday, 1963. 264 pp. $4.50; Macfadden-Bartell, paper 60¢. 61–12512. (SH)
Did you know that we all dream 5 or 6 times a night and that the average dream lasts 20 minutes? Have you heard about the doctor who suspended himself in lukewarm water for several hours to investigate the hallucinations

reported by some jet pilots and truck drivers? This highly readable book reports on modern investigations of dreams and what these common but misunderstood phenomena mean.

Freud, Sigmund. *The Interpretation of Dreams*. (Tr. by James Stachey.) Basic, 1955. xxxii+736 pp. $10.00; Avon, paper $1.65. 55–10432. (SH–C)
This important work is not only a classic in psychoanalytic history, but is also the basis for contemporary thinking and the impetus for many current theories. Freud's coverage is all-encompassing: dream-work, psychology of dream-processes, dream interpretation, and material and sources of dreams. Thorough bibliographies and appendices supplement the text.

155 PSYCHOLOGY—DIFFERENTIAL AND GENETIC

Bonner, Hubert. *Psychology of Personality*. Ronald, 1961. vi+534 pp. $7.00. 61–5656. (C)
Presents various points of view and typologies concerning personality and covers its formation, organization, and dynamics. Includes such factors as organic foundations, human groups, culture, language, motivation, the attitude-value complex, character and moral behavior. For the student who has had basic instruction.

Harris, Dale B. (ed.). *The Concept of Development: An Issue in the Study of Human Behavior*. U. of Minnesota, 1967. x+287 pp. $6.75, paper $2.75. (C)
Essays from 17 contributors representing psychology, philosophy, natural sciences, medical sciences, and the humanities which discuss and evaluate developmental concepts from the viewpoints of the several disciplines.

Lazarus, Richard S. *Personality and Adjustment*. (Foundations of Modern Psychology.) Prentice-Hall, 1963. x+118 pp. illus. $4.50, paper $1.95. 63–7440. (C)
Presents the main issues in this often-confusing area of psychology, and the ways they have been resolved. Deals with the nature of "adjustment" and its relation to conflict and mental health, and the basic aspects of the study of personality, including its fundamental qualities and structures, development, assessment, and theoretical framework.

Lidz, Theodore. *The Person: His Development Throughout the Life Cycle*. Basic, 1968. xxii+574 pp. $10.00. 68–9443. (C)
Traces the development of man as a person from the earliest days of infancy to the termination of his life. It is a synthesis of man's growth as a human being, designed particularly for medical students who need to know their patients as complex individuals rather than mere biological organisms. It is worthwhile reading for all mature readers, and indispensable background for anyone who works with other people.

Michael, Donald N. *The Next Generation: The Prospects Ahead for the Youth of Today and Tomorrow*. Random, 1965. xxvi+218 pp. $4.95; paper (Vintage V273) $1.65. 64–20020. (SH–C)
A thoughtful and provocative book that raises important questions and problems concerning education, employment, family life, social adjustment, and international relations in a world that will inevitably become more heavily

populated, much more sophisticated, and increasingly automated. It demonstrates convincingly that a laissez-faire attitude and the complacent preservation of the *status quo* will not work. This outstanding work will appeal to all serious readers. Educators and guidance personnel should read and digest it.

155.9 SOCIAL PSYCHOLOGY

Bacon, M., and M. B. Jones. *Teen-Age Drinking.* Crowell, 1968. 228 pp. $5.95. 67–13590. (SH–C)
The authors have combed the available literature, both technical and popular, to present a reasonable sociological overview of teenage drinking. They have found many interesting details that should come as reassurance to parents—the finding, for example, that few teens drink anything stronger than beer, and that young people have very strong taboos about drinking and driving. This should be read by guidance counselors, parents, and teenagers themselves.

Bohannan, P. *Love, Sex, and Being Human.* Doubleday, 1969, xiv+144 pp. illus. $3.95. 68–54238. (JH–SH)
This lucid discussion of the biology and morality of sex is addressed mainly to teen-agers, and while it takes up the biological and moral issues separately, the continuity is excellent. Stresses moral responsibility and moral decision without taking a definite stand. Highly recommended as a springboard for discussion by all age groups.

Lambert, W. W., and W .E. Lambert. *Social Psychology.* (Foundations of Modern Psychology Series.) Prentice-Hall, 1964. viii+120 pp. $4.50, paper $1.95. 64–10868. (C)
An introduction to the branch of psychology that experimentally studies individuals in their social and cultural settings. Deals with the socialization process, perceiving and judging social events, social attitudes and interaction, group settings, and the relations of culture to social psychology.

Lindemann, Bard. *The Twins Who Found Each Other.* (Intro. by Amram Scheinfeld) Morrow, 1969. 288 pp. illus. $6.95. 76–85876. (C)
This is the story of a pair of identical twins who were separated at birth, grew up in radically different environments, and who were reunited at age twenty-four. To the trained psychologist or sociologist, this represents a detailed case history, to the casual reader it is an intensely fascinating story of human behavior.

Munn, Norman L. *The Evolution and Growth of Human Behavior* (2nd ed.). Houghton Mifflin, 1965. x+525 pp. illus. $8.95. 55–14268. (SH–C)
Developmental psychology is interested in the origin and growth of psychological processes in species and in individual organisms, including man. This text is a good introduction to the biological aspect of psychology and, although technical, it will be rewarding for the student who is well versed in biology and the scientific method.

Mussen, Paul H. *The Psychological Development of the Child.* (Foundations of Modern Psychology Series.) Prentice-Hall, 1963. xv+109 pp. illus. $4.50, paper $1.95. 63–18023. (C)
A well-written account of modern child psychology explaining psychological

development from birth to adulthood. Discusses the general principles and biological foundations of development, then deals specifically with the development of higher mental processes such as perception and thinking, of personality, and of social behavior.

Noshpitz, Joseph D. *Understanding Ourselves: The Challenge of the Human Mind.* (Challenge Books.) Coward-McCann, 1962. 114 pp. illus. $2.60. 63–15551. (JH)

A forceful explanation of modern depth psychology at the junior-senior high level, with an integrity of style refreshing to the reader. Discusses some of the great psychologists of the 20th century and their contributions; the psychological development of the child and adolescent; the mechanisms that control our feelings; and suggestions for dealing with such teen-age problems as college entrance, dating, and careers.

Offer, D. *The Psychological World of the Teen-Ager: A Study of Normal Adolescent Boys.* Basic, 1969. xiv+286 pp. $7.95. 73–78465. (C)

This in-depth research study presents numerous insights on normality. The study relies on a variety of techniques: interviews, questionnaires, and projective tests, and therefore some sophistication is needed to interpret the findings. Areas such as the importance of parents, athletic coaches and teachers; peer relations; emotions; self insight; and reaction to the many traumas of adolescence are reported.

Richardson, Stephen A., and Alan F. Guttmacher (eds.). *Childbearing—Its Social and Psychological Aspects.* Williams & Wilkins, 1967. xvii+334 pp. $8.50. 67–18156. (SH–C)

Considers effects on childbearing that arise from psychological, physiological, sociological, and cultural sources. Anyone, from the intelligent layman to the most advanced professional, should enjoy this lucid presentation.

Rogers, Dorothy. *Issues in Adolescent Psychology.* Merideth, 1969. xv+614 pp. $5.50. 69–12326. (C)

This is an accumulation of 46 articles covering 6 general areas of adolescent psychology: general issues; stage theory vs. adolescence; self-actualization; peer relations; developing values and social consciousness; and cultural variants. The articles reported deal with contemporary issues and are preceded by a brief commentary and preview.

Splaver, Sarah. *Your Personality and You.* Messner, 1965. 190 pp. $3.95; LB $3.64. 65–22251. (SH)

The book recognizes the problems of adolescents and attacks these problems from the point of view of a well-trained and experienced psychologist. There is discussion of topics such as grief, anger, fear, love, jealousy, defense mechanisms, escape mechanisms, neurotic behavior, cheating, popularity, dating, communications with parents and teachers, and vocational guidance. There are references to similar writings for adolescents and to a series of problem playlets for presentation. *The Socio-Guidrama Series.* Not a "do-it-yourself" course in psychotherapy. High school students will enjoy reading this book and discussing it with adult leaders and with each other.

Schulz, D. A. *Coming up Black: Patterns of Ghetto Socialization.* Prentice-Hall, 1969. xiv+209 pp. $5.95. 69–15340. (SH–C)

The Preface discloses: "This study attempts to overcome the middle-class myopia often experienced in looking by paying close attention to the process of growing up (coming up) in the ghetto of a mid-western city. The problems faced by children struggling to reach maturity here seem not too far different from those faced by other children in New York, Washington, or Los Angeles." The author sheds light on conditions of ghetto living and poses problems that must be solved. Excellent index.

156 COMPARATIVE PSYCHOLOGY

Burtt, Harold E. *The Psychology of Birds: An Interpretation of Bird Behavior.* Macmillan, 1967. 242 pp. illus. $5.95. 67–12222. (SH–C)
Analyzes the how and why of bird behavior through a discussion of interesting topics which include sensation and perception, motivation, instinctive behavior, migration, learning, social behavior, communication, personality, and intelligence. An interesting and scholarly book.

Freedman, Russell, and James E. Morris. *How Animals Learn.* Holiday House, 1969. 159 pp. illus. $3.95. (JH)
Investigations of animal behavior, traditionally called "comparative physiology," also are comprehended under the newer term "ethology" which is the subject of this excellent young people's book. Classical experiments by such well known scientists as Lorenz and Pavlov are mentioned, but many ongoing studies of the present are introduced. The great dividend in this book is the treatment of problem solving and the description of open ended investigations in which the student may engage. These projects are genuine learning and investigative activities that will require several weeks or even months if they are developed and reported adequately. Bibliography is excellent. Indexed.

Klopper, Peter H. *Habitats and Territories: A Study of the Use of Space by Animals.* Basic, 1969. (Basic Topics in Comparative Psychology.) x+117 pp. illus. $3.95. 76–78471. (C)
Considers the question of why animals are not randomly distributed in space and what physical, biological, and social factors influence the observed distribution of living organisms in nature. Gives an excellent summary of scientific work on the heterogeneity of habitat, species development in relation to habitat, selection of habitats, the nature and evolution of territoriality, and adaptations to geographical localities.

Lilly, John Cunningham. *The Mind of the Dolphin: A Nonhuman Intelligence.* Doubleday, 1967. xix+310 pp. illus. $5.95. 67–10417. (C)
A partial account of current research with dolphins. Thought-provoking and stimulating, this report points out that conventional research methods are inadequate to study the dolphin mind and dolphin vocalizations, and points out the necessity for a new approach.

Lorenz, Konrad Z. *King Solomon's Ring.* Crowell, 1952. xix+202 pp. illus. $6.95; paper (Apollo) $1.95. 52–7373. (JH)
An outstanding naturalist and keen observer discusses the subject of animal intelligence and communication. The title is inspired by I Kings IV:33 in which King Solomon communicated with animals. Lorenz believes that animals can talk to men, if the men "understand their language."

Lorenz, Konrad Z. *On Aggression.* (Tr. by Marjorie Kerr Wilson.) Harcourt, Brace, 1966. xiv+306 pp. $5.75. 66–12369. (C)

Deals with the question: Why do animals of the same species, even including human beings, fight each other? The triggering mechanisms that release aggression in such varied animals as fishes, birds, and wild and domestic mammals are described. The author discusses stimulus situations that he believes elicit aggression in human societies, and he suggests how such conditions can be controlled to avoid undesirable aggression. The author expresses his belief in the power of human reason and in the possibility of developing a thoroughgoing control of active hostility between human beings, and even between human societies or nations.

McGaugh, James L., Norman M. Weinberger, and Richard E. Whalen (eds.). *Psychobiology: The Biological Bases of Behavior.* (Readings from *Scientific American.*) Freeman, 1967. 382 pp. illus. $10.00, paper $5.45. 67–12182. (SH–C)

Instinct, imprinting, peck order, learning, memory, and intelligence in an array of animals from sticklebacks to elephants are treated in this collection of 45 articles reprinted from *Scientific American.* Articles treating motivation, sensation, perception, emotion and their neurophysiological bases include studies on frogs, monkeys, cats, rats, and people. Well illustrated.

157 ABNORMAL PSYCHOLOGY

Felix, Robert H. *Mental Illness: Progress and Prospects.* Columbia U., 1967. 110 pp. $4.75. 67–20278. (C)

This transcript of four lectures delivered at Columbia University offers a historical survey of the rise of the mental health movement. The lectures focus on the development of community mental health clinics, a topic on which Dr. Fleix speaks with great authority. The style is elegant, readable, and enjoyable. Useful as collateral reading in courses in public health and psychology.

Hall, Bernard H. (ed.). *A Psychiatrist for a Troubled World: Selected Papers of William C. Menninger, M.D.* Viking, 1967. xxii+871 pp. $12.50. 67–11268. (SH–C)

It contains 81 papers reflecting the multifaceted interests and activities of the author over the 43 years of his exceedingly productive professional career. One central and dominant theme runs through the entire book, namely, the author's insistent and relentless effort to gain understanding for the mentally-ill patient and to translate this understanding into concrete financial and legislative support. The selections are classified under the broad headings of contributions to the art and science of medicine, to military psychiatry, to social and community psychiatry and to his leadership as a spokesman for the mentally ill.

Kluver, Heinrich. *Mescal and Mechanisms of Hallucinations.* U. of Chicago, 1966. xviii+108 pp. $3.95; paper (Phoenix PSS531) $1.50. 66–20593. (C)

The two papers that comprise this book are extremely timely and should be of interest to all who find this topic intriguing. Most of the speculation about the mechanism of action is couched in psychological, not neurological, terms.

Rycroft, Charles (ed.). *Psychoanalysis Observed.* Coward-McCann, 1966. 1965 pp. $4.50. 67–15275. (C)
This book is a rare treat: a discussion of psychoanalysis which is neither blind defense nor rabid attack. The contributors—three analysts, a social anthropologist, and a "theologian and industrial scientist"—evaluate analysis as theory, as therapy, and as ideology. The authors assume that their public has more than a passing acquaintance with psychoanalytic theory. Not intended to be an undergraduate text, but for the interested student (and certainly for the teacher) of personality, abnormal psychology, and other relevant subjects. Invaluable supplementary reading.

333 CONSERVATION (See also 631.4 SOIL CONSERVATION)

Allen, Durwald L. *Our Wildlife Legacy* (rev. ed.). Funk & Wagnalls, 1962. x+422 pp. illus. $6.50 (text ed. $5.00). 62–7980. (SH)
Considers conservation of living natural resources in relation to the often conflicting interests of sportsmen and fishermen, the administrators, and public officials. Dr. Allen communicates his keen insights into how renewable resources can be managed successfully and used for satisfying other interests. He also deals with questions of predator control and artificial propagation and rearing.

Callison, Charles H. (ed.). *America's Natural Resources* (rev. ed.). Ronald, 1967. vii+220 pp. $5.00. 67–14482. (C)
Provides background information to the problems of conservation with chapters written by prominent authorities. A general review of natural resource conservation that stresses the importance of the environment for the well-being of American society.

Clepper, Henry (ed.). *Origins of American Conservation.* Ronald, 1966. x+193 pp. $4.50. 66–14161. (JH–SH–C)
An excellent condensation of information dating from early settlement of the United States to the present time. It deals with pleasant as well as unpleasant annals of American history. The great resources of plant and animal life and their influence on soil and on mankind are considered. The destruction of buffalo as a wild animal; disappearance of the passenger pigeon; increase in deer population, and the surges and recessions in numbers of various animals of land, air and water are discussed. Will appeal to a wide range of readers.

Darling, F. Fraser, and John P. Milton (eds.). *Future Environments of North America.* Nat. Hist. Press, 1966. xv+767 pp. $12.50. 66–20989. (C)
The preservation and development of natural habitats is central to this unique collection of authoritative writings which epitomizes the breadth of the conservation movement. An excellent reference for students and professionals in conservation and ecology.

Douglas, William O. *A Wilderness Bill of Rights.* Little, Brown, 1965. 192 pp. illus. $5.95, paper $1.95. 65–21350. (JH–SH–C)
Lays before the reader the sad testimonial of the previous and current practices of land use by commercial, private and Federal interests. *The Bill of Rights* guarantees certain rights to minorities. Why not a *Wilderness Bill of Rights* to assure that wilderness areas and sanctuaries will be supported and maintained? A plan for and discussion of the necessary ingredients of a

Wilderness Bill of Rights makes up the last half of the volume. Ideally suited for those with serious interests in conservation of natural resources.

Frome, Michael. *Strangers in High Places: The Story of the Great Smokey Mountains.* Doubleday, 1966. ix+394 pp. illus. $5.95. 66–17401. (SH–C)
Chronicles the current problems of use and abuse of a national park in the face of a highly mobile, increasing population and the gradual but inevitable loss of isolation. In view of the increasing concern and involvement of commercial, private, and public interests along with the social and political implications, this book is especially timely.

Gabrielson, Ira N. *Wildlife Conservation* (2nd ed.). Macmillan, 1959. 244 pp. illus. maps. $5.95. 41–51524. (JH–SH)
An analysis of wildlife conservation problems, with specific reference to fishes, migratory birds and fur-bearing mammals, with development of the thesis of wise utilization of resources consistent with restoration and maintenance at optimal abundance levels.

Gabrielson, Ira N. *Wildlife Management* (2nd ed.). Macmillan, 1959. xii+274 pp. illus. $5.95. Agr 51–501. (JH)
A noted authority examines the American wildlife resources used for recreation and for other human purposes, taking into consideration the need for these wild populations to exist along with necessary societal activities—farming, lumbering, dams and highways, and other projects. Although the statistics are out of date, the basic principles and conservation methods discussed and recommended are still sound.

Helfman, Elizabeth S. *Rivers and Watersheds in America's Future.* McKay, 1965. x+244 pp. illus. $4.95; LB $4.19. 65–22964. (SH–C)
Useful information about dams, watersheds, and conservation in the United States. Great watersheds such as those drained by the Colorado, Columbia, Mississippi and Tennessee Rivers and their tributaries are discussed in detail. Information concerning location and construction of dams is accompanied by a discussion of the significance of water control to the people, agriculture, industry and the wildlife of the area. Well written and adequately illustrated. Will appeal to many nonspecialists.

Laycock, George. *The Sign of the Flying Goose: A Guide to the National Wildlife Refuges.* Nat. Hist. Press, 1965. 299 pp. illus. map. $5.95. 65–10598. (JH–SH–C)
Today there are more than 280 national wildlife refuges, administered by the U. S. Fish and Wildlife Service, marked with the sign of the flying goose. Many species of birds and mammals would be extinct if it were not for the protection afforded by these refuges. Here is a history of their establishment and a directory for those who wish to visit them.

Lillard, Richard G. *Eden in Jeopardy; Man's Prodigal Meddling with His Environment: The Southern California Experience.* Knopf, 1966. xii+329+vi pp. illus. $6.95. 66–10744. (SH–C)
Sound discussions of significant changes in population, water supply, agriculture, industry, architecture, transportation, topography, recreation, and culture. Occasionally the author shows a lack of scientific background, but there are

no serious errors. The facts and thoughts about man's interaction with nature presented with numerous historical references will be useful and revealing.

Park, Charles F., Jr. (Margaret Freeman, Collaborator.) *Affluence in Jeopardy: Minerals and the Political Economy.* Freeman, Cooper, 1968. xii+368 pp. illus. $9.50. 68–30737. (SH–C)
This up-to-date analysis of the economic and political problems of mineral resources begins by posing the questions: "Are the earth's mineral resources adequate to support a universally affluent society?," or, "Are they even adequate to support a continuance of today's level of affluence?" The very broad background of the book makes it an informative work for general reading as well as a basic resource for students of economics, political science, sociology, and the applied sciences.

Smith, Guy-Harold (ed.). *Conservation of Natural Resources* (3rd ed.). Wiley, 1965. xi+533 pp. illus. $10.95. 65–21449. (SH–C)
The 19 contributors are well-known specialists. Each is careful to consider the basic tenets of resource conservation. The two basic resources, land and water, receive most of the effort and space. Land reclamation, agriculture, the public domain, and land-use for food production, industry, for recreation, and for the urban sprawl are closely scrutinized. Forests, mineral fuels, and fisheries are carefully considered. A major fault: only scattered, minimal references are made to air pollution.

U.S. Department of Agriculture. *Outdoors USA.* (Ed. by Jack Hayes). U.S. Govt. Printing Office, 1967. xxxix+408 pp. illus. $2.75. (JH–SH–C)
Reports the magnitude, diversity, wonders, and resources of the great outdoors and of the recreational opportunities they afford. This is an anthology of a number of small works dealing with major themes on the big woods, water, beautification, and the countryside.

U.S. Department of the Interior. *The Third Wave . . . America's New Conservation.* (Conservation Yearbook No. 3.) U.S. Govt. Printing Office, 1967. 128 pp. illus. $2.00. (JH–SH)
This is the third in the series of conservation yearbooks. It attempts to explain the ecological emphasis which endeavors to understand and relate man to the natural resources. An excellent collection of color photography supplements the text.

Walsh, John, and Robert Gannon. *Time is Short and the Water Rises.* Dutton, 1967. 224 pp. illus. $6.95. 67–11380. (JH–SH–C)
Reports the rescue of wildlife from the rising waters caused by the building of the Afobaka Dam. An excellent account of the Surinam region, its Bushnegro population, and the effect of civilization on wildlife and on an undeveloped region.

338.19 FOOD SUPPLY

Dumont, René, and Bernard Rosier. *The Hungry Future.* Praeger, 1969. 271 pp. $6.95. 69–11861. (SH–C)
The authors predict that 1980 will be a point of no return in food production, and advocate world-coordinated trade, the extensive application of technically

advanced agricultural methods, intensified use of fertilizers, land reform, the development of new sources of nutrition, and birth or population control.

371.4 GUIDANCE AND COUNSELING

Many scientific and technical organizations are deluged week-by-week with letters from students requesting information on careers in science and technology, to fulfill a class assignment, or because of a personal interest in exploring various vocational fields. Since most of the letters indicate that the students have not been directed to do any study in libraries on their own initiative, and are unaware of major works that should be available in all school, college and public libraries, the following suggestions are made concerning three indispensable library acquisitions. These works are for general orientation. Reading in encyclopedias, biographies, trade books and textbooks in a good science library collection such as that suggested in this annotated list will provide far more useful and worthwhile material than ephemeral pamphlets and reprints obtained by correspondence which seldom have sufficient depth and detail.

The Encyclopedia of Careers and Vocational Guidance. William E. Hopke (Editor-in-chief). Vol. 1: *Planning your Career;* Vol. 2: *Careers and Occupations.* Ferguson, 1967. xvi+752 pp.; xvi+784 pp. $21.65 set. (JH–SH–C)
A good supplement to the *Occupational Outlook Handbook.* Vol. 1 has various chapters written by specialists consisting of general articles on careers, followed by descriptions of the various career fields in general. Vol. 2 describes specifically this myriad of jobs, trades, occupations and professions, including information on definition, history, nature of work, requirements, opportunities for experience, advancement, outlook, earnings, working conditions, and social and psychological factors.

Occupational Literature (1964 edition). Gertrude Forrester (comp.). H. W. Wilson Co., 1964. 675 pp. Price on application. 64–22814. (C)
The most complete annotated bibliography, so far as it is humanly possible for a dedicated specialist to compile one, of most of the vast array of occupational and vocational literature. It covers all aspects of career guidance and is an indispensable source book for all librarians, vocational counsellors, and interested educators. It should be available in school and college libraries as a source book and collateral reading guide for students.

Occupational Outlook Handbook. Prepared by the Bureau of Labor Statistics, U. S. Department of Labor. Department of Labor, Bulletin 1550, 1968–69 edition, 1968. (Purchase from Superintendent of Documents, U. S. Government Printing Office, Washington, D. C. 20540). 754 pp. $4.25 (paper). (JH–SH–C)
This guide, revised biennially, is an authoritative condensed guide to every skill, trade, occupation and profession. It provides information on duties, education, training, demand, opportunities for advancement, compensation, location of work, fringe benefits, etc. This is the basic reference to which all students should be referred as they begin their career studies. (The *Handbook* is updated by a periodical *Occupational Outlook Quarterly* available from the same source at $1.50 per year.)

390 CULTURAL ANTHROPOLOGY

Ardley, Robert. *The Territorial Imperative: A Personal Inquiry into the Animal Origins of Property and Nations.* Atheneum, 1966. xii+390 pp. illus. $6.95. 66–23572. (SH–C)

The scientific facts presented here are told in the expert prose of an experienced playwright. His basic thesis is that man, like many other animals, will guard his "territory" against all members of his own kind. Includes a 245-item bibliography.

Benedict, Ruth. *Patterns of Culture.* Houghton Mifflin, 1934. xvii+291 pp. $4.00; paper (Sentry 8, 1961) $1.95, also NAL (Mentor MD89) 50¢. 34–37639. (SH)

The first part of this anthropological classic introduces the concept "culture" and tells some ways in which it is studied. The middle chapters describe the three cultural groups the book is based on—the Zuni Indians of New Mexico, the Dobuans of Melanesia, and the Kwakiutl of Vancouver Island—and points out contrasts among them, while the last two chapters deal with the nature of society and the place of the individual in culture, drawing on findings from the groups studied.

Bohannan, Paul James. *Social Anthropology.* Holt, Rinehart and Winston, 1963. viii+421 pp. illus. $8.50. 63–13322. (C)

A well-written, up-to-date text on the institutions of culture. Includes kinship, social agreements and contracts, economies and political systems, and a concluding chapter on social and cultural change.

Brown, Ina Corinne. *Understanding Other Cultures.* Prentice-Hall, 1963. viii+184 pp. $4.25, paper $1.95. 63–9783. (SH)

"The notion that if people would just get to know one another they would all be friends and everything would be all right is as dangerous as it is sentimental . . . we have to learn to get along with people who are different and are likely to stay that way." With this in mind, the author describes all areas of cultural studies, from the nature of culture itself, through its various parts—physical subsistence, family groups, art, religion, the life cycle—to a discussion of the problems of culture change, both for individuals and for whole societies.

Brundage, Burr Cartwright. *Lords of Cuzco: A History and Description of the Inca People in Their Final Days.* U. of Oklahoma, 1967. viii+458 pp. illus. $6.95. 67–15576. (SH–C)

A social history that stresses the daily life and habits of the ruling class: the Inca nobles of Cuzco. Presents a complete picture of the Incas: their personal characteristics, religion and ritual, power structure, and customs.

Casagrande, Joseph B. (ed.). *In The Company of Man.* Harper & Row, 1960. xviii+540 pp. illus. $6.50; paper (Torchbooks Tb3047) $2.95. 60–5731. (C)

A basic tool of anthropology is the use of informants—members of a culture who tell the anthropologist about social structure, what some particular ceremony means, why someone's wives are fighting, and other important information for the field worker. In this collection of 20 articles, such anthropologists as Clyde Kluckhohn, Margaret Mead, and Cora DuBois tell of informants they have used in their diversified field work; the results are intriguing descriptions of the cultures they studied and warm, informative portraits of the individuals who guided them.

Dalton, George (ed.). *Tribal and Peasant Economics: Readings in Economic Anthropology.* Nat. Hist. Press, 1967. xv+584 pp. $2.95 (paper). (C)
A collection of 29 previously published selections from anthropological, sociological, and economic sources that record aspects of the traditional economic systems of various tribal groups as they existed before industrialization and the associated developments occurred. There are several contributions each from Africa, Asia, Oceania, Europe and the Americas. Selections from each geographic area are concluded with articles relating to social and economic change. Numerous footnotes and an index.

Douglas, Mary, Gerald Barry, J. Bronowski, James Fisher, and Julian Huxley. *Doubleday Pictorial Library of Man in Society: Patterns of Human Organization.* (Illus. by Hans Erni.) Doubleday, 1964. 367 pp. illus. $15.00. 64–12288. (SH–C)
An illustrated overview of aspects of anthropology and sociology, abundantly illustrated in color and monochrome, that traces the beginnings of human society, discusses physical differences, organization, social role, personal and family relations, religion, employment and personal wealth, politics, law, education, and other activities. Appendices include map, glossary, reading list, and a description of the work of anthropologists and sociologists.

Fabre-Luce, Alfred. *Men or Insects? A Study of Population Problems.* (Tr. by Robert Baldick.) Horizon, 1965. 160 pp. $5.00. 65–26724. (SH–C)
An eminent French sociologist attacks the fables of the population problem. To a great extent, the book is a moralistic essay confronting the Western mind with the backwardness of its views on abortion, contraception, euthanasia, and the social values of suburbia. More than this, the author shows how Western and Eastern cultures view the same problems in different ways. Thus, no one culture can plan for another, or dictate solutions that are not compatible with local customs and beliefs. Facts are not presented for their own value, but are used to illustrate the complexity of the problems. Occasional glib political formulations, stereotype asides, and somewhat superficial interpretations of historical writings can be overlooked.

Folsom, Franklin. *Science and the Secret of Man's Past.* Harvey House, 1966. 192 pp. illus. $5.00; LB $4.79. 66–14178. (JH–SH)
Recounts the engaging story of man's attempt to learn more about his own past. It is concerned primarily with the development of techniques for assigning dates to events in man's cultural and biological evolution, rather than a discussion of the development of man's culture *per se*. Well researched and consistently clear and accurate.

Gennep, Arnold Van. *The Rites of Passage.* U. of Chicago, 1960. xxvi+198 pp. $5.00; paper (Phoenix P64) $1.95. 59–14321. (C)
Originally written in French in 1908, this description and comparison of the rituals that accompany changes in social position in different cultures has become a classic in anthropology. Discusses and gives examples of rites and ceremonies that accompany birth and pregnancy, aspects of childhood, social

puberty and initiation, betrothal and marriage, funerals and other social phenomena.

Goldschmidt, Walter. *Man's Way: A Preface to the Understanding of Human Society.* Holt, Rinehart and Winston, 1959. 255 pp. paper $2.95. 59–5926. (C)

Describes the evolution of cultures and the nature of culture itself, using modern findings and ideas from anthropology and other behavioral sciences. Various anthropological theories are discussed, along with the "social imperatives"—characteristics necessary if a society is to exist—and the actual mechanisms and changes of social evolution. For students with background in cultural anthropology.

Hall, Edward T. *The Silent Language.* Doubleday, 1959. 240 pp. $4.50; Fawcett (Premier R204, 1961), paper 60¢. 59–6359. (SH)

Discusses the role of nonverbal behavior and manners in intercultural communication, and describes for the layman some reasons for cross-cultural conflict, particularly in international affairs and business. An appendix gives a theory of culture, and presents the components of culture in chart form in a "Map of Culture."

Hays, Hoffman R. *From Ape to Angel: An Informal History of Social Anthropology.* Knopf, 1958. xxii+440+xv pp. illus. $7.95; Putnam's (Capricorn 247), paper $2.65. 58–7713. (SH)

A popular history of cultural anthropology that shows its growth through the expeditions and other work of major contributors, from the classical evolutionists to modern psychological investigators and culture-theorists.

Herskovits, Melville Jean. *Cultural Anthropology.* Knopf, 1955. xvi+569+xxxiv pp. illus. $6.50. 55–5171. (C)

A soundly organized, well-written introductory text. The five sections cover: the setting of culture, including biological and environmental factors; aspects of culture such as social organization, technology, religion, language; the nature of culture itself and its relation to the individual and to a society; cultural structure and dynamics, including its origin and mechanisms for change; and a conclusion on anthropology in the world today. Large bibliography and reading list.

Hunt, Robert (ed.). *Personalities and Culture: Readings in Psychological Anthropology.* Doubleday, 1967. xxi+434 pp. $6.95, paper $2.50. 67–10405. (C)

A comprehensive collection of articles on cross-cultural studies of personality based on empirical data testing interdependence of the psychic and sociocultural variables. For those interested in the scientific knowledge of human behavior, Hunt's collection is one of the best.

Kardiner, Abram, and Edward Preble. *They Studied Man.* World, 1961. 287 pp. illus. $5.00; paper (Meridian M192) $2.25, also NAL (Mentor MT486) 75¢. 61–5808. (SH)

While anthropologists study man and his culture, it is culture that produces the anthropologists. This interesting book traces the lives and work of ten well-known figures who studied man, from Charles Darwin and James Frazer to Ruth Benedict and Sigmund Freud. A fascinating account of the rise of cultural anthropology and the interaction of men and culture that brought it about.

Keesing, Felix. *Cultural Anthropology: The Science of Custom.* Holt, Rinehart and Winston, 1958. xxv+477 pp. illus. $8.95. 58–6432. (C)
Virtually all aspects of this broad subject are presented as a series of 84 problems, each one a short article on a particular concept. Some of the section titles are: "Culture and people: some basic concepts," "Culture and biological heritage," "Theories relating to culture, society and personality," and "Stability and change." The writing is clear and the examples and questions in the text make it thought-provoking as well.

Kluckhohn, Clyde. *Mirror for Man: The Relation of Anthropology to Modern Life.* McGraw-Hill, 1949. xi+313 pp. $5.00; Fawcett (Premier T316), paper 75¢. 49–7012. (SH)
"Anthropology holds up a great mirror to man and lets him look at himself in his infinite variety." Kluckhohn's mirror is a highly polished one and this anthropological classic, despite its age, gives a clear view of custom, language, race, personality, and a general reflection on the way anthropologists look at things. A good basic introduction.

Lewis, Oscar. *The Children of Sanchez.* Random, 1961. xxxi+500 pp. $8.95 (text ed. $5.60); paper (Vintage V280) $2.95. 61–6270. (C)
An engrossing study of Mexican poverty culture, based on recorded interviews with five members of the Sanchez family, by the well-known anthropologist who has made several other studies of Mexican cultures. Deals with all aspects of life there, including kinship, religion, social institutions, and the problems of cultural change being brought about by economic and technological development.

Linton, Ralph. *The Tree of Culture.* Knopf, 1955. 629 pp. $9.75 (ext ed. $7.25); Random (abridged ed., Vintage V-76, 1959, vii+261+viii pp.), paper $1.45. 55–5173. (C)
An overflow of the elements and growth of culture. Briefly describes the development of *Homo sapiens* and the basic inventions and turning points of culture, then gives a world-wide geographical description of culture areas and civilizations. A good introduction to the evolutionary relationships between cultures and a sound reference work.

Lisitzky, Gene. *Four Ways of Being Human: An Introduction to Anthropology.* Viking, 1956. 303 pp. illus. $5.00; LB $4.53; paper (Compass C128, 1960) $1.45. 56–14304. (JH)
Clear introductory accounts of four widely-scattered cultures: the Semang of the tropical rain forests of Malay and Siam, the Polar Eskimos, the Maioris of New Zealand, and the Hopis of the Arizona desert.

Mead, Margaret. *And Keep Your Powder Dry: An Anthropologist Looks at America* (rev. ed.). Morrow, 1965. xxx+340 pp. $5.00; paper (Apollo) $2.50. 65–11491. (SH–C)
The original was published in 1942 representing the author's effort "to present the culture and character of my own people in a way they would find meaningful and useful in meeting the harsh realities of war." A new chapter has been added in which Dr. Mead presents a discussion of America and Amer-

icans from the end of World War II up to the present. Concludes with a revised 1942 bibliography, a new bibliographical note, and a 1965 bibliography.

Mead, Margaret. *Family*. (Photographs by Ken Heyman.) Macmillan, 1965. 208 pp. 29 cms. $10.00. 65–21158 (SH–C)
A most magnificent work made up of individual sections, each beginning with the text and followed by several pages of superb photography. One reads a section of text for the background and the mood, and then patiently enjoys the photographs in which he can read much more than words can tell about warm personal relationships that occur among members of families everywhere. An unparalleled exhibition.

Mead, Margaret. *People and Places*. World, 1959. 318 pp. illus. $5.95. 59–11544. (JH)
An absorbing introduction to the elementary aspects of anthropology and what it knows about man. Includes both general discussions of anthropology and the nature of man, with the author drawing on her extensive experience in field work and reading, and a description of five widely differing cultures. Good illustrations, many in color.

Rowley, Charles D. *The New Guinea Villager: The Impact of Colonial Rule on Primitive Society and Economy*. Praeger, 1966. 225 pp. $7.50. 66–15452. (C)
A record of a changing culture or cultures on the exotic islands of New Guinea is presented in this study of eight decades of changes that have affected two million New Guineans due to the contact with and rule by the "white man." As a record of these important developments, this book is useful and interesting to read.

White, Leslie A. *The Science of Culture*. Farrar, Straus & Giroux, 1949. xx+ 444 pp. $6.00; Grove (Evergreen E-105, 1958), paper $2.95. 49–11015. (C)
A collection of the author's essays from a variety of journals, some of which are near-classics in anthropology ("Energy and the evolution of culture," "Man's control over civilization: an anthropocentric illusion," "The definition and prohibition of incest"), and all of which are stimulating and thought-provoking. White's ideas provide a contrast to those of some other leading anthropologists.

410 LINGUISTICS

Barry, Gerald, J. Bronowski, James Fisher, and Julian Huxley. *The Doubleday Pictorial Library of Communication and Language*. Doubleday, 1965. 367 pp. illus. $15.00. 65–17419. (JH–SH–C)
This comprehensive survey attempts to answer the questions as to how and what animals communicate, how alphabets begin and languages grow, what social changes we owe to the mass media, the social responsibilities of an advertiser, what actors, dancers and musicians communicate to their audiences, what makes effective television, what special information do pictures and symbols carry, what is brainwashing, how do computers translate, how do teaching machines work, how do relay satellites work? The book also examines techniques of teaching and learning. Beautifully illustrated with appropriate graphic aids, but no topic is treated in depth.

Hildum, Donald C. (ed.). *Language and Thought.* (Insight Series.) Van
Nostrand, 1967. vi+200 pp. illus. $1.95 (paper). 67–5062. (C)
A collection of authoritative articles by leading linguists and psychologists
on research relating language to thought. Although quite technical, this
work might be used by advanced undergraduates for research and collateral
reading.

Hockett, Charles F. *Course in Modern Linguistics.* Macmillan, 1958. xi+621
pp. $8.50. 58–5007. (C)
A comprehensive introduction to the modern aspects of linguistic science.
Discusses phonology, the pattern and production of sounds in speech;
grammatical systems and syntax; morphemes and the formation of word-units;
linguistic change; the prehistory of language; and the use of writing and
literature. Students who hope to specialize in cultural anthropology need
the background in this book.

Lenneberg, Eric H. (ed.). *New Directions in the Study of Language.* M.I.T.
Press, 1964. ix+194 pp. $5.00; paper $2.45. 64–8088. (SH–C)
Although it does not cover all extant ways of thinking about and studying
language, it does provide an interesting variety of approaches to the study
of language phenomena. Topics include biological disposition for language,
communication at the sub-human level, taboos and puns, the development of
language in children, and the relationship between language acquisition and
intellectual ability. Collateral reading for both high school and college-
level psychology students.

Potter, Simeon. *Language in the Modern World.* Peter Smith, 1960. 221 pp.
$3.00; Penguin (Pelican A470), paper 95¢. 60–3222. (C)
An intriguing introduction to language and how linguists study it. Demonstrates basic linguistic concepts such as the "phoneme" and "morpheme,"
describes how langauges are classified into families and groups, and shows
some of the ideas that modern science is forming on the relation between
language and thought. A comprehensive glossary defines basic linguistic
terminology from "allophone" through "umlaut" and beyond.

Sapir, Edward. *Language: An Introduction to the Study of Speech.* Harcourt,
Brace, 1921. ix+242 pp. paper (Harvest HB7) $1.25. 21–20134. (C)
This classic in anthropological linguistics is the only book that Sapir wrote
for the general reader. It covers all aspects of language from the elements
of speech and linguistic structure through a discussion of language in relation
to race and culture.

Singh, Jagjit. *Great Ideas in Information Theory, Language and Cybernetics.*
Dover, 1966. ix+338 pp. illus. $2.25 (paper). 66–20417. (C)
Covers the definitions of language, problems of language transmission, computers, the human brain, and certain aspects of human intelligence dealing
with information theory. Little mathematics is required.

Whorf, Benjamin Lee. *Language, Thought, and Reality.* (Ed. by John B. Carroll.) M.I.T. Press, 1956. xi+278 pp. $8.00; paper $1.95. 56–5367. (C)
Whorf grasped a new relationship between human language and human thinking with his theory that language can shape our thoughts. This selection of
his writings includes well-known work on American Indian languages as
well as articles on psychological linguistics and the concepts that underlie the

famous Sapir-Whorf hypothesis. Though difficult in parts, it will be a rewarding book for anyone interested in language and modern concepts of how we think.

500 SCIENCE IN GENERAL

Asimov, Isaac. *Adding a Dimension.* Doubleday, 1964. xi+202 pp. $4.50. 64–15340. (SH)
Asimov, in his brisk and informal style, discusses topics in mathematics, physics, biology, chemistry, and astronomy in 17 unorthodox essays on the history of science. He presents the scientific achievements of the past to provide background for each topic. In discussing imaginary and very large numbers, the speed of light, or evolution, he assumes that his readers are well informed. Highly readable, omitting just enough so that the reader can make his own discoveries.

Asimov, Isaac. *The New Intelligent Man's Guide to Science.* Basic, 1965. xvi+864 pp. illus. $12.50. 65–23045. (SH–C)
The second edition of a solid introduction to the modern world of science and technology. This careful revision incorporates those important discoveries and developments of the last few years not mentioned in the original. The total wordage has been increased by 20 percent, though compacted from two volumes into one. A valuable book for everybody. Good bibliography.

Asimov, Isaac. *Only a Trillion.* Abelard-Schuman, 1958. 195 pp. $3.50. 57–9947. (JH)
Subtitled "Speculations and explorations on the marvels of science," this popular account communicates to the layman concepts of the incredibly small and, contrarily, the immensity beyond comprehension of the phenomena with which scientists deal.

Barber, Bernard, and Walter Hirsch (eds.). *The Sociology of Science.* Free Press, 1962. viii+662 pp. $9.50. 62–15357. (SH–C)
An outstanding collection of professional writings that provides an understanding of the nature of science and scientific research in relation to the social organization and institutions of modern times. Also deals with the social image of the scientist. Particularly good for collateral reading by high school and college students and non professional adults.

Bluemle, Andrew (ed.). *Saturday Science.* Dutton, 1960. 333 pp. illus. $5.95. 60–8944. (SH)
Based on a series of lectures to a group of superior high school students by scientists of the Westinghouse Research Laboratories. The first and major part deals with principles and the second part with some of the techniques. Good bibliography. Excellent collateral reading.

Calder, Nigel (ed.). *The World in 1984* (2 vols.). Penguin, 1965. 215, 205 pp. illus. 95¢ ea. 65–2870. (JH–SH)
The credentials of the contributors are impressive. The range and variety of topics is evident in the following sample: "Fundamental science," "Natural resources," "Computers," "Surface transport," "Human mind," "Domestic life," "Education," "Government," and "Leisure and the arts." Contributors were asked to forecast "on the basis of known possibilities and trends, rather

than to speculate freely." Thus many of the predictions may turn out to be somewhat conservative. The brevity of the articles is a net virtue, but it prevents the inclusion of much of the evidence upon which the author's predictions presumably were based. Selections from these books will serve teachers and students of many subjects as collateral or basic reading.

Cockcroft, John (ed.). *The Organisation of Research Establishments.* Cambridge U., 1966. 275 pp. illus. $11.50. 65–14353. (SH–C)
A collection of papers on the organization and management of a group of establishments in the United Kingdom ranging from those in the physical sciences, atomic energy, and high-energy physics through railroads and aircraft to cotton, agriculture, pharmacology, medicine, and psychiatry. To these are added descriptions of the international center for high-energy physics, CERN, and the Bell Telephone Laboratories in the U.S.A. A valuable analysis of the elements which go into making a good research center.

Cohen, I. Bernard. *Science, Servant of Man.* Little, Brown, 1948. xiv+362 pp. $5.00. 48–3708. (SH–C)
The purpose is to give the layman a cogent conception of the manner in which science operates, of the need for scientific literacy as a process of a broad education, and why science is the essence of human civilization.

Colborn, Robert (Editor-in-chief). *Modern Science and Technology.* Van Nostrand, 1965. xiv+746 pp. illus. $22.50. 65–7696. (SH–C)
A collection of 81 articles which have previously appeared in the periodical, *International Science and Technology.* Each topic is treated by an authority in the field. The depth of treatment varies from popular science to the level of a good scientific journal, but most students can pursue profitably any one of the areas included. The articles have been grouped under general headings of physics, chemistry, technology, space science, earth science, and mathematics.

Commoner, Barry. *Science and Survival.* Viking, 1966. 150 pp. $4.50, paper $1.35. 66–16068. (SH–C)
The author makes a strong plea, while there is yet time to prevent biological disaster, to both scientist and layman to examine the implications of current scientific activities and their technological applications. He analyzes in depth the responsibilities of scientists, to science and to the public, with respect to the decision-making process at a number of levels. Problems of insecticides, air and water pollution, nuclear and biological warfare, weather, *et al.* are presented clearly. The focus is on the obligation of mankind to preserve a biosphere that will support human life. In addition to its obvious value as a social tract, the book yields the unintentional bonus of depicting the genuinely objective scientific mind at work. Useful for students in philosophy, sociology and political science as well as science majors.

Conant, James Bryant (gen. ed.). *Harvard Case Histories In Experimental Science* (2 vols.). Harvard, 1957. xvi+639 pp. diagrs. $10.00 set. 57–12843. (SH–C)
Each volume contains a detailed account of four outstanding developments in science that have had far-reaching effects on the course of future scientific research, on later technological development and on the advancement of human

knowledge and welfare. These books provide the reader with a real appreciation of the scientist at work.

Deming, Richard. *The Police Lab at Work.* Bobbs-Merrill, 1967. 125 pp. illus. $3.95. 66–29903. (JH)
The basic methods used in crime detection are summarized by examining its history, methods, and applications. This primer is written in nontechnical language that will be easily understood by the young reader.

Dubos, René. *The Dreams of Reason: Science and Utopias.* Columbia U., 1961. xii+167 pp. illus. $6.00, paper $1.75. 61–11753. (C)
Another lucid approach to the "myth" of two cultures which maintains that science cannot be disassociated from the total human experience. The relations between science and society are examined.

Fox, Russell, Max Garbuny, and Robert Hooke. *The Science of Science: Methods of Interpreting Physical Phenomena.* Walker, 1963. 243 pp. illus. $5.95. 63–20273. (SH)
Three scientists of the Westinghouse Research Laboratories present an exposition of scientific methods and show their potentialities and limitations with the aid of examples of actual projects. Particularly good is the discussion of measurement as a standard language of science, and as an extension of the human senses.

Froman, Robert. *Wanted: Amateur Scientists.* McKay, 1963. xv+102 pp. illus. $3.75; Hale, 1966, LB $2.88. 63–19797. (JH)
This is not, as some might infer from the title, a book about science fair projects. It does indicate, however, the important role that amateurs can and do play in bona fide research projects—how personal hobbies may assist research workers.

Gamow, George. *Matter, Earth, and Sky* (2nd ed.). Prentice-Hall, 1965. xiv+624 pp. illus. $11.00 (text ed. $9.75). 65–13813. (SH–C)
A textbook in introductory physical science useful in freshman physics courses, in beginning college courses in "physical science" and possibly in advanced high school courses that follow the usual introductory course. The book has the virtue of a strong emphasis upon "modern" physics and the related physical sciences. The informal sketches may appeal to some readers although many teachers object to such "light treatment." Styling, typography, illustrations, and editorial features are commendable.

Gamow, George. *One Two Three . . . Infinity: Facts and Speculations of Science* (2nd ed.). Viking, 1961. xii+340 pp. illus. $5.00, paper $1.65. 61–3926. (SH)
A most absorbing adventure into a melange of such things as atoms, stars, genes, the fourth dimension, the days of creation, whether one can bend space—all for the purpose of giving the reader a true insight into the microscopic and macroscopic components of the universe.

Gray, Dwight E., and John W. Coutts. *Man and His Physical World* (3rd ed.). Van Nostrand, 1958. x+682 pp. illus. $8.95. 58–8613. (SH)
Another comprehensive survey text covering astronomy, chemistry, physics and geology which emphasizes methods, history and scientific theory. Good enrichment material for high school students.

Greenberg, Daniel S. *The Politics of Pure Science*. New American Library, 1967. xiii+303 pp. $7.95. 68–15277. (SH–C)
Greenberg traces the transformation of science from its purity before World War II when science was nonpolitical, ignored by government, and impoverished, to its present highly political status in which it consumes a significant percentage of the total Federal budget.

International Science and Technology, The Editors of. *The Way of the Scientist: Interviews from the World of Science and Technology*. Simon & Schuster, 1966. 382 pp. illus. $8.95. 66–27585. (SH–C)
Edited interviews with some 50 prominent contemporary men of science, engineering, government and business, originally published in *International Science and Technology*. Gives the imaginative reader the feeling he is there. One finds fascinating glimpses into how these men view their responsibilities, how they discovered their life work, and their creative processes. Much of the discussion deals with the interface between science/technology and government. Useful as collateral reading to help humanize science and relate it to social and economic problems. No index.

Isaacs, Alan. *Introducing Science*. Basic, 1963. 244 pp. illus. $5.95. Penguin (Pelican A562, 1962), paper $1.25. 64–13752. (SH–C)
The salient points of chemistry and physics are organized around the concepts of matter and energy. All discussions are succinct and explanations are minimized. Accordingly, persons who have a good general background in science will find the book ties together a wealth of information; those with little background may find it uninformative. The author's intended audience is the adult general reader and the nonscience student. However, the teacher of general science will find the book helpful in presenting general principles to younger students.

Jones, Bessie Zaban. *The Golden Age of Science: Thirty Portraits of the Giants of 19th-Century Science by Their Scientific Contemporaries*. Simon & Schuster, 1966. xxxiii+659 pp. $12.00. 66–20254. (SH–C)
Primarily a reference work, this serves as an excellent introduction to the period of science characterized as an age of synthesis and professionalism. The introduction by Everett Mendelsohn is admirable and should be read with care by science-minded high school students.

Life, The Editors of. *Life Nature Library*. Time, Inc., 1961–67. illus. $6.60 ea. (JH–SH–C)
Bates, Marston. *The Land and Wildlife of South America*, 1964. 200 pp.
Beiser, Arthur. *The Earth*, 1962. 192 pp.
Bergamini, David. *The Land and Wildlife of Australia*, 1964. 198 pp.
Bergamini, David. *The Universe*, 1966. 192 pp.
Bourliere, Francois. *The Land and Wildlife of Eurasia*, 1964. 198 pp.
Carr, Archie Fairly. *The Land and Wildlife of Africa*, 1964. 200 pp.
Carr, Archie Fairly. *The Reptiles*, 1967. 128 pp.
Carrington, Richard. *The Mammals*, 1967. 128 pp.
DeVore, Irven, and Sarel Eimerl. *The Primates*, 1965. 200 pp.
Edey, Maitland A. *A Guide to the Natural World (Index)*, 1965. 210 pp.
Engel, Leonard. *The Sea*, 1961. 190 pp.
Farb, Peter. *Ecology*, 1963. 192 pp.

Farb, Peter. *The Forest,* 1961. 192 pp.
Farb, Peter. *Insects,* 1962. 192 pp.
Farb, Peter. *The Land and Wildlife of North America,* 1966. 200 pp.
Howell, Francis Clark. *Early Man,* 1965. 200 pp.
Leopold, A. Starker. *The Desert,* 1961. 192 pp.
Ley, Willy. *The Poles,* 1962. 192 pp.
Milne, Louis J., and Margery Milne. *The Mountains,* 1962. 192 pp.
Moore, Ruth E. *Evolution,* 1962. 192 pp.
Ommanney, Francis Dorones. *The Fishes,* 1963. 192 pp.
Peterson, Roger Tory. *The Birds,* 1967. 128 pp.
Ripley, Sidney Dillon. *The Land and Wildlife of Tropical Asia,* 1964. 200 pp.
Tinbergen, Niko. *Animal Behavior,* 1965. 200 pp.
Went, Frits W. *The Plants,* 1963. 195 pp.
This series is divided into four sections: major habitats, kinds of living things, major land units, and the physical and behavioral evolution of man, and are interrelated by means of bridging volumes. The texts, written by prominent authorities, are supplemented with rich color photographs and diagrams. As with the *Life Science Series,* these volumes are designed for the layman and student. Each volume is meritorious in its own right, but the series as a whole provides a wealth of information. All libraries should own both series.

Life, The Editors of. *Life Science Library.* Time, Inc., 1963–1969. 200 pp. ea. illus. $6.60 ea. (JH–SH–C)
Bergamini, D. *Mathematics,* (Vol. 14), 1963.
Bowen, E., and W. Owen. *Wheels,* (Vol. 25), 1967.
Claiborne, R., and S. Goudsmit. *Time,* (Vol. 22), 1966.
Clark, A. C. *Man and Space,* (Vol. 13), 1964.
David, W., and L. Leopold. *Water,* (Vol. 23), 1966.
Eimerl, S., and R. V. Lee. *The Physician,* (Vol. 17), 1967.
Furnas, C. C., and J. McCarthy. *The Engineer,* (Vol. 5), 1966.
Haggerty, J. J., and W. Sebrell. *Food and Nutrition,* (Vol. 7), 1967.
Lapp, R. E. *Matter,* (Vol 15), 1969.
Leonard, J., and C. Sagan. *Planets,* (Vol. 18), 1966.
Lewis, E. V. *Ships,* (Vol. 20), 1965.
Margenau, H., and D. Bergamini. *The Scientist,* (Vol. 19), 1964.
Mark, H. F. *Giant Molecules,* (Vol. 8), 1966.
Modell, W., and A. Lansing. *Drugs,* (Vol. 3), 1967.
Nourse, A. E. *The Body,* (Vol. 1), 1964.
O'Brien, R. *Machines,* (Vol. 12), 1969.
Pfeiffer, J. *The Cell,* (Vol. 2), 1964.
Pines, M., and R. Dubos. *Health and Disease,* (Vol. 10), 1965.
Rudolph, M., and C. Mueller. *Light and Vision,* (Vol. 11), 1964.
Stevens, S. S., and F. Warshofsky. *Sound and Hearing,* (Vol. 21), 1965.
Stever, H. G., and J. J. Haggerty. *Flight,* (Vol. 6), 1965.
Tanner, J. M., and G. R. Taylor. *Growth,* (Vol. 9), 1965.
Thompson, P. D., and R. O'Brien. *Weather,* (Vol. 24), 1968.
Wilson, J. R. *The Mind,* (Vol. 16), 1969.
Wilson, M. *Energy,* (Vol. 4), 1967.
Index. *A Guide to Science,* (Vol. 26), 1969.

Each of the 26 volumes in this collection covers a specific area of science in a manner suitable to educate the layman. The text of each volume is written by prominent and respected authorities in the particular field and is supplemented with excellent color photographs and drawings. Each chapter contains a picture essay of the major theme of that chapter. A list of books for further references and an exhaustive index is appended to each volume. The series is ended with a guide and index which shows how the 26 volumes can be used as supplemental reading for science courses. On the whole each volume is highly recommended for school, public, and college libraries.

Lovell, Bernard, and Tom Margerison (eds.). *The Explosion of Science: The Physical Universe.* Meredith, 1967. 216 pp. illus. 35 cm. $24.95. 67–12219. (SH–C)
Seven studies by eminent British scholars comprise this exposition of the revelations and applications of modern astronomy and astrophysics, space exploration, molecular and nuclear physics, and modern chemical research and industry. Colored and monochrome illustrations are a major feature in an accurate, interesting, and useful volume for nonprofessional readers which is well worth the high price. Bibliography.

Mattfeld, Jacquelyn A., and Carol G. Van Aken. *Women and the Scientific Professions.* M.I.T. Press, 1965. 185 pp. $6.95, paper $2.95. 65–26968. (SH–C)
Papers presented at a symposium cover the commitment required of women entering the scientific professions, the role and status of women in science and engineering, the case for and against the employment of women, and suggestions for improving the status of women.

Nobel Foundation (eds.). *Nobel: The Man and His Prizes.* American Elsevier, 1962. x+690 pp. $12.50. 62–20556. (SH)
A facsimile of Nobel's will, a biographical sketch, the story of the establishment of the Nobel Foundation, and an account (1901–1961) of the prizes and prize-winners in literature, medicine and physiology, chemistry, physics, and the peace prize; concluding with a list of all prize-winners. An indispensable collateral reading and reference work for all school and public libraries.

Science Service (Comp.). *Science News Yearbook 1969/70.* Scribner's 1969. xx+434 pp. illus. $9.95. 69–17043. (SH–C)
A collection of 55 articles concerning newsworthy scientific developments or happenings in 1968. They are grouped under the headings of biomedicine, space, astronomy, physics and chemistry, earth, engineering and technology, environment and ecology, behavioral and social sciences, and science policy. This should be a useful book for school and public libraries for general orientation in current subject matter.

Silverberg, Robert. *Scientists and Scoundrels: A Book of Hoaxes.* Crowell, 1965. x+251 pp. illus. $3.95. 65–11649. (JH–SH)
A collection of 13 scientific hoaxes that illustrate how the scientists' search for "truth" has gone awry. Some of the hoaxes were the work of deceptive charlatans. But others were not intentional: the scientists grossly misinterpreted their evidence. Instructive reading for anyone. Indexed.

Untermeyer, Louis. *Makers of the Modern World.* Simon & Schuster, 1962. xix+809 pp. $7.95, paper $2.95. 54–12364. (SH)
Brief biographical sketches of 92 representatives of various learned pursuits,

including pure and applied science, who have influenced most greatly our modern culture, art, and technology.

Walker, Marshall. *The Nature of Scientific Thought*. Prentice-Hall, 1963. vii+ 184 pp. diagrs. $4.50, paper $2.25. 63–9519. (C)
The author uses his background as chemist, physicist and mathematician to demonstrate the unity of all sciences. "It will show the beginner where he can go, and it will show the specialist where he has been." A useful cultural background study for thinking readers.

Weinberg, Alvin M. *Reflections on Big Science*. M.I.T. Press, 1967. ix+182 pp. $5.95, paper $1.95. 67–14205. (C)
Anyone interested in science-government relations and resource allocation should read this book. The imbalance between social and scientific criteria that determines resource allocation for research is completely and authoritatively examined. Weinberg also deals with the promise, the choices, and the institutions of Big Science. This volume will stimulate informed, nonpolemic scientific discussion, and perhaps spark intelligent public demands.

Weisskopf, Victor F. *Knowledge and Wonder: The Natural World as Man Knows It* (rev. ed.). Doubleday, 1962. 222 pp. illus. $4.95, paper $1.45. 62–15935. (SH)
Beginning with a discussion of space and time and the dating of universal events, this eminent physicist lays a foundation for discussing properties common to all phenomena. A PSSC collateral reading book.

501 PHILOSOPHY OF SCIENCE

Benjamin, A. Cornelius. *Science, Technology, and Human Values*. U. of Missouri, 1965. x+296 pp. illus. $7.00. 65–10698. (SH–C)
An explanation of what science is and how it works from the philosophical point of view. The author provides many suggestions for further reading. Any reader will gain an introduction to the language and concepts of the philosophy of science. Although useable by students in senior high school, the book more nearly fits an introductory college course.

Bronowski, J. *Insight: Ideas of Modern Science*. Harper & Row, 1964. 108 pp. illus. $5.95. 64–19950. (SH–C)
Presents modern science as a "web of ideas" relevant to the humanities and contemporary life in general. Commences with a discussion of beginnings: the beginning of reason and imagination in a young child, the evolution of the elements and the universe, the state of organic homogeneity eventually leading to more complex and diverse organisms. The book concludes with discussions by prominent artists and scientists showing the interrelationship of modern science to modern society. Written clearly and well illustrated with photographs. Aimed at the nonspecialist. No index or bibliography.

Casserley, J. V. *In the Service of Man*. Regnery, 1967. ix+204 pp. $4.95. 67–28493. (SH–C)
Casserley, a cultural conservative, tries to estimate the nature and necessity of the impending social, political and economic changes. He reviews values of Western Civilization and considers how these changes can be negotiated without threatening the values concerned with preserving our identity.

Crick, Francis. *Of Molecules and Men.* U. of Washington, 1966. xv+99 pp. $3.95, paper $1.95. 66–26994. (SH–C)
The nature, formation, and function of proteins and nucleic acids in viruses and cells are clearly stated in layman's terms. In time, Crick states, biological problems such as mechanisms of control, cell division and morphogenesis, will be understood in exact physical and chemical terms. Concludes with a stimulating discussion of such controversial issues as present-day teaching of religion, what students should study in science, what is man, and what will be the new culture. It will be useful in stimulating discussions in science and liberal arts courses including philosophy and theology, and in core courses on the nature of man.

Davies, J. T. *The Scientific Approach.* Academic, 1965. x+100 pp. illus. $5.50, paper $2.75. 65–27734. (SH–C)
An interesting, nontechnical primer on the philosophy of science. Gives the layman insight and perspective into rational, scientific approach, without extending itself beyond the elementary. Many useful examples, diagrams, and photographs supplement the text.

Priestley, John B. *Man and Time.* Doubleday, 1964. 319 pp. illus. 28 cm. $14.95, paper, 1968, $1.25, Dell. 64–22384. (C)
Summarizes theories and concepts of time and its measurement from the viewpoints of history, philosophy, the sciences, and the humanities, with about a third of the text devoted to Priestley's own speculations. Illustrations, some colored, from manuscripts and works of art. More appropriate for leisurely reading than reference purposes.

McCain, Garvin, and Erwin M. Segal. *The Game of Science.* Brooks/Cole, 1969. xiv+178 pp. illus. $2.95. (paper). 69–11975. (SH–C)
Sheds light on the frequently misunderstood and often misrepresented domain called science. The authors provide an interesting and direct introduction to the rules and concepts of the science game and comment on the slippery ideas of scientific law, explanation, theory and the general nature of scientific inquiry. They also scrutinize the specific types of scientific practitioners, and attempt to relate science to broader aspects of social change and the problems of social ethics, free will, and related ideas.

503 SCIENCE DICTIONARIES, ENCYCLOPEDIAS, AND DIRECTORIES

American Men of Science; A Biographical Directory (11th ed., 1965–1969). Edited by The Jacques Cattell Press. *The Physical and Biological Sciences* (6 vols.); *The Social and Behavioral Sciences* (2 vols.). Bowker. $25.00 per vol.; $200.00 per set. (Semiannual supplements.) (SH–C)
Brief biographical sketches on about 150,000 scientists who have at least attained the doctorate with continued activity in scientific work, or who have demonstrated research abilities of high proficiency as evidenced by publications, or attainments. Data for each entry included: name, address, field of specialization, place and date of birth, degrees, positions held, memberships and research specialties, honors received, etc.

Eggenberger, David I. (Exec. ed.). *McGraw-Hill Basic Bibliography of Science and Technology*. McGraw-Hill, 1966. 738 pp. $19.50. 66–14809. (SH-C)
A selected and annotated listing of some 8,000 primary reference sources. Bibliographic entries, including brief descriptive annotations, are listed under subject headings corresponding to the titles of articles in the *McGraw-Hill Encyclopedia of Science and Technology*. The unique arrangement will be more useful to students and research workers than standard comprehensive bibliographies arranged according to standard library classification schemes. It will not be as useful to librarians as an acquisition guide as standard bibliographies and book lists. A necessary purchase for senior high school, public, college, and technical libraries.

Eggenberger, David I. *McGraw-Hill Encyclopedia of Science and Technology* (2nd ed.). McGraw-Hill, 1966. 15 vols. illus. $2.95 (discount to schools and libraries). 65–26484. (SH–C)
The most complete and authoritative scientific and technological reference work in print. More than 7200 articles by over 2000 professional contributors are included. It is needed where more advanced reference material is required than is found in standard general encyclopedias. It should be purchased by all public, college and university libraries, and by senior high schools with good curricula in the sciences and mathematics. The set is upated by a yearbook which also provides a resume of major developments of the previous year.

Eggenberger, David I. (Exec. ed.) *McGraw-Hill Modern Men of Science* (Vols. I and II). McGraw-Hill, 1966, 1968. vi+620, xiv+679 pp. $19.50 each. ($12.50 each to subscribers of *McGraw-Hill Encyclopedia of Science and Technology*). 66–14808. (SH–C)
No other comparable source contains so much valuable information on the current and recent happenings in science and technology, related by and to the men who are in the forefront of modern scientific enterprise. Volume I contains 426 brief biographies of scientists who have won recognition since 1940. Volume II extends coverage to another 420 individuals, updates previous entries, has a composite table of contents to both volumes, and an analytical, classified index.

Feldman, Robert J. (Managing Editor). *Cowles Encyclopedia of Science and Technology* (rev. ed.). Cowles Education Corp., 1969. 639 pp. illus. $17.50. 69–17310. (JH–SH)
A substantial revision of the 1967 edition. A much needed section on mathematics has been included, the articles are arranged in a more logical sequence, and the entire volume appears to be quite current up to the early fall of 1968.

Ford, Charles A. (Editor-in-chief). *Compton's Dictionary of the Natural Sciences* (2 vols.). Compton, 1966. 879 pp. illus. $34.95 set; school and library ed., $24.95. 65–27264. (JH–SH)
The scope is the natural sciences *per se:* earth sciences, astronomy, botany, and zoology. It is more than a dictionary because the definitions, each accompanied by an illustration usually in color, are precise enough to provide informative detail. Should have very strong appeal to younger students

and nonspecialist adults as a first source of information. Excellent charts and tables provide such information as animal and plant classification, characteristics of the earth, information on meteorological phenomena, characteristics of common metallic elements, astronomical information, etc. Glossary of scientific terms. A useful reference set for elementary and secondary school libraries, public libraries, and home libraries.

Kondo, Herbert (Editor-in-chief). *The Book of Popular Science* (10 vols.). Grolier, 1970. illus. $89.50. (discount to schools and libraries). 78–88108. (JH)

An outstanding set for general background reading in all phases of science and technology, and also useful for reference with the aid of the index in Volume 10. Outstanding contributors have prepared authoritative and informative articles organized in 14 departments: "The universe; The earth; Life; Plant life; Animal life; Man; Health; Mathematics; Matter and energy; Industry; Transportation; Communication; Science through the ages; and Projects and experiments." Bibliography. Kept current by extensive annual revision.

Scientific American Resource Library. Freeman, 1969. 15 vols. illus. 29 cm. $150 per set; $10 per volume. (SH–C)

(1) *Readings in the Earth Sciences.* (Introd. by William H. Matthews III.) 2 vols. 622 pp. 69–15600.

(2) *Readings in the Life Sciences.* (Introd. by Marston Bates.) 7 vols. 2832 pp. 71–87739.

(3) *Readings in the Physical Sciences and Technology.* (Introd. by Isaac Asimov.) 3 vols. 1171 pp. 78–87738.

(4) *Readings in Psychology.* (Introd. by Harry F. Harlow.) 2 vols. 829 pp. 73–80078.

(5) *Readings in the Social Sciences.* (Introd. by Harry F. Harlow.) 1 vol. 360 pp. 78–87222.

Hundreds of offprints of articles published in *Scientific American* since 1948 have been carefully selected to make up this series of reference works which will be a boon to readers who are searching for some of the most authoritative and significant articles in the various differences of the pure, applied and behavioral sciences. Each offprint is serially numbered and those numbers are used in the indexes. In each volume of the five series there are separate author and topic indexes for every volume in the series, and also there is a cumulative topic index for the entire series which includes references to related articles in the other subject series. The topical indexes, carefully organized under collective subject headings make the set valuable for collateral reading and reference by students and their teachers.

Tressler, Arthur G. (Exec. ed.). *Science Year: The World Book Science Supplement, 1965–66–67–68–69.* Field, 1965–69. illus. $6.95 (to owners of *World Book Encyclopedia*, $5.95). 65–21776. (JH–SH–C)

This outstanding yearbook, first published in 1965, provides insights into the current status and progress of science and technology. Each volume is begun with a photographic essay, followed by an opening report of general scientific importance. A section on special reports of current and lasting interest is included. A "Science file" contains numerous articles and reports on the latest scientific developments, precedes a list of major scientific awards and prizes, their recipients, and a necrology of notables, which closes each volume. Essential for all libraries.

Van Nostrand's Scientific Encyclopedia (4th ed.). Van Nostrand, 1968. ix+ 2008 pp. 2,000 illus. 30.5 cm. $42.75. 68–20922. (JH–SH)
A well-known, authoritative one-volume encyclopedia of science and technology that is useful in laboratories and classrooms, and libraries. Now current in all topics.

504 SCIENCE—ADDRESSES, ESSAYS, AND LECTURES

Asimov, Isaac. *Is Anyone There?* Doubleday, 1967. xiii+320 pp. $5.95. 67– 12879. (JH–SH)
A collection of 37 essays on science dealing with the question of man in the universe. Some of the essays are humorous, others deal with well-polished facets of modern science, but each is valuable in itself.

Bernstein, Jeremy. *A Comprehensible World: On Modern Science and Its Origins.* Random, 1967. 269 pp. $4.95. 67–12740. (SH–C)
A collection of popular essays concerning science, scientists, and the world of physics. Includes exciting personal insights of many famous physicists, as well as clearly defined concepts, book reviews, and science fiction highlights.

Bronowski, Jacob. *The Common Sense of Science.* Harvard, 1953. 154 pp. $2.75; Random (Vintage), paper 95¢. A53–9924. (SH)
A classic little book by a distinguished scientist and man of letters who believes that there is no broad chasm between the sciences and the humanities. He discusses the essential nature of science and its inescapable involvement in all human activity.

Conant, James B. *Modern Science and Modern Man.* Columbia, 1952. 111 pp. $3.50; Doubleday (Anchor A10, 1959), paper 95¢. 53–35926. (C)
Discusses four important topics of interest to the layman: "Science and technology in the last decade," "The changing scientific scene, 1909–1950," "Science and human conduct," and " Science and spiritual values."

Jones, Howard Mumford, and I. Bernard Cohen. *A Treasury of Scientific Prose:A Nineteenth-Century Anthology.* Little, Brown, 1963. 372 pp. $7.50. 62–10533. (SH)
A well-known English professor and a noted historian of science have selected items of exquisite expository prose from the works of well-known scientists. The substitution of this book for some of the unappealing and traditional assignments in high school English classes should enliven interest and broaden students' perspectives.

Kepes, Gyorgy (ed.). (1) *Education on Vision;* (2) *The Nature and Art of Motion;* (3) *Structure in Art and in Science;* (4) *The Man-Made Object;* (5) *Module, Proportion, Symmetry, Rhythm;* (6) *Sign, Image, Symbol.* (Vision + Value Series.) Braziller, 1965, 1966. 233, 195, 189, 230, 233, 281 pp. 28 cms. illus. Trade ed. $12.50 ea,; student ed. $8.50 ea. 65–10807 (vols. 1–3), 66–13046, 66–13044, 66–13045. (SH–C)
Six handsome volumes of illustrated essays. The authors—physicists, paint-

ers, architects, designers, educators, and museum directors—cover a wide range of viewpoints showing the relevancy of many disciplines to the solution of problems of visual communication in our time. This set of outstanding books cannot fail to appeal to thoughtful readers interested in any aspect of science, the arts, or the humanities. They are worthwhile acquisitions for all secondary school, academic, and public libraries.

Lakoff, Sanford A. *Knowledge and Power: Essays on Science and Government.* Free Press, 1966. x+502 pp. $9.95. 66–23079. (SH–C)
The association of science and government, resting upon a base of tremendous economic development, is the subject of this fascinating book of selected essays by 15 celebrated writers. Part 1 contains essays on "Cases and controversies" to illustrate instances where bureaucracy and science have conflicted. Part 2 deals with the government's role in the promotion and financing of science. Here, special attention is given to the President's Science Advisory Committee. Part 3 deals with an overview of knowledge and power—of policies, programs, and prognostications concerned with improvements in the relation between science and government.

Medawar, P. B. *The Art of the Soluble.* Methuen (dist. by Barnes & Noble), 1967. 160 pp. $4.95. (SH–C)
Nine essays on what a scientist is and what kind of reasoning leads to scientific discovery. This is an exposition on scientific research and method.

Selye, Hans. *From Dream to Discovery: On Being a Scientist.* McGraw-Hill, 1964. xiv+419 pp. $6.95. 63–23306. (SH–C)
This is not an autobiography, but it is a series of reflections on his life's work. Although it is not an easy book to digest, prospective young scientists need to read and reflect on its main thesis in order to appraise their own personal attributes and to understand fully the "how to" of science: behavior, working, reading, thinking, writing, and speaking.

Shapley, Harlow. *Beyond the Observatory.* Scribner's, 1967. 222 pp. $4.50. 67–14493. (SH–C)
A collection of refreshing and stimulating essays for all students and laymen, prepared by the former director of the Harvard Observatory. They concern all humanity and encourage one to seek a better understanding of the universe of which he is a part and of his inescapable role therein.

Stratton, Julius A. *Science and the Educated Man: Selected Speeches of Julius A. Stratton.* M.I.T. Press, 1966. viii+186 pp. $5.00. 66–29173. (SH–C)
Contains many wise observations about the nature and place of science in our culture, the difference between science and engineering, and about significant elements of our modern society. Shows the breadth and depth of the concerns of an intelligent and articulate individual.

Weaver, Warren. *Science and Imagination: Selected Papers of Warren Weaver.* Basic, 1967. xvi+295 pp. $5.95. 67–28325. (SH–C)
A collection of the author's essays on a wide range of topics, all related to the broad aspects of science. Essays deal with areas such as the characteristics of science and scientists, science versus technology, scientific

freedom, science and religion, and the imperfections of science. Much of the misunderstanding about science can be reduced by reading this work.

506 SCIENTIFIC ORGANIZATIONS

Carmichael, Leonard, and J. C. Long. *James Smithson and the Smithsonian Story.* Putnam's, 1965. 316 pp. illus. $7.95. 65–20672. (JH–SH)
A commemorative history of the Smithsonian Institution and an illustrated description of its current organization and diversified activities by Dr. Carmichael, the seventh Secretary; together with a carefully researched biography of James Smithson by Dr. Long. Appendices list the officers and staff (as of 1965), a bibliography of principal sources, and an index. This work is a long needed and authoritative contribution to the history of science.

Cochrane, Rexmond C. (James R. Newman, Consultant.) *Measures for Progress: A History of the National Bureau of Standards.* U.S. Dept. of Commerce, National Bureau of Standards, 1966. xxv+703 pp. illus. $5.25. (Purchase from Supt. of Documents, U.S.G.P.O., Washington, D.C. 20402.) (SH–C)
This thorough and scholarly history of the origin and development of a key scientific agency of the United States Government, involved in highly skilled scientific and technical work on all phases of the physical sciences, has been written from the point of view of the men who established the agency and of the people who accomplished its first 50 years of service. One learns here of the improvements in standards in the basic international qualities of length, mass, time, temperature, electrical current, and luminous intensity. Contributions of the Bureau in scientific research and development also are related. Indexed. All university, college, technical, large high school and public libraries should own it.

Gurney, Gene. *The Smithsonian Institution.* Crown, 1964. 129 pp. illus. $3.95. 64–23813. (JH–SH)
A pictorial story of the building, exhibits and activities of America's greatest and most widely-known scientific organization. Introduces the Arts and Industries Building, the Museum of Natural History, the Museum of History and Technology, the National Air Museum, three world-famous art galleries, the National Zoological Park, and the Astrophysical Laboratory at Cambridge, Massachusetts. It is a useful general guide and memento.

Karp, Walter. *The Smithsonian Institution.* Smithsonian Institution Press (dist. by Random), 1968. 128 pp. illus. $2.95. 65–61775. (JH–SH)
A beautifully illustrated history of the Smithsonian Institution, prepared by the editors of *American Heritage.* The scope and diversity of its activities and the Institution's all-encompassing research in science, industry, and the arts are reported.

Museums Directory of the United States and Canada (2nd ed.). Smithsonian Institution Press (published with American Association of Museums), 1965. xvii+1039 pp. $7.50. 61–9712. (JH–SH–C)
A complete reference for all types of museums and related institutions that

lists name, address, officers and department heads, founding date, major collections, special holdings, activities, hours, admission charge, etc.

National Academy of Sciences. *Scientific and Technical Societies of the United States.* National Academy of Sciences, Publication 1499, 1968. 221 pp. 28.6 cm. $12.50. 27–21604. (C)
This eighth edition of the directory has been streamlined to include only membership societies devoted to a particular discipline and those societies within the United States. Each entry consists of name, address, principal officers, brief history, statement of purpose, details of membership and meetings, and listings of special prizes and awards, and publications. An annual supplement will update this edition.

507 SCIENCE—STUDY AND TEACHING

Ashford, Theodore Askounes. *The Physical Sciences: From Atoms to Stars* (2nd ed.). Holt, Rinehart and Winston, 1967. xi+736 pp. illus. $11.50. 67–11739. (SH–C)
An excellent, revised, text for a college survey course of the physical sciences, involving an interdisciplinary treatment of fundamentals of physics, chemistry, astronomy, geology, and mathematics. Useful as collateral reading for secondary school students.

Booth, Verne H. *Physical Science: A Study of Matter and Energy* (2nd ed.). Macmillan, 1967. xvi+742 pp. illus. $8.95. 67–16052. (SH)
Although intended for a college survey course, this is a valuable reference for teachers of general science and for secondary school students with a background in mathematics.

Brooks, Stewart M. *Integrated Basic Science* (2nd ed.). Mosby, 1966. 507 pp. illus. $8.50; lab. manual workbook (1964), $4.75. (SH–C)
A general and integrated orientation for the nonspecialist that provides a basic survey of the essentials of physics and chemistry and builds thereon a survey of microbiology, human biology, and pathology.

Barrett, Raymond E. *Build-It-Yourself Science Laboratory.* Doubleday, 1963. xvi+340 pp. diagrs. $4.95. 62–12920. (JH–SH)
A detailed guide to equipment and facilities for individual student research at home, or in a school that does not have a standard well-equipped laboratory. A separate section is provided for each major scientific discipline, to which is appended a list of suggestions for student investigations.

Barzun, Jacques, and Henry F. Graff. *The Modern Researcher.* Harcourt, Brace, 1957. xiii+386 pp. $8.25 (text ed. $5.95), paper $1.95. 57–10615. (SH)
An indispensable guide and manual for anyone engaged in investigation, research, bibliographic searching, abstracting and writing reports and manuscripts, irrespective of the field of study or interest. High school English and science teachers might profitably employ this manual in the instruction of all college-bound students.

Edelson, Edward. *Parents' Guide to Science.* Crowell, 1966. 212 pp. illus. $4.95. 66–14935. (SH–C)
This is the first major attempt to present a truly informative and integrated

account of the new developments in education that have resulted from the work of the Physical Science Study Committee, the Chemical Bond Approach, the Chemical Education Materials Study, the Biological Sciences Curriculum Study, the Earth Sciences Curriculum Project, the School Mathematics Study Group, the University of Illinois Committee on School Mathematics, and other groups—most of which have been financed by the National Science Foundation. This compact and informative work is ideal reading for teachers, for parents of students who are being trained under the new curricula, parents of children in schools that still "do things the old way," and school board members in such school systems. Glossary and reading list.

Holton, Gerald. *Introduction to Concepts and Theories in Physical Science.* Addison-Wesley, 1952. xviii+650 pp. illus. $10.00. 52–76666. (SH–C)
The title is fully descriptive, for this book does explain theories, concepts, and principles lucidly for those with a background of mathematics through elementary calculus. A good reference for superior students.

Joseph, Alexander, et al. *Teaching High School Science: A Sourcebook for the Physical Sciences.* Harcourt, Brace, 1961. xxx+674 pp. illus. $8.95. 61–11840. (SH)
An indispensable guide and reference work for teachers which will also be very useful to students conducting their own investigations at home or in school.

Krauskopf, Konrad, and Arthur Beiser. *The Physical Universe.* McGraw-Hill, 1960. viii+536 pp. illus. $8.50 (text ed. $7.95) ; instructor's manual $1.50; study guide $2.75. 60–6959. (SH–C)
Presents the major concepts of physical science—physics, chemistry, astronomy, and geology. A condensation and up-to-date version of a well-known text that has stood the test of 20 year's wide acceptance. Stresses the continuous and evolutionary nature of scientific endeavor.

Moore, A. D. *Invention, Discovery, and Creativity.* (Science Study Series S60). Doubleday, 1969. 178 pp. illus. $4.95; $1.45 (paper) (SH–C)
Abundant opportunities for personal experimentation, self-analysis, and independent thinking and action, that help the young adult to discover and understand just what qualities a creative and inventive scientist and engineer should possess are found within this book. It also contains suggestions as to how the young person can develop his potentiality for original thought, discovery, and invention. An appendix tells how devices and processes may be patented and lists America's greatest inventors. There is an excellent bibliography and a good index.

Stong, C. L. *The Scientific American Book of Projects for the Amateur Scientist.* Simon & Schuster, 1960. xxii+584 pp. illus. $5.95. 60–14286. (JH–SH)
The most explicit how-to-do-it book currently in print which will acquaint students with the essentials of the scientific method and assist them in beginning their own research studies. Examples in all major disciplines.

509 HISTORY OF SCIENCE

Bettex, Albert W. *The Discovery of Nature.* Simon & Schuster, 1965. 380 pp. illus. 32 cm. $24.95. 65–25050. (SH–C)
A magnificent volume featuring paintings, woodcuts, drawings, etchings,

charts, portraits and diagrams selected from many sources with accompanying legends and a text that give historical highlights of man's discovery and gradual enlightenment concerning the natural world. The approach is through a series of historical essays which emphasize the work of a limited number of the chief contributors to advancement of knowledge.

Boas, Marie. *The Scientific Renaissance 1450–1630.* (The Rise of Modern Science Series.) Harper & Row, 1966. xi+376+11 pp. illus. $6.95, paper $2.45. 62–8615. (SH–C)
This reprint of a 1962 original makes more widely available an outstanding contribution to the history of science. This was a period of world exploration, of important discoveries in all of the emerging scientific disciplines, and covered the dawn of "the scientific method" as represented by the experimental and inductive approach. All of the discussions are based primarily on original sources which are cited in the bibliography and notes.

Brumbaugh, Robert S. *Ancient Greek Gadgets and Machines.* Crowell, 1966. xiv+152 pp. illus. $4.95. 66–25430. (SH–C)
Explores the relationship between ancient Greek technology and the history of classical thought. Illustrated with over 30 drawings and 17 photographs of ancient gadgets and artifacts. A clear and sometimes humorous account of these gadgets and their effect on Greek philosophy. A witty introduction and excellent annotated bibliography add much to the book.

Crombie, Alistair C. *Augustine to Galileo: The History of Science, A.D. 400–1650.* Harvard U. 1953. xv+436 pp. illus. $5.50. 54–1443. (SH)
This important historical work covers the period between the ancient classical tradition that began with the rise of Christianity and the early beginnings of the modern world of science; a period of transition from mythology and speculation, first to the recording and interpretation of observations, then on to the development of the experimental method. The central theme is that the history of science is the history of thought.

Dampier, William. *A History of Science and its Relations with Philosophy and Religion* (4th ed.). Cambridge U., 1949–1966. xxvii+527 pp. $9.50, paper $3.45 (1966). 49–7999. (SH)
A comprehensive account of the origin, development, and accomplishments of science as representative of the greatest achievements of the human mind. Thus the influence of the developments in science on human thought and action is portrayed in a scholarly and thoughtful work.

Dampier-Whetham, William Cecil. *The Recent Development of Physical Science* (6th ed.). John Murray, London, 1927. xvi+313 pp. illus. $4.00. 65–29925. (SH–C)
Although listed in a trade publication among the "new books" published in 1966, this is a reissue or reprinting of the 1927 edition. It is a standard work valuable for reference.

Davis, Watson. *The Century of Science.* Duell, Sloan & Pearce, 1963. vi+313 pp. illus. $5.95. 63–16823. (JH)
The founding director of *Science Service* and the "father" of the science fair movement in America wrote this verbal and photographic panorama of

man's principal scientific and technological achievements during the past 60 years in some 30 fields of endeavor. An exciting reading adventure for students and laymen.

Debus, Allen G. *The English Paracelsians.* (The Watts History of Science Library.) Watts, 1966. 222 pp. illus. $5.95. 66–15981. (C)
An excellent evaluation of the contributions of Paracelsus and his followers. It is an American printing of an earlier English edition (Oldbourne Book Co., Ltd., 1965). The American printing is not as well bound and one plate has been deleted. Written in a concise and easy-to-read style and followed by an extensive listing of notes. These notes plus the excellent bibliography enhance its use as a reference.

Fortune, Editors of. *Great American Scientists.* Prentice-Hall, 1961. xii+144 pp. $3.95. 61–6215. (JH)
A succinct account of the physicists, chemists, astronomers and biologists of the 20th century whose efforts have placed America in a position of scientific leadership.

Gillispie, Charles Coulston. *The Edge of Objectivity: An Essay in the History of Scientific Ideas.* Princeton, 1960. 562 pp. $7.50, paper $2.95. 60–5748. (SH)
Lectures to Princeton students have been refined and enlarged in this portrayal of intellectual barriers faced by many of the outstanding figures in science, how these barriers were broken down by some scientists, and how others failed. Although the younger student will find the going hard, at times, the knowledge and insights derived will reward him.

Gilman, William. *Science: U.S.A.* Viking, 1965. xii+499 pp. $7.95. 65–23995. (SH–C)
Presents the development of science and technology since World War II and the present position of these "Siamese twins" in America's social, economic, and political structure. Assesses the value of scientific research both in money and in human happiness. Significant present-day research is well described, despite a few minor errors. The style is journalistic, the writing smooth and the tone serious without being ominous. Humor and human frailties are not neglected. Recommended collateral reading for all young people considering a career in the pure or applied sciences.

Halacy, D. S., Jr. *Science and Serendipity: Great Discoveries by Accident.* (Illus. by Frank Fretz.) Macrae Smith, 1967. 155 pp. $3.95. 67–26974.(JH)
Eleven examples of chance discoveries by famous scientists and inventors that have or will change our lives are given. A discussion of the background surrounding and the consequences of the discovery make for a thorough coverage.

Hogben, Lancelot. *Science for the Citizen* (rev. ed.). Norton, 1957. 1146 pp. illus. $12.50. 56–10849. (SH)
A well-known compendium of history of science, scientific method, and exposition of salient concepts and principles aimed at the general reader. The occasional interposition of philosophic considerations make it more than a book for informational reading.

Hoyt, Edwin P. *A Short History of Science: Vol. 1—Ancient Science; Vol. 2—Modern Science.* John Day, 1965, 1966. 256, 286 pp. illus. $4.95, $5.95. 64–20704. (JH–SH)
Volume 1 tells of the earliest beginnings of science in the Fertile Crescent and the Far East and concludes in the Middle Ages. It is chronologically arranged emphasizing more the persons and events than tracing the evolution and progressive development of concepts, ideas, and techniques. Volume 2 begins with the Renaissance and terminates in our present-day period. Here the arrangement is by subject matter in separate chapters. Each volume has a bibliography and an index. Interesting general background reading for the student and layman but lacks sufficient scientific depth and introspection to qualify as a reference source.

Jaffe, Bernard. *Men of Science in America* (2nd ed.). Simon & Schuster, 1958. xli+715 pp. illus. $6.95. 58–59443. (SH)
A narrative account of the development of a science in America related in terms of the lives and achievements of 20 outstanding men beginning with Harriot (1560–1621) and concluding with Fermi (1901–1954). Concludes with a general discussion of the future of American science and a bibliography.

Jones, Bessie Zaban. *The Golden Age of Science: Thirty Portraits of the Giants of 19th-Century Science by Their Scientific Contemporaries.* Simon & Schuster, 1966. xxxiii+659 pp. $12.00. 66–20254. (SH–C)
Thirty biographies of scientists, written by their contemporaries, were selected from the *Annual Reports of the Smithsonian Institution* to portray the life of scientists and the state of scientific research in the Nineteenth Century. The introduction by Everett Mendelsohn should be read with care by high school students interested in the sciences. Although the accounts were prepared by scholars of another century and may be difficult reading, they are packed with solid information.

McLanathan, Richard. *Images of the Universe; Leonardo da Vinci: The Artist as Scientist.* Doubleday, 1966. 192 pp. llus. $4.50. AC 66–10368. (SH–C)
Superb illustrations! The accompanying text is excellent in this fascinating and unusual glimpse of the mind and talent of Leonardo da Vinci. The book is concerned with him as a scientist, technologist, and engineer and is illustrated profusely with his drawings and sketches. Only the barest biographical details are given, but it does convey the almost limitless understanding of da Vinci as an inquiring scientist. Several pages of "Important dates in Leonardo's life and times," a short bibliography, and index.

Moulton, Forest Ray, and Justus J.. Schifferes (eds.). *The Autobiography of Science* (2nd ed.). Doubleday, 1960. xxxiii+748 pp. $6.95. 60–15193. (SH)
An anthology of the original writings of early precursors of science, of philosophers who gave birth to the first scientific concepts, of those whose labors laid the foundations of modern science and technology, and finally of some of our 20th century savants. A good introduction to history of science.

Newman, James R. *Science and Sensibility* (2 vols.). Simon & Schuster, 1961. 372, 309 pp. illus. $10.00 set. 61–12869. (SH)
These volumes are made up of essays or analytical reviews by the author, published in *Scientific American* and *The New Republic* mainly during the

period 1947–1960. They are good reading for the student or layman who wishes to improve his scientific literateness by becoming acquainted with some of the leading scientists and their accomplishments.

Sarton, George. *A History of Science.* Vol. 1: *Ancient Science Through the Golden Age of Greece;* Vol. 2: *Hellenistic Science and Culture in the Last Three Centuries B.C.* Harvard, 1952, 1959. xxvi+626 pp., xxvi+554 pp. illus. $12.00 ea. 52–5041. (C)
The author planned a complete historical survey of science in 8 volumes, but he died before the second volume was printed so the set will not be completed. Although written primarily for professionals, students can read them as the best account of the genesis of science in pre-Christian civilizations.

Sarton, George. *Six Wings: Men of Science in the Renaissance.* Indiana, 1956. xiv+318 pp. illus. $7.50. 56–11998. (SH–C)
The title is based on an allusion to the seraphim (Isaiah 6:2) and the "Wings" of this book are as follows: (1) Exploration and education; (2) Mathematics and astronomy; (3) Physics, chemistry, technology; (4) Natural history; (5) Anatomy and medicine; (6) Art and science. The illustrations are contemporary portraits of the scientists discussed.

Shapley, Harlow, Samuel Rapport, and Helen Wright (eds.). *The New Treasury of Science* (6th ed.). Harper & Row, 1965. xv+762 pp. illus. $10.00. 65–20439. (SH)
The authors have carefully examined all of the major literature of science to distill, in this one volume, outstanding pieces of writing that afford a broad panorama of man's knowledge of the universe and all that therein exists.

Singer, Charles. *A Short History of Scientific Ideas to 1900.* Oxford, 1959. xviii+525 pp. illus. $10.50, paper $2.95. 59–4952. (SH)
"These pages were written to give an elementary idea of how science came to occupy its distinctive position in the life of our own time. For this it is necessary to have some knowledge also of the Civilizations within which the science of the past has several times waxed and waned."—Introd. Since reading and enjoyment require only an elementary science background, this book is ideal for collateral reading.

Taton, René (ed.). *History of Science:* (1) *Ancient and Medieval Science;* (2) *Beginnings of Modern Science;* (3) *Science in the Nineteenth Century;* (4) *Science in the Twentieth Century.* (Tr. by A. J. Pomerans.) Basic, 1963, 1964, 1965, 1966. xx+552, xx+667, xxi+623, xxiv+638 pp. illus. $17.50 ea. 63–21689. (SH–C)
Surveys the history of science from its prehistoric beginnings to modern times. All fields, from mathematics through physical and biological sciences to medicine, are covered. Major developments are noted; minor ones usually not. No picture of a coherent and orderly development emerges, so the four volumes serve best as reference sources. Factually reliable and authoritative. Excellent photographs and illustrations. Bibliographies, indexes.

Wiener, Philip P., and Aaron Noland (eds.). *Roots of Scientific Thought: A Cultural Perspective.* Basic, 1957. x+677 pp. $10.00. 57–12414. (C)
An anthology of contributions of various writers that trace basic ideas, concepts, and modes of thought in the development of science. It deals with the

development of the dominant scientific ideas from the time of the ancient Greeks to modern cosmologists, and the approach is interdisciplinary.

Weaver, Warren. *Scene of Change: A Lifetime in American Science.* Scribner's, 1969. 226 pp. illus. $6.95. 79–85247. (SH–C)
Not only does Weaver portray his long, rich, and varied experience as a scientist who refused confinement in an ivory tower for a life dedicated to the advancement of science in general, but he also reflects his many associations with some of the world's top scientists. The book ends with two essays which are valuable contributions to the philosophy of science and which reveal the warmth, depth and sincerity of a great scholar.

510 MATHEMATICS

Adler, Irving. *Magic House of Numbers.* John Day, 1957. 128 pp. illus. $3.75; LB $3.49, NAL (Signet P2117), paper 60¢. 57–5978. (JH)
Mathematical curiosities and riddles hold interest, but more than mere entertainment makes this primer an excellent introduction to the whole basis of our number system.

Adler, Irving. *The New Mathematics.* John Day, 1958. 187 pp. illus. $4.50; NAL (Signet P2099, 1960), paper 60¢. 58–12132. (SH)
Advanced mathematical concepts are presented on a relatively elementary level by expanding the number system from natural numbers to integers, to rational numbers, to real numbers, to complex numbers.

Ahrendt, Myrl H. *The Mathematics of Space Exploration.* (Holt *Library of Science.*) Holt, Rinehart and Winston, 1965. 160 pp. illus. $2.95, paper $1.96. 65–23270. (JH–SH)
The purpose is to present in simple form enough of the applications of mathematics in the space program to enable the student with two or three years of secondary school mathematics (algebra and logarithms) to form an appreciation and understanding of the role of mathematics in the exploration of space. The book is structured about topics in space science rather than about topics in mathematics, an arrangement which gives the book continuity as an introduction to space science. The student who enjoys solving simple equations will find the book delightful. Teachers of mathematics, physics and general science will find the "space flights" a rich source of illustrative problems with an up-to-date flavor.

Bakst, Aaron. *Mathematics: Its Magic and Mastery* (3rd ed.). Van Nostrand, 1967. 842 pp. illus. $8.75. 67–23114. (JH–SH)
A comprehensive presentation of the basic mathematical concepts through algebra, geometry, and trigonometry. Entertaining text and examples are combined to demonstrate these concepts. Interest is further stimulated with numerous mathematical brain-teasers. Updated with new material on coordinate systems, elementary mathematics on space vehicles, and game theory.

Beck, Anatole, Michael N. Bleicher, and Donald W. Crowe. *Excursions into Mathematics.* Worth, 1969. xxi+449 pp. illus. $10.75. 68–57963. (SH–C)
Six topics: Euler's formula and related topics, perfect numbers, area, geometries, game theory, and number theory are rigorously developed and

clearly explained in an unsophisticated manner. Mathematical induction is developed in each of the first three chapters and employed in each of the last three.

Court, Nathan A. *Mathematics in Fun and in Earnest.* Dial, 1958. 250 pp. illus. $4.75; NAL (Signet), paper 60¢. 58–11431. (SH)
A series of independent essays concerning mathematics in culture, its philosophical and sociological aspects, and its role as a basic science. The concepts of infinity and the axiomatic method in mathematics are discussed at length.

Dantzig, Tobias. *Number: The Language of Science* (4th ed.). Macmillan, 1954. 340 pp. illus. $6.95; Doubleday (Anchor A67), paper $1.25. 54–8531. (SH)
Uses the historical method "to bring out the role intuition has played in the evolution of mathematical concepts." Number theory is viewed as symbolic of and a product of ideas, avoiding the purely technical and mechanical aspects of the procedures involved.

Dörrie, Heinrich. *100 Great Problems of Elementary Mathematics.* (Tr. by David Antin.) Dover, 1965. x+393 pp. illus. $2.00 (paper). 65–14030. (SH–C)
Problems are clearly stated and their solutions can be followed by the patient reader. The author divides his selections into six groups: arithmetical, planimetric, concerning conic sections and cycloids, stereometric, nautical and astronomical, and extremes. Each group has both simple and complex problems. Simultaneously, many worthwhile ideas, definitions and structures are introduced where appropriate. Should be in every secondary school library and will prove valuable to any math club.

Dubisch, Roy. *The Nature of Number: An Approach to Basic Ideas of Modern Mathematics.* Ronald, 1952. xi+159 pp. illus. $5.50. 52–6182. (SH)
Explores the new ideas about mathematics which are the basis for the expanding subject matter of pure mathematics, concentrating on the abstractions from ordinary arithmetic which lead to modern mathematics.

Freund, John E. *A Modern Introduction to Mathematics.* Prentice-Hall, 1956. xvi+543 pp. illus. $9.25. 56–7146. (SH)
Postulate thinking is augmented and directed by emphasizing basic number theory and the abstract nature of mathematics. Although the traditional topics such as algebra, trigonometry, analytic geometry, and introductory calculus are covered, the emphasis throughout is on basic concepts and ideas.

Fujii, John N. *An Introduction to the Elements of Mathematics.* Wiley, 1961. 312 pp. illus. $6.95. 61–15397. (SH)
A modern approach to the "new" mathematics is presented, stressing ideas and concepts rather than computation. Symbolic logic and argument, set theory, the function concept, and probability are among those topics discussed.

Glenn, William H., and Donovan A. Johnson. *Invitation to Mathematics.* Doubleday, 1962. 373 pp. illus. $6.50. 62–8932. (JH)
An introduction at the junior-high level to the pure science itself, to measurement, graphing, short cuts in numerical computation, computing devices, and statistics.

Grossman, Israel, and Wilhelm Magnus. *Groups and Their Graphs*. (New Mathematical Library, No. 14.) Random, 1964. vii+195 pp. illus. LB $2.95, paper $1.95. 64–8512. (SH–C)

The concept of a group is developed painstakingly. By the end of Chapter 4 the conscientious reader finds himself with a good understanding of groups. The authors use this foundation to discuss such topics as generators, subgroups, mappings, permutation groups, etc. Ideally suited for the senior high and junior college level. Also serves as excellent background reading for elementary and secondary school teachers.

Haag, Vincent H., and Edwin A. Dudley. *Introduction to Secondary Mathematics* (2 vols.). Heath, 1967. Vol. 1: ix+306 pp. illus. $4.68; Vol. 2: ix+339 pp. illus. $4.80. (JH–SH)

Originally published in 1964 and 1965. There appears to have been no revision of the text, but two new chapters have been added on "Experiments in geometry" and "Functions—an epilogue." Incorporates both "modern" content and a discovery approach. Carefully structured exercises are designed to lead the student to an understanding of new ideas. The computational skills in arithmetic typical of general mathematics courses are covered. The conceptual flavor is that of the SMSG texts. Two- and three-dimensional geometry is introduced by a nonmetric approach. Though the geometry is intuitive, there is stress on the use of accurate terminology. A chapter on probability is an interesting and less standard beginning.

Henkin, Léon, et al. *Retracing Elementary Mathematics*. Macmillan, 1962. 418 pp. illus. $6.95. 62–7986. (SH–C)

Intended primarily for teachers or prospective teachers, and for laymen, providing not only a review of traditional techniques, but a basic training in rational and irrational numbers, theory of groups, set theory, axiomatic logic, etc.

Hogben, Lancelot. *Mathematics for the Million* (3rd ed.). Norton, 1951. xiv+ 697 pp. illus. $8.95, paper 95¢. 51–8025. (JH)

A survey of the entire field of mathematics, from arithmetic to calculus. Basic number theory is liberally interspersed with practical applications of geometry, trigonometry, and algebra.

Hooke, Robert, and Douglas Shaffer. *Math and Aftermath*. Walker, 1965. xii+ 233 pp. illus. $5.95. 64–23896. (SH–C)

Samples of the applications of mathematics to such varied areas as astronomy, suspension cables, clocks, violins, aerodynamics, traffic problems, factory production, and reliability testing. The mathematics ranges through calculus, linear programming, probability, and statistics. The style is informal but, in places, the mathematics is nontrival.

Kasner, Edward, and James Newman. *Mathematics and the Imagination*. Simon & Schuster, 1940. 380 pp. illus. $4.50, paper $1.95. 40–27575. (SH)

Includes an explanation of the "new" mathematical terminology, various geometries, many fascinating puzzles, and an introduction to the calculus. Footnote references to original sources and concrete proofs of the problems discussed make this treatment of interest to the more erudite mathematician.

Katsoff, Louis O., and Albert J. Simone. *Foundations of Contemporary Mathematics*. McGraw-Hill, 1967. xiv+533 pp. illus. $9.50. 67–12327. (SH–C)

Covers topics in finite mathematics, calculus, linear algebra, and the applications of mathematics to linear programming and decision theory at an elementary level. Represents a successful marriage of pure and applied mathematics. Useful for both high school and college students.

Kline, Morris. *Mathematics and the Physical World.* Crowell, 1959. 482 pp. illus. $6.00; Doubleday, paper $1.95. 59–5252. (JH)
Mathematics is viewed in its relationship to the physical sciences as the basic tool of research scientists. The role of mathematics in the study of nature is surveyed in all phases of the subject from arithmetic to calculus.

Korn, Granino A., and Theresa M. Korn. *Manual of Mathematics.* McGraw-Hill, 1967. 391 pp. illus. $6.95. 66–23624. (SH–C)
Even though the coverage is somewhat thin, this is a tight-packed compendium of definitions, theorems, and formulas of undergraduate engineering mathematics. For the college mathematics major there is no other more complete work.

Land, Frank. *The Language of Mathematics.* Doubleday, 1963. 264 pp. illus. $4.95. 63–8772. (SH)
Explores the basis of our number system, including notation, units of measure, and calculations. Algebra, geometry and trigonometry are treated in their relation to the everyday world, while statistics are viewed in their application to the social sciences.

Lieber, Lillian R., and Hugh G. Lieber. *The Education of T. C. Mits.* Norton, 1944, 230 pp. illus. $6.95. 44–40067. (SH)
Amusing, yet philosophical; humorous, yet serious—mathematics is placed in its proper perspective as the basis for all scientific thought. The scientific method is presented by the use of elementary mathematical concepts.

Lieber, Lillian R. *Infinity.* Holt, Rinehart and Winston, 1953. 359 pp. illus. $5.00. 53–5355. (SH)
Pure number theory and the calculus are presented in the author's inimitable style. The interrelationship among the various mathematical disciplines is emphasized.

Lieber, Lillian R., and Hugh G. Lieber. *Mits, Wits, and Logic* (3rd ed.). Norton, 1960. 240 pp. illus. $4.50. 60–12023. (SH)
A thoroughly original approach to logical thinking which emphasizes the mathematical aspects. Elementary algebra and geometry are requisites.

Lieber, Lillian R. *Take a Number: Mathematics for the Two Billion.* Ronald, 1946. 221 pp. illus. $3.75. 47–1469. (SH)
The basic theory of elementary mathematics through geometry and algebra is presented in an intriguing manner—delightfully illustrated.

Maxwell, E. A. *Fallacies in Mathematics.* Cambridge U., 1959. 95 pp. illus. $3.50, paper 95¢. 59–16217. (SH)
Contains relatively advanced fallacious proofs: the fallacies are pointed out and the reader encouraged to learn through the errors of others. A fairly strong background in geometry and trigonometry and a little calculus are prerequisites for complete understanding.

Menninger, K. W. *Mathematics in Your World.* Viking, 1962. 291 pp. illus. $5.00. 62–8867. (JH)
Not a mathematics book, but a narrative account of the role that mathematics plays in human everyday life. This is for the student and average layman who would like a little "recreation" with his learning.

National Research Council. *The Mathematical Sciences: A Collection of Essays.* (Ed. George A. W. Boehm.). MIT Press, 1969. x+271 pp. illus. $3.95 (paper) 69–12750. (C)
This is a collection of essays on mathematics and its relation to the sciences and social sciences that range from the very abstract to the academic to the very concrete with contemporary applications. For the most part the essays should be intelligible to a mathematically bright high school student or any college student with one year of theoretical mathematics.

Newman, James R. *The World of Mathematics* (4 vols.). Vol. I: *Men and Numbers;* Vol. II: *World of Laws and the World of Chance;* Vol. III: *Mathematical Way of Thinking;* Vol. IV: *Machines, Music and Puzzles.* Simon & Schuster, 1956. 2535 pp. illus. $25.00, set; paper $9.95, set ($2.95; $2.75; $2.45; $2.25). 55–10060. (C)
An encyclopedic anthology of mathematics containing excerpts from original works with commentary by the author. This remarkable work leaves no aspect of the science unturned, but emphasis is placed on mathematical history, mathematics and the physical sciences, probability and statistics, mathematical logic, and mathematics in culture.

Newsom, Carroll V. *Mathematical Discourses: The Heart of Mathematical Science.* Prentice-Hall, 1964. 125 pp. $5.95. 64–10138. (C)
"The mathematizing of science has led to revolutionary changes in our civilization. . . . Thus a person in our time can hardly be characterized as educated unless he has some understanding of the fundamental concepts of mathematical sciences, their meaning, and their proper utilization."—Preface.

Niven, Ivan. *Mathematics of Choice or How to Count Without Counting.* (New Mathematical Library, No. 15.) Random, 1965. xi+202 pp. illus. LB $2.95, paper $1.95. 65–17470. (SH–C)
Collateral reading book prepared under the direction of SMSG. The opening chapters provide a careful discussion of permutations, combinations, and binomial coefficients. It is a book of problems, with background material for attacking them. Numerical answers, as well as sketches of solutions to the more difficult problems, are included. Should be available to capable high school and junior college students. Also valuable as background reading for teachers introducing work on probability in lower grades.

Ore, Oystein. *Graphs and Their Uses.* (New Mathematical Library, No. 10). Random, 1963. 131 pp. illus. LB $2.95, paper $1.95. 63–9345. (SH)
Shows how the theory of graphs may be employed in recording athletic statistics; biological and chemical experiments; mathematical problems; and diverse other practical situations. Another SMSG book.

Polya, George. *Mathematics and Plausible Reasoning.* Vol. 1: *Induction and Analogy in Mathematics;* Vol. 2: *Patterns of Plausible Inference.* Princeton U., 1954. 190, 280 pp. illus. $9.00 set (Vol. 1, $5.50; Vol. 2, $4.50). 53–6388. (C)
Through the psychology of creative mathematics the author shows how to attack a new problem and the trains of thought that may lead to a solution; the methods of deduction and analogy. The second volume uses the material

in the first as a basis and then goes forward to discuss questions involving psychology and philosophy, as well as mathematics, and their applications.

Reid, Constance. *From Zero to Infinity: What Makes Numbers Interesting* (3rd ed.). Crowell, 1965. 145 pp. illus. $3.95; paper (Apollo, 3rd ed., 1961) $1.95. 55–9200. (JH)
Each of the ten digits, zero through nine, is discussed individually, showing its historical development, its unique characteristics, and its particular usefulness in the everyday world.

Rényi, Alfréd. *Dialogues on Mathematics.* Holden-Day, 1967. 100 pp. illus. $4.95, paper $2.50. 67–13839. (SH–C)
Presents the foundations of mathematical thought, the birth of applied mathematics, and the important role of the mathematical method in discovering the laws of nature. This should be made available to and read by all high school students interested in mathematics.

Rosenthal, Evelyn B. *Understanding the New Math.* Hawthorn, 1965. 240 pp. illus. $4.95; Fawcett (Crest R784), paper 60¢. 65–1954. (JH–SH–C)
This is a book for parents to teach them about the "new math." It is also for teachers who are still not too proficient; lastly, it is good for junior-high students to work at. Definitions are good. Some history of mathematics is injected and adds interest and clarity. Questions and answers help the reader test his progress. Not intended for classroom use, but as a background and resource book.

Salkind, Charles T. (comp.). *The MAA Problem Book II.* (New Mathematical Library, No. 17.) Random, 1966. vi+112 pp. illus. $1.95 (paper). 66–15479. (SH–C)
A very useful compilation of the contest problems with suggested elementary solutions of the Annual High School Contests of the Mathematical Association of America, 1961–1965. Problems have been thoughtfully selected and graded and will present a challenge to even the high-ability student. A classified index enables the reader to refer quickly to examples of a special type of problem.

Sloan, Robert W. *An Introduction to Modern Mathematics.* Prentice-Hall, 1960. 73 pp. illus. $4.95. 59–15391. (C)
Mathematics based upon logical inference from axioms is treated through a study of such topics as directed numbers, set theory, the concept of the function, and logical truth tables.

Stein, Sherman K. *Mathematics: The Man-Made Universe.* Freeman, 1963. 316 pp. illus. $6.50. 63–7786. (C)
Mathematics is revealed in its true nature, as a product of the human mind, and this book presents selected topics from number theory, topology, set theory, geometry, and analysis.

Steinhaus, Hugo. *Mathematical Snapshots* (2nd ed.). Oxford U., 1960. 328 pp. illus. $7.50. 60–5104. (SH)
A visual excursion into the entire field of elementary mathematics, running the gamut of exercises from simple tricks and puzzles to more advanced problems. Emphasis is on learning through visual aids.

510.3 MATHEMATICS—DICTIONARIES AND ENCYCLOPEDIAS

James, Robert C., and Edwin F. Beckenbach (eds.). *James & James Mathematics Dictionary* (3rd ed.). Van Nostrand, 1968. vii+446 pp. illus. $13.50. 59–8655. (SH–C)
Combines features of both a dictionary and encyclopedia to present a correlated concept of mathematical concepts. Defines more than 8000; some 800 in excess of 2nd (1959) edition. Pronunciation of terms is marked. Lists of symobls and abbreviations, and miscellaneous tables are included. (Multilingual indexed edition also available.)

The Universal Encyclopedia of Mathematics. (Foreword by James R. Newman.) Simon & Schuster, 1964. 715 pp. illus. $8.95. 63–21086. (SH–C)
A translation of a good German reference book of mathematics that is " . . . designed to serve the needs of high school and college students, . . . encompasses many branches of mathematics from arithmetic through the calculus and includes a collection of essential formulae and tables. While the higher branches such as group theory or algebraic topology are not treated, the coverage within the limits indicated is succinct and to the point. . . ."— Foreword.

510.7 MATHEMATICS—STUDY AND TEACHING (See also 511–519)

Adler, Irving. *Groups in the New Mathematics: An Elementary Introduction to Mathematical Groups through Familiar Examples.* (Illus. by Ruth Adler and Ellen Viereck.) John Day, 1967. 274 pp. $7.95. 67–24635. (JH–SH)
An introduction to elementary group theory which should prove of most value to the casual general reader. Covers topics found in a one semester course in abstract algebra.

Allendoerfer, Carl B., and Cletus O. Oakley. *Principles of Mathematics* (2nd ed.). McGraw-Hill, 1963. xi+539 pp. $8.95; teacher's manual $2.00. 63–12123. (SH–C)
Intended for students who have completed intermediate algebra and a first course in trigonometry; therefore it is appropriate for advanced secondary school students or college freshmen. It is based on an introduction of much of the modern in mathematics and the treatment is enlivened with practical illustrations.

Brixey, John Clark, and Richard Vernon Andree. *Fundamentals of College Mathematics* (rev. ed.). Holt, Rinehart and Winston, 1961. xiv+750 pp. illus. $10.50; teacher's guide, paper $2.95; tables 50¢. 61–7863. (SH)
Presents an integration of college algebra, trigonometry, basic statistical reasoning, analytic geometry and elementary calculus suitable for the college freshman or the advanced secondary school student.

Cooke, Nelson M. *Basic Mathematics for Electronics* (2nd ed.). McGraw-Hill, 1960. 679 pp. illus. $10.75 (text ed. $7.50). 59–14441. (SH)
Beginning with a review of arithmetical computations, followed by elementary algebra, trigonometry, and logarithms, the text relates mathematical processes directly to electrical and electronic applications.

Courant, Richard, and Herbert Robbins. *What is Mathematics?* Oxford U., 1941. xix+521 pp. illus. $9.00 (text ed. $7.00). 41–25632. (SH)

Pure mathematical theory is presented to the educated layman, including such advanced topics as non-Euclidean geometry, topology, algebraic set theory, mathematical functions, and the concept of the limit. Emphasis is upon abstract reasoning and rigorous proofs.

Crowdis, David G. and Brandon W. Wheller. *Introduction to Mathematical Ideas.* McGraw-Hill, 1969. xii+352 pp. illus. $7.95. 68–27505. (SH–C)

For the non-mathematically oriented student, this book is a boon. Topical areas include numbers, numerals, and symbols; digital computers; symbolic logic, Boolean algebra; and probability and statistics. For the more advanced reader, excellent references are given for a greater challenge.

Dinkines, Flora. *Elementary Concepts of Modern Mathematics.* (Part I: *Elementary Theory of Sets,* paper $2.45; Part II: *Introduction to Mathematical Logic,* paper $1.45; Part III: *Abstract Mathematical Systems,* paper $1.45.) Appleton-Century-Crofts, 1964. x+457 pp. illus. $6.50 (hardbound Parts I, II, & III). 64–18143. (SH–C)

Provides basic material for branches of mathematics which have been newly developed or have found new applications in present-day science and technology, including automation systems and high-speed digital computers. Each of the three major parts of the book is relatively independent of the others and requires no more preparation than elementary algebra and plane geometry. Concepts are lucidly developed in a straightforward and detailed manner with numerous examples, illustrations, and exercises. Answers to exercises are included at the end of each part of the book.

Ferrar, W. L. *Mathematics for Science.* Oxford, 1965. xi+328 pp. illus. $4.80. 65–9644. (SH–C)

Designed to provide mathematical background for students of the physical sciences. The first part of the book for secondary schools covers essentially precalculus topics: trigonometry, complex numbers, inequalities, analytic geometry through conics and polar coordinates, and vectors through both scalar and vector products. The second part develops what constitutes two or three semesters of college calculus. Appropriate for the secondary school and junior college library, as a supplementary text for a capable class at either level, and—perhaps most important—as a teacher reference.

Hemmerling, Edwin M. *Elementary Mathematics.* (Technical Education Series.) McGraw-Hill, 1965. xiv+361 pp. illus. $5.95. 65–18746. (JH–SH)

Designed for students with a weak background who plan to study intermediate algebra, trigonometry or technical mathematics. Basic material is presented with many illustrative examples as well as a sufficient number of problems to fix the ideas and techniques in the mind of the student. The text will prove useful in technical schools that have many students with weak or marginal backgrounds. Even though the author has not followed the accepted modern curriculum, he has produced a useful text along traditional lines.

Hight, Donald W. *A Concept of Limits.* Prentice-Hall, 1965. xii+138 pp. illus. $4.95. 66–10183. (SH–C)

An excellent background on "Limits." Assumes a knowledge of high school algebra and geometry and was designed as a teacher-training text. Also valuable as a supplementary text for those high school courses intended to provide a good basis for calculus. Numerous exercises; answers for odd-numbered exercises are included, and answers for even-numbered ones are available in a separate booklet. Begins with historical accounts of limits, and progresses through limits of sequences and functions, continuity, and proofs of theorems, to the development and applications of a unifying concept of a generalized limit.

Jaeger, Chester George, and Howard Maile Bacon. *Introductory College Mathematics* (2nd ed.). Harper & Row, 1962. xvii+423 pp. illus. $7.95. 62–7139. (C)

The concept of the function is the thread which binds the text. Topics in algebra are interwoven throughout, while plane trigonometry is rigorously treated. The basic ideas of calculus are clearly put forth, but the treatment is generally superficial.

Kemeny, John G., J. L. Snell, and G. L. Thompson. *Introduction to Finite Mathematics.* Prentice-Hall, 1957. 372 pp. illus. $9.95. 57–7294. (SH)

Modern mathematics; i.e., set theory, probability theory, vectors and matrix algebra, linear programming and game theory is presented on an elementary level by confining the survey to the consideration of finite problems.

Kinsolving, May Risch. *Set Theory and the Number Systems.* International Textbook Co., 1967. xii+154 pp. $5.95. 67–12105. (SH–C)

Introduces the interested neophyte to formal mathematics through the development of set theory, natural numbers, integers, rational numbers, irrational numbers, and complex numbers. An interest and ability for abstract thinking is required.

Kline, Morris. *Mathematics: A Cultural Approach.* Addison-Wesley, 1962. 701 pp. illus. $9.75. 61–10970. (C)

Intended as a guide for a one-year terminal and cultural college course for nonscience or nonmathematics majors, which might be used also in secondary schools. The stress is on the practical application of mathematical ideas and concepts.

Mancill, Julian D., and Mario O. Gonzalez. *Contemporary Mathematics.* Allyn and Bacon, 1966. xii+590 pp. $12.65; text ed. $9.50. 66–10509. (SH–C)

Written as a two-semester course for the general liberal arts student in mathematics. Pays attention both to basic structure and operational facility. There is a unique organization of the material into two distinct sections: basic algebra and analytic geometry, and introduction to the calculus with trigonometry. Interwoven are careful introductory developments of matrices, probability, statistics, and linear programming. The final section on the elements of the calculus leads up to the treatment of areas, volumes, and maxima and minima problems. Student exercises are graded and there is a complete set of answers to problems.

National Council of Teachers of Mathematics. *Enrichment Mathematics for High School* (28th yearbook). NCTM, 1963. ix+388 pp. diagrs. $4.00, paper $2.50. 63–14060. (SH)

A valuable book for secondary school libraries, for the teacher working with talented students, or for the student who wishes to explore on his own. Contains a list of mathematics books for school libraries recommended by NCTM.

Niles, Nathan O. *Algebra and Trigonometry*. Wiley, 1965. x+399 pp. illus. $7.95. 65–16422. (SH)
An appropriate text for a one-semester course in the senior year of high school. It begins with set theory and the axioms of the real numbers and follows with a study of analysis of elementary polynomial, exponential, logarithmic, and trigonometric functions. Modern notation and terminology are used throughout and the diagrams are unusually clear. Excellent selection of problems. Should not be considered as a replacement for a comprehensive elementary analysis text. As a preparation for further work in mathematics, the book is entirely satisfactory.

Nunz, Gregory J., and William L. Shaw. *Electronic Mathematics*. Vol. 1: *Arithmetic and Algebra;* Vol. 2: *Algebra, Trigonometry, and Calculus* (2 vols. in 1). McGraw-Hill, 1967. xvii+358, 418 pp. illus. $6.00 ea.; set, $9.95. 67–15433.(C)
A wide range of the mathematics underlying electronic engineering is covered in this two-volume set. Highly recommended for a two-year college electronics course.

Rees, Paul K. *Principles of Mathematics*. Prentice-Hall, 1965. xiii+383 pp. illus. $8.25. 65–16592. (SH–C)
A revised and enlarged edition of a 1959 work, *Freshman Mathematics*. A book of pre-calculus mathematics and brief introduction to the calculus based on a direct approach with the absence of profound proof or discussion. The presentation of a principle is followed by worked-out examples which serve as models for the non-complicated problems that follow. Assumes previous training in elementary algebra, trigonometry and analytical geometry. The inclusion of topics such as income and property tax, statistics, charts, sets, and annuities suggest its usefulness to liberal arts and social science students. Tables, answers to exercises, and an analytical index.

Richardson, Moses. *Fundamentals of Mathematics* (3rd ed.) Macmillan, 1966. xx+603 pp. illus. $8.95. 65–20177. (SH–C)
This new edition of one of the best general surveys of mathematics at the advanced high school or elementary college level has been rewritten to conform with current trends in improving mathematics curricula. Especially suitable for social science students or for prospective secondary school teachers.

Schaff, William L. *Basic Concepts of Elementary Mathematics* (2nd ed.). Wiley, 1965. xix+384 pp. illus. $7.95. 64–25883. (SH–C)
An excellent example of "general mathematics" that is comprehensive and appropriate for reference or enrichment at the senior-high level, also for preservice training of elementary teachers. Its primary use might be a general mathematics course for liberal arts freshmen. The range is encyclopedic and hence lacks depth. Problems and a good bibliography are provided with each chapter.

Sharp, Henry, Jr. *Modern Fundamentals of Mathematics*. Prentice-Hall 1968. ix+390 pp. illus. $8.95. 68–13022. (C)
A standard college textbook with an integrated presentation by fundamental concepts. A successor to the author's *Modern Fundamentals of Algebra and Trigonometry* (1961).

LORETTE WILMOT LIBRARY
NAZARETH COLLEGE

Vance, Elbridge P. *Fundamentals of Mathematics.* Addison-Wesley, 1960. x+469 pp. $9.50. 60–5165. (SH)
A text which presents mathematics as a logical system and as a unification of basic ideas of algebra, trigonometry, analytic geometry, and introductory calculus—which usually are taught separately.

510.71 MATHEMATICAL RECREATIONS

Bakst, Aaron. *Mathematical Puzzles and Pastimes* (2nd ed.). Van Nostrand, 1954. vi+206 pp. illus. $5.95. 54–8490. (JH)
Amusing mathematical recreations are presented without sacrificing mathematical principles. Examples are taken from pure number theory, algebra, geometry, and trigonometry.

Ball, W. W. Rouse. *Mathematical Recreations and Essays.* Macmillan, 1960. xvi+418 pp. illus. $6.00, paper $1.95. 39–27626. (JH)
Mathematical fallacies, puzzles and games are presented to the mathematically inclined reader. Of special interest are the chapters on chessboard recreations, cryptography, and cryptanalysis.

Dudeney, Henry Ernest. *536 Puzzles & Curious Problems.* Scribner's, 1967. xii+428 pp. illus. $7.95. 67–15488. (JH–SH–C)
A staggering collection of Dudeney's best puzzles, arranged by problem type arithmetic, algebraic, geometric, combinatorial, topological, game, and others. Answers are included. Many a challenge to even the most skilled puzzle fiend.

Friend, J. Newton, *Numbers: Fun & Facts. More Numbers: Fun & Facts.* Scribner's, 1954, 1961. 208, 201 pp. $3.50 ea. 54–8690, 61–13364. (SH)
Collections of mathematical curiosities, puzzles and problems requiring only a knowledge of elementary algebra for manipulation, but requiring real ingenuity for solution.

Gamow, George, and Marvin Stern. *Puzzle-Math.* Viking, 1958. 119 pp. illus. $3.50. 58–5402. (SH)
A collection of mathematical problems involved in human situations, complete with solutions. Most of the puzzles concern the application of strict mathematical logic and probability.

Gardner, Martin. *The Scientific American Book of Mathematical Puzzles & Diversions.* Simon & Schuster, Book 1, 1964. 178 pp. illus. $4.50, paper $1.45. 59–9501. Book 2, 1961. 253 pp. illus. $4.95, paper $1.45. 61–12845. (SH)
Collections of puzzles and brain-teasers compiled partially from the author's column in the *Scientific American.* Provides the mathematically inclined student not only with hours of amusement and enjoyment, but also with a stimulating introduction to modern mathematical thought.

Glenn, William H., and Donovan A. Johnson. *Exploring Mathematics on Your Own.* Doubleday, 1960. 303 pp. illus. $4.95. 61–6195. (JH)
A book of recreational mathematics that includes an introduction to recent developments as well as a consideration of classical mathematical concepts.

Golomb, Solomon W. *Polyominoes.* Scribner's, 1965. 182 pp. illus. $5.95. 64–24805. (SH–C)
These puzzles are problems in combinatorial geometry—different ways of combining geometric figures. Theorems are included, together with answers. Twelve pentominoes are provided in a pocket in the back of the book and the problem compendium in the text challenges the reader to fit them together in various ways. An ideal book for recreation and enrichment for those with appropriate background.

Kurschak, Jozsef. *Hungarian Problem Book* (2 vols.). (Ed. by G. Hajox, G. Neukomm, and J. Suranyi; tr. by Elvira Rapaport.) Nos. 11 and 12, New Mathematical Library. Random, 1963. 111, 120 pp. illus. LB $2.45 ea., paper $1.95 ea. 63–16149. (SH)
SMSG collateral study books for secondary students based on the Eotvos Contests in elementary mathematics open to students in their last year of high school since 1894. Winners of these contests often have become internationally famous scientists. Problems have been selected from the 1894–1928 contests.

Northrop, Eugene P. *Riddles in Mathematics: A Book of Paradoxes.* Van Nostrand, 1944. 262 pp. illus. $4.50. 44–3524. (SH)
Over two hundred mathematical fallacies and paradoxes—drawn from every branch of mathematics from arithmetic to calculus—are presented, including an appendix of complete solutions.

Salkind, Charles T. *The Contest Problem Book.* (New Mathematical Library No. 5.) Random, 1961. 154 pp. illus. LB $2.95, paper $1.95; 61–13843. (SH)
A compilation of problems from the Annual High School Contests of the Mathematical Association of America, with solutions.

510.78 COMPUTERS

Adler, Irving. *Thinking Machines: A Layman's Introduction to Logic, Boolean Algebra and Computers.* John Day, 1961. 189 pp. illus. $4.50; NAL (Signet P2065), paper 60¢. 61–5924. (SH)
Provides the basic foundation in mathematics for an understanding of the principles and operation of computers, and explains the programming and operation of computers.

Baron, Robert C., and Albert T. Piccirilli. *Digital Logic and Computer Operations.* McGraw-Hill, 1967. xii+330 pp. illus. $13.50. 67–17196. (SH–C)
This simple introduction to computer logic could be used for high school instruction and for collateral reading for teachers of arithmetic and of the new mathematics. Coverage includes: number systems, computer logic, internal computer organization, input-output equipment, computer organization, and an introduction to computer programming. Examples are given in detail and the text is clearly presented.

Davidson, Charles H., and Eldo C. Koenig. *Computers: Introduction to Computers and Applied Computing Concepts.* Wiley, 1967. xii+596 pp. illus. $10.95. 67–19447. (C)
This comprehensive text provides an excellent introduction to the general

nature of computers, computer programming, specific computational procedures, and areas of future growth in computer technology. An excellent introduction to computers.

Diebold, John. *Man and the Computer: Technology as an Agent of Social Change.* Praeger, 1969. xi+157 pp. $5.95. 70–75237. (SH–C)
This is based on five of the author's speeches and articles on the place of the computer in society, and the problems of our adaptation in a period of constant change. Should be especially useful as background reading for anyone interested in trends in contemporary society.

Harris, Charles O. *Slide Rule Simplified* (2nd ed.). American Technical Soc., 1961. 278 pp. illus. $3.75. 61–10984. (SH)
A complete manual of instruction on the proper techniques of slide rule manipulation. Numerous illustrative examples and sample problems afford the student ample opportunity to gain proficiency.

Hull, T. E. *Introduction to Computing.* Prentice-Hall, 1966. xi+212 pp. illus. $10.60 (text ed. $7.95). 66–22085. (SH–C)
A well-rounded discussion of the use of the digital computer to solve problems. The general nature of the computer is explained and then a number of chapters are devoted to teaching the common computer language, FORTRAN. Additional topics are flow charts, debugging, numerical techniques, algorithms, automata and languages. Exercises at the end of each chapter. A very up-to-date selected bibliography. A high school student eager to learn about using (not building) a computer will profit from this book.

Jordain, Philip B., and Michael Breslau. *Condensed Computer Encyclopedia.* McGraw-Hill, 1959. xv+605 pp. illus. $14.50. 68–25654. (SH–C)
Achieves the goal of defining computer terms clearly and meaningfully for the nonspecialist. Computer terms are divided into three categories, generic terms, specific terms, and specialized terms. An excellent bibliography is included and is broken into three sections, systems, computers, and programming.

Lewis, Alfred. *The New World of Computers.* Dodd, Mead, 1965. 79 pp. illus. $3.00; LB $2.79. 65–11157. (JH)
Accurately and adequately explains the basic principles of computer mathematics, engineering, and electronics to the junior high school student. With the aid of excellent photographs, he describes many applications. Also describes the challenges and opportunities which lie ahead and the education and skills that will be required, indicating to young readers the training they must have to work in the "computer world."

McCalla, Thomas Richard. *Introduction to Numerical Methods and FORTRAN Programming.* Wiley, 1967. xiii+359 pp. illus. $8.95. 66–28745. (C)
Although designed for a one-semester numerical methods course, the text is so readable that it can be used by the intelligent student without an instructor. It is a clear, extremely comprehensive guide to a wide range of simple and advanced computational techniques. The programming examples are well done and the range of numerical techniques covered is very complete.

Saxon, James A., and Wesley W. Steyer. *Basic Principles of Data Processing.* Prentice-Hall, 1967. xvi+278 pp. illus. $8.95. 67–16391. (SH–C)
This easy-to-understand introduction to the basics of electromechanical and computerized data processing systems is appropriate for text, collateral read-

ing, or reference. Topics include electric accounting machines, supplementary equipment, number systems, electronic data processing, input-output devices, flow charting, elementary programming (including FORTRAN and CO-BOL), computer applications, and future developments. The brief index is adequate. Review questions are both well placed and well chosen to put proper emphasis on the material covered.

Sippl, Charles J. *Computer Dictionary and Handbook.* Sams, 1966. 766 pp. illus. $12.95. 66–21405. (C)

An exceptionally complete dictionary of terms using what the author calls "full concepts explanations." Appendices are useful not only to the layman but to people in computing who would like a capsule description of an area within the computer field. Very useful to students and those in the computer industry who require a good general reference.

Stark, Peter A. *Digital Computer Programming.* Macmillan, 1967. xv+525 pp. illus. $9.95. 67–16057. (SH–C)

General features and uses of computers are described in this carefully written book that treats its subject simply and clearly. It is a good first text. Both machine language and symbolic languages are treated (FORTRAN IV is used). Hundreds of examples and problems showing programs and their actual printed outputs are given. Simple applications to numerical computations and file work stressed rather than more sophisticated.

Stuart, Fredric. *FORTRAN Programming.* Wiley, 1969. xix+353 pp. illus. $7.95. 68–30922. SBN 471–83477–7. (SH–C)

Written both as a text and as a reference, this is designed for use with any computer for which FORTRAN is available. Material is presented in a logical manner, is accompanied by 120 practice exercises from a variety of disciplines, and is supplemented with helpful appendices.

510.8 MATHEMATICAL TABLES

Beyer, William H. (ed.). *Handbook of Tables for Probability and Statistics.* Chemical Rubber Co., 1966. xv+502 pp. $16.00. 66–17301. (C)

In recognition of the vast amount of research currently conducted in the fields of theoretical and applied statistics, as related to the pure, behavioral, social, and applied sciences, the publisher has produced this collection of tables that will serve as a reference work for students, research and professional workers. The numerous and diverse tables, formulas and other materials are organized under 13 major headings: (1) "Probability and statistics," (2) "Normal distribution," (3) "Binomial, Poisson, hypergeometric, and negative binomial distributions," (4) "Student's T-Distribution," (5) "Chi-square distribution," (6) "F-distribution," (7) "Order statistics," (8) "Range and studentized range," (9) "Correlation coefficient," (10) "Nonparametric statistics," (11) "Quality control," (12) "Miscellaneous statistical tables," (13) "Miscellaneous mathematical tables." Previously available only from a large number of sources.

Comrie, Leslie J. (ed.). *Barlow's Tables of Squares, Cubes, Square Roots, Cube Roots and Reciprocals of all Integers up to 12,500* (4th ed.). Chemical Pub. Co., 1957. xii+258 pp. $5.95. 47–6963. (JH)

These tables were first published in 1814, and in 1840 a new edition was prepared and its plates were used for 90 years. In 1940 new plates were made and the tables revised. A standard work known to all statisticians and research workers.

Selby, S. M. (ed.). *Handbook of Tables for Mathematics* (3rd ed.). Chemical Rubber Co., 1968. 1050 pp. $19.50. 62–15661. (SH–C)
Contains all standard tables found in the *CRC Handbook of Chemistry and Physics* plus such specialized topics as spherical harmonics, the binomial and Poisson distributions, and information on elliptic planetary orbits.

510.9 MATHEMATICS—HISTORY

Aaboe, Asger. *Episodes from the Early History of Mathematics.* (New Mathematical Library, No. 13.) Random, 1964. 133 pp. illus. LB $2.95, paper $1.95. 63–21916. (SH)
Four episodes from the early history of mathematics were selected to show some notion of their proper setting; and the mathematical content can be understood by anyone who has studied high school algebra and geometry: (1) "Babylonian mathematics"; (2) "Early Greek mathematics"; (3) "Archimedean mathematics"; and (4) "Ptolemy's table of chords." Answers to problems provided.

Bell, Eric Temple. *Development of Mathematics* (2nd ed.). McGraw-Hill, 1945. xi+637 pp. $9.50. 45–10599. (SH)
Main trends in the evolution of mathematical thought over the past 6000 years are presented in this narrative. Familiarity with mathematics in general is a prerequisite for complete comprehension.

Bell, Eric Temple. *Mathematics: Queen and Servant of Science.* McGraw-Hill, 1951. xx+437 pp. illus. $7.50, paper $2.95. 51–9241. (SH)
Both pure and applied mathematics receive consideration in this account of the evolution of modern mathematical thought. Though not a history of mathematics, this comprehensive work contains a wealth of information on the personalities and theories of the great mathematicians.

Bell, Eric Temple. *Men of Mathematics.* Simon & Schuster, 1937. xi+ 590 pp. illus. $7.50, paper $2.25. 37–27177. (SH)
A series of biographical sketches of some of the great mathematicians throughout history. Although their contributions to the science are discussed, the gist of each story lies in the personalities of these men and the circumstances surrounding the realization of their goals.

Bochner, Salomon. *The Role of Mathematics in the Rise of Science.* Princeton U., 1966. x+386 pp. $9.00 66–10550. (SH–C)
Scholarly essays are presented in a collection designed to describe the role and nature of mathematics in the rise of Western intellectuality. A minor part of the volume is devoted to biographical sketches of over 150 mathematicians, scientists, and philosophers from the time of Thales of Miletus to Albert Einstein. Some of the chapters could be very profitably studied by an advanced secondary school student; whereas others require a much more sophisticated background. Clear, concise, and very well documented. Serves well as a reference and a source of discursive reading.

Eves, Howard. *An Introduction to the History of Mathematics* (rev. ed.). Holt, Rinehart and Winston, 1964. 460 pp. illus. $9.95. (SH)
A chronological treatment of the development of "elementary" mathematics through the beginnings of calculus. Appreciation of the concepts presented is gained through the working of problem studies at the end of each chapter.

Gauss, Carl Friedrich. *Disquisitiones Arithmeticae*. (Tr. by Arthur A. Clarke, S.J) Yale, 1966. xx+472 pp. $12.50, paper $2.95. 65–22318. (SH–C)
This work of genius is by a man acclaimed to be one of the three greatest mathematicians ever to have lived. The value of such a book in a library is essentially historical. Because it is a book with such fruitful ideas it rates as a first choice of an "original" work to be included in a high school library. The translation reads well, having been "edited" to present-day terminology.

Hogben, Lancelot. *Mathematics in the Making*. Doubleday, 1961. 320 pp. illus. $9.95. 61–5067. (SH)
Excellent illustrations are integrated with a lucid text to provide an historical treatment of the development of modern mathematics, from elementary arithmetic through introductory calculus.

Kline, Morris. *Mathematics in Western Culture*. Oxford, 1953. 484 pp. illus. $9.50, paper $2.50. 53–9187. (SH)
Discusses the contributions which mathematics has made to Western life and thought in such areas as philosophy, the physical and social sciences, religion, literature, and the arts.

Kramer, Edna E. *The Main Stream of Mathematics*. Oxford, 1951. 321 pp. illus. $7.75; Fawcett (Premier T130, 1961), paper 75¢. 51–2067. (SH)
Combines a survey of the history of mathematics with practical applications of all aspects of elementary mathematics through introductory calculus, including a discussion of relativity, probability, and infinity.

Logsdon, Mayme I. *A Mathematician Explains* (2nd ed.). U. of Chicago, 1936. 189 pp. illus. $4.00, paper (Phoenix PSS502) $1.50. 36–8177. (SH)
Elementary mathematics through trigonometry is treated on an historical basis, pointing up the advances in this area through practical problems. The chapters on calculus and analytic geometry are more theoretical, emphasizing concepts rather than historical significance.

Muir, Jane. *Of Men and Numbers: The Story of the Great Mathematicians*. Dodd, Mead, 1961. 249 pp. illus. $3.50, paper (Dell) 50¢. 60–14795. (JH)
Contains succinct, informative biographical sketches of some of the more famous mathematicians, emphasizing their contributions to the pure science rather than personal anecdotes.

Terry, Leon. *The Mathmen*. McGraw-Hill, 1964. 222 pp. illus. $3.95. 64–7740. (JH–SH)
A well-written brief examination of the mathematical contributions of the early Greeks with number theory and geometry pleasantly mixed. Contains vignettes about Greek personages and tells of the interrelation between personalities and math contributions. Although the mathematics doesn't flow logically and sequentially, it could be blended nicely into the school curriculum.

Valens, Evans G. *The Number of Things: Pythagoras, Geometry and Humming Strings.* Dutton, 1964. 189 pp. illus. $4.95. 64–10695. (JH)
The origins of geometry, musical scales and annotation, and cosmology can be traced to Pythagoras and his followers, as explained in this informative and entertaining item of recreational reading.

Van Der Waerden, B. L. *Science Awakening.* (Tr. by Arnold Dresden.) Oxford, 1961. 306 pp. illus. $7.50; Wiley (Science Eds.), paper $2.65. 61–19368. (C)
A comprehensive study of the history of Greek mathematical thought as influenced by the Egyptian and Babylonian cultures—developed with the unifying theme that all science and technology is based on mathematics and physics. A scholarly work of merit.

Wilder, Raymond L. *Evolution of Mathematical Concepts: An Elementary Study.* Wiley, 1968. xviii+224 pp. $8.00. 68–28508. (SH–C)
It is the author's contention that the growth of mathematics has been an integral part of the growth of Western culture for several millenia. This is demonstrated by tracing the evolution of geometry and the number concept from earliest civilization up to the present. Educational aspects of mathematics are considered, and an extensive bibliography and a good index are included.

511 ARITHMETIC

Asimov, Isaac. *Realm of Numbers.* Houghton Mifflin, 1959. 200 pp. $3.50; Grosset & Dunlap (Hale), paper $1.92, also Fawcett R333 60¢. 59–7480. (JH)
For the reader with a background in simple mathematics acquired in elementary school, this is an introduction of mathematical principles and their evolution. The discussion begins with finger counting and progresses through square root, logarithms, rational and irrational numbers and on to infinity.

Brumfiel, Charles F., Robert E. Eicholz, and Merrill E. Shanks. *Fundamental Concepts of Elementary Mathematics.* Addison-Wesley, 1962. xi+340 pp. $8.50. 62–9399. (C)
An introduction to the mathematical concepts underlying the traditional techniques of computation, which includes an analysis of arithmetic and presents intuitive algebra and geometry.

Brumfiel, Charles F., Robert E. Eicholz, Merrill E. Shanks, and P. G. O'Daffer. *Principles of Arithmetic.* Addison-Wesley, 1963. x+373 pp. illus. $7.50. 63–12468. (JH)
Useful for those whose mathematical background for algebra courses needs bolstering, and for those planning to teach elementary mathematics; exercises including reasoning and development of intuitive skills.

Fujii, John N. *Numbers and Arithmetic.* Blaisdell, 1965. xi+559 pp. illus. $8.75; teacher's manual $1.50. 65–17960. (JH–SH)
An excellent source book for students and teachers. It covers all fundamental number concepts and operations of arithmetic. The inclusion of vocabulary lists and reference sources, use of problems of historical as well as modern, but practical, orientation make this an excellent treatment of the real number system and its operations for the pre-college level. The supplementary material of the teacher's manual suggests concern for methods of presentation and instruction, and provides suggested test material.

Hacker, Sidney G., Wilfred E. Barnes, and Calvin T. Long. *Fundamental Concepts of Arithmetic*. Prentice-Hall, 1963. 271 pp. $7.95 63-20708. (SH-C)
The content was developed and tested for NSF-sponsored in-service training of elementary teachers, is useful for college students intending to teach, and will be useful to individuals for review and practical exercises.

Peterson, John A., and Joseph Hashisaki. *Theory of Arithmetic* (2nd ed.). Wiley, 1967. xiv+337 pp. illus. $8.50. 67-12569. (C)
A text at the college level on general mathematics for teachers. Strongly recommended for those interested in self-study, this has been widely used for in-service programs for elementary and junior high school teachers.

512 ALGEBRA

Allendoerfer, Carl B., and Cletus O. Oakley. *Fundamentals of College Algebra*. McGraw-Hill, 1967. xiv+446 pp. illus. $7.50; instructor's manual $1.75. 66-22291. (SH-C)
Designed to present the topics contained in a traditional college algebra course with a more modern treatment. For students of business and economics there are applications of matrices, some discussion of supply and demand functions, simple game theory and linear programming. Lends itself nicely to high school and college courses designed to prepare students for analytic geometry and calculus. A wealth of graded exercises.

Allendoerfer, Carl Barnett, and C. O. Oakley. *Fundamentals of Freshman Mathematics* (2nd ed.). McGraw-Hill, 1965. xiii+586 pp. illus. $8.95; instructor's manual $1.75. 64-7924. (SH-C)
Bridges the "gap" between intermediate algebra and analytic geometry and calculus; useful for review, self study, or reference. Revision influenced by recommendations of CUPM.

Andree, Richard V. *Selections from Modern Abstract Algebra*. Holt, Rinehart and Winston, 1958. xii+212 pp. $8.95. 58-6799. (SH)
A college-level textbook used successfully with advanced senior high students and in many summer institutes for teachers. It provides a basic introduction to abstract algebra needed by physical scientists and engineers.

Arnold, Bradford H. *Logic and Boolean Algebra*. Prentice-Hall, 1963. 144 pp. illus. $10.60 (text ed. $7.95). 62-19100. (C)
An introduction to Boolean algebra beginning with fundamental logic and basic ordered sets and concluding with computer applications.

Asimov, Isaac. *Realm of Algebra*. Houghton Mifflin, 1961. 230 pp. $3.50; Fawcett R334, paper 60¢. 61-10637. (JH)
Introduces the junior-high student to the theory behind algebraic manipulations from the most basic concepts of algebra to the more refined considerations of quadratic, cubic and simultaneous equations, concluding with a chapter on "putting algebra to work."

Birkhoff, Garrett, and Saunders Maclane. *A Brief Survey of Modern Algebra* (2nd ed.). Macmillan, 1965. vii+279 pp. $7.95. 65-27327. (SH-C)
A new short version of a classic, *Survey of Modern Algebra* (3rd ed.; Macmillan, 1965), which first appeared 25 years ago. Either version is still probably the best introduction to the subject because it never loses sight of the concrete origins of the abstract ideas or of their applications in other fields.

Clearly written with many examples and exercises. Although no more than high school algebra is assumed, the book as a whole would be heavy going for most high school seniors or college freshmen.

Bryne, J. Richard. *Number Systems: An Elementary Approach*. McGraw-Hill, 1967. xiii+291 pp. illus. $6.95. 67–16299. (JH–SH)
Introduces algebraic abstractions and the simple ideas of number theory. A clear and understandable discussion on number systems written for mature college students with at least one year of college mathematics. An excellent text for teacher training and reference.

Cameron, Edward A. *Algebra and Trigonometry* (rev. ed.). Holt, Rinehart and Winston, 1965. x+338 pp. $8.95. 65–14884. (SH–C)
A modern treatment (basic number, set, and function concepts) of algebra and trigonometry. Recommended as a text for first-year college students or specially qualified high school students who plan to follow a career in the mathematical sciences or engineering. It will also prove useful both as a collateral reference text for high school mathematics teachers and as a primer for professional workers.

Campbell, Hugh G. *An Introduction to Matrices, Vectors, and Linear Programming*. (Appleton-Century Mathematics Series.) Appleton-Century-Crofts, 1965. xiv+244 pp. $6.50. 65–16820. (JH–SH)
Comparable to the *Introduction to Matrix Algebra* text of SMSG, with which the author claims similar objectives and level of presentation. Each text will appeal to different groups principally on the basis of style. Definitions are emphasized and examples to illustrate new ideas are frequent. There is more complete coverage of special matrices, inverses and elementary transformations than in the SMSG text. Although this would serve best as a classroom text, its use by individual students with occasional teacher consultation is strongly recommended where students are interested.

Clarkson, Donald R., Edwin C. Douglas, Arthur W. Eades, Joyce F. Olson, and Elizabeth Glass. *Algebra and Trigonometry*. (The Prentice-Hall Modern Mathematics Series.) Prentice-Hall, 1966. 437 pp. illus. $5.80. (SH)
An integrated text reflecting recommendations from various mathematics study groups. Includes introductory treatment of vectors, mathematical systems, and conic sections in addition to traditional concepts, with emphasis on the function concept.

Crouch, Ralph, and David Beckman. *Fundamental Mathematical Structures: Algebraic Systems*. Scott, Foresman, 1966. 304 pp. illus. $5.56. (SH–C)
Carefully and well written. The mathematics is formal and consistently presented. The four chapters, "Sets, relations and functions," "Groups," "Rings and integral domains," and "Fields," indicate that this is ground common to a great many contemporary texts, with which this presentation compares favorably. Aimed at high school and lower division college, but it is a good reference for students and teachers working with these ideas at any level.

Crouch, Ralph, and David Beckman. *Fundamental Mathematical Structures: Linear Algebra*. Scott, Foresman, 1965. 432 pp. illus. $6.24. (SH–C)
The text is intended to serve as a transition from high school to college mathematics, and for use by 12th grade students who have completed a traditional mathematics program. It introduces abstract reasoning based on an axiomatic approach and presents material useful in college courses in physics

and engineering. Considers vector spaces and subspaces, linear independence and bases, linear transformations, matrices, and systems of linear conditions for equality. A booklet of specially drawn geometric figures with a view that gives a three-dimensional effect is included in a pocket.

Davis, Philip J. *The Lore of Large Numbers.* (New Mathematical Library, No. 6.) Random, 1961. 165 pp. illus. LB $2.95; paper $1.95 61–13842. (JH)
Deals with the intriguing aspects of very large numbers which are necessary for assessing the magnitudes of the inordinately large and the infinitesimally small. Intended for senior highs as collateral reading, but can be used by some junior highs.

Drooyan, Irving, Walter Hadel, and Frank Fleming. *Elementary Algebra; Structure and Skills* (2nd ed.). Wiley, 1969. xii+390 pp. illus. $7.50. 69–11185. (JH–SH)
This text was written for the beginning student of algebra and may be used for a one semester course with advanced pupils or a one year course for the average type pupils. Each chapter includes a summary and a review. Answers to these reviews (which make good self-tests) are given for all items in the back of the text. Answers to all odd numbered exercises are also included. Appendices include symbols used.

Dubisch, Roy. *Introduction to Abstract Algebra.* Wiley, 1965. ix+193 pp. $6.95. 65–16405. (SH–C)
A well-written "gradual introduction to the basic concepts of abstract algebra." Excellent development of the real number system alternates with chapters on abstract topics that can be illustrated by the familiar number systems being developed. Good supplementary reading in a college calculus course.

Eves, Howard, and Carroll V. Newsom. *An Introduction to the Foundations and Fundamental Concepts of Mathematics* (rev. ed.). Holt, Rinehart and Winston, 1965. xv+398 pp. illus. $9.95. 65–13241. (SH–C)
A minor revision of a significant mathematics text. It is a technical book, providing insight into mathematical foundations including algebraic structure, axiomatics, set theory and logic, motivated by an historical survey of the inherent logical difficulties of synthetic geometry from Euclid to Hilbert. Careful explanations and excellent exercises help to clarify the content. The book can best be studied under the guidance of a competent mathematician. Given this kind of support, many high-ability senior high school students could study the text with profit.

Hart, William L. *Contemporary College Algebra and Trigonometry.* Heath, 1967. xvi+488 pp. illus. $9.50. 67–22727. (C)
Contemporary treatment is given to sets, function, domain, range, inequalities, complex numbers, induction, probability, linear equations, quadratic equations, etc. Frequent historical notes add interest and there is an abundance of well-selected problems.

Hillman, Abraham P., and Gerald L. Alexanderson. *Algebra Through Problem Solving.* (Topics in Contemporary Mathematics Series.) Allyn and Bacon, 1966. vii+129 pp. Text ed. $9.95, paper $3.95. 66–16281. (SH–C)
Not to be confused with a two-volume text for first and second year algebra students in high school having a similar title. Intended for use by students who have completed the material usually included in the first two algebra

courses, whether they be high school students, undergraduates in colleges, or teachers of mathematics. The topics include mathematical induction, determinants and inequalities, among others. The problems have been arranged by topics. Many are non-routine and quite interesting.

Holberg, Charles J. A., Jr., and John F. Devlin. *Fundamental Mathematical Structures: Elementary Functions.* Scott, Foresman, 1967. 672 pp. illus. $6.04. (SH–C)
The advanced algebra text of the series which introduces the function concept, real and complex number fields, the Cartesian plane, real-valued functions of a single real variable, constant and linear functions, quadratic and polynomial functions, circular functions, exponential functions and logarithmic functions. Answers and necessary tables are appended.

Hunter, John. *Number Theory.* Wiley, 1965. ix+149 pp. $3.75. 65–24412. (SH–C)
The first three chapters on number systems and algebraic structures, division and factorisation properties, and congruences, contain the basic concepts in number theory. Other chapters introduce the theory of algebraic congruences and quadratic residues, the representation of integers by binary quadratic forms and something on Diophantine equations and Fermat's Last Theorem. All theorems are proved, often in such a way as to introduce ideas that appear in other branches of mathematics. Should be particularly useful to high school and college teachers who need a more rigorous presentation of the subject.

Levi, Howard. *Elements of Algebra* (4th ed.). Chelsea, 1960. 161 pp. $3.95. 59–6891. (SH)
Emphasizes the theoretical aspect of the algebra rather than the manipulative by constructing the natural number system, the integers, the reals, and the rationals and developing the algebra appropriate to each of these systems.

Lovaglia, Anthony R., and Gerald C. Preston. *Foundations of Algebra and Analysis: An Elementary Approach.* Harper & Row, 1966. xi+516 pp. illus. $9.50. 66–15673. (SH–C)
Although intended for a full-year pre-calculus course, portions may be used for shorter courses in college algebra and trigonometry, foundations of number systems for teachers, and general education courses. The text is mathematically sound, the level of discourse is high and the pace is brisk. Answers to problems, an index of symbols, and an index of subjects.

Niven, Ivan. *Numbers: Rational and Irrational.* (New Mathematical Library, No. 1.) Random, 1961. 136 pp. illus. LB $2.95, paper $1.95. 61–6226. (SH)
Supplemental study book for secondary students dealing with the number system, which is one of the basic structures of mathematics. The early chapters are fairly easy, but the later chapters are advanced material for ambitious students.

Niven, Ivan, and Herbert S. Zuckerman. *An Introduction to the Theory of Numbers* (2nd ed.). Wiley, 1966. 250 pp. $8.95. 60–10322. (C)
A good introduction to the theory of numbers which begins with a presentation of the basic concepts followed by more specialized materials in the three final chapters. A large number of problems of great variety is included.

Olds, C. D. *Continued Fractions.* (New Mathematical Library, No. 9.) Random, 1962. 162 pp. LB $2.95, paper $1.95; Singer (student ed.) 90¢. 61–12185. (SH)

Collateral book for secondary students beginning with the expansion of rational fractions into continued fractions, and the expansion of irrational numbers into infinite continued fractions. For the mathematically ambitious.

Ore, Oystein. *Number Theory and its History.* McGraw-Hill, 1948. 370 pp. illus. $7.95. 48–8825. (C)

Combines historical chapters on counting, recording, and properties of numbers with basic number theory. Amongst those topics selected for discussion are: prime numbers, Diophantine problems, congruences, and Euler's theorem.

Payne, Joseph N., Floyd F. Zamboni, and Francis G. Lankford, Jr. *Algebra One.* Harcourt, Brace and World, 1969. ix+566 pp. illus. $4.35. (JH)

A pure "new math" text for high school freshman algebra. There are "now try this" exercises with answers furnished, checkpoint exercises, more challenging problems, chapter reviews, chapter tests, and cumulative reviews at the end of every third chapter. A teacher's edition is available.

Pedoe, Daniel. *A Geometric Introduction to Linear Algebra.* Wiley, 1963. xi+224 pp. $7.50. 63–20637. (C)

Based on the author's introductory courses on linear algebra at Purdue University. He believes that the study of this subject provides a good introduction to the notion of a mathematical proof. Some examples are "worked" in the text and there are additional exercises for the reader.

Richardson, M. *College Algebra* (3rd ed.). Prentice-Hall, 1966. xvii+605 pp. illus. $9.25. 65–21801. (SH–C)

A revision of a well-established text for college freshmen or high school teachers. The author has added appropriate material from the modern curriculum. A reasonable treatment of linear algebra, a better-than-usual treatment of the number system and the theorems on partial fractions. There is a broad spectrum of problems from the trivial to the more challenging ones. A happy combination of rigorous presentation, a wealth of traditional valuable topics, and many exercises. Useful tables and answers to odd-numbered problems.

Rose, Israel H. *Algebra: An Introduction to Finite Mathematics.* Wiley, 1963. 489 pp. illus. $7.50. 63–8058. (SH–C)

The author has designed this book for advanced secondary students, particularly those who have been using SMSG programs, to provide articulation between secondary training and college undergraduate courses. Useful also for teacher-training courses.

Sah, Chih-Han. *Abstract Algebra.* Academic, 1967. xiii+342 pp. $9.75. 66–29641. (C)

Every serious undergraduate student of mathematics should be familiar with the basic material from algebra contained in this book. The text is clearly, but concisely, written so the reader will have to do some work to get through it. Examples and exercises are liberally scattered through the text. A particularly good feature occurs in the final chapter where the reader will find a proof of the fundamental theorem of algebra which is almost completely algebraic in nature.

Shields, Paul C. *Elementary Linear Algebra.* Worth, 1968. x+349 pp. illus. $8.95. 67–31678. (C)

The proper approach to beginning linear algebra, relying heavily on geometry and systems of linear equations to motivate the student, as well as the historical approach to the subject. The author utilizes some modern generalizations without overwhelming the student with recondite abstraction. Chapter 4, on linear differential operations, is a superlative introduction to differential equations. Physical science and engineering students, who usually get a cookbook differential equations course, now have available a correct treatment of the subject.

Swokowski, Earl W. *Algebra and Trigonometry.* Prindle, Weber & Schmidt, 1967. x+485 pp. illus. $8.95. 67–11929. (SH–C)

Discusses the algebra of polynomials, systems of linear equations and inequalities, complex numbers, series, and sequences. A background in elementary algebra is helpful.

Van Engen, Henry, Maurice L. Hartung, Harold C. Trimble, Emil J. Berger, and Ray W. Cleveland. *Fundamental Mathematical Structures: Algebra.* Scott, Foresman, 1966. 607 pp. illus. $5.20. (JH–SH)

Intended as a ninth grade algebra text, this one is written in the modern vein, stressing the "why" as well as the "how." Sets, logic, and modern notation are employed. The presentation is excellent. The pace is fast and the conceptual material advanced. A special feature is a set of 15 sections, labelled "Special challenges," on such topics as number theory, mathematical induction, and groups. Highly recommended for above-average teachers and students only.

Zippin, Leo. *Uses of Infinity.* (New Mathematical Library, No. 7) Random, 1962. 151 pp. illus. LB $2.95, paper $1.95. 61–12187. (SH)

A SMSG supplemental book that shows the various aspects of the infinity concept, gives examples of infinite processes and provides the reader with a grasp of the fundamental notions used in the several mathematical disciplines.

513 GEOMETRY

Adler, Irving. *A New Look at Geometry.* (Illus. by Ruth Adler.) John Day, 1966. 414 pp. $7.95; NAL (Signet Y3225, 1967), paper $1.25. 66–15089. (C)

Successfully explains the evolution of geometry throughout history in order to demonstrate how a synthesis of ideas and concepts has produced our modern geometry. Ties together all of geometry in a logical fashion and hence, is a useful book for reference and collateral reading.

Barnett, Raymond A., and John N. Fujii. *Vectors.* Wiley, 1963. 132 pp. illus. $4.50. 63–11427. (SH–C)

A supplementary book useful to advanced high school students and to students of physics and engineering courses, for which basic geometry and trigonometry courses are prerequisites.

Barr, Stephen. *Experiments in Topology.* Crowell, 1964. 210 pp. illus. $3.75. 64–10866. (C)

Topology is one of the fascinating bypaths of mathematics. It is concerned initially with continuity of such things as space and shape, but eventually leads into other kinds of continuity beyond space. Anyone who knows a few rudiments of algebra and geometry will be intrigued by this book.

Beckenbach, Edwin, and Richard Bellman. *An Introduction to Inequalities.* (New Mathematical Library, No. 3.) Random, 1961. 133 pp. illus. LB $2.95, paper $1.95. 61–2228. (SH)

An introduction to inequalities beginning with a dissection of the relation "greater than" and the meaning of "absolute values" of numbers, including descriptions of some unusual geometries. Recommended collateral reading for secondary students.

Blackett, Donald W. *Elementary Topology: A Combinatorial and Algebraic Approach.* (Textbooks in Mathematics Series.) Academic, 1967. ix+224 pp. illus. $9.50. 66–30139. (C)

Designed to introduce undergraduates to topology through a discussion of elementary surfaces, covering surfaces, winding numbers, vector fields, networks, and manifolds. The best currently available undergraduate introduction to algebra and combinatorial methods in topology.

Coxeter, H. S. M., and Samuel L. Greitzer. *Geometry Revisited.* (New Mathematical Library, No. 19.) Random, 1967. xiv+195 pp. illus. LB $2.95, paper $1.95. 67–20607. (SH–C)

Designed to carry the student beyond the customary beginning course in plane geometry by acquainting him with some of the "treasures" of older geometry that are often neglected, and by acquainting him with some of the advances in its development. Particularly interesting is the consideration of collinearity and concurrence, transformations, and the introductions to inversive geometry and projective geometry. The orbits of artificial satellites and four-dimensional geometry of the space-time continuum are two examples that illustrate that geometry is more useful to the scientist than at any previous time.

Coxeter, H. S. M. *Introduction to Geometry.* Wiley, 1961. xiv+443 pp. illus. $9.25. 61–11175. (C)

This comprehensive book is an excellent and somewhat rigorous exposition of the entire subject which also shows its uses in algebra, analysis, cosmology, kinematics, crystallography, topology and even in botany.

Cundy, H. Martyn, and A. P. Rollett. *Mathematical Models* (2nd ed.). Oxford, 1961. 286 pp. illus. $6.50. 61–65743. (SH)

Contains descriptions of and detailed instructions for the construction of mathematical models, including dissections, loci and envelopes produced by folding, drawing, and stitching, and various polyhedra. Excellent for illustrating the principles of plane and solid geometry.

Diggins, Julia E. *String, Straightedge, and Shadow: The Story of Geometry.* Viking, 1965. 160 pp. illus. $5.00; LB $4.53. 64–13597. (JH–SH)

The author takes her readers—basically in fact, partly in fantasy—through the ancient world from the Stone Age to the age of Euclid by means of a delightful explanation of the history and uses of mathematics, with emphasis on geometry. One may glean appreciation for the culture, geography and color of Egypt, Greece and Mesopotamia, and learn to know better the personalities of mathematicians such as Thales and Pythagoras. Mathematically, the book is acceptably correct. Simple experiments using only string, straightedge and shadow are suggested; each leads to the proof (not given) of the geometric principle involved. Should be read by every junior and senior high school student. Teachers could use the book and the "Suggestions for further reading" to enrich their classroom activities.

Dodes, Irving Allen. *Geometry.* Harcourt, Brace, 1965. xii+560 pp. illus. $4.50. 65–1433. (SH)

A new textbook with a tremendous sweep—a fusion of plane and solid geometry that extends to trigonometry and analytic geometry. Some classes may not be able to cover the entire book in a year. It contains a wealth of problems and good illustrations. The methods and principles of formal logic are explained and used as are many kinds of proof: arithmetic, algebraic, inductive, set-theoretic, direct and indirect, formal and informal. Teacher's manual available.

Dorwart, Harold L. *The Geometry of Incidence.* Prentice-Hall, 1966 xvii+156 pp. illus. $6.95. 66–11240. (SH–C)

A noteworthy attempt to revive interest in geometry, specifically in projective geometry. It is not intended as a textbook but is an introduction and a guide to further study. By selecting several fundamental concepts and theorems from projective geometry, together with their historical backgrounds, the reader is given a fascinating treatise. May be used for collateral reading by honor students in senior high, or by mathematics majors in college. Extensive bibliography.

Fischer, Irene, and Dunstan Hayden. *Geometry.* Allyn and Bacon, 1965. ix+582 pp. illus. $5.28. 65–1813. (SH–C)

Recommendations of the College Entrance Examination Board in the *Report of the Commission on Mathematics* (New York, 1959) have been followed in this new text that treats plane and solid geometry simultaneously. While the subject matter is comparable to that offered in conventional courses, the spirit of the approach is unique and modern. The teacher who uses this text will need time for careful preparation. The text will be primarily appropriate for the better high school student, but is a good supplementary reference for all students.

Fishback, W. T. *Projective and Euclidean Geometry.* Wiley, 1962. ix+244 pp. $8.95. 62–20162. (C)

Developed especially as a foundation text for secondary teachers attending NSF-sponsored institutes which proceeds from elementary analytic geometry to a consideration of linear algebra and then to the development of projective geometry.

Hilbert, David, and S. Cohn-Vossen. *Geometry and the Imagination.* (Tr. by P. Nemenyi.) Chelsea, 1952. 357 pp. illus. $7.50. 52–2894. (SH)

Stresses the "intuitive understanding" of geometric principles rather than the abstract, logical relationships, though this latter approach is utilized in the chapter on topology.

Jones, Burton W. *Elementary Concepts of Mathematics* (2nd ed.). Macmillan, 1963. xvii+350 pp. illus. $7.25. 63–7216. (SH)

Pure number theory is presented to introduce the student to the abstract reasoning processes which are used to explain various geometries and topology. Practical applications of the number theory are stressed in the discussions of graphs, averages, permutations, combinations, and probability.

Kazarinoff, Nicholas D. *Geometric Inequalities.* (New Mathematical Library, No. 4.) Random, 1961. 132 pp. illus. LB $2.95, paper $1.95. 61–6229. (SH) "Geometric inequalities are especially appealing because their statements can be easily grasped; at the same time they provide an excellent introduction to creative mathematical thought and to the spirit of modern mathematics." —Preface.

Kelly, Paul J., and Norman E. Ladd. *Fundamental Mathematical Structures: Geometry.* Scott, Foresman, 1965. 600 pp. illus. $5.96. (SH) Several unusual features: First, a clear distinction is made between informal material, designed to help the student understand the text, and the formal material itself. This is done by indenting the paragraphs of informal explanations along the left-hand margin. Second, some of the proofs are set up in columnar style and others in paragraphs so that students may familiarize themselves with both forms. Third, 44 anaglyphs that correspond to certain three-dimensional drawings in the book help the student visualize difficult three-dimensional figures. Appendices discuss special topics such as induction, the nine-point circle, the Fermat point, and the Simson line. Many of the more than 450 theorems and corollaries could not be proved because of space limitations.

Meserve, Bruce E. *Fundamental Concepts of Geometry.* Addison-Wesley, 1955. ix+321 pp. illus. $8.95. 55–7374. (SH) Presents geometry as a logical system based upon postulates and undefined elements, and conveys an appreciation of the historical evolution of geometrical concepts and of the relation of Euclidean geometry to the space in which we live.

Reid, Constance. *A Long Way from Euclid.* Crowell, 1963. 292 pp. illus. $5.00. 63–18418. (SH) "It is the hope of the author that the reader of this book will be able to glimpse through his own misty memories of Euclid's geometry the outline of some of the more imposing edifices of modern mathematics."—Preface.

Schuster, Seymour. *Elementary Vector Geometry.* Wiley, 1962. 213 pp. illus. $5.95. 62–10933. (SH) Develops vector algebra as a mathematical tool in geometry. In addition, vector analysis is applied in such areas as analytic geometry and trigonometry.

Smail, Lloyd L. *Analytic Geometry and Calculus.* Appleton-Century-Crofts, 1953. 714 pp. illus. $8.00. 52–13696. (SH–C) One of many standard textbooks integrating the study of analytic geometry and calculus, an approach which is used most often in schools of science and engineering.

Tuller, Annita. *A Modern Introduction to Geometries.* Van Nostrand, 1967. xiii+201 pp. illus. $7.50. 67–148. (SH–C) Illustrates two primary principle approaches to geometry: the study of a body of theorems deduced from a set of axioms, and the study of the invariant theory of a transformation group. Excellent for the advanced high school student.

Wernick, William, and Bruce R. Vogeli. *Analytic Geometry.* Silver Burdett, 1968. 278 pp. illus. (Teacher's edition 294 pp.) $4.72, LB $3.54; teacher's edition $4.72; solution key $3.60, LB $2.70. (C) For the student with an above-average background in general mathematics, this modern text will be valuable as a reference. The modern approach is emphasized throughout. Lacks somewhat in tables of standard symbols, equations, trigonometric identities, and reduction formulas.

Yaglom, I. M. *Geometric Transformations.* (New Mathematical Library, No. 8.) Random, 1962. 113 pp. illus. LB $2.95, paper $1.95. 62–18330. (SH)

A SMSG collateral reading book which is concerned with transformations of the plane (isometries) that plays a fundamental role in the group-theoretic approach to geometry.

514 TRIGONOMETRY

Bettinger, Alvin K., Gerald A. Hutchison, and John A. Englund. *A Modern Approach to Algebra and Trigonometry.* International Textbook Co., 1966. xii+ 420 pp. illus. $7.95. 66–20526. (SH–C)

Presents an excellent background for a rigorous calculus course. A first rate text for a one year high school or a freshman college course. Areas covered include: trigonometric functions of angles and real numbers, complex numbers, polynomials, iterative methods, sequences, and matrices.

Wooton, William, Edwin F. Beckenbach, and Mary P. Dolciani. *Modern Trigonometry.* Houghton Mifflin, 1966. viii+427 pp. illus. $5.60. 65–8580. (SH–C)

The approach is from circular functions, with domains over the sets of real numbers, to trigonometric functions whose domains are sets of angles. Related subject matter includes: two-dimensional vectors; basic elements of complex numbers; basic properties of matrices and their use; and development of the expression of the circular functions as infinite series. The treatment is rigorous but so well motivated and carefully executed that it will suit average as well as superior students. There are chapter summaries and tests, essays on applications of trigonometry, a list of symbols and a very complete set of tables. A special teachers' edition is available.

515 DESCRIPTIVE GEOMETRY

Burger, Dionys. *Sphereland: A Fantasy about Curved Spaces and an Expanding Universe.* (Tr. by Cornelie J. Rheinboldt. Crowell, 1965. x+208 pp. illus. $4.95. 65–21409. (SH–C)

Flatland, a Romance of Many Dimensions, by A Square (written by Edwin A. Abbott; available in paperbound reprint from Dover, or Barnes & Noble) was the inspiration for this fantasy. " . . . the author of *Sphereland* tells of the grandson of the Square, a Hexagon, who is confronted by even greater problems that can be understood only by assuming that the plane he lives in is curved and, even more confounding, that his two-dimension world is expanding."—Preface. The first chapter is a summary of *Flatland.* Although an interesting and nonexisting fantasy, it helps the reader gain an insight into the problems of outer space. Indexed.

Chinn, W. G., and N. E. Steenrod. *First Concepts of Topology: The Geometry of Mappings of Segments, Curves, Circles, and Disks.* (New Mathematical Library, No. 18.) Random, 1966. viii+160 pp. LB $2.95; paper $1.95. 66–20367. (SH–C)

A good overview of topology is provided by using as a basic goal the proofs of "intermediate value" theorems for continuous functions in dimensions 1 and 2. Significant topological concepts are developed within this context; included are open and closed sets, completeness, compactness, homeomorphism, fixed point properties, winding numbers of curves, and vector fields. Several applications of independent interest are developed. Valuable supplementary reading at a variety of levels of instruction.

Paré, Eugene George, Robert O. Loving, and Ivan L. Hill. *Descriptive Geometry* (3rd ed.). Macmillan, 1965. viii+383 pp. illus. $6.50. 65–15774. (C)
Established as an effective method of teaching graphic communication and providing orientation to geometric concepts. Solution illustrations show the necessary steps for easy construction and are intended to teach fundamentals as well as to "introduce new engineering experiences." Abstract and practical problems at the end of each chapter. Full-size self-testing problems are provided, with solutions in an appendix.

516–517 ANALYTICAL GEOMETRY AND CALCULUS

Adams, Lovincy J. *Applied Calculus.* Wiley, 1963. ix+278 pp. illus. $6.95. 63–17471. (C)
A brief text presenting a large number of concepts and emphasizing their practical applications to mechanical, civil, electrical and electronic engineering.

Agnew, Ralph Palmer. *Calculus: Analytical Geometry and Calculus with Vectors.* McGraw-Hill, 1962. xii+738 pp. illus. $11.00. 61–18624. (C)
The book attaches primary importance to basic concepts which are the stones of the foundation on which applications of the calculus are based. The historical and philosophic aspects of the subject are considered also.

Apostol, Tom M. *Calculus* (2nd ed.). Vol. 1: *One-Variable Calculus, with an Introduction to Linear Algebra.* Blaisdell, 1967. xx+666 pp. illus. $11.50. 67–14605. Vol. 2: *Linear Algebra, Calculus of Several Variables with Applications to Probability and Vector Analysis.* Blaisdell, 1969. (Pagination and price not available.) (C)
An outstanding and comprehensive work designed for a two-year course, but valuable for reference and background. The new edition will serve a wider group than the 1961–62 edition, but still will have its greatest appeal to teachers for use with well-motivated students. Includes some easier exercises and more applications to physics and engineering than the first edition, but still retains the historical and philosophical approach.

Begle, Edward G. *Introductory Calculus with Analytic Geometry.* Holt, Rinehart and Winston, 1954. 304 pp. illus. $8.95. 54–6596. (C)
Treats calculus as a branch of mathematics rather than a mere adjunct of the physical and engineering sciences, to give the student a better understanding of the basic concepts of the subject than is usually done in an introductory course.

Bell, Stoughton, J. R. Blum, J. Vernon Lewis, and Judah Rosenblatt. *Introductory Calculus: With Algebra and Trigonometry.* Holden-Day, 1966. xxiv+309 pp. illus. $8.50. 66–15006. (SH–C)
Covers intuitive calculus of functions of one variable and gives a brief intro-

duction to calculus of functions of several variables. The student gets a rapid introduction to the main ideas of calculus. The authors are careful to point out what has been omitted instead of trying to make the student believe that more has been proved than actually has been. In effect, the authors assume as axioms many facts that will be theorems in the next course. Students interested primarily in calculus as a tool for other subjects will find this a useful text.

Britton, Jack R., R. Ben Kriegh, and Leon W. Rutland. *University Mathematics I & II* (2 vols.). Freeman, 1965. xiii+662, xii+650 pp. illus. $9.50 ea. 65–11880. (SH)
Mainly concerned with calculus, but includes some algebra, analytic geometry, linear algebra, and probability, as well as a chapter on Laplace transforms. Well balanced between intuition and rigor, and between technique and theory, the text would work well as a continuation of the modern approach to elementary and secondary school mathematics, provided that adequate time is available for a rather detailed course.

Dresden, Arnold. *Solid Analytical Geometry and Determinants.* Dover, 1964. v+310 pp. $2.00 (paper). 64–8267. (SH–C)
Written a generation ago, this classic covers useful and interesting material which has almost disappeared from the mathematics presently taught in high school and college. Its early chapters introduce the study of solid analytical geometry with sections on determinants and matrices, and on the solution of systems of linear equations. Although the author calls the remainder of the book a standard treatment of the subject, he consistently refers to the concepts learned in the first two chapters. Can be used by anyone who desires a fuller understanding of analytical geometry at the late high school or early college level.

Friedrichs, K. O. *From Pythagoras to Einstein.* (New Mathematical Library, No. 16.) Random, 1965. 88 pp. illus. LB $2.95, paper $1.95. 65–24963. (SH–C)
For readers with a background in elementary physics and Euclidean geometry. A somewhat abstract offering that will disclose the parts played by energy and momentum in the description of the physical universe. The first three chapters on Euclidean geometry and vectors are fairly easy, but from then on the text is progressively difficult.

Greenberg, Daniel A. *Mathematics for Introductory Science Courses: Calculus and Vectors.* Benjamin, 1965. xiv+214 pp. illus. $6.50, paper $2.95. 65–17012. (SH–C)
A summary of pre-college mathematics and an informal introduction to calculus and vectors. The ground covered is essentially that of the standard introductory or analysis texts of two decades ago, many of which are still available. Perhaps the greatest value is its informality. Too many introductory college mathematics courses stress formalism so much that broader understanding of fundamental ideas is lost. This book provides a more superficial but also a more directly applicable view of its subject matter. Recommended for high school and junior college libraries.

Hummel, James A. *Introduction to Vector Functions.* (Addison-Wesley Series in Mathematics.) Addison-Wesley, 1967. x+372 pp. illus. $9.75. 67–17259. (SH–C)
Formally treats various dimensions of vector analysis. Clearly written, this text requires knowledge of advanced calculus for comprehension. Suitable for mathematics majors.

Johnson, R. E., and F. L. Kiokemeister. *Calculus with Analytic Geometry* (3rd ed.). Allyn and Bacon, 1964. 798 pp. $15.95 (text ed. $11.95). 64–14271. (C)
A revision of a well-known textbook that incorporates changes dictated by recent curriculum revisions. It is an alternative to the book by George B. Thomas, Jr., bearing a similar title.

Kells, Lyman M. *Analytic Geometry and Calculus* (2nd ed.). Prentice-Hall, 1963. 628 pp. $12.65 (text ed. $9.75). 63–7063. (C)
Develops understanding on the intuitive level, then on the logical level. Helps the student to master concepts by applying them to make his own formulas and evaluate his results. Alternative to the George B. Thomas, Jr., or the Johnson and Kiokemeister textbooks—all recommended by referees.

Kline, Morris. *Calculus: An Intuitive and Physical Approach* (Parts 1 and 2). Wiley, 1967. xiii+574, xii+415 pp. illus. $9.95, $8.95. 66–26748. (SH–C)
Geometrical, physical, and heuristic arguments are used rather than formal mathematical proofs, and great emphasis is placed on the relationship between calculus and physical science, both by using physics to motivate the mathematics and by using the mathematics to solve physical problems. Applications in other sciences are not considered. The presentation is attractively informal and quite leisurely. The author's insistence that calculus is a tool for calculating and not a body of theorems makes this book more suitable for beginners, especially for students of the physical sciences, than many current "rigorous" ones.

Rainville, Earl D., and Phillip E. Bedient. *Elementary Differential Equations.* (4th ed.). Macmillan, 1969. xiv+466 pp. illus. $8.95. 69–10274. (SH–C)
Strong support for that mathematician's statement, "God organizes the universe with differential equations," is found in this, *the* standard text for differential equations. Its obvious position is in the third or fourth semester of the standard calculus sequence, but Chapters I–III may also be used in the 12th grade with beginning calculus students.

Reichmann, W. J. *Calculus Explained.* Van Nostrand, 1964. viii+331 pp. illus. $5.95. 64–57190. (SH–C)
The careful progression of ideas and techniques in this collateral reading book requires that the reader have only slight acquaintance with algebra and trigonometry. The writing is strong on general principles and basic definitions. The slight British flavor does not interfere with the easy and conversational style and the author's intuitive approach.

Roberts, A. Wayne. *Introductory Calculus.* Academic, 1968. xiii+527 pp. illus. $8.75. 68–14643. (SH–C)
An attractive, concise presentation of calculus, including infinite series and the

elements of differential and integral calculus of several variables. This text follows many of the recommendations of the Committee on the Undergraduate Program in Mathematics (CUPM).

Sagan, Hans. *Integral and Differential Calculus: An Intuitive Approach.* Wiley 1962. ix+329 pp. illus. $6.95. 62–17469. (SH)
This text is developed on an historical plan in which physical and geometric applications are interwoven to provide motivation for introducing new mathematical concepts.

Sawyer, Walter W. *What is Calculus About?* (New Mathematical Library, No. 2.) Random, 1961. 118 pp. illus. LB $2.95, paper $1.95. 61–6227. (SH)
Intended as required collateral reading for secondary school students to inform them concerning the role of the calculus in the development of mathematics and modern technology as an indispensable tool of the pure and applied sciences.

Stanaitis, O. E. *An Introduction to Sequences, Series, and Improper Integrals.* (The Mathesis Series.) Holden-Day, 1967. vii+210 pp. illus. $7.95. 66–17896. (C)
More appropriate for the teachers or others who have not studied calculus recently. It could be used for self study by those with sufficient background and could be a useful tool to a calculus instructor. It is at the post-calculus level. It starts with fundamentals and contains, in part, subject matter covered in the usual calculus course, but with such precision and clarity that it may give the reader new insights. Definitions and theorems are carefully worded, there are plenty of worked-out examples, a good supply of problems, and proofs are carefully labeled as such. Answers to odd-numbered problems are provided.

Stein, Sherman K. *Calculus in the First Three Dimensions.* McGraw-Hill, 1967. xiv+613 pp. illus. $9.95. 65–28827. (C)
The general student will be well served by this book written for a one-year course in calculus and advanced mathematics. With the introduction of each new concept a problem is presented, the solution of which is facilitated by the concept. Also, in addition to the normal applications to the physical sciences, there are numerous applications to biology, economics, psychology, and other fields of interest to the general student. A teacher's manual is available.

Taylor, Angus E. *Calculus with Analytic Geometry.* Prentice-Hall, 1959. 762 pp. illus. $15.00 (text ed. $11.95). 59–11305. (C)
One of a number of standard college texts combining the studies of calculus and analytic geometry. It is favored by almost all of the advisors who evaluated mathematics books for this list.

Thomas, George B., Jr. *Calculus and Analytic Geometry* (3rd ed.; 2 parts). Addison-Wesley, 1960. 1010 pp. illus. Part 1, $8.75; Part 2, $6.75; 1 volume ed. $12.50. 60–5015. (SH–C)
An excellent undergraduate college text for the science or engineering major which will be quite useful as a reference source for high school students who have been introduced to calculus. Emphasis is upon theoretical considerations.

Zelinsky, Daniel. *A First Course in Linear Algebra.* (Academic Press Textbooks in Mathematics Series.) Academic, 1968. viii+266 pp. illus. $6.50. 67–31043. (C)

Functional, algebraic, and geometric aspects of vectors, matrices, and linear transformations are introduced in parallel to provide background for calculus of several variables, following the CUPM approach. This is also suitable for self study; supplemented with problem sets and well-executed drawings.

517.5 SET THEORY

Anderson, Kenneth W., and Dick Wick Hall. *Sets, Sequences, and Mappings: The Basic Concepts of Analysis.* Wiley, 1963. 191 pp. illus. $5.95. 63–9394. (C)
For the person who has mastered a basic calculus course, this work introduces a modern approach to set theory, topology, and continuous functions.

Breuer, Joseph, *Introduction to the Theory of Sets.* (Tr. by Howard Fehr.) Prentice-Hall, 1958. 108 pp. illus. $7.95. 58–8673. (C)
A treatment of set theory, which is now being introduced in some new mathematics courses for elementary school students; suitable for advanced high school students, college undergraduates and teachers. Develops properties and principles which then are used to deal with sets and collections of abstract entities.

Kemeny, John G., et al. *Finite Mathematical Structures.* Prentice-Hall, 1959. 487 pp. illus. $14.35 (text ed. $11.00). 59–12841. (C)
Similar to the author's *Introduction to Finite Mathematics,* this text presents set theory and linear algebra at a higher level of sophistication with illustrative examples slanted toward the physical sciences.

May, Kenneth O. *Elements of Modern Mathematics.* Addison-Wesley, 1959. xvi+607 pp. $11.75. 59–7545. (C)
Introduces the ideas and notation of logic and sets, and proceeds to use them for the purpose of laying a foundation of mathematical concepts and skills that will prepare the student for the serious study of statistics and statistical methods.

Stoll, Robert R. *Introduction to Set Theory and Logic.* Freeman, 1963. xiv+474 pp. illus. $9.50. 63–8995. (C)
The foundations of mathematics are introduced through emphasis on standard mathematical systems, including a consideration of the role of logic in connection with axiomatic theories. The author says that the mathematics presented was directly stimulated by investigations pertaining to the real number system.

Whitesitt, J. Eldon. *Boolean Algebra and its Applications.* Addison-Wesley, 1961. 182 pp. illus. $8.50. 61–5027. (C)
Boolean algebra theory is viewed in the abstract as the natural algebra of sets, and in its practical applications in circuit theory and symbolic logic.

519.9 PROBABILITY AND STATISTICS (For Statistical Tables see 510.8)

Adler, Irving. *Probability and Statistics for Everyman: How to Understand and Use the Laws of Chance.* John Day, 1963. 256 pp. illus. $5.95; NAL (Signet T2850), paper 75¢. 62–16293. (C)
An interesting book for the student or adult who has a good basic knowledge

of algebra. Since chance invades many facets of a man's existence and experience, this new approach to a study of probability through the modern use of the set theory is timely.

Alder, Henry L., and Edward B. Roessler (4th ed.). *Introduction to Probability and Statistics* (3rd ed.). Freeman, 1968. 289 pp. illus. $7.00. 62–10143. (SH) A basic text on probability theory and statistical analysis with applications to such diversified fields as the agricultural sciences, psychology, geology, and the medical sciences.

Alterman, Hyman. *Numbers at Work: The Story and Science of Statistics.* (Illus. by F. W. Taylor.) Harcourt, Brace, 1966. 280 pp. $5.95. 65–12610. (SH–C) Presents in a fresh light such basic topics as the various kinds of averages, the preparation and use of graphs and charts, and the important areas of probability and sampling, exploring them from their simple beginnings to their use in solving a wide variety of complicated problems. Closes with a discussion of the electronic computer.

Campbell, R. C. *Statistics for Biologists.* Cambridge U., 1967. xi+242 pp. illus. $7.50, paper $2.45. 67–21955. (C) Concentrating on the exposition of principles of statistics rather than procedures, this text proceeds to show the need for biological statistics, the logic of statistical reasoning, decision theory, application to the normal distribution, and analysis of variance. Lucid examples, list of references, and tables are included.

Chakravarti, I. M., R. G. Laha, and J. Roy. *Handbook of Methods of Applied Statistics.* Vol. 1: *Techniques of Computation, Descriptive Methods, and Statistical Inference;* Vol. 2: *Planning of Surveys and Experiments.* Wiley, 1967. xiv+460, x+160 pp. illus. Vol. 1, $12.95; Vol. 2, $9.00. 66–26737. (C) A very thorough handbook with descriptions of a multivariate of topics. The first volume deals with computational techniques, inference, and decision making. The second volume examines sampling, design, and factor analysis. New ideas are presented rigorously with many examples. A useful reference for students of statistics.

Diamond, Solomon. *The World of Probability: Statistics in Science.* (Science & Discovery Series.) Basic, 1964. vii+193 pp. $5.95. 64–23454. (SH–C) The binomial, geometric and negative distributions are developed in this largely verbal presentation which requires little mathematical background beyond elementary algebra. No problems, nor are there exercises for each assignment. Simple derivations are given in appendices. May be used effectively as collateral reading for selected high school students and will serve well as a text for college students beginning statistical studies.

Freund, John E. *Modern Elementary Statistics* (3rd ed.). Prentice-Hall, 1967. x+432 pp. $9.25. 66–29560. (SH) The purpose of this book is to acquaint beginning students in the natural and social sciences with the fundamentals of modern statistics that will be useful in their research. Illustrations and exercises are distributed among various fields of application.

Hoel, Paul G. *Elementary Statistics* (2nd ed.). Wiley, 1966. ix+351 pp. illus. $8.50. 66–11523. (SH–C)
A textbook of elementary statistical methods emphasizing sampling methodology and theory, and including such additional topics as correlation, regression, analysis of variance, and an introduction to time series and index number analysis.

Huff, Darrell. *How to Lie with Statistics*. Norton, 1954. 142 pp. illus. $3.95, paper $1.95. 53–13322. (JH–SH)
"Averages and relationships and trends and graphs are not always what they seem." Extremely interesting and amusing, the author warns the reader of the misleading nature of many statistical presentations.

Huff, Darrell. *How to Take a Chance*. Norton, 1959. 173 pp. illus. $4.50, paper $1.25. 58–13953. (SH)
An enjoyable excursion into probability considerations in the everyday world. Profusely illustrated.

Lordahl, Daniel S. *Modern Statistics for Behavioral Sciences*. Ronald, 1967. xii+365 pp. illus. $8.00. 67–14485. (C)
Helps to prepare students of the behavioral sciences to apply statistical methods to research. Considers descriptive statistics, scale transformations, random sampling, analysis of variance, Chi square, probability, and correlation.

Mack, C. *Essentials of Statistics for Scientists and Technologists*. Plenum, 1967. vii+174 pp. illus. $5.95. 67–17769. (C)
A brief but rigorous introduction to fundamental statistical methods and tests. Topics in descriptive statistics include significance tests, normal and Poisson distributions, regression, and correlation. This may be the best book in its class.

Mendenhall, William. *Introduction to Probability and Statistics* (2nd ed.). Wadsworth, 1967. xiii+393 pp. illus. $8.95. 67–19290. (SH–C)
Begins with some mathematical background information and then proceeds through descriptive procedures, probability rules, binomial and normal distributions, etc. The mathematics is minimal and intuitive discussions are frequently presented.

Mosteller, Frederick, et al. *Probability: A First Course*. Addison-Wesley, 1961. 319 pp. illus. $5.00. 61–4119. (SH)
Requiring mathematics through intermediate algebra, this text presents elementary probability theory for finite sample spaces, a section on random variables, binomial distribution, and some statistical applications of probability.

Ostle, Bernard. *Statistics in Research* (2nd ed.). Iowa State U., 1963. 487 pp. $10.50. 63–7545. (C)
This text presents the techniques of modern statistics as statistical methods per se, with examples from varied fields of application; emphasis is placed upon presenting the material in such a way as to provide a thorough understanding of the limitations of various techniques. Covers elements of both descriptive and mathematical statistics, statistical methods, and design and analysis of experiments.

Reichmann, W. J. *Use and Abuse of Statistics.* Oxford, 1962. 336 pp. illus. $6.00. 62–3388. (SH)

Designed for the general reader who wants to learn something about statistical methods and their application and yet does not want to plough through a textbook. It may serve as an introduction to the subject, or as collateral reading for a student of beginning statistics.

Snedecor, George W. *Statistical Methods Applied to Experiments in Agriculture and Biology* (6th ed.). Iowa State U., 1967. 593 pp. illus. $9.95. (C)

This work has been a fundamental text for beginners in statistics as well as a manual for those using statistics in biological research over a great many years. This edition features a chapter on "Design and analysis of samplings," by William G. Cochran. Each chapter begins with explanations for beginners, then follow formulae and technical explanations.

Throp, Edward O. *Elementary Probability.* Wiley, 1966. x+152 pp. illus. $4.95. 66–16130. (SH–C)

Contains discussions of equally likely alternatives, combinatorial analysis, sets, functions, probability measures on finite sample spaces, relative frequencies, Bayes' rule, random variables, expectation, dispersion, Chebychev's inequality, Poisson distribution, Borel sets and continuous density functions. Exercises at the end of each section, and most solutions are given. The book may be used for a one-semester introductory course, and as a supplementary text for two-year calculus courses or statistics courses for social scientists. The author succeeds at making "each successive step seem to be the natural consequence of the preceding material."

Walker, Helen M., and Joseph Lev. *Statistical Inference.* Holt, Rinehart and Winston, 1953. 510 pp. illus. $10.50. 52–13908. (C)

The intuitive approach is implemented to introduce various statistical methods and techniques. Many of the techniques presented are illustrated in a great variety of practical applications.

Weaver, Warren. *Lady Luck: The Theory of Probability.* Doubleday (Anchor S30), 1963. 392 pp. illus. $1.45 (paper). 63–8759. (SH)

A PSSC book written because of the author's conviction that the type of thinking about problems which one learns in probability theory is of the highest importance. Should be required reading for every college-bound secondary school student.

Williams, John D. *The Compleat Strategyst* (rev. ed.). McGraw-Hill, 1965. 234 pp. illus. $6.95. 53–9007. (SH)

This primer on the theory of games of strategy (as opposed to games of chance) analyzes game situations and available strategies with the emphasis placed on two-person, zero sum games.

520 ASTRONOMY

Page, Thornton, and Lou Williams Page (eds.). *Sky and Telescope Library of Astronomy.* Macmillan, 1965–1969. $7.95 ea. (SH–C)

Vol. I: *Wanderers in the Sky: The Motions of Planets and Space Probes,* 1965. xiv+338 pp. illus.

Vol. II: *Neighbors of the Earth,* 1965. xvi+336 pp. illus.
Vol. III: *The Origin of the Solar System,* 1966. xiv+336 pp. illus.
Vol. IV: *Telescopes,* 1966. xiv+338 pp. illus.
Vol. V: *Starlight,* 1967. xiv+337 pp. illus.
Vol. VI: *The Evolution of Stars,* 1967. xi+334 pp. illus.
Vol. VII: *Stars and Clouds of the Milky Way,* 1968.
Vol. VIII: *Galaxies of the Universe,* 1969.
This unique, eight volume, library of astronomy is the result of careful compilations of articles by distinguished authors that have appeared in *Sky and Telescope, The Sky,* and, *The Telescope.* The editors' skillful commentary relates to and connects the articles, published from 1931. The series portrays the historical development of astronomical theories and their recent changes and refinements. Each volume is thoroughly illustrated with black-and-white photographs and line drawings, and is well appendixed and indexed. A helpful glossary aids the layman.

Ronan, Colin A. *Astronomers Royal.* Doubleday, 1969. xii+224 pp. illus. $5.95. 69–10364. (SH–C).
This detailed history of astronomy in the United Kingdom, from the solar system to the universe, and from reflector telescope to radio-dish, is unusual in that the author does not confine himself to the great names, but includes the contributions made by minor figures as well. The lay reader will find interesting the parallel development of the clock and telescope. The accurate measurement of time and the checking of this accuracy by movement of the stars are inextricably linked.

Smart, W. M. *The Riddle of the Universe.* Wiley, 1968. x+228 pp. illus. $5.00. 68–25830. (SH–C)
An account of the history and development of astronomy and astronomical principles. Subjects such as the shape of the earth, structure of the galaxy, stellar motions, the application of spectroscopy to solution of astronomical problems, the age of the universe and theories of its origin, the origin of the elements, and quasars, are treated lucidly and expertly. Photographs are excellent. Recommended for beauty and clarity of presentation as well as a reference book. Good index.

Whipple, Fred L. *Earth, Moon, and Planets.* (3rd ed.). (Harvard Books on Astronomy) Harvard, 1968. viii+297 pp. illus. $7.25. 68–21987. (SH–C)
Material from recent investigations of the Moon, Venus, and Mars is included in this third edition. Appendices include Bode's Law, planetary configurations, planetary data, star charts, planet finder, and the age of the moon. This book continues to be the one that should be available for reference, as well as for leisure-time reading by the astronomy buff.

520.3 ASTRONOMY—ENCYCLOPEDIAS

Rudaux, Lucien, and G. De Vaucouleurs. *Larousse Encyclopedia of Astronomy* (2nd ed.). Prometheus (dist. by Putnam's), 1962. ix+506 pp. illus. $17.50. 60–27693. (SH)
A basic reference work, well illustrated, that is so complete, authoritative, and

up to date that it is necessary for all secondary school, academic and public libraries. It is topically arranged and has an analytical index.

520.7 ASTRONOMY—STUDY AND TEACHING

Abell, George. *Exploration of the Universe.* Holt, Rinehart and Winston, 1964. 646 pp. illus. $11.95. 64–10120. (SH–C)
This new and comprehensive introductory text is recommended as the "foundation book" of the astronomy collections in all libraries, and for personal libraries of interested nonspecialists; valuable for instruction, background reading, and general reference.

Baker, Robert H. *An Introduction to Astronomy* (7th ed.). Van Nostrand, 1968. viii+364 pp. illus. text ed., $7.50. (SH)
Since 1935 this has been the classic and short introduction to astronomy, which has been kept up to date by several revisions. Students with a good mathematical foundation will prefer the author's more comprehensive textbook *Astronomy* (8th ed.).

Baker, Robert H. *Astronomy* (8th ed.). Van Nostrand, 1964. 557 pp. illus. $8.95. 59–8367. (C)
Since the publication of the first edition in 1930, this textbook has built a reputation for unexcelled clarity and completeness, and the policy of frequent revision has kept it strictly up to date. Those desiring a briefer text may prefer the author's *Introduction to Astronomy* (7th ed.).

Davidson, Martin. *Elements of Mathematical Astronomy with a Brief Exposition of Relativity.* (Revised by Cameron Dinwoodie.) (3rd ed.). Macmillan, 1962. 276 pp. illus. $9.25. 62–5897. (C)
The student who has read one or more good general introductions to astronomy needs the discipline imparted by this book if he desires to understand astronomical research.

McLaughlin, Dean B. *Introduction to Astronomy.* Houghton Mifflin, 1961. viii+ 463 pp. illus. $8.95. 61–16200. (C)
The author considers that "facts and figures" should constitute only the beginning, not the end, of one's knowledge of astronomy. The philosophy of measurement, observation and experiment must be learned and one must know how to apply rational processes to observed facts.

Moore, Patrick (ed.). *A Handbook of Practical Amateur Astronomy.* Norton, 1964. 254 pp. illus. $5.95. 64–10570. (SH)
Contributions from very experienced amateur astronomers in the British Astronomical Association. Many people spend years in perfecting a useable small telescope, then wonder what to do with it. This book suggests things that can be done with such amateur equipment.

Moore, Patrick. *Naked-Eye Astronomy.* (The Amateur Astronomer's Library.) Norton, 1965. 253 pp. illus. $6.50. 65–27466. (JH–SH)
After a brief orientation to the sky and explanation of why the stars change from season to season, Moore describes the stars one can see without a telescope or binoculars during each of the 12 months of the year. Information

for viewing the planets, the sun, moon, and artificial satellites. Practical and useful.

Paul, Henry E. *Outer Space Photography for the Amateur* (3rd ed.). Chilton, 1967. 156 pp. illus. $5.95. 67–21698. (JH–SH)
More than 160 of the best amateur astronomical photographs illustrate this concise and stimulating guide. Constant reference to supplies and additional information is most useful.

Paul, Henry E. *Telescopes for Skygazing* (2nd ed.). Chilton, 1965. 157 pp. illus. $4.95. 65–26425. (JH–SH–C)
Hints and suggestions on the various types of telescopes, their relative advantages and disadvantages, and the points to be considered in building, assembling, or buying astronomical equipment. Methods of testing optical components and the interpretation of the results of such tests are described and illustrated. A cross-referenced equipment and supplier list should be very handy. Chapter on astrophotography points out the details of equipment, film, development and exposure. Also a chapter on the use of binoculars in astronomy.

Roy, Archie E. *The Foundations of Astrodynamics*. Macmillan, 1965. xiv+385 pp. $10.95 illus. 65–15172. (SH–C)
Clearly and lucidly presents the ideas, principles, methods and results achieved by investigators in the various branches of astrodynamics. Intended for junior and senior college students and requires a mathematical background of spherical trigonometry, the calculus, differential equations and vector notation. However every chapter contains carefully prepared expository material which can be read with profit by any good student. A senior high school library can well afford a few carefully selected books, like this one, which may be beyond the grasp of the general student but which could offer the able student a glimpse of the actual details of scientific endeavor. Exceptionally well written.

Sidgwick, J. B. *Amateur Astronomer's Handbook* (2nd ed.). Macmillan, 1961. 580 pp. illus. $12.75. 55–29173. (SH)
This volume is a handbook of methods and techniques to be used with its companion, *Observational Astronomy for Amateurs*, by the same author. It fills the gap between amateur literature and technical works for professionals.

Sidgwick, J. B. *Observational Astronomy for Amateurs*. Macmillan, 1955. 358 pp. illus. $10.75. 55–57280. (SH)
Having mastered the author's *Handbook*, the amateur astronomer is now ready to use this book to give him background for his studies and observations of the heavens.

Van de Kamp, Peter. *Principles of Astrometry, With Special Emphasis on Long-Focus Photographic Astrometry*. Freeman, 1967. vii+227 pp. illus. $7.00. 66–22077. (C)
A very well-written text in an area for which few texts exist. The introductory part covers the necessary topics in spherical astronomy. The last part deals with Gaussian error analysis. The main body of the work is an admirable and lucid account of the way in which long-focus astrometry can be made to yield the maximum amount of information.

520.9 ASTRONOMY—HISTORY

Abetti, Giorgio. *The History of Astronomy*. (Tr. by Betty Burr Abetti.) Abelard-
Schuman, 1952. 338 pp. illus. $6.00. 52–13485. (SH)
A working astronomer has given an insight into the personalities of his predecessors and contemporaries. A useful and basic historical account for the
general reader.

Caspar, Max. *Kepler*. (Tr. by C. Doris Hellman.) Abelard-Schuman, 1960. 401
pp. $7.50; Macmillan (Collier BS55), paper $1.50. 59–5797. (C)
The best and most authoritative biography ever written of Johannes Kepler
(1571–1630) who was poet, theologian, scientist, and mathematician. Best
known for his laws of planetary motion, he also did pioneering work in the
calculus, and founded the study of optics.

De Santillana, Giorgio. *The Crime of Galileo*. U. of Chicago, 1955. xvi+389
pp. illus. $6.50, paper (Phoenix P40) $2.25. 55–7400. (C)
The exciting story of Galileo's bitter conflicts with religious and political
leaders that arose because of his discoveries and the publication of his *Di-
alogue on the Great World Systems*.

De Vaucouleurs, Gerrard. *Discovery of the Universe*. Macmillan, 1957. 328 pp.
illus. $5.25. 57–10015. (SH)
Man's ideas and comprehension of the solar system and of the galaxies beyond have evolved by a gradual process of speculation, investigation, verification, through the procession of a long and intriguing series of historical events
narrated herein.

Geymonat, Ludovico. *Galileo Galilei: A Biography and Inquiry into His Philosophy of Science*. (Tr. by Stillman Drake.) McGraw-Hill, 1965. xii+260
pp. $6.50; paper $2.95. 64–8549. (SH–C)
The biographical details of this scholarly and readable work serve as a
framework for a consideration of Galileo's complex philosophy of science
which must be understood and appreciated to evaluate his achievements. Appended historical notes greatly enhance the value of the book. High school
students should read it in preference to other and superficial biographies of
Galileo.

Ley, Willy. *Watchers of the Skies: An Informal History of Astronomy from
Babylon to the Space Age*. Viking, 1963. xiii+528 pp. illus. $8.50. 61–8386.
(SH)
A popular science historian and prophet of the space age has presented a
compact history of astronomy from its beginnings up to the unsolved present-
day problems for which contemplated space explorations may provide at
least partial solutions.

Moore, Patrick. *The Picture History of Astronomy* (2nd ed.). Grosset & Dunlap,
1964. 253 pp. illus. $6.95; LB $6.25. 62–13365. (JH)
The story of astronomy from the earliest recorded attempts of man to understand heavenly phenomena to recent man-made satellites and the development of radio-astronomy. Included are the names and pictures of outstanding individuals who have contributed to man's knowledge of the universe,
with adequate descriptions of their contributions. Mathematical descriptions

are avoided. The appendix includes the index, a 3-page list of illustrations, and a 2-page compendium of landmarks.

Pannekoek, Antonie. *A History of Astronomy.* Interscience (dist. by Wiley), 1961. 521 pp. illus. $12.50. 61–66763. (C)
Several well-known professional astronomers have stated that this is the best general history of astronomy currently in print. It is authoritative, yet not technical, and relates the intellectual, technological and social background to the astronomy of each historical period.

Pickering, James S. *Famous Astronomers.* Dodd, Mead, 1968. 128 pp. illus. $3.50. 68–12812. (JH)
Rather dramatic descriptions of the lives of the major astronomers of antiquity, from Aristotle through Herschel, are given. The author makes some assumptions to obtain continuity in the lives of the astronomers. Very readable and appropriate for young readers.

Richardson, Robert S. *The Star Lovers.* Macmillan, 1967. x+310 pp. illus. $6.95. 67–16714. (JH–SH–C)
A collection of informal biographies of 16 scientists who were either astronomers or worked closely with astronomy. The combination of personal interest, scientific reliability, and easy communication between author and reader makes this work widely acceptable.

Shapley, Harlow (ed.). *Source Book in Astronomy, 1900–1950.* Harvard, 1960. xv+423 pp. illus. $10.00. 60–13294. (C)
A collection of 69 papers, nonmathematical in style, by well-known authorities which deal with a great variety of topics and thus portray the advancement of astronomical knowledge during the first half of this century.

Shapley, Harlow. *Through Rugged Ways to the Stars.* Scribner's 1969. 180 pp. illus. $6.95. 68–57085. (SH–C)
Shapley tells of his multi-faceted career as astronomer and scientist in a witty and anecdotal fashion. Unknown to many, Shapley was integral in the shaping and development of the National Science Foundation, UNESCO, and other organizations. The only valid criticism is that the book is too brief—many more stories could be told by this most remarkable man.

Struve, Otto, and Velta Zebergs. *Astronomy of the 20th Century.* Macmillan, 1962. 544 pp. illus. $12.50. 62–21206. (SH)
Covers the same period as Shapley's *Sourcebook,* but is an historical account which can be read with relish by laymen. Particularly valuable is the portrayal of the development of modern astrophysics.

Thiel, Rudolph. *And There Was Light: The Discovery of the Universe.* (Tr. by Richard and Clara Winston.) Knopf, 1957. xv+415+vii pp. illus. $7.95; NAL (Mentor MT290), paper 75¢. 57–13059. (SH)
Those who like historical novels will relish this book written in charming narrative style which presents a completely factual account of the adventure and discovery of astronomy. Not a reference book, but good background reading.

Toulmin, Stephen, and June Goodfield. *The Fabric of the Heavens: The Development of Astronomy and Dynamics.* Harper & Row, 1965. 285 pp. illus. $1.95 (paper). 62–7916. (SH–C)
Drawing upon historical, literary, and scientific sources, the authors trace

the progress of man's knowledge of the universe and its laws through the representative figures of Aristotle, Aristarchos, Ptolemy, Copernicus, Tycho Brahe, Kepler, and Newton. A fundamental and readable contribution to the history and philosophy of science.

521 THEORETICAL AND CELESTIAL MECHANICS

Danby, John M. A. *Fundamentals of Celestial Mechanics*. Macmillan, 1962. xiii+ 348 pp. illus. $9.25. 62–7032. (C)
This is a book for the astronomy student with a strong and comprehensive mathematical background which will provide him with a modern treatment of celestial mechanics. It will be useful in college libraries and in large public libraries with advanced science collections. It has some reference value for mathematics majors.

522.2 OBSERVATORIES AND INSTRUMENTS

Howard, Neale E. *Standard Handbook for Telescope Making*. Crowell, 1959. x+326 pp. illus. $8.95. 59–12503. (SH)
A standard handbook giving detailed directions for the construction of an 8-inch telescope. The author has had many years of experience in teaching young people to build their own telescopes and following construction details, he tells how to use optical instruments.

Keene, George T. *Star Gazing with Telescope and Camera* (2nd ed.). Chilton, 1967. 128 pp. illus. $3.50. 67–25847. (SH–C)
Gives the serious amateur detailed assistance in selecting binoculars and telescopes, in building and using a telescope, and in using a camera with or without a telescope for celestial photography. A must for anyone who owns or plans to build or buy a telescope.

Lovell, Sir Bernard. *The Story of Jodrell Bank*. Harper & Row. 1968. xvi+265 pp. illus. $5.95. (SH–C)
The author planned and supervised construction of the first massive radio telescope in the world, the famous Jordrell Bank installations. This account was based on the author's diary and recollections of personal, financial, and technical details in constructing the world's largest (250-foot diameter) fully steerable radio receiver and later a transmitter to *Pioneer V*. Professional scientists will find the story fascinating and will undoubtedly recognize vignettes similar to their own experiences. Index is excellent.

Miczaika, G. R., and William M. Sinton. *Tools of the Astronomer*. Harvard U., 1961. viii+294 pp. $7.75. 60–13299. (C)
A guide to the instruments and techniques employed by modern astronomers which will be of interest to amateur and professional astronomers and to students in general.

Texereau, Jean. *How to Make a Telescope*. Interscience (dist. by Wiley), 1957. 191 pp. illus. $4.95; Doubleday (Anchor N31), paper $1.45. 57–9156. (SH)
Perhaps the most popular and worthwhile book for the amateur on this subject currently in print. The author built his first successful 10-inch telescope at

the age of 20 and has been building them ever since, and herein he communicates the results of this experience.

Wood, Frank Bradshaw. *Photoelectric Astronomy for Amateurs.* Macmillan, 1963. ix+223 pp. illus. $8.95. 63–8638. (SH)
During the past 30 years the photoelectric photometer has developed from a temperamental instrument of limited use to a successful and dependable instrument of diversified usefulness, and this is that story.

Woodbury, David O. *The Glass Giant of Palomar.* Dodd, Mead, 1953. 385 pp. illus. $6.00. 39–27835. (JH)
An interesting popular account of the development, design and construction of the first of the "giant telescopes," the 200-inch reflector at Palomar. Although mainly of historical interest it merits a place in school libraries.

523 DESCRIPTIVE ASTRONOMY

Alter, Dinsmore, et al. *Pictorial Astronomy* (2nd ed.). Crowell, 1963. 312 pp. illus. $8.95 (text ed. $6.00). 63–14545. (JH)
One of the most widely used basic books for the beginning student or amateur astronomer. Divided into sections dealing with the earth, the moon, eclipses, the planets, comets and meteors, and the stars and nebulae.

Bernhard, Hubert J., et al. *New Handbook of the Heavens* (2nd ed.). McGraw-Hill, 1948. xi+360 pp. illus. $6.50, paper $2.75; NAL (Signet T3132) 95¢. 50–56429. (JH)
This authoritative guide for students and laymen is one of the most widely used introductions to astronomical science. Contains a glossary and other informative appendices and is well indexed.

Del Rey, Lester. *The Mysterious Sky.* Chilton, 1964. 183 pp. illus. $3.95. 64–10698. (SH)
Facts about the atmosphere, stars and space are arranged in a narrative with a great deal of historical reference as well as much recent theoretical and observational material. The 18 chapters cover the atmosphere, planets, sun, stars, galaxies and the origin of the universe, and describe such phenomena as ball lighting, auroras, cosmic rays, Van Allen Belts, meteors, lunar volcanoes, sunspots, gravity of planets, variable stars, measurements of the universe, age of the earth and universe.

Gallant, Roy A. *Exploring the Universe* (rev. ed.). (Illus. by Lowell Hess.) Doubleday, 1968. 65 pp. $3.95. 68–11959. (JH)
The first edition received the Thomas Alva Edison Foundation Award as the best science book of the year for young people. Traces the progress of man's knowledge of the universe from the ancient Chinese to the present. Completely revised and updated with new illustrations.

Hoyle, Fred. *Astronomy.* Doubleday, 1962. 320 pp. illus. $12.95; deluxe ed., $14.95. 62–14108. (SH)
With the aid of magnificent colored illustrations, the author presents an historical summary of man's investigations of the universe, as well as a descriptive account of astronomy and of its instruments and techniques.

Inglis, Stuart J. *Planets, Stars, and Galaxies* (2nd ed.). Wiley, 1967. 474 pp. illus. $8.50. 61–5672. (SH)
Presents the general field to the student of liberal arts who does not have a sufficient grounding in mathematics to absorb a more sophisticated presentation, and yet feels he must understand the "mysterious" universe.

King, Henry C. *Our World in Space: An Easy Guide to the Universe.* Macrae Smith, 1964. 94 pp. illus. $3.25. 64–23916. (JH)
Presents a brief summary of man's knowledge and observations as he has peered outward into space. Fully aware that he is addressing readers with little or no background in astronomy, the author uses some approximations which may be technically debatable, but which, in these circumstances, are both adequate and sensible.

MacGowan, Roger A., and Frederick I. Ordway, III. *Intelligence in the Universe.* (Illus. by Harry H-K Lange.) Prentice-Hall, 1966. xiii+402 pp. $13.95. 66–18343. (C)
A speculative analysis of the problem of extra-solar intelligence that surveys existing literature and assesses man's scientific, technological, and social progress. The authors develop their own interpretations of established facts. An interesting and provocative presentation, certain to accelerate controversy.

Moore, Patrick. *Amateur Astronomy.* Norton, 1968. 328 pp. illus. $6.95. 68–10882. (SH)
First published in 1957 under the title, *The Amateur Astronomer,* this popular guide has been revised and brought up to date (as of September 1967 when the manuscript was completed). The Director of the Armagh Planetarium in Ireland has included data obtained from probes and flybys, quasars and radio telescope among his new material. The guide is useful for readers with or without observational equipment. Miscellaneous useful appendices, bibliography, and index.

Olcott, William T. (rev. and ed. by R. Newton and Margaret W. Mayall.) *Field Book of the Skies* (4th ed.). Putnam's 1954. 482 pp. illus. $5.00. 54–8707. (JH)
An introduction to astronomy for the beginner and amateur, and a reference handbook for the professional. Some portions are not strictly up to date, but this does not impair its usefulness.

Peltier, Leslie C. *Starlight Nights: The Adventures of a Star-Gazer.* Harper & Row, 1965. x+236 pp. illus. $4.95. 65–20992. (JH–SH–C)
The story of a scientist, a devoted amateur. It is the story of a man in love with all of nature, and the heavens in particular. Cogent comment on subjects ranging from berry-picking to flying saucers. Highly entertaining reading. Although strictly autobiographical, the budding astronomer can learn much from it of a practical nature.

Sciama, D. W. *The Unity of the Universe.* Doubleday, 1959. 288 pp. illus. $3.95, paper (Anchor A247) 95¢. 59–11608. (SH)
Written for the layman, presenting the development of our present observational picture of the universe and a view of various theories concerning the functioning of the universe. Expounds lucidly on basic theories of inertia and relativity.

523.013 ASTROPHYSICS

Asimov, Isaac. *The Universe: From Flat Earth to Quasar*. Walker, 1966. 308 pp. illus. $6.50. 66–22515. (SH–C)

A thrilling sketch of man's search for understanding the mysteries of space, time, and the incredible dimensions of astronomy. Detail is minimal, Asimov touches lightly on many topics; the scale of the universe, astrophysical explanations of the evolution and life cycles of the universe, and finally the question of the origin of the universe. Readers should have some knowledge of astronomy.

Hoyle, Fred. *Frontiers of Astronomy*. Harper & Row, 1955. xvi+360 pp. illus. $7.50, paper 95¢; also NAL (Signet) 75¢. 55–6582. (SH)

Although important new observations have been made since this book was written, it contains much fundamental information on how major discoveries in physics, particularly in nuclear physics, and improved observational techniques have increased our knowledge of the universe.

Hynek, J. Allen, and Norman D. Anderson. *Challenge of the Universe*. (A Vistas of Science Book.) McGraw-Hill, 1962. 143 pp. illus. $2.50; LB $2.63; NSTA, paper 50¢. 62–13888. (JH)

The "challenge" is explained in terms of facts, theories and speculations on astrophysics, celestial mechanics and information on relativity. A dividend is incorporated in the form of amateur projects and experiments.

Jastrow, Robert. *Red Giants and White Dwarfs: The Evolution of Stars, Planets and Life*. Harper & Row, 1967, xii+176 pp. illus. $5.95. 66–20765. (SH–C)

From the background of his direct involvement in basic research programs so fundamental to modern astrophysical and astronautical research, Dr. Goddard has described how gravity, electromagnetic and nuclear forces act on matter, synthesize elements in the interiors of stars, and how stars, planets and other astronomical objects are formed from these elements. Also described are the processes by which organic materials are created and become life forms. The possibility of life forms beyond present human experience, and the evolution and future of the universe as a whole are also treated.

Merrill, Paul W. *Space Chemistry*. U. of Michigan, 1963. 166 pp. illus. $5.00, paper $1.95. 60–15776. (C)

A unique presentation of the present state of our knowledge of the chemistry of the universe which is intended for the general reader, who will find it somewhat difficult yet intellectually rewarding.

Moore, Patrick, and Francis Jackson. *Life in the Universe*. Norton, 1962. ix+140 pp. illus. $3.95. 61–13044. (C)

An astronomer and a research microbiologist deal with the general question of life on the other planets and are exponents of the deterministic viewpoint. An extensive bibliography is appended. An elementary knowledge of astronomy is needed by the prospective reader.

Wright, Helen. *Explorer of the Universe: A Biography of George Ellery Hale*. Dutton, 1966. 480 pp. illus. $10.00. 66–11542. (SH–C)

Extremely well written, thoroughly documented biography. Hale's achievements in advancing science in general are shown to be as important as his astronomical discoveries which were great contributions to the founding of the "new science," modern astrophysics. Recommended as collateral reading in the history of astronomy.

523.016 RADIO ASTRONOMY

Allen, Tom. *The Quest: A Report on Extraterrestrial Life.* Chilton, 1965. xii+ 333 pp. illus. $4.95. 65–13926. (JH–SH)
Recounts the meeting at Green Bank, West Virginia, where the topic of extraterrestrial life was given respectability. Arrives at the number of civilizations which may be trying to communicate with us. Explores the mysterious radio signals which appeared 40 years ago, and for which there is still no satisfactory explanation. The criteria for the development and perpetuation of life are then developed. Describes the instrumentation contemplated in our search for life on Mars. An appendix, hard-fact notes pertaining to various chapters, an excellent bibliography, and a satisfactory index.

Ehrensvard, Gosta. *Man on Another World.* (Tr. by Lennart and Kajsa Roden.) U. of Chicago, 1965. vii+182 pp. illus. $5.95. 65–17287. (SH–C)
Three main considerations: (1) A discussion of manned expeditions to the moon and local planets, and the possible establishment of bases; (2) The classic question of indigenous life on Mars or other possible planets in the Milky Way, and (3) The reception and transmission of radio signals for exchanging information with possible conscious life on these possible planets. Scientifically sound, neither overly optimistic nor pessimistic.

Lovell, Alfred Charles Bernard, and Joyce Lovell. *Discovering the Universe.* Harper and Row, 1964. 136 pp. illus. $4.95; LB $4.43. 64–12700. (JH)
The Director of Jodrell Bank and his wife have written this book to explain the fundamentals of astronomical instruments, radio astronomy, and space exploration to visitors. This is a good book for the curious layman.

Smith, Alex G. *Radio Exploration of the Sun.* Van Nostrand, 1967. viii+159 pp. illus. $1.95 (paper). (SH–C)
Presents in a beautiful and understandable manner how solar eclipses are used to analyze the sun, and how the earth is affected by solar activity as seen through the eyes of the radio astronomer.

523.1 COSMOLOGY

Abetti, Giorgio. *Stars and Planets.* (Tr. by V. Barocas.) American Elsevier, 1966. 341 pp. illus. $12.50. 66–12963. (SH–C)
A somewhat unique discussion of stars and planets which illustrates the growth of theory and interpretive ideas concerning the origin of the solar system. The frequent use of astronomical history is a valuable characteristic which makes the subject matter more palatable to those with little background in astronomy.

Alfven, Hannes. *Worlds-Antiworlds: Antimatter in Cosmology.* Freeman, 1966. 103 pp. $3.50. 66–27947. (C)
Explains (to readers who lack special training in physics or astronomy)

the author's extensions to the original cosmological theory of Professor O. Klein. The idea is based on known physical laws, particularly the anti-matter notion of symmetry that every proton is matched by an anti-proton somewhere in the metagalaxy and every electron is matched by a positron. Can be profitably and enjoyably read by the scientifically-trained as well as the erudite layman. Some of the answers to the age-old question, "How did it all come to be?" may lie within this theory and hence it deserves wide-spread attention.

Asimov, Isaac. *To the Ends of the Universe*. Walker, 1967. 136 pp. illus. $3.95. 67–23096. (JH–SH)
The various contents of the universe beyond our solar system are described. The concepts, historical development, and names of astronomers, are presented together with theories and discoveries. Understanding the subject matter in each chapter does not necessarily depend on knowledge of the information given in previous chapters.

Bondi, Hermann. *The Universe at Large: Views of Cosmology*. Doubleday (Anchor S14), 1960. 154 pp. illus. $1.25 (paper). 60–13501. (SH)
A Science Study Series book that explains the author's view of the nature of scientific theory, his idea of the expansion of the universe, and of theories of cosmology and tests thereof.

Edgeworth, K. E. *The Earth, the Planets, and the Stars—Their Birth and Evolution*. Chapman and Hall (dist. by Barnes & Noble), 1961. xiii+193 pp. illus. $4.50. (C)
For the reader with some basic knowledge of astronomy, this discussion of planetary evolution will be enlightening. The various theories that have been proposed are discussed as well as some original notions of the author.

Gamow, George. *The Creation of the Universe* (rev. ed.). Viking, 1961. xii+147 pp. illus. $5.75; paper (Compass C7) $1.25. 61–7387. (SH)
The "Big Bang" hypothesis and the "Steady State" hypothesis are explained and the pro and con arguments presented for the layman. Actual photographs and "Gamovian" sketches assist the exposition.

Gamow, George. *Gravity: Classic and Modern Views*. (Science Study Series.) Doubleday (Anchor S–22), 1962. 157 pp. illus. $1.25 (paper). 62–8840. (SH)
With his characteristic popular and whimsical style, Gamow presents the classical and modern notions and concepts of gravity. He concludes with a discussion of "unsolved" problems of gravity.

Hawkins, Gerald S. *Splendor in the Sky*. Harper & Row, 1961. 292 pp. illus. $7.95. 61–10222. (SH)
An inspired account of the historical evolution of astronomy as a science, the formation and characteristics of the Earth, the stars, meteors and meteorites, concluding with a discussion of the theories of cosmology.

Lovell, Alfred Charles Bernard. *The Individual and the Universe*. Harper & Row, 1959. x+111 pp. illus. $3.00; NAL (Mentor), paper 60¢. 59–6326. (SH)
Another series of lectures which deal primarily with the various past and present theories of the origin and evolution of the universe; basic knowledge for the lay reader.

Lyttleton, R. A. *Man's View of the Universe*. Little, Brown, 1961. 108 pp. illus. $4.75. 61–5739. (SH)

What is known and what is conjectured about the astronomical universe are summarized for the layman. The book clears up some misconceptions that frequently appear in beginning textbooks and popular writings.

McVittie, G. C. *Fact and Theory in Cosmology*. Macmillan, 1962. 190 pp. illus. $5.95. 62–5858. (C)

A presentation of rather difficult material in a form intelligible to nonspecialists. Fills the gap between elementary discussion and technical monographs.

Martin, Charles-Noël. *The Universe of Science*. Hill and Wang, 1963. 208 pp. illus. $3.95. 63–18482. (SH)

A French journalist, who is also a nuclear physicist, describes in an integrated narrative of four sections the development of the universe. The sections are: 1. The Infinitessimal; 2. Life; 3. The Planet Earth; 4. Infinities.

Messel, H., and S. T. Butler (eds.). *The Universe and its Origin*. St. Martin's, 1964. 147 pp. illus. $3.75. 64–9999. (SH–C)

Fairly well-coordinated lectures by three astrophyisicists: G. Gamow discusses the time and distance scales of the universe and the evolution of the universe with special emphasis on relativity theory; B. J. Bok's lectures on the evolution of stars and clusters, the role of interstellar dust and gas, and the universe of galaxies are superb; T. Gold lectured on the origin of the solar system, the structure of the earth, and the origin of lunar features; C. B. A. McCusker has a brief but well-organized statement of the origin of cosmic rays. A fairly good background in astronomy must be assumed if the full significance of the lectures is to be appreciated.

Pfeiffer, John. *From Galaxies to Man*. Random, 1959. xii+234 pp. illus. $4.95. 59–12308. (SH)

A discussion which attempts to synthesize the views of many scientists on the origin and evolution of the universe and of life on the earth.

Shapley, Harlow. *Of Stars and Men: The Human Response to an Expanding Universe* (rev. ed.). Beacon Press, 1964. 134 pp. illus. $12.00. 64–20491. (C)

A stimulating, beautifully illustrated and very readable essay on current theories of the origin of the solar system, the origin of life on earth, and the possibilities of life on other planets. The treatment is inspiring and thought-provoking. Highly recommended for collateral reading and classroom discussion.

Shapley, Harlow. *The View from a Distant Star: Man's Future in the Universe*. Basic, 1963. ix+212 pp. $4.95; Dell, paper 75¢. 63–18673. (C)

A most eminent astronomer addresses himself to the question of man's future in the universe, after having led the reader on an exploratory quest into the evolution of galaxies, of life, and of mankind. He has very provocative thoughts that should stimulate the reader's thinking.

523.2 SOLAR SYSTEM

Kuhn, Thomas S. *The Copernican Revolution: Planetary Astronomy in the Development of Western Thought*. Harvard U., 1957. xviii+297 pp. illus. $5.75; Random (Vintage), paper $1.45. 57–7612. (C)

The text reveals that the Copernican Revolution is of significance at the present day, for understanding the processes that led to the Revolution provides a perspective for the evaluation of our own scientific beliefs. Recommended for those seeking *knowledge,* rather than *information.*

523.3 MOON

Alter, Dinsmore. *Pictorial Guide to the Moon.* Crowell, 1967. 199 pp. illus. $8.95. 67–18396. (SH–C)

An interesting history of the development of lunar observations and theory, this guide book shows the moon and its characteristics throughout the lunar month. Outstanding photographs, illustrations, and diagrams richly supplement the text.

Baldwin, Ralph B. *The Measure of the Moon.* U. of Chicago, 1963. xix+488 pp. illus. $13.50. (C)

Baldwin develops the thesis that most of the craters on the moon are of external or meteoric origin rather than the result of volcanic action. The early chapters consider terrestrial meteorite craters, and the remainder is devoted to discussions of the shape, surface, structures, history, and atmosphere of the moon. Tabular data support the discussion, and references are supplied at the end of each chapter. A challenging book but the reader does not need astronomical training.

Gamow, George. *The Moon* (rev. ed.). Abelard-Schuman, 1959. 127 pp. illus. $3.00. 59–7280. (JH)

Describes and explains the moon, its orbit, gravitation, origin, chemical make-up, atmosphere and topography; also discusses gaps in our knowledge that should be filled before moon travel can be attempted.

King, Henry C. *The World of the Moon.* Crowell, 1966. 125 pp. illus. $4.95. 67–10374. (JH–SH)

With each new lunar probe, theory of and information about the moon is altered. This is a popular, accurate, and scholarly account of the moon. Appropriate emphasis is given to all sides of conflicting theories and ideas. Timely examples and touches of humor maintain reader interest.

523.4 PLANETS

Alexander, A. F. O'D. *The Planet Uranus.* American Elsevier, 1965. 316 pp. illus. $12.75. 65–22522. (SH–C)

A historical approach; the early chapters deal with the discovery of Uranus and its basic characteristics. Later chapters recount the dogged work of astronomers in wringing newer information from their observations. Although this might lack the appeal of an exciting novel, it brings to the reader a feel for one fundamental aspect of the work of the astronomer as he overcomes one difficulty after another. An excellent index and a thorough bibliography, with many useful illustrations.

Armitage, Angus. *William Herschel.* Doubleday, 1962. xii+158 pp. illus. $3.95. 63–7962. (JH)

A biography of the astronomer who rose to fame suddenly in 1781 upon his discovery of the planet Uranus, one of the most dramatic events in the history of astronomy.

Ley, Willy, and Wernher Von Braun. *The Exploration of Mars.* (Illus. by Chesley Bonestell.) Viking, 1956. x+176 pp. 28 cm. $5.95. 56–7596. (JH) Although the contents are dated they still provide an elementary account of the preparations necessary for an exploratory expedition to Mars. A hypothetical trip is described in detail. Knowledge about Mars is reviewed as well as theories concerning the possibility of life on the planet. Illustrations include reproductions of 16 paintings in color.

Moore, Patrick. *Guide to Mars* (3rd ed.).)Macmillan, 1960. 124 pp. illus. $4.50. 57–418. (SH) Moore provides a summary of what is known about the Red Planet—its deserts, atmosphere, dark areas, icy polar caps, its two tiny moons and its mysterious "canals."

Moore, Patrick. *The Planet Venus* (3rd ed.). Macmillan, 1961. 151 pp. illus. $4.95. 59–65190. (SH) A fascinating account of the theories and observations which have been made of the planet Venus. The author points out the pros and cons for each hypothesis and concludes with his point of view.

Sagan, Carl, Jonathan Norton Leonard, and The Editors of *Life. Planets.* Time, Inc., 1966. 200 pp. illus. $3.95; LB $4.95. 66–22436. (JH–SH) A clearly-written, dramatically-illustrated, and up-to-date summary of our knowledge of the planets. Each chapter is paralleled with a "picture essay" consisting of astronomical photographs, diagrams, color photos, and highly imaginative drawings. There is a general focus on the origin of life (on earth and elsewhere) and a refreshing open-mindedness in discussing speculations and interpretations of all sorts. Requires little background in science and mathematics. Useful appendix, references for further reading, complete index.

Whipple, Fred L. *Earth, Moon and Planets* (rev. ed.). Harvard, 1963. 278 pp. illus. $6.50. 63–17216. (SH) The Director of the Smithsonian Astrophysical Observatory has summarized information about the planets and their moons in a layman's book, and includes data obtained by radio, radar, high-altitude balloon, rocket flights and space capsule.

523.5 METEORS AND COMETS

Heide, Fritz. *Meteorites.* (Tr. by Edward Anders and Eugene R. DuFresne.) U. of Chicago, 1964. x+144 pp. illus. $6.50, paper (Phoenix) $1.95. 63–20906. (SH) Consists of three chapters: the first deals with phenomena associated with the fall of a meteorite; the second describes the matter of the meteorite itself, and although the author is a mineralogist he emphasizes other studies; the third chapter is on the ages and origin of meteorites. A good companion for Brian Mason's, *Meteorites* (Wiley).

Mason, Brian. *Meteorites.* Wiley, 1962. xii+274 pp. illus. $8.95. 62–17466. (C) A comprehensive account of the subject which emphasizes extensive researches, hypotheses and theories, and interpretation of actual data. Includes a list of meteorites in the U.S.A. and a scholarly bibliography.

Nininger, H. H. *Arizona's Meteorite Crater: Past-Present-Future.* Amer. Meteorite Museum, 1956. 232 pp. illus. $3.75, paper $2.25. 57–826. (JH)

An account of the controversies that have raged for many years concerning the greatest of the authenticated meteorite craters in North America.

Richardson, Robert S. *Getting Acquainted with Comets*. McGraw-Hill, 1967. 306 pp. illus. $7.50. 66–24480. (SH–C)
This narrative covers the history, important facts, and theory behind comets. Presented in a simple, clear format. Designed for science students, it will hold the atttention of those with only a passing interest as well.

Watson, Fletcher G. *Between the Planets* (rev. ed.). Harvard, 1956. vi+228 pp. illus. $5.00. 56–6526. (SH)
Limited to a discussion of the wanderers of the heavens: asteroids, comets, meteors, and meteorites. Includes a discussion of radio studies of meteors.

532.7 SUN

Abetti, Giorgio. *Solar Research*. Macmillan, 1962. 173 pp. illus. $5.95. 63–320. (SH)
Covers observational details of the sun and also explores the physical processes that underlie them. Includes discussions of sunspots, radio observations, and solar effects on terrestrial phenomena.

Gamow, George. *A Star Called the Sun*. Viking, 1964. 224 pp. illus. $5.75; Bantam, paper 75¢. 64–13594. (JH)
The author has dedicated this to his earlier book, *The Birth and Death of the Sun* (1940), of which it is a revision and modernization. An easy and interesting introduction to all aspects of solar lore, knowledge, and research.

Kiepenheuer, Karl. *The Sun*. U. of Michigan, 1959. 160 pp. illus. $5.00, paper $1.95. 59–7294. (SH)
The known facts of solar physics are ably summarized and therefore it is a good introduction to that subject. Contains straight scientific material—no speculative theories.

Menzel, Donald H. *Our Sun*. Harvard, 1959. ix+350 pp. illus. $7.50. 59–12975. (SH)
A book on the sun, light and atomic energy. The author has interesting hypotheses on some of the phenomena of our sun and he discusses all of the important aspects of solar research.

Zirin, Harold. *The Solar Atmosphere*. Blaisdell, 1966. ix+502 pp. illus. $15.00. 65–21458. (C)
Woven through and around the mathematical and technical core designed for professionals and students is a very readable expository account of the physical processes at work in the sun. There is thus much of interest to the general reader as well. Describes the corona, chromosphere, photosphere, prominences, solar activity, and flares. Amply illustrated with diagrams and excellent photographs.

523.8 STARS

Baade, Walter. *Evolution of Stars and Galaxies*. (Ed. by Cecilia Payne-Gaposchkin.) Harvard, 1963. xiii+321 pp. $6.75. 63–9547. (C)
Based on recorded lectures delivered by the author, and prepared for publication by the editor after the author's death. The author was an astronomer

at the Mount Wilson and Palomar Observatories and the book presents historical perspectives and original ideas.

Bok, Bart J., and Priscilla F. Bok. *The Milky Way* (3rd ed.). Harvard, 1957. vi+269 pp. illus. $5.50. 56–11279. (JH)
Written in the form of a travelogue, the authors take the reader on a tour of the Milky Way; includes a discussion of radio astronomy.

Howard, Neale E. *The Telescope Handbook and Star Atlas.* Crowell, 1967. ix226 pp. illus. $10.00. 66–14939. (JH–SH)
Coverage includes topics on the sky and coordinates, the solar system, double stars, variables, galaxies, comets, aurorae, and photography. Accurate and up to date, this is a complete summary for the beginning student and amateur astronomer. An atlas of the whole sky, tables, bibliography, index and glossary are included.

King, Henry C. *Pictorial Guide to the Stars.* Crowell, 1967. 167 pp. illus. $8.95. 67–12404. (SH–C)
An excellent companion to Alter's *Pictural Guide to the Moon* (listed under 523.3), this work provides an up-to-date survey of the stars that will be useful to astronomers and general readers. A brief glossary and bibliography of books of various levels of difficulty are included.

Kruse, W., and W. Dieckvoss. *The Stars.* U. of Michigan, 1957. 202 pp. illus. $5.00, paper $1.95. 57–7745. (SH)
With the aid of more than 100 illustrations the authors explain how to determine the speed, position and temperature of stars with a telescope, how stars emanate light, and the origins of galaxies, nebulae, etc.

Meadows, A. J. *Stellar Evolution.* Pergamon, 1967. vii+169 pp. illus. $4.50, paper $1.95. (JH–SH–C)
Describes a classification of stars based on observational characteristics such as luminosity, surface temperature, and size. Shows how these properties of stars can be explained in terms of mass, chemical composition, origin, and age. The text is simple but will best be understood by those with some background in physics.

Menzel, Donald H. *A Field Guide to the Stars and Planets.* Houghton Mifflin, 1963. xiv+397 pp. $4.95. 63–7017. (JH)
A complete pocket guide to the sky at night with 48 sky maps and 54 charts of the fainter stars, each with a paired guide containing proper names. Appendices contain a glossary and useful tabular information.

Moore, Patrick. *A Guide to the Stars.* Norton, 1960. 222 pp. illus. $4.95. 60–7584. (JH)
Reviews the few physical facts known about the stars, but the constellations described are in British skies; there are explanations of binaries, variables, clusters, nebulae, and outer galaxies. A good book for the beginner.

Page, Thornton (ed.), et al. *Stars and Galaxies: Birth, Ageing, and Death in the Universe.* Prentice-Hall, 1962. xiii+163 pp. illus. $3.95, paper $1.95. 62–19406. (SH)
Many questions concerning the stars, galaxies, origin and evolution of the universe are answered or discussed by seven authorities. All authors were

selected by the American Astronomical Society as the best persons to present current developments.

Shapley, Harlow. *Galaxies* (rev. ed.). Harvard, 1961. vii+186 pp. illus. $5.00. 61–7393. (C)
This engrossing and up-to-date volume contains detailed descriptions and analyses of galaxies, metagalaxies, and the expanding universe. The reader should have had a course in astronomy, or developed an elementary background by self-study before he attempts it.

Shapley, Harlow. *The Inner Metagalaxy.* Yale, 1957. xiii+204 pp. illus. $6.75. 57–6877. (SH)
Summarizes three decades of pioneering research on the problems of the galaxies. Consists of explorations of the north and south galactic polar zones, in the canopy galaxies, and along the borders of the Milky Way. Also deals with the peripheral position of the earth in its own galaxy. Good for all students and amateurs.

Zadde, Arthur J. *Making Friends with the Stars.* (Revised by Theodore A. Smits.) Barnes & Noble, 1963. xiii+144 pp. illus. $3.50, paper $1.25. 63–21769. (JH)
A manual for the beginner and the amateur which conveys the elementary principles of the subject, and assists him in becoming a "star watcher" by means of unique star maps.

523.9 SATELLITES

Sandner, Werner. *Satellites of the Solar System.* American Elsevier, 1965. 151 pp. illus. $6.50. 65–22523. (JH–SH–C)
Interesting stories concerning contributions to the knowledge of the 31 satellites of the solar system that have been made by a great many astronomers. The treatment is nonmathematical. For anyone who wants a résumé of the basic facts and lore concerning the satellites of the solar system.

525 PLANET EARTH (see also 550 EARTH SCIENCES)

Asimov, Isaac. *The Double Planet* (rev. ed.). Abelard-Schuman, 1966. 158 pp. illus. $4.00. 60–13922. (JH)
The fascinating story, begun in ancient times and ended in our present decade, of how man has probed the secrets of the Earth and the Moon. Appendices contain useful statistics and historical notes.

Beiser, Arthur. *Our Earth: The Properties of Our Planet, How They Were Discovered and How They Came into Being.* Dutton, 1959. 123 pp. illus. $2.95. 59–5825. (SH)
An introduction to a brief history of geophysics dealing with fundamental topics of evolution of earth features, transformations of the earth's crust and the causative agents. Appended tables of useful data.

Stumpff, Karl. *Planet Earth.* U. of Michigan, 1959. 191 pp. illus. $5.00, paper $1.95. 59–5266. (SH)
Geophysics is introduced to the reader through information on the movements,

structure, size and shape of the earth, and its relationship to the whole of the universe. Research methods and the concepts of time and measurement are ably explained.

526   GEODESY AND CARTOGRAPHY

Brown, Lloyd A. *Map Making: The Art that Became a Science*. Little, Brown, 1960. 217 pp. illus. $4.95. 60–9338. (JH)
An outstanding survey of the development of cartography from earliest times to the beginning of the twentieth century. The discussion takes into account discovery and geography, measuring instruments (including the compass), clocks, navigation by the stars and related topics. Authenticated with reproductions of old prints.

Greenhood, David. *Mapping* (rev. ed.). U. of Chicago, 1963. 289 pp. illus. $6.00, paper (Phoenix) $2.95. 63–20905. (JH)
Explains how to know good maps; how to make your own maps; how to do you own surveying for a map with the use of inexpensive devices; how to collect and take care of maps. An expanded and detailed revision of a book first published in 1944 under another title.

Rayner, William H., and Milton O. Schmidt. *Elementary Surveying* (4th ed.). Van Nostrand, 1963. 483 pp. illus. $7.50. 63–1318. (C)
Designed for a beginning course in surveying and mapping for civil engineering students; also, to provide surveying and mapping knowledge for students in geology, geography, forestry, limnology, landscape architecture, and agriculture. Contains information on instruments and photogrammetry, and there are mathematical tables in the appendices.

Robinson, Arthur, and Randall D. Sale. *Elements of Cartography* (3rd ed.). Wiley 1969. 413 pp. illus. $10.95. 69–19232. (C)
The fascinating story of the development and science of a form of communication vital to many physical, biological, and social scientists holds a compelling intellectual interest for many thoughtful people. Includes discussions of new methods for map construction, reproductions, design, and computer technology.

Tannenbaum, Beulah, and Myra Stillman. *Understanding Maps: Charting the Land, Sea, and Sky*. McGraw-Hill, 1957. 144 pp. illus. $3.50; LB $3.06. 57–8632. (JH)
The science and technique of reading and making maps is presented in an elementary and entertaining fashion for the novice.

Tooley, R. V. *Maps and Map-Makers* (2nd ed.). Crown (Bonanza), 1952. xii+ 140 pp. illus. $3.95. 53–6245. (SH)
A history of maps and of map making presented chiefly through the achievements of outstanding experts. Each chapter is devoted to a particular country or region and there are regional lists of the most important maps.

529   CHRONOLOGY

Asimov, Isaac. *The Clock We Live On* (rev. ed.). Abelard-Schuman, 1965. 172 pp. illus. $3.50. 65–12072. (JH–SH)

Discusses the techniques used to measure time, and the advantages and disadvantages of the earth, moon, and stars for these purposes. This is good reading for anyone wishing to learn about the origins of the day, days of the week, the month, and the year.

Harrison, Lucia Carolyn. *Sun, Earth, Time and Man.* Rand McNally, 1960. xiv+287 pp. illus. $4.50. 60–8992. (SH)
The study of chronology is grounded in a study of Earth-Sun relationships, a consideration of the subdivisions known as latitude and longitude, and of the problems involved in man's attempts to break the continuous flow of time into measurable units. The position of the sun in the sky and its dependence on time, season, and the observer's latitude are discussed in detail.

Quill, Humphrey. *John Harrison: The Man Who Found Longitude.* (Foreword by Sir Richard Wooley.) Humanities, 1966. xiv+255 pp. illus. $10.00. 67–12246. (SH)
The only biography of Harrison, this fills a necessary place in the history of navigation and invention. Harrison built the first successful marine chronometer. This work serves as a background volume or for collateral reading.

Wright, Lawrence. *Clockwork Man: The Story of Time, Its Origins, Its Uses, Its Tyranny.* Horizon, 1969. 260 pp. illus. $7.95. 69–14997. (SH–C)
Tells of the origin of the human need for time, the history of timekeeping, the innovations and improvements in time-keeping methods, and the regulations imposed on temporal routines and activities of people. The book provides good general background and material for discussion.

530 PHYSICS

Adler, Irving. *The Wonders of Physics: An Introduction to the Physical World.* (Illus. by Cornelius DeWitt.) Golden Press, 1966. 165 pp. $4.95. 66–13405. (JH–SH)
Superior in its breadth, clarity, concern for current research, and for its attractive illustrations. Begins by asking 25 questions that might occur to a thoughtful observer at the beach. At the end of the book the author returns to these same questions and shows the reader that he has now acquired the insight to answer any that may previously have puzzled him. The reader has been treated to a rapid but meaningful survey of a wide range of physics.

Bell, Raymond M. *Your Future in Physics.* Richards Rosen, 1967. 138 pp. illus. $4.00. 67–18710. (SH–C)
A guide for high school college students considering a career in physics. Describes educational opportunities, employment possibilities, and career development guidance. Highly recommended for counselors, science teachers, and libraries.

Brown, Sanborn C. *Count Rumford: Physicist Extraordinary.* (Science Study Series.) Doubleday (Anchor S28), 1962. xv+178 pp. illus. $1.25 (paper). 62–14130. (SH)
A superior biography of Benjamin Thompson (Count Rumford), who founded the Royal Institution and who "launched" Sir Humphry Davy. He made many important contributions to physics and laid the foundation for the modus operandi of industrial technology.

Messel, H., and S. T. Butler (eds.). *Atoms to Andromeda.* Pergamon, 1966.
301 pp. illus. $3.50 (paper). 66–22726. (SH–C)
A series of 6 lectures given to outstanding four-year high school students:
(1) "Cosmic rays," (2) "Light from the stars," (3) "Radio waves," (4)
"Plasmas," (5) "Digital computers," and (6) "Theoretical physics." Some
parts may be difficult for the average high-school student, but the work gives
a good insight to those fields in the forefront of physics. Diagrams are good.
Minimum of mathematics.

Oppenheimer, J. Robert. *The Flying Trapeze: Three Crises for Physicists.*
Oxford, 1964. ix+65 pp. $2.75. 64–56486. (SH–C)
This work is based on three lectures. The first two, "Space and time" and
"Atom and field", tell of the impact of relativity, and later of nuclear physics,
on Newtonian physics. For these the reader needs an elementary physics
background. The third lecture, "War and the nation," addressed more to
the general reader, is an historical résumé of the development of nuclear
power and the involvement of scientists in government policy. Good reading
for thoughtful students.

Pollack, Philip. *Careers and Opportunities in Physics.* Dutton, 1961. 159 pp.
illus. $3.75. 61–6025. (JH–SH)
Discusses various specialties in physics—electronics, nuclear physics, meteorological research, aeronautics, biophysics, geophysics, astrophysics, physics
teaching—and provides information on qualifications, educational requirements, and monetary rewards in each field.

Rogers, Eric M. *Physics for the Inquiring Mind.* Princeton, 1960. 778 pp. illus.
$12.50. 59–5603. (SH–C)
An unparalleled, comprehensive exposition of the "methods, nature and
philosophy of physical science." Developed from an undergraduate course for
the non-physicist, this survey includes a study of matter, motion and force,
astronomy, molecules and energy, electricity and magnetism, and atomic and
nuclear physics. No library should be without it and every physics teacher
needs his own copy.

Rothman, Milton A. *The Laws of Physics.* Basic, 1963. 254 pp. illus. $5.95.
63–20022. (SH)
An elementary exposition of the principles and concepts of classical physics,
which proceeds on this foundation to deal with the elements of nuclear
physics. Recommended for those who find Rogers', *Physics for the Inquiring
Mind,* too substantial fare.

Shamos, Morris H. (ed.). *Great Experiments in Physics.* Holt, Rinehart and
Winston, 1959. viii+370 pp. illus. $7.95 (text ed.). paper $5.95. 59–5751.
(SH)
Consists of excerpts from the classic accounts of experiments of 24 great
physicists, compiled and edited with biographical sketches and marginal notes
to amplify and clarify the original texts.

Shortley, George, and Dudley Williams. *Elements of Physics: For Students of
Science and Engineering* (4th ed.; 2 vols. in 1). Prentice-Hall, 1965. 924+
xxiv+viii pp. illus. $12.95; $7.95. 65–19190. (SH–C)
The fourth edition of this well-known text follows the pattern of the previous

editions. The student should have had or be taking a college course in mathematics which includes the notions of the calculus. The British gravitational units are employed in mechanics; the meter-kilogram-second-ampere (MKSA) system is used in other sections of the text. A *rigorous, complete* text that deals with basic principles and their applications. Should be on library shelves of colleges and high schools for reference.

Stewart, Alec T. *Perpetual Motion: Electrons and Atoms in Crystals.* (Science Study Series.) Doubleday (Anchor S39), 1965. xii+146 pp. illus. $1.25 (paper). 64-19244. (SH-C)
A brief yet reasonably comprehensive discussion of the properties of crystals with particular emphasis on the properties that can be deduced from theories about the locations and behavior of electrons and atoms in crystals. The presentation is lucid, but this does not mean that it is easy reading.

Sutton, Richard M. *The Physics of Space.* Holt, Rinehart and Winston, 1965. 176 pp. illus. $2.95, paper $1.96. 65-23281. (JH-SH-C)
Fascinating, lucid, refreshingly *correct* in its scientific explanations. Almost all of the text is in easy (although rigorous) English, but there are a few simple algebraic expressions. There are many diagrams, a few good photographs, 11 appendices, a bibliography, and a good index. Many simple experiments are described. Almost all of the book could be followed by motivated junior high school students. Even college students can profit by reading it.

Westphal, Wilhelm H. *Physics Can Be Fun.* Hawthorn, 1965. 207 pp. illus. $4.95. 65-22908. (JH-SH)
Considers a number of common everyday activities and observations which can be readily and interestingly explained in terms of the physical principles involved. Neither a textbook on physics nor a collection of unrelated explanations. Chapters on energy, inertia, friction, sound, electricity, weights and measures, buoyancy, heat, and time. Assumes no previous knowledge of physics.

White, Harvey E. *Physics: An Exact Science* (5th ed.). Van Nostrand, 1966. 597 pp. illus. $5.96. (SH)
One of the very best and most lucid presentations of physics for the advanced high school student. Topics on modern atomic and nuclear physics are discussed at length. Chapter summaries, questions and problems facilitate understanding and comprehension. An alternative for the PSSC text if a more traditional survey course is desired.

530.03 PHYSICS—ENCYCLOPEDIAS

Besançon, Robert M. (ed.). *The Encyclopedia of Physics.* Reinhold, 1966. xii+ 832 pp. illus. $25.00. 65-29253. (SH-C)
Concise, accurate and well-written information about carefully selected branches of physics. Useful to college students, able high school physics students, teachers, librarians, scientists, and engineers. Short, introductory articles are provided on physics, the history of physics, measurement, symbols, units and nomenclature. Other articles deal with phenomena at the boundaries

between physics and other disciplines. The articles are explanations at the technical level. Mathematics is used as needed but does not replace verbal explanation.

Michels, Walter C. (Editor-in-Chief). *International Dictionary of Physics and Electronics*. Van Nostrand, 1961. 1355 pp. illus. $27.85. 61–2485. (C)
An encyclopedic dictionary of terms, valuable as a reference work to the undergraduate physics or electrical engineering major and to the professional engineer. It should be in all large public libraries and college libraries; high school libraries do not need it unless the school offers physics instruction beyond a basic course.

Susskind, Charles (ed.). *The Encyclopedia of Electronics*. Reinhold, 1962. xxi+ 974 pp. illus. $25.00. 62–13258. (SH)
A comprehensive encyclopedia of electronics arranged alphabetically. Each topic is well explained, many being supplemented by charts and diagrams. Electronic entries are more technical and in greater detail than comparable entries in Van Nostrand's *International Dictionary of Physics and Electronics.*

530.07 PHYSICS—STUDY AND TEACHING

Alonso, Marcelo, and Edward J. Finn. *Fundamental University Physics*. Vol. I: *Mechanics;* Vol. II: *Fields and Waves*. Addison-Wesley, 1967. xvi+xv+965 pp. $8.75 ea. 66–10828; Vol. III: *Quantum and Statistical Physics,* 1968. ix+ 598 pp. $8.75. 68–10828. (C)
This three-volume set was designed for a three (more realistically four) semester course in general college physics requiring the calculus. The authors eschew the conventional arrangement of general physics in favor of (1) mechanics, (2) interactions and fields, (3) waves, (4) quantum physics, and (5) statistical physics. Mechanics occupies all of Volume I; interactions and fields with waves, Volume II. The quantum mechanics and statistical physics are treated in Volume III.

Arons, Arnold B. *Development of Concepts of Physics*. Addison-Wesley, 1965. xx+972 pp. illus. $14.75. 65–10924. (SH–C)
The material is a useful combination of physics, mathematics, history of physics, and excerpts from writings of note in the sciences and philosophy. Approximately one-half of the text is devoted to the field of mechanics, including the laws of motion as applied to astronomy. The problems throughout require thought and careful analysis. List of references for supplementary reading and study appears at the end of each chapter. An introduction to recent developments in physics and to special relativity is included. Appendices contain useful reference material.

Atkins, Kenneth Robert. *Physics*. Wiley, 1965. 754 pp. illus. $9.95. 65–12719. (C)
A textbook for a two-semester introductory (and probably terminal) course in general college physics for liberal arts students. Slanted in the direction of modern physics. The treatment of textual material often involves analytic thinking and will require intense concentration on the part of the student.

It does not seem likely that the text will be found suitable for college freshmen and sophomores of less than good aptitude who have not had strong high school preparation.

Baez, Albert V. *The New College Physics: A Spiral Approach.* Freeman, 1967. xviii+739 pp. illus. $11.75. 67–12180. (C)
Designed for liberal arts and education majors this text represents a new approach of presenting an introductory, terminal physics course. The illustrations are of high quality and the text has good appendices. A valuable addition to existing literature.

Beiser, Arthur. *Modern Technical Physics.* Addison-Wesley, 1966. 706 pp. illus. $10.75. 66–10829. (SH–C)
The explanations of the physical phenomena are lucid; the illustrations are selected wisely and are well done; about 200 solved problems are included to illustrate the transition from abstract principle to practical application, and approximately 1,000 problems are presented with answers to the odd-numbered ones. There is a separate study guide. Conventional presentation, starting with a brief introduction to mathematical principles (largely algebra, trigonometry and simple vector addition and substraction), and progressing through mechanics, wave motion, heat and thermodynamics, electromagnetism, optics and modern physics. Suited for physics classes in secondary schools, colleges, and technical institutes.

Bitter, Francis. *Mathematical Aspects of Physics: An Introduction.* (Science Study Series S32) Doubleday, 1963. 188 pp. illus. $1.25 (paper). 63–18044. (SH)
The purpose of this PSSC book is to enable the reader to recognize the repeated mathematical patterns of nature and to give him a true appreciation of the laws of physics and how they are formulated.

Blackwood, Oswald H., William C. Kelly, and Raymond M. Bell. *General Physics* (3rd ed.). Wiley, 1963. 685 pp. illus. $9.50. 63–9395. (SH–C)
A distinguished elementary college text or a reference and collateral study book for the secondary school student whose mathematics background includes trigonometry. Each chapter concludes with review questions, problems and recommended reading.

Bulman, A. D. *Model-Making for Physicists.* Crowell, 1968 [c. 1963]. 184 pp. illus. $4.50. 68–17077. (JH)
Offers detailed instructions for building 25 pieces of apparatus using common materials. Rewards the builder who sees something that he had made work well and understands why it works, in terms of physical principles.

Efron, Alexander. *Exploring Modern Physics.* Hayden, 1968. xxiii+855 pp. illus. $7.90. 67–13869. (SH–C)
High school physics is presented here with care and respect for the intellectual maturity and serious intent of many students. Chapters are clearly organized, usually with a historical introduction, with excellent diagrams and many recent photographs, and with illustrative problems and solutions; they conclude with review questions and problems. An excellent text for the serious high school student, and appropriate even for some college level physical science courses.

Ford, Kenneth W. *Basic Physics*. Blaisdell, 1968. xxii+968 pp. illus. $11.75. 68–10415. (C)
"The book is organized in the main about the great theories of physics: mechanics, thermodynamics, electromagnetism, relativity, and quantum mechanics. The organization, the choice of individual topics, and the various recurring themes have all been governed by one idea: to help the student come to understand physics as a human activity dedicated to organizing a certain part of experience in the simplest, most economical, most general, and most satisfying way."—Preface. Useful for reference and supplemental reading.

Maleh, Isaac. *Mechanics, Heat, and Sound*. (Merrill Physical Science Series). Merrill, 1969. xii+159 pp. illus. $3.95; $1.95 (paper). 68–21655. (SH–C)
Each chapter has an introduction, review and set of problems and questions. Extensive line drawings supplement the text. Compares favorably with appropriate sections from other texts for physical science survey courses.

Miller, Franklin, Jr. *College Physics* (2nd ed.). Harcourt, Brace, 1967. xix+ 715 pp. illus. $10.50. 67–12525. (SH–C)
Interesting and helpful as a supplement to high school and college courses as well as a text for a nonscience physics course. Attractive features include concept check list, thought questions, and graded problems. Only simple algebra is required.

Orear, Jay. *Fundamental Physics* (2nd ed.). Wiley, 1967. xix+472 pp. illus. $8.95. 66–26754. (C)
A thoroughly revised edition of a well-written, completely modern, general physics text that contains many new photographs and carefully worked problems and examples. Little calculus is required.

Physical Science Study Committee. *Physics* (2nd ed.). Heath, 1965. 686 pp. illus. (Price not available.) 65–24096. (SH)
This is the high school text of the PSSC, which presents physics "not as a mere body of facts but basically as a continuing process by which men seek to understand the nature of the physical world." This revision includes revised and improved problems, clarity of photographs, and streamlining of chapter content. Recommended as the best textbook for secondary school physics. Should be in all libraries for review and reference.

Sears, Francis Weston, and Mark W. Zemansky. *University Physics*. Addison-Wesley, 1970. 716 pp. illus. $13.25. 70–93991. (SH–C)
Mechanics, heat, sound, electricity and magnetism, light, and atomic physics are well covered in the fourth edition of this standard college text. Useful for collateral reading for the advanced high school student.

Warren, John W. *The Teaching of Physics*. Butterworths, 1965. x+130 pp. illus. $4.25. 65–8232. (SH–C)
Written as a result of the author's belief that the traditional interpretation of elementary physics for teaching purposes contains errors of fact and absurdities in logic. Succeeds in stimulating thought on the part of the reader and in raising questions in his mind that will usually result in clearer understanding. Focuses attention on important ideas in physics and often presents

the material in a fresh manner. This book will be of most value to teachers who are concerned lest their presentation of ideas in physics lead to misconceptions. It will be of use as a reference to the most discerning high school and elementary college student of physics. Stimulates thought rather than memorization.

Weber, Robert L., Kenneth V. Manning, and Marsh W. White. *College Physics* (4th ed.). McGraw-Hill, 1965. vi+710 pp. illus. $10.50. 64-25177. (SH–C)
This fourth edition of a widely-used text is suitable for the first year of college both for those students who intend to go into a technical field and for those pursuing a liberal arts course. Also useful reference and supplementary reading by advanced and well-qualified secondary school students. Elementary algebra and trigonometry are sufficient mathematical background. The text has four main sections: "Physics of particles and aggregates of particles," "Physics of fields," "Wave physics," and "Quantum physics."

Weidner, Richard T., and Robert L. Sells. *Elementary Classical Physics.* Vol. 1: *Mechanics, Kinetic Theory, Thermodynamics;* Vol. 2: *Electromagnetism and Wave Motion.* Allyn and Bacon, 1965. xv+xiv+1242 pp. illus. $10.50 ea. (text ed. $7.95 ea.); combined ed. $17.95 (text ed. $13.50 ea.). 65-10891. (C)
Presents physics as the fundamental experimental science whose purpose is to explain the behavior of the physical universe. Miscellaneous appendices. Problems in the text with answers to odd-numbered problems in the appendices. Useful for reference and collateral study. (See also *Elementary Modern Physics* by the same authors under "539 Molecular and Nuclear Physics".)

White, Harvey E. *Modern College Physics* (5th ed.). Van Nostrand, 1966. 765 pp. illus. $10.75. (SH–C)
The well-known basic college physics text for science students. Unites the elementary principles of classical physics with the contemporary concepts associated with atomic and nuclear physics. The text contains numerous illustrations from the biological and medical sciences as well as from engineering. Useful for reference by secondary school students.

530.09 PHYSICS—HISTORY

Born, Max. *My Life and My Views.* (Introd. by I. Bernard Cohen.) Scribner's, 1968. 216 pp. $4.95. 68-12510. (SH–C)
A collection of essays by a Nobel Prize winner in physics that deals with the problems of our technological culture, of keeping peace, and of the future of Western civilization. Excellent reading for a class in intellectual history, philosophy, or general science.

Cooper, Leon N. *An Introduction to the Meaning and Structure of Physics.* Harper & Row, 1968. 746 pp. illus. $13.95. 68-12925. (SH–C)
A remarkable book with an apt title that will be a useful reference for teachers and students at high school and beginning college levels. It provides an acquaintance with the great leaders in physics by explaining what they accomplished (640 BC—1967 AD) and its relevance to the whole of civilization.

Childs, Herbert. *An American Genius: The Life of Ernest Orlanda Lawrence.* Dutton, 1968. 576 pp. illus. $12.95. 68–12456. (JH–SH–C)
An excellent biography of one of America's chief contributors to the great age of discovery and advancement in physics, 1925–1960. Lawrence's achievements include: invention of the cyclotron, electromagnetic separation of uranium isotopes, and founding of the Lawrence Radiation Laboratory. This work is well written, intensely interesting, and highly reliable.

Dibner, Bern. *Wilhelm Conrad Röntgen and the Discovery of X Rays.* (Immortals of Science Series.) Watts, 1968. 149 pp. illus. $3.95; LB $2.96. 68–10434. (JH–SH)
A short, interesting biography of Röntgen, written by an electrical engineer and historian of science. Useful as collateral reading in science courses at the high school level.

Fermi, Laura, and Gilberto Bernardini. *Galileo and the Scientific Revolution.* Basic, 1961. ix+150 pp. illus. $4.50; Fawcett, paper 60¢. 61–7486. (SH)
A remarkable and invigorating work in the field of science biography as was the hero, Galileo, in his hey-day. It adheres to facts and is a model for worthwhile science biographies.

Gamow, George. *Biography of Physics.* Harper & Row, 1961. x+338 pp. illus. $6.50; paper (Torchbooks) $2.45. 61–6433. (SH)
An historical survey of the development of theoretical physics, with equal emphasis upon both biographical data and substantive scientific contributions.

Irwin, Keith Gordon. *The Romance of Physics.* (Illus. by Anthony Ravielli.) Scribner's, 1966. 240 pp. $4.95; LB $4.37. 66–24489. (JH–SH)
A survey of major developments in physics by means of sketches of the outstanding works, the contributions of those who made the basic studies of liquids, gases, temperature, and the atmosphere; those who developed the disciplines of electricity and magnetism; the men who contributed to our knowledge of heat, light, energy, and power; and finally, three titans who founded modern nuclear physics. Appendices consist of a chronology, a glossary, suggested readings, and an index.

Koestler, Arthur. *The Watershed: A Biography of Johannes Kepler.* (Science Study Series.) Doubleday (Anchor S16), 1960. 280 pp. illus. $1.45 (paper). 60–13537. (SH)
Describes Kepler's efforts to improve the techniques of observational science and his outstanding contributions to techniques of observational science and his outstanding contributions to astronomy and man's changing vision of the universe.

MacDonald, D. K. C. *Faraday, Maxwell and Kelvin.* (Science Study Series.) Doubleday (Anchor S33), 1964. 143 pp. illus. $1.25 (paper). 64–11313. (SH)
In three essays, the author relates the backgrounds, personalities, and scientific achievements of three outstanding scientists.

Rabi, I. I. et al. *Oppenheimer.* Scribner's, 1969. x+90 pp. illus. $5.95. 68–57086. (SH–C)
This collection of essays about J. Robert Oppenheimer constitutes an inter-

esting and enlightening description of the man's work and his personality. Oppenheimer's concern for humanity is revealed throughout the book. No sophisticated knowledge of physics is required for comprehension, just a feeling for the field of physics in general and of some of the discoveries and advances mentioned.

Riedman, Sarah R. *Trailblazer of American Science: The Life of Joseph Henry.* Rand McNally, 1961. 224 pp. illus. $3.50. 61–6844. (JH)
Joseph Henry, the first Secretary of the Smithsonian Institution, was truly one of America's great pioneers of science. Since this is the only general biography available it is listed for reading by younger students. Meanwhile the hope is recorded that another author will produce a biography of real merit and substance.

Rozental, S. (ed.). *Niels Bohr: His Life and Work as Seen by His Friends and Colleagues.* North-Holland (dist. by Wiley), 1967. 355 pp. illus. $9.00; paper, $5.95. (SH–C)
A prestigious group of more than a score of contributors have recorded their own experiences and used firsthand course materials to produce an indispensable anthology and source book for anyone interested in Niels Bohr and the history of atomic physics. A most scholarly work.

Scott, William T. *Erwin Schrödinger: An Introduction to His Writings.* U. of Massachusetts, 1967. xv+175 pp. $6.50. 66–28117. (C)
Not only an excellent evaluation of Schrödinger's own life and immensely profitable work, this biography also recounts the parallel development of the investigations and explorations of his fellow physicists.

Seeger, Raymond J. *Men of Physics: Galileo Galilei, His Life and His Works.* Pergamon, 1966. xi+286 pp. illus. $6.00, paper $3.50. 66–23858. (SH–C)
An excellent, readable presentation of the man, his times and his physics. The first portion deals with Galileo's life, and is skillfully written with an objectivity befitting a scientific book. The second section deals with selections from the major works of Galileo. Collateral reading useful to students of physical science, and to students of any discipline who are interested in gaining insight into that period of history.

Thompson, George. *The Inspiration of Science.* Oxford U., 1962. x+150 pp. illus. $4.00. 62–161. (SH)
Taking examples from physics, Sir George shows how the scientist works and thinks; he discusses the scientific method and describes how certain discoveries actually were made. Short biographical sketches of eleven physicists enhance its charm.

Thompson, George Paget. *J. J. Thomson: Discoverer of the Electron.* (Science Study Series S48.) Doubleday, 1966. xvi+215 pp. illus. $1.45 (paper). 65–10821. (SH–C)
An excellent perspective of J. J. Thompson's achievements and his directorship of the Cavendish laboratory, written clearly and objectively by his son. Essential collateral reading for all elementary physics students and is a useful source book for historians of physics.

Williams, L. Pearce. *Michael Faraday.* Basic, 1965. xvi+531 pp. illus. $12.50. 65–19542. (SH–C)

In all probability this book will take its place as the standard biography of Faraday. The author has not only written a biography of one of the greatest experimental scientists of the 19th century, but has also brilliantly illuminated the period during which Faraday lived and worked. For those who wish to be informed about the origins of modern physical science, no better source can be recommended.

530.1 PHYSICS—THEORIES

Barnett, Lincoln. *The Universe and Dr. Einstein* (2nd ed.). Sloane, 1957. 127 pp. illus. $4.00; NAL (Mentor MP435), paper 60¢. 52–1367. (SH)
An elementary exposition of Einstein's work by an outstanding science writer. Includes an annotated list of recommended supplementary reading.

Einstein, Albert, and Leopold Infeld. *The Evolution of Physics: The Growth of Ideas from Early Concepts to Relativity and Quanta.* Simon & Schuster, 1938. x+319 pp. illus. $4.50, paper $1.45. 33–27272. (SH)
An account of the growth of ideas in physical science from the earliest concepts to the more abstruse theories of modern times. The authors trace the progress of the mechanical view of the universe propounded by classical physicists, the decline of that view, and progress to the more satisfactory explanations of modern science.

Faraday, Michael. *On the Various Forces of Nature.* Crowell, 1961. xxx+155 pp. illus. $2.75. 61–10486. (SH)
A reprint of the famous Christmas Lectures at the Royal Institution delivered in 1859. These lectures deal with fundamental physical concepts: gravitation, cohesion, chemical affinity, heat, magnitism, electricity, and the correlation of physical forces.

Gamow, George. *Mr Tompkins in Paperback.* (Illus. by the author and John Hookham.) Cambridge, 1965. 186 pp. $1.95 (paper). 65–20791. (SH–C)
A combined reprint of the author's popular expositions of relativity and the quantum theory for the layman. (*Mr Tompkins in Wonderland,* 1940; and *Mr Tompkins Explores the Atom,* 1944). The text has been expanded and updated to include information on fission and fusion, the steady state, and problems concerning elementary particles.

Koslow, Arnold (ed.). *The Changeless Order.* Braziller, 1967. viii+328 pp. illus. $7.50. 67–24207. (C)
The author has compiled the writings of physicists and philosophers concerning the central ideas of physics: space, time, motion, and conservation. Gives more than the usual attention to the cumulative aspects as well as to positive achievements of the physical sciences.

530.11 RELATIVITY

Bergmann, Peter G. *The Riddle of Gravitation.* Scribner's, 1968. xvi+270 pp. illus. $7.95. 68–11537. (C)
As one of Einstein's students, Bergmann is qualified and is a recognized authority. His new presentation on the 52-year-old theory of relativity (con-

cerned primarily with gravitation) is arranged in three sections: "Newtonian Physics and Special Relativity," "General Relativity," and "Recent Developments." Of most value to graduate students of mathematics and physics.

Coleman, James A. *Relativity for the Layman: A Simplified Account of the History, Theory, and Proofs of Relativity.* Macmillan, 1958. x+127 pp. illus. $2.75; NAL (Signet P2049), paper 60¢. 59–13781. (SH)
Written for those with little training in mathematics, physics, or astronomy, the book presents a brief history of scientific events preceding the theory or relativity and contains a simple account of the theory itself.

Eddington, Arthur. *Space, Time and Gravitation: An Outline of the General Relativity Theory.* Cambridge, 1966. vi+218 pp. $7.50, paper $1.35. 66–7834. (SH–C)
Sir Arthur wrote for those without much technical knowledge of the subject, to explain the basics of the general and special theories of relativity, using a minimum of mathematics. An appendix contains mathematical notes for the more advanced reader. This is a verbatim reprint of the 1920 original.

Gardner, Martin. *Relativity for the Million.* Macmillan, 1962. 182 pp. illus. $6.95. 62–21214. (JH)
An excellent, nonmathematical exposition of Einstein's Theory of Relativity —what it is, how it was derived, and what implications it holds. Profound concepts are handled in a clear and straightforward manner; effective illustrations accompany the text.

Kacser, Claude. *Introduction to the Special Theory of Relativity.* Prentice-Hall, 1967. vii+229 pp. illus. $2.95 (paper). 67–16387. (C)
A special relativity theory for a liberal arts physics course. Little mathematics is required. There are several worked examples, many problems, and an excellent bibliography. Useful for collateral reading.

530.12 QUANTUM MECHANICS

Emmerich, Werner, Milton Gottlieb, Carl Helstrom, and William Stewart. *Energy does Matter.* Walker, 1964. 250 pp. illus. $5.95. 64–19376. (SH)
Staff members of the Westinghouse Research Laboratories, representing physics, mathematics, and engineering, explore the subject of energy for the layman by telling how energy is involved in his own field of interest and research and how it is used to increase human understanding and control of nature.

Guillemin, Victor. *The Story of Quantum Mechanics.* Scribner's, 1968. xvi+332 pp. illus. $8.95. 68–17354. (SH–C)
The purpose of this book is to detail the history of atomic and nuclear physics and its antecedents. It will be exciting and helpful to bright high school students who have had a course in physics, and to college students and adults as well. History of classical physics; the quantum-mechanical theory of atoms, atomic nuclei, subatomic particles and their interactions; and the relation of physics theories to those of philosophy, ethics, and religion is covered. An excellent glossary is included.

Holden, Alan. *The Nature of Solids*. Columbia U., 1965. v+241 pp. illus. $7.50.
65–22156. (SH–C)
Solid state physics has become one of the most exciting fields of science
during the past 20 years and in consequence most scientific and engineering
activities require some knowledge of the physico-chemical properties of
solids. This book meets the need for a simple, nonmathematical, lucidly-
written discourse on the solid state. The author has the gift of clearly explaining theories and concepts and connecting them with the historical background.

Kittel, Charles. *Introduction to Solid State Physics* (3rd ed.). Wiley, 1967. 648
pp. illus. $12.95. 66–21055. (C)
This most widely-used textbook in solid state physics has undergone considerable revision. Intended for use by advanced undergraduate and graduate students, the coverage is quite comprehensive—differential equations, classical
mechanics, thermodynamics, atomic physics, and quantum mechanics are required.

Moore, Walter J. *Seven Solid States: An Introduction to the Chemistry and
Physics of Solids*. Benjamin, 1967. xii+224 pp. illus. $10.00, paper $3.95.
67–19435. (C)
Extremely interesting and well written, the book discusses the solid state of
matter, in particular the solid state of sodium chloride, gold, silicon, steel,
nickel oxide, ruby and anthracine. It further covers the crystal structure and
theories on the nature of forces which bond together atoms, molecules, and
ions. Then it describes in some depth what we know of each of the above
solids. It asks and answers many provocative questions about the solid state.

531 MECHANICS

Cohen, I. Bernard. *The Birth of a New Physics*. (Science Study Series.)
Doubleday (Anchor S10), 1960. 200 pp. illus. $1.25 (paper). 60–5918.
(SH)
Traces the development of mechanics from the Greeks to Newton, including
discussions of the work of Aristotle, Ptolemey, Copernicus, Kepler, Galileo,
and Newton, the latter in terms of present-day nonclassical physics.

Froman, Robert. *Baseball-istics: The Basic Physics of Baseball*. (Illus. by Sam
Salant.) Putnam's, 1967. 128 pp. illus. LB $3.49. 67–14804. (JH)
A unique coverage of the basic tenets of physics for young readers, that deals
with principles of physics and at the same time provides sophisticated belly-
laughs. A play-by-play account of the 7th game of the Year 2000 World
Series that concurrently explains the fundamental laws of the physics of
motion.

Gamow, George. *Gravity*. (Science Study Series S22.) Doubleday, 1962. 157
pp. illus. $1.25 (paper). (SH)
Emphasizing Newton and his concepts of Universal Gravitation, presents both
classic and modern views—from Galileo's experiments to post-Einsteinian
speculations concerning relation of gravity to other physical phenomena.

Harrison, George Russell. *The Conquest of Energy*. Morrow, 1968. xv+297 pp.
illus. $6.95. 68–8401. (SH–C)

Both theory and practical applications are considered in this unified picture of energy and its control. This is not a text or reference book, but a skillful discussion of what physical scientists have learned about energy. The book is a sequel to *Atoms in Action* (3rd ed.; Verry, 1947) written to give "the layman a unified picture of certain recent scientific developments of great social importance."

Shockley, William, and Walter A. Gong. *Mechanics.* Merrill, 1966. x+213 pp. illus. $3.95. 66–28253. (SH–C)
An entertaining and successful effort to explain fundamental concepts of mechanics and scientific method to nonmathematicians. Public, high school, and college libraries will find this an important addition.

Sterland, E. G. *Energy into Power: The Story of Man and Machines.* Nat. Hist. Press, 1967. 252 pp. illus. $5.95. 67–16902. (JH–SH)
The basic themes and objectives that seem paramount in the history of technology; man's search for new ways in which to utilize energy and for new sources of energy are described in this informative and attractive text. Mathematical background required is minimal.

Taylor, Edwin F. *Introductory Mechanics.* Wiley, 1963. xvii+423 pp. illus. $9.95. 63–11454. (C)
Since mechanics is a branch of physics fundamental to engineering and technology, the prospective professional needs a specialized knowledge of mechanics beyond that obtained in a general college physics course. This work is recommended for that purpose.

531.5 MASS, GRAVITY, BALLISTICS

Valens, Evans G. *The Attractive Universe: Gravity and the Shape of Space.* (Photographs by Bernice Abbott.) World, 1969. 188 pp. $5.95; LB $5.28. 68–14702. (EA–JH–SH)
Excitement is an apt word to describe the experience of reading this book, which is commended to all science students and their teachers, particularly because of its relevance in this age of space exploration and voyages to the moon. The mathematics requires introductory algebra; those who haven't studied it can understand much of the book anyway. Topics covered include: properties of falling objects, planetary orbits, Kepler's laws, tidal forces, Galileo's work with the pendulum, the weight of a planet, and the shape of space.

532 MECHANICS OF FLUIDS

Boys, Charles V. *Soap Bubbles and the Forces Which Mould Them.* (Science Study Series S3.) Doubleday, 1959. 156 pp. illus. $1.25 (paper); also Dover, paper 95¢; and Crowell, 1962. xxii+280 pp. illus. $3.95. 59–9612. (JH–SH)
Based on three lectures delivered at London in the Christmas Holidays of 1889–1890. It is both delightful reading and good science, for the experiments described are simultaneously entertaining and educational. The Crowell edition has the advantages of better paper, larger type and an index.

Rosenfeld, Sam. *Science Experiments with Water*. (Illus. by John J. Floherty, Jr.) Harvey House, 1965. 190 pp. $5.00; LB $4.79. 65–19983. (JH–SH)
An interesting combination of history, do-it-yourself experiments with water, explanations, and problems for the reader to think about. The approach is to pose a problem, usually one familiar to the reader, and then to try some simple experiments which will help explain an answer or give insight into the problem. If the reader actually performs the experiments and takes part in the action, he should have a rewarding experience.

533 MECHANICS OF GASES

Thompson, Paul D. *Gases and Plasmas*. (Introducing Modern Science Series.) Lippincott, 1966. 168 pp. illus. $4.25. 66–10888. (JH–SH)
An elementary summary of facets of modern research on gases and plasmas, notable for its breadth, clarity and enthusiasm. Accurate and relatively up to date. The historical perspective, and the presentation of how science and technology intermix and cross-fertilize each other are interesting features that will attract nonspecialist readers.

534 SOUND

Backus, John. *The Acoustical Foundations of Music*. Norton, 1969. xiv312 pp. illus. $7.95. 68–54957. (SH–C)
Serving a twofold educational purpose, this book introduces the music student or musician to the physical principles dealing with his field. With no more difficult mathematics than elementary algebra, the author discussed physical quantities, and their measurement, waves and wave propagation, vibrations and resonance, physiology of the human ear and the reception of sounds; also the characteristics of musical sounds, and the basic principles of acoustics as related to auditoriums, halls and rooms. Valuable for study and reference because it brings together material that is available only piecemeal in many sources.

Benade, Arthur H. *Horns, Strings and Harmony*. (Science Study Series S11.) Doubleday, 1960. 271 pp. illus. $1.45 (paper). 60–10663. (SH)
After presenting some of the physics necessary for understanding musical vibrators, the author discusses how the structure of the human ear determines our hearing of music, and concludes by examining various musical instruments.

Culver, Charles A. *Musical Acoustics* (4th ed.). McGraw-Hill, 1956. xiii+305 pp. illus. $8.50. 56–6953. (SH–C)
Science and music students, and others with only a cultural interest, will enjoy this logical and authoritative, yet nonmathematical treatment of acoustics which discusses the physics of sound as well as the musical aspects.

Griffin, Donald R. *Echoes of Bats and Men*. (Science Study Series S4.) Doubleday, 1959. 156 pp. illus. $1.25 (paper). 59–12051. (SH)
An account of the discovery that bats and porpoises hunt and navigate by a system of ultrasonic echo location similar to sonar. This discussion branches into such areas as radar, sonar, acoustics, and echoes.

Josephs, Jess J. *The Physics of Musical Sound.* Van Nostrand, 1967. xiii+165 pp. illus. $1.75 (paper). 67–574. (C)
Considerable basic information is given on musical science, however, the reader should have a background in college physics. The content is highly readable and reliable, and is supplemented with useful sketches, graphs, tables, and adequate bibliography.

Kinsler, Lawrence E., and Austin R. Frey. *Fundamentals of Acoustics* (2nd ed.). Wiley, 1962. vii+524 pp. illus. $12.95. 62–16151. (C)
The equations and laws of acoustics are developed from the fundamental principles underlying the generation, transmission and reception of sound waves. Then these principles are applied to a limited number of engineering fields. Includes information on ultrasonic sonar transducers. The reader needs a good foundation in mathematics.

Kock, Winston E. *Sound Waves and Light Waves.* (Science Study Series S40.) Doubleday, 1965. xvi+165 pp. illus. $1.25 (paper). 65–13088. (SH–C)
The similarities between sound and light or radio waves are used to explain the behavior of each type. It begins with definitions and descriptions, then proceeds to such topics as diffraction, wave and group velocities, microwave lenses, wave guides, delay lines, etc. The presentation is nonmathematical. Valuable to one who wishes only a glimpse of the physical phenomena involved.

535 LIGHT

Basford, Leslie, and Joan Pick. *The Rays of Light: Foundations of Optics.* (Foundations of Science Library: The Physical Sciences.) Sampson Low, Marston (dist. by Ginn), 1966. 128 pp. illus. $5.25. 66–17978. (JH–SH)
Descriptions of image formation, lens blooming, Fresnel's biprism, phase contract microscopes and other applications will hold reader interest. Some topics are treated very simply. Other topics are treated in rather high-level language but a student who reads *Popular Science* should have little trouble, despite occasional British spelling and expressions. The illustrations are colorful and accurate.

Dogigli, Johannes. *The Magic of Rays.* (Tr. and ed. by Charles Fullman.) Knopf, 1961. 275 pp. illus. $5.95. 60–15673. (SH)
Describes the theoretical investigations on and practical applications of such phenomena as cosmic rays, light, ultraviolet waves, infrared waves, alpha and beta radiation, and others. Although not current, it has historical and background value for students.

Gregg, James R. *Experiments in Visual Science: For Home and School.* Ronald, 1966. viii+158 pp. illus. $5.00. 66–12385. (JH–SH)
Consists of an introduction and 39 experiments. There are, in addition, a glossary, bibliography, and index. The introductory section discusses physical optics, basic anatomy and physiology of the visual mechanism (correctly emphasizing the importance of the brain in visual perception), the equipment needed to perform the experiments, and comments on interpretation of results. The simple experiments require very little in the way of equipment.

More complex ones need some slide projectors and (optionally) a variable-speed motor.

Jaffe, Bernard. *Michelson and the Speed of Light.* (Science Study Series S13.) Doubleday, 1960. 197 pp. $1.25 (paper). 60–13533. (SH)
A charming biographical account of an eminent scientist whose many accomplishments included studies of the speed of light.

Kogan, Philip. *The Unseen Spectrum: Foundations of Electromagnetic Radiation.* (Foundations of Science Library; The Physical Sciences.) Sampson Low, Marston (dist. by Ginn), 1966. 112 pp. illus. $5.25. 66–17979. (JH–SH)
The major divisions deal with light and spectra, the unseen spectra, man-made waves and radiation from atoms and outer space. The reader who wants to know a little about wave-guides, lasers, or the workings of a radiation film badge can get useful information in clear, well-illustrated form. The level is mainly popular technology; however, many students are seeking this level today since textbooks currently avoid many applications of science.

Ruechardt, Eduard. *Light: Visible and Invisible.* U. of Michigan, 1958. 201 pp. illus. $4.50, paper $1.95. 58–5904. (C)
A technical presentation of the physical nature of light, including a discussion of such aspects as light waves, the speed of light, light diffraction, wave interference, polarization, scattering, and dispersion.

535.6 COLOR

Burnham, Robert W., Tandall M. Hanes, and C. James Bartleson. *Color: A Guide to Basic Facts and Concepts.* Wiley, 1963. 249 pp. illus. $9.95. 63–18620. (SH)
This book resulted from a study sponsored by the Inter-Society Color Council of the basic elements of color education and consists of basic facts including physics and physiology of color vision, colorimetry, color systems, and color aesthetics, with bibliography.

Evans, Ralph M. *An Introduction to Color.* Wiley, 1948. x+340 pp. illus. $16.00. 48–7620. (C)
A comprehensive study of color which, for complete understanding, sprawls across the three major disciplines of physics, physiology, and psychology for its scientific basis; then, in practical application, spreads through many aspects of art and technology.

536.5 PHYSICS—TEMPERATURE

Castle, Jack, Jr., Werner Emmerich, Robert Heikes, Robert Miller, and John Rayne. *Science by Degrees: Temperature from Zero to Zero.* Walker, 1965. 229 pp. illus. $5.95. 64–23985. (SH–C)
Five scientists of the Westinghouse Research Laboratories have collaborated in explaining the present understanding of temperature, and of heat and cold. Covers the succeeding range of scientific discovery and experience of temperatures ranging from absolute zero to the temperature of the sun. Ex-

cellent collateral reading for students and laymen with some basic understanding of the physical sciences.

Hall, J. A. *The Measurement of Temperature.* Barnes & Noble, 1966. viii+96 pp. $3.75 (paper). (C)
A good introduction to the subject of temperature measurements. The topics covered are temperature scales, resistance thermometers, thermocouples, optical pyrometers, expansion thermometers, calibration of instruments, and temperature measurement in practice. The references cited are generally up to date and sufficient in number to give the student sources of additional material on items of interest.

McClintock, Michael. *Cryogenics.* Reinhold, 1964. 270 pp. illus. $12.00. 64–16625. (C)
A condensed, interesting and descriptive account of the important subject of extreme low-temperature research. The first chapter deals with the basics of cryogenics; the next five deal with phenomena that occur at low temperatures; and the final chapter presents some ways in which cryogenics is receiving practical applications.

MacDonald, D. K. C. *Near Zero: The Physics of Low Temperature.* (Science Study Series S20.) Doubleday, 1961. 116 pp. $1.25 (paper). 61–16716. (SH)
A study of the basics of low-temperature physics and of the behavior of materials at very low temperatures. The physicist tries but cannot reach absolute zero, and the reason is explained.

Meetham, A. R. *The Depth of Cold.* Barnes & Noble, 1967. 173 pp. illus. $4.50. (JH–SH–C)
Provides an introduction to low temperatures as they affect everyday life as well as science and technology. Coverage is complete: nature of cold temperatures, methods of obtaining these extremes, and uses of low temperatures are included. Interesting and well illustrated.

Mendelssohn, K. *The Quest for Absolute Zero: The Meaning of Low Temperature Physics.* McGraw-Hill, 1966. 256 pp. illus. $4.95, paper $2.45. 65–23829. (SH–C)
This readable account of low temperature physics begins with the liquefaction of acetylene by Cailletet. The story is carried through the 19th century and into the 20th century, bringing in not only the names, experiments and theories of the pioneers of low-temperature physics and engineering, but also their personalities, lives and trials. The gas laws, the laws of thermodynamics, quantum theory and the principle of "determinancy" are discussed within the terms of the behavior of gases. The names and the lives of the scientists who participated in the experimental theoretical work are interwoven into the background of theoretical physics. Well illustrated.

537 ELECTRICITY

Basford, Leslie, and Joan Pick. *Lightning in Harness: Foundations of Electricity.* (Foundations of Science Library; The Physical Sciences.) Sampson Low, Marston (dist. by Ginn), 1966. 127 pp. illus. $5.25. 66–17976. (JH–SH)
The bases of electrical theory are handled along with a variety of techno-

logical developments. The topics are treated individually and there is some overlap of ideas between sections; the presentation format is more typical of an encyclopedia than a text. Illustrations are colorful and prominent.

Bishop, Calvin C. *Fundamentals of Electricity*. Chilton, 1960. ix+230 pp. illus. $6.95. 59–9639. (SH–C)
A textbook useful to nonspecialists who wish to understand electrical circuits and principles of electrical equipment and devices. The final sections survey electronic and atomic theory.

Dart, Francis E. *Electricity and Electromagnetic Fields*. (*Merrill Physical Science Series*.) Merrill, 1966. ix+101 pp. illus. $3.95; paper $1.75. 66–16412. (SH–C)
Basic concepts of electricity and major applications of electrical theory. Contains excellent discussion questions that should promote fruitful classroom activity. Slight ability to manipulate algebraic equations is needed, although the student will need to think about the implications of an algebraic formula. Will stimulate a desire for additional study in depth.

Dunsheath, Percy. *Giants of Electricity*. Crowell, 1967. xii+200 pp. illus. $4.50. AC 67–10471. (JH–SH)
Describes the lives of 13 great men who pioneered in the discovery and elucidation of electrical phenomena. Presents biographies and explanations of what they accomplished and the many barriers that had to be overcome.

Graham, Kennard C., et al. *Fundamentals of Electricity* (5th ed.). American Technical Soc., 1960. 342 pp. illus. $6,00; study guide, paper $1.50. 60–6659. (SH)
Basic treatment of electrical science with a minimum of mathematics. Graphic analysis is well integrated in the text and a glossary of electrical terms is included.

Noll, Edward M. *Audels: Practical Science Projects in Electricity/Electronics*. Sams, 1966. 472 pp. illus. $4.95. 66–21937. (SH)
Can be used effectively for home study or in a practical high school course. It leads the student gradually from basic circuitry and components to AM and FM radio. With each discussion a simple experiment is offered to allow the student to build and understand the principles involved.

Singer, Bertrand B. *Basic Mathematics for Electricity and Electronics* (2nd ed.). McGraw-Hill, 1965. vii+600 pp. illus. $7.50. 64–20534. (JH–SH)
Presents lucidly and forcefully the physical ideas of electricity described with simple mathematics. Uses primarily linear equations, simultaneous equations and the elements of trigonometry. Summaries and reviews are plentiful. Designed either for self study or class use. Answers are given to the odd-numbered problems in the chapters.

Turner, Rufus P. *Basic Electricity* (2nd ed.). Holt, Rinehart and Winston, 1963. xvi+412 pp. illus. $9.95. 63–10789. (SH)
A fundamental book designed especially for students in vocational courses; hence explanations are clear, mathematical derivations are avoided unless essential for clarity, and practical applications are stressed. Authoritative and not superficial.

537.2 ELECTROSTATICS

Moore, A. D. *Electrostatics: Exploring, Controlling, and Using Static Electricity.*
(Science Study Series.) Doubleday, 1968. xxii+240 pp. illus. $4.95, paper
$1.45. 68–11781. (SH–C)
A retired professor of electrical engineering communicates to young scientists his fascination with electrostatics, teaching basics as well as sug-
gesting open-ended experiments. There are detailed instructions for making
a Kelvin generator and four Dirods. Annotated bibliography and index.

537.5 ELECTRONICS

Bennett, Alan, Robert Heikes, Paul Klemens, and Alexei Maradudin. *Electrons
on the Move.* Walker, 1964. x+229 pp. illus. $5.95. 64–23295. (SH–C)
Explores the properties of electrons, both free and in matter, and gives
particular emphasis to vacuum tubes and solid-state devices. Some recent
developments are discussed such as "picture tubes" and masers and lasers.
The reader should know fundamentals of algebra. Appendices include a
reference list, a list of symbols, biographical notes on all scientists mentioned,
and the index.

De Waard, Hendrik, and David Lazarus. *Modern Electronics: A Practical Guide
for Scientists and Engineers.* Addison-Wesley, 1966. ix+358 pp. $10.50. (C)
Although it presupposes a knowledge of elementary physics and passive DC
and AC electrical circuits, the book affords the non-electronically-minded
students and scientific workers an excellent opportunity to obtain an intuitive
"feel" for circuit action in vacuum tube and solid state devices. The mathematics is kept at basic algebraic level with a few necessary exceptions.

Kogan, Philip, and Joan Pick. *The Cathode Ray Revolution: Foundations of
Electronics.* (Foundations of Science Library; The Physical Sciences.)
Sampson Low, Marston (dist. by Ginn), 1966. 128 pp. illus. $5.25. 66–
14641. (JH–SH)
The emphasis is clearly on electronics technology, but there are good explanations of the many common devices. Detection, amplification and transmission of radio frequencies are discussed for vacuum-tube and solid-state
circuits. Sections on frequency modulation, television and the oscilloscope are
also included. A unique feature is the use of photographs of basic circuits
wired to match a schematic diagram. Profusely diagrammed in color. A
good starter for the young person curious about the principles of electronic
devices.

Middleton, Robert G., and Milton Goldstein. *Basic Electricity for Electronics.* Holt,
Rhinehart and Winston, 1966. x+694 pp. illus. $10.95. 66–10519. (SH–C)
Intended for junior college, community college and technical institute students. Only algebra and trigonometry are required. A brief historical survey
of electricity is followed by an introduction to basic physical concepts.
Electric circuit analysis is then systematically developed; magnetic circuits are
also discussed. Latter portions deal with transformers, motors, meters, vacuum
tubes, and semiconductor devices.

Pierce, John R. *Electrons and Waves: An Introduction to the Science of Electronics and Communication.* (Science Study Series S38.) Doubleday, 1965. 226 pp. illus. $1.25 (paper). 64–25265. (JH–SH–C)
Beginning with a résumé of electronics in the world, traces the principles and applications of electronic technology. The reader needs basic algebra, and preferably trigonometry, for complete comprehension.

Pierce, John R. *Quantum Electronics: The Fundamentals of Transistors and Lasers.* (Science Study Series S44.) Doubleday, 1966. xvi+ 138 pp. illus. $1.45 (paper). 66–11750. (SH–C)
A well-written, comprehensive introduction to four domains of the modern electronics technology which have revolutionized scientific work and daily life: transistors and lasers, noise and masers, radio astronomy, and satellite communications. The emphasis is on the ideas of physics which made these devices of technology possible. Many basic concepts are explained. Indexed.

Spielman, Harold S. *Electronics Source Book for Teachers* (3 vols.). Hayden, 1965. 1609 pp. in all. illus. $48.00 set. 65–14296. (SH–C)
A veritable encyclopedia of information in the broad subject area of electronics, arranged in progressive developmental order of subject matter. Will be invaluable for those who must plan and teach various courses in electronics in secondary schools or technical institutes. Particularly useful for teachers who lack extensive background and practical experience in electronic technology. Supplementary reference for students.

538 MAGNETISM

Bitter, Francis. *Magnets: The Education of a Physicist.* (Science Study Series S2.) Doubleday, 1959. 155 pp. $1.25 (paper). 59-9611. (SH)
The autobiography of a M.I.T. professor whose productive career has included extensive explorations of theoretical and applied problems of magnetism, and, most recently, magnetic aspects of the nuclei of atoms.

Sootin, Harry. *Experiments in Magnetism and Electricity.* Watts, 1962. x+244 pp. illus. $3.95. 62–7421. (JH)
Gives details of approximately 200 experiments that can be performed by an amateur with simple, inexpensive equipment. Each experiment is prefaced with scientific background information.

539 MOLECULAR, ATOMIC, AND NUCLEAR PHYSICS

Amaldi, Ginestra. *The Nature of Matter: Physical Theory from Thales to Fermi.* U. of Chicago, 1966. 332 pp. $5.95. 66–6791. (SH–C)
A nontechnical treatment of the 20th century physics of atoms, nuclei, and elementary particles. Up-to-date features include photographs of particle tracks in an emulsion, a Wilson chamber, a bubble chamber, and a spark chamber. Current topics such as weak interactions and neutrino physics are discussed. Thorough index, lack of references.

Anderson, William R., and Vernon Pizer. *The Useful Atom.* World, 1966. 191 pp. illus. $5.75. 63–18466. (JH–SH)

An organized view of the many current uses of atomic energy, and some predictions of future expansion. Includes an illuminating view of the structure of atoms, the problems of developing a controlled-fusion process, and applications of nuclear power.

Andrade, E. N. da C. *Rutherford and the Nature of the Atom.* (Science Study Series S35.) Doubleday, 1964. 218 pp. illus. $1.25 (paper). 64–11734. (SH) Andrade was one of Rutherford's assistants at the University of Manchester when he conducted his investigations of radioactivity that won him the Noble Prize in 1908. His biography of Rutherford is outstanding.

Asimov, Isaac. *Inside the Atom* (2nd ed.). Abelard-Schuman, 1966. 223 pp. illus. $4.00. 66–11053. (JH) Explains that all matter is composed of atoms, and then discusses atomic nuclei, isotopes, radioactivity, electrostatic generators, newly-discovered particles, and future possibilities for the use of atomic power.

Asimov, Isaac. *The Neutrino: Ghost Particle of the Atom.* Doubleday, 1966. xiii+223 pp. illus. $4.95; Dell, paper $1.95. 66–17073. (SH–C) Scientists first suggested the existence of the neutrino because something was needed to balance their energy equations. Neutrinos—without electrical charge and mass—travel with the speed of light and are produced inside the sun and other stars. The subject is introduced with excellent background, covering momentum, energy, atomic structure, mass-energy, electric charge, and antiparticles. The final three chapters tell of the postulation and detection of neutrinos and discuss neutrino astronomy. A lucid and intriguing presentation for anyone with a rudimentary physics background.

Beiser, Arthur. *Concepts of Modern Physics.* McGraw-Hill, (2nd ed.). 1967. 357 pp. illus. $9.50. 63–13154. (C) Intended for those with a foundation in elementary classical physics and the calculus to give them a basic understanding of atomic and nuclear physics. The latter part of the book deals primarily with ideas rather than experimental detail.

Beiser, Germaine, and Arthur Beiser. *The Story of Cosmic Rays.* Dutton, 1962. 126 pp. illus. $2.95. 62–10383. (JH) Elementary introduction to cosmic rays and their effects on the earth and its atmosphere; includes a chapter on the use of balloons in cosmic ray research.

Biquard, Pierre. *Frederic Joliot-Curie: The Man and His Theories.* (A Profile in Science.) Eriksson, 1966. 192 pp. illus. $5.00. 65–24212. (SH–C) A short and sympathetic account of Frederic Joliot's life by a fellow scientist. We meet not only a brilliant scientist who helped bring in the atomic age, but also a sensitive man who was concerned with the social implications of science and technology. An interesting account of this Nobel Laureate's contribution to nuclear physics: the research on the neutron, the discovery of artificial radioactivity, the experiments leading to nuclear fission.

Bohr, Niels. *Atomic Physics and Human Knowledge.* Wiley, 1958. viii+101 pp. illus. $3.95, paper (Science Editions) $1.45. 58–9002. (C) A collection of addresses and articles by the famous Danish physicist and winner, in 1957, of the $75,000 Atoms for Peace Award. They are valuable contributions to the philosophy of atomic physics.

Boorse, Henry A., and Lloyd Motz (eds.). *The World of the Atom* (2 vols.). Basic, 1966. xxvi+xx+1, 873 pp. in all. illus. $35.00 set. 66–11693. (SH–C)
This two-volume work takes the reader through the whole fascinating story of atoms from antiquity to the present day by means of more than 100 landmark classic papers. Each paper is introduced with a biographical sketch of the author which also includes a discussion of the setting in which his scientific work was done. Some require considerable background in mathematics and physics. A magnificent anthology, carefully selected and arranged, and beautifully printed.

Cohen, Bernard L. *The Heart of the Atom: The Structure of the Atomic Nucleus.* (Science Study Series.) Doubleday, 1967. x+107 pp. illus. $3.95, paper $1.25. 67–12871. (SH–C)
The nucleus, the "atomic heart," is the source of all matter and energy. The physics of the nucleus is introduced through a discussion of its structure, its principal properties, and large-scale applications. Explains fundamentals of "nuclear vocabulary."

Davis, George E. *Radiation and Life.* Iowa State, 1967. xvi+344 pp. illus. $7.50. 66–24398. (SH–C)
Provides a background in nuclear physics and engineering for students and nonspecialist adults, including a necessary introduction to algebraic operations, logarithms, exponential notation, and graphical presentations. Covers nuclear radiations and their known effects on organic and inorganic matter, as well as types of nuclear accelerators, reactors, and weapons. The effects of nuclear radiations on biological organisms are mentioned.

Fermi, Laura. *Atoms in the Family: My Life with Enrico Fermi, Architect of the Atomic Age.* U. of Chicago, 1954. ix+267 pp. illus. $5.00, paper (Phoenix P58) $1.95. 54–12114. (SH)
A wife of an eminent physicist presents an interesting account of the life and work of her famous husband, winner of the Nobel Prize and of the Congressional Medal of Merit, who led the research team that brought about the first self-sustaining chain reaction.

Frisch, O. R. *Working with Atoms.* Basic, 1966. 96 pp. illus. $3.95. 66–18154. (JH–SH)
Relates the story of the efforts that led from the discovery of radium to the first self-sustaining nuclear chain reaction. The narrative includes all of the significant events, logically strung together to form a complete account of this fruitful era. The author was a major contributor to this great effort. Excellent collateral reading material, is indexed, lacks references.

Gardner, Martin. *The Ambidextrous Universe.* Basic, 1964. x+294 pp. illus. $5.95. 64–23953. (SH–C)
A unique discussion of the structure of matter in the universe. The author explains its contents: "After examining the nature of mirror reversals in one, two and three dimensions, followed by an interlude on left and right in magic and the fine arts, we plunge into a wide-ranging exploration of the

left-right symmetry and asymmetry in the natural world. The exploration culminates in an account of the fall of parity and an attempt to relate its fall to some of the deepest mysteries in modern physics." An excellent book for collateral reading.

Glasstone, Samuel. *Sourcebook on Atomic Energy* (3rd ed.). Van Nostrand, 1967. vii+883 pp. illus. $9.25. 67–29947. (SH–C)
The most widely-used general reference book on atomic and nuclear science, prepared under the sponsorship of the U. S. Atomic Energy Commission. Completely revised and updated, including bibliographies. Essential for school, college, and public libraries.

Gottlieb, Milton, Max Garbuny, and Werner Emmerich. *Seven States of Matter.* (A Westinghouse Search Book.) Walker, 1966. vii+247 pp. illus. $5.95. 66–13265. (SH–C)
An unusual ammount of varied information on (1) "Of matter and spaces," (2) "Structure in transition," (3) "Substance of starts," (4) "Matter is degeneracy," and (5) "Some generalizations." While a subject is introduced at a very elementary level, part of the treatment requires considerable prior knowledge. Best suited to the physics or chemistry teacher for supplementing his own information.

Harvey, Bernard G. *Introduction to Nuclear Physics and Chemistry.* Prentice-Hall, 1962. x+370 pp. illus. $12.95 (text ed.). 62–13510. (C)
The more important nuclear phenomena are described in the first part in a nearly nonmathematical manner, and the second part describes experimental equipment and methods.

Hughes, Donald J. *The Neutron Story.* (Science Study Series S1.) Doubleday, 1959. 158 pp. illus. $1.25 (paper). 59–6910. (SH)
A substantial but surprisingly understandable account of neutron physics— their discovery, disintegration and atomic usefulness; highly readable.

Jaworski, Irene D., and Alexander Joseph. *Atomic Energy: The Story of Nuclear Science.* Harcourt, Brace, 1961. 258 pp. illus. $4.95. 60–13702. (JH)
A simple, bird's-eye view of elementary particles and nuclear reactions, with complete descriptions of experiments that can be conducted with relatively simple apparttus.

Kogan, Philip. *The Cosmic Power: Foundations of Nuclear Physics.* (Foundations of Science Library; The Physical Sciences.) Sampson Low, Marston (dist. by Ginn), 1966. 128 pp. illus. $5.25. 66–17980. (JH–SH)
The topics are interesting and related to any study of atomic energy. Radioactivity is treated historically, nuclear study is well handled, and the specific problem of detecting particles is carefully discussed. The illustrations are detailed and representative of scenes in modern British installations and the history of atomic science.

Lapp, Ralph E., and Howard L. Andrews. *Nuclear Radiation-Physics* (3rd ed.). Prentice-Hall, 1963. xiii+413 pp. illus. $11.50. 63–13280. (C)
A thorough and systematically arranged textbook, involving the use of calculus for the serious student with the necessary foundation courses in physics, chemistry and mathematics.

Lapp, Ralph E. *Roads to Discovery*. Harper & Row, 1960. 191 pp. illus. $3.95.
60–5965. (SH)
A very simple historical approach to nuclear physics that begins with Roent-
gen's discovery of X-rays, continuing with the working of Becquerel, the
Curies, Rutherford, Bohr, and concluding with the discoveries of the rare
elements.

Millikan, Robert Andrews. *The Electron*. (Ed. and introd. by Jesse W. M.
DuMond.) U. of Chicago, 1963. lxii+268 pp. illus. $6.00, paper (Phoenix
PSS523) $2.45. 63–20910. (SH)
A description of the isolation, measurement, and determination of some of
the properties of the electron, which was published first in 1917 and revised
in 1924. These researchers won for Millikan the Nobel Prize for Physics in
1923. The editor was an associate of Millikan and his introduction pays
tribute to the learned man and his scientific endeavors.

Oldenberg, Otto. *Introduction to Atomic and Nuclear Physics*. McGraw-Hill,
1967. x+414 pp. illus. $9.95. 67–10877. (C)
A basic introductory book for readers with a background of preliminary
physics and general chemistry. Since the detailed theory of atomic structure
is based on experimental evidence, the book stresses the relation between
theory and observed fact.

Romer, Alfred. *The Restless Atom*. (Science Study Series S12.) Doubleday,
1960. 198 pp. $1.25 (paper). 60–10681. (SH)
A stimulating nonmathematical account of the development of man's knowledge of atomic particles from 1890 to 1916.

Rusk, Rogers D. *Introduction to Atomic and Nuclear Physics*. Appleton-Century-
Crofts, 1964. xiv+470 pp. illus. $9.00. 64–20992. (C)
The methods and concepts underlying the fields of quantum physics and
relativity, with attention to practical applications, are explained for students
of the sciences and engineering who have completed a basic college physics
course and introductory calculus. Problems and recommended readings conclude each chapter. More advanced than White's text on the same subject.

Sproull, Robert L. *Modern Physics: The Quantum Physics of Atoms, Solids and
Nuclei* (2nd ed.). Wiley, 1963. xiv+630 pp. illus. $11.50. 63–11452. (C)
The author has attempted to tread the "middle ground" between simplified
introductory narrative and the highly mathematical treatment of advanced
books. The approach here is through the application of elementary quantum
mechanics. Intended primarily for undergraduate engineers.

Swartz, Clifford E. *The Fundamental Particles*. Addison-Wesley, 1965. viii+
152 pp. illus. $3.95, paper $2.95. 65–10929. (SH–C)
Presents the fundamental concepts and procedures selected from the vast body
of knowledge of particle physics. Systematically arranged, carefully illustrated. Good for orientation and collateral reading by advanced high school
students, their teachers, and recommended for use as part of a comprehensive
first-year physics course for college students.

Weidner, Richard T., and Robert L. Sells. *Elementary Modern Physics*. Allyn
and Bacon, 1960. xi+513 pp. illus. $13.25 (text ed. $9.95). 60–9402. (C)

For students with a knowledge of elementary physics and introductory calculus, presents a sequential account of the relativity and quantum theories, of atomic and nuclear structure, and of some topics in molecular and solid-state physics. Of particular background value for science and engineering majors. (See also *Elementary Classical Physics* by the same authors under "530.07 Physics—Study and Teaching".)

White, Harvey E. *Introduction to Atomic and Nuclear Physics*. Van Nostrand, 1964. 560 pp. illus. $9.75. 64–9067. (C)
An introductory text suitable for those who have completed a first college course in classical physics; requires no knowledge of mathematics beyond algebra and trigonometry. Features simple explanations of basic principles and practical applications.

540 CHEMISTRY

Chedd, Graham. *Half-Way Elements*. (Science Study Series). Doubleday, 1969. 192 pp. illus. $2.45. (SH–C)
Recounts the nature and certain applications of the group of chemical elements known as the metalloids. Easy to read and illustrated with schematic diagrams and pictures this is recommended for collateral reading on the general subject of metalloids.

Esterer, Arnulf. *Your Career in Chemistry*. Messner, 1964. 190 pp. illus. $3.95; LB $3.64. 64–11820. (JH)
Surveys the opportunities for employment and describes the work of chemists and chemical engineers in popular terminology; reviews academic requirements; contains bibliography. Similar to Pollack's book which is preferable, except for younger readers.

Pollack, Philip. *Careers and Opportunities in Chemistry*. Dutton, 1960. xii+147 pp. illus. $3.95. 60–5978. (JH)
The author of successful career guidance books in other science fields is concerned here with chemistry and chemical engineering. Duties, academic requirements, accredited institutions, and a bibliography are included. Preferable to the more recent book by Esterer listed above.

Vaczek, Louis. *The Enjoyment of Chemistry*. Viking, 1964. 243 pp. $6.00. 64–11193. (SH)
"And what Mr. Vaczek has done is to take his readers by the hand, to lead them through the channels of thought which the human scientific collectivity has followed for several thousand years. . . . it will make it possible for all intelligent persons to share in the intellectual enjoyment of those who probe in the structure of matter and the machinery of life."—Foreword by René Dubos.

540.03 CHEMISTRY—DICTIONARIES AND ENCYCLOPEDIAS

Clark, George L., and Gessner G. Hawley (eds.). *The Encyclopedia of Chemistry* (2nd ed.). Reinhold, 1966. xxi+1144 pp. illus. $25.00. 66–22807. (JH–SH–C)

Completely revised, with special emphasis given to chemical bonding, molecular orbitals, new developments in instrumental analysis, and recent advances in chemical technology. An invaluable reference that includes excellent narrative articles.

Hey, D. H. (ed.). *Kingzett's Chemical Encyclopedia: A Digest of Chemistry & Its Industrial Applications* (9th ed.). Van Nostrand, 1966. xi+1097 pp. illus. $32.95. 67–2338. (SH–C)

A standard reference proving information on the meaning of chemical terms, the nature of chemical operations, and properties of chemical substances and their industrial applications. Invaluable to students, scientists, and technologists. Deals with basic chemistry, chemical processes and products, medicinal products, and raw materials.

Lange, Norbert Adolph (ed.). *Handbook of Chemistry* (10th ed.). McGraw-Hill, 1967. xiv+2001 pp. illus. $12.00. 34–34381. (SH–C)

Revised and expanded, this volume covers a large amount of useful data: physical constants of chemical compounds and natural substances, the usual tables, plus some items on first aid and safety, qualitative analysis, and others. Similar to Chemical Rubber Company's *Handbook of Chemistry and Physics*.

Stecher, Paul G., M. Windholz, and D. S. Leahy (eds.). *The Merck Index: An Encyclopedia of Chemicals and Drugs* (8th ed.). Merck, 1968. xii+1713 pp. $15.00. 68–12252. (C)

Contains descriptive monographs of over 9,500 chemicals and drugs arranged alphabetically by nonproprietary name, illustrated with some 5,000 chemical structures. Cross-index of names containing 42,000 entries for locating chemical descriptions by proprietary, chemical, generic and nonproprietary names. Useful tables and appendices. An indispensable reference for all college libraries, large public libraries, and specialists of the chemical, pharmaceutical and medical professions.

Van Nostrand's International Encyclopedia of Chemical Science. Van Nostrand, 1964. 1331 pp. $32.50. 64–1609. (SH–C)

A dictionary of chemical terms, compounds, reactions, etc., including appropriate related physics and mathematical terminology. Also contains a glossary of foreign language equivalents. A good companion for Reinhold's *Encyclopedia of Chemistry*.

Weast, R. C. (Editor). *CRC Handbook of Chemistry and Physics: A Ready-Reference Book of Chemical and Physical Data*. Chemical Rubber Co., 1969. 2057 pp. $22.50. 13–11056. (SH–C)

The Golden Anniversary Edition of the *CRC Handbook of Chemistry and Physics* reflects, as has each revision, current advances in chemistry and physics, and in the constants, tables, formulas and mathematics associated therewith. Indespensible as a reference for senior high school, college, public and professional libraries.

540.1 CHEMISTRY—EARLY THEORIES

Burland, C. A. *The Art of the Alchemists*. Macmillan, 1968. 224 pp. illus. $9.95. 68–10841. (SH–C)

In alchemy the roots of scientific method and the inductive approach arose and were nurtured. Burland reports the evolution of alchemy in this informative historical account. This would be most meaningful to students who have had an elementary chemistry course, one in ancient history, and one in philosophy.

540.7 CHEMISTRY—STUDY AND TEACHING

Brescia, Frank, John Arents, Herbert Meislich, and Amos Turk. *Fundamentals of Chemistry: A Modern Introduction.* Academic, 1966. xv+816 pp. illus. $8.95; lab manual $4.95. 65–26049. (SH–C)
Attempts presentation at "a more comprehensive and rigorous level that interests and challenges, but does not dumbfound the student." The coverage of topics is ambitious, but the book is well organized. Two appendices summarize basic physical concepts and mathematical operations used. Can serve quite effectively for supplementary study for advanced high school students.

Brown, Theodore L. *General Chemistry.* Merrill, 1963. xvi+512 pp. illus. $11.95 (text ed. $8.95). 63–14202. (C)
A general chemistry text which presents a systematic view with references to experimental procedures and contains good discussions of quantum theory, gas kinetic theory, atomic and molecular structures, thermodynamics and thermochemistry, fuel cells, and crystal field theory. A worthwhile reference work.

Buttle, J. W., D. J. Daniels, and P. J. Beckett. *Chemistry: A Unified Approach.* Butterworth, 1966. viii+538 pp. illus. $7.95. 66–2848. (SH–C)
This British text approximates the U.S.A. *CHEM Study* or *Chemical Bond Approach* texts in difficulty. One important goal is stated in the authors' words: "It is . . . with the awareness that the theories and principles quoted are unlikely to stand the test of time, that the reader should approach this book." This is a remarkable précis. A tremendous amount of material is condensed. Every word is made to count but without loss of readability.

Chemical Bond Approach Project. *Chemical Systems.* (Prepared under a grant from National Science Foundation.) McGraw-Hill, 1964. 772 pp. illus. $7.96. 63–15889. (SH)
An alternative to *Chemistry: An Investigative Approach,* both of which are courses for secondary students developed by chemists and teachers of merit. This course is designed to provide an understanding of the constitution of matter, the nature of chemical reactions, interactions of electrostatic charges, fundamentals of atomic theory, etc. Available also are a laboratory manual and a teacher's manual that guide the student through chemical research experiences.

Cotton, F. Albert, and Lawrence D. Lynch. *Chemistry: An Investigative Approach.* (Teacher's Guide included. NY: Houghton Mifflin, 1968. xi+660 pp. illus. $6.96. (SH)
Cotton and Lynch have revised and extensively rewritten an earlier work, *Chemistry—An Experimental Science,* produced by the Chemical Educational Material Study (Freeman, 1963). The revisions have been substantial, though the general format of the previous, highly successful predecessor has been

retained. Experiments are performed by the student before the text discusses the ideas which they illustrate. Thus, the empirical nature of chemistry is stressed, and the fruitful interplay of theory and experiment is more fully appreciated. Equipment required for the experimentation is relatively simple, well within the normal purchasing requirements of secondary schools. Illustrations are numerous and excellent. The book also contains several tables of constants, a laboratory supplement section, a glossary, and a good index. The book should provide an interesting and challenging study of chemistry for every secondary school student, particularly those who will pursue the field more intensely in their college studies.

Gray, Charles A. *Explorations in Chemistry*. Dutton, 1965. 224 pp. illus. $5.95. 65–12180. (JH–SH)

Neither a textbook nor a laboratory manual, but a set of experiments and explanations designed to teach the student the major principles and concepts of chemistry, and lay the foundation for original research. The book discusses setting up a home laboratory, fundamental chemistry concepts, basic experiments and the theories behind them, discussions of each group of the chemical elements and experiments to explain their behavior, and suggestions for original research based on the knowledge developed. Excellent bibliography and index. Valuable to all amateur researchers, and to elementary and junior high teachers. Indispensable for all school and public libraries.

Gray, Harry B., and Gilbert P. Haight, Jr. *Basic Principles of Chemistry*. Benjamin, 1967. xviii+595 pp. illus. $10.95. 67–19433. (C)

A college chemistry text oriented towards structure and reactivity. The approach represents a realignment in teaching procedure and approach. Picture essays—adaptations of words and pictures to clarify concepts—are a helpful innovation.

Hoffman, Katherine Blood. *Chemistry for the Applied Sciences*. Prentice-Hall, 1963. xiii+429 pp. illus. $9.95. 63–7544. (SH)

Designed for students with an algebra background. Traces first the historical development of chemistry, its methods and applications, and then proceeds with a thorough portrayal of the chemistry of matter. Good advanced study and reference for secondary students.

Mahan, Bruce H. *University Chemistry*. Addison-Wesley, 1965. xii+660 pp. illus. $9.75. 65–12119. (SH–C)

A college textbook designed for those who have had a good modern chemistry course in high school. The first part might be considered fundamentals of chemistry in which the important principles of modern chemistry are discussed in depth. The second part deals with descriptive inorganic and organic chemistry. An understanding of the chemical principles in this textbook will prepare the serious science student for further study in chemistry, biology, or physics. Can be profitably used in high school chemistry courses as a reference.

Masterton, William L., and Emil J. Slowinski. *Chemical Principles*. Saunders, 1966. xxiv+668 pp. illus. $8.75. 66–12422. (SH–C)

Deals with descriptive material rather than with theoretical concepts, and is suitable for a terminal chemistry course and as supplemental reading beyond a high school or college freshman general chemistry course. Rich in interesting details, numerical examples, and review questions and problems.

Mortimer, Charles E. *Chemistry: A Conceptual Approach.* Reinhold, 1967. xii+692 pp. illus. $9.50. 67–18707. (C)
A general chemistry textbook definitely above the current average quality. The sequence of topics leaves something to be desired and occasionally illogical situations arise. The text is well appended.

Pauling, Linus. *College Chemistry* (3rd ed.). Freeman, 1964. xxiv+832 pp. illus. $8.75. 64–12953. (SH–C)
Intended for college students, but also an outstanding supplementary resource for good secondary students. This revision has incorporated much new material, particularly in modern aspects of chemistry.

Quagliano, James V. *Chemistry* (2nd ed.). Prentice-Hall, 1963. xvi+897+xxvii pp. illus. $10.95. 63–11096. (SH)
A comprehensive introductory college text, that makes a useful reference for secondary school students who wish to go beyond the elementary information in their assigned textbooks, or who may be interested in special problems.

Sanderson, R. T. *Inorganic Chemistry.* Reinhold, 1967. xii+430 pp. illus. $14.00. 67–21189. (SH–C)
A noble attempt at unifying and systematizing the field of inorganic chemistry. Useful in specific courses in inorganic chemistry and as a reference in first year college chemistry courses.

Sanderson, R. T. *Principles of Chemistry.* Wiley, 1963. x+626 pp. illus. $8.95. 63–11447. (SH–C)
Although intended as an introductory college textbook, the easy, informal and warm narrative style make it a good general background and review source for the nonspecialist. It also would entice secondary students as collateral reading.

Siebring, B. Richard. *Chemistry.* Macmillan, 1967. viii+743 pp. illus. $8.95. 67–18455. (SH)
Modern introductory chemistry is systematically covered at the senior high level in a straightforward manner, with numerous illustrative examples. Pertinent exercises, including numerical problems, are given at the end of each chapter. Answers to the even-numbered problems are in an appendix.

Smith, R. Nelson. *Chemistry: A Quantitative Approach.* Ronald, 1969. xiii+639 pp. illus. $10.50; $4.25 (Lab. Manual). 71–75642. (C)
Intended for use in a freshman chemistry course which includes quantitative analysis. Attention is paid to such topics as confidence limits, separation by precipitation, and the calibration of instruments. The main divisions are measurement, equilibrium, structure, and synthesis. The general approach is highly pragmatic and the author's style is conversational and clear.

540.9 CHEMISTRY—HISTORY

Asimov, Isaac. *A Short History of Chemistry.* (Science Study Series.) Doubleday, 1965. xiii+263 pp. illus. paper $1.45. 65–10641. (SH–C)

Over 300 scientists are mentioned, along with their contributions. The development is primarily chronological. Only readers with considerable prior understanding of chemical principles are likely to appreciate the significance of each event, to understand the importance of experimental evidence, and to be able to sort out the more important steps.

Carrier, Elba O. *Humphrey Davy and Chemical Discovery.* (Immortals of Science Series.) Watts, 1965. 161 pp. illus. LB $3.95. 65–11751. (JH–SH)
A well-researched biographical account for young people. The final chapter, "Assessment of Davy's contributions," and an index round out a worthwhile book.

Chilton, Thomas H. *Strong Water—Nitric Acid: Its Sources, Methods of Manufacture and Uses.* M.I.T. Press, 1968. viii+170 pp. illus. $7.50. 67–16496. (SH–C)
Examines the history and uses of the common and important chemical reagent—HNO_3. A historical work, not a how-to-do-it manual. Recommended for secondary and college students in their first general chemistry course.

Clark, Ronald W. *JBS: The Life and Work of J.B.S. Haldane.* Coward-McCann, 1969. 326 pp. $6.95. 68–11875.
Recounts the many accomplishments and personal eccentricities of Haldane in a manner that sustains interest and enjoyment. Haldane the scientist is seen at work, but not at the expense of seeing Haldane the man as well.

Clements, Richard. *Modern Chemical Discoveries* (rev. ed.). Dutton, 1963. xv+278 pp. illus. $5.95. 63–2487. (SH)
A summary of the principal developments in chemistry and chemical technology during the first six decades of this century. Although British-oriented as to industrial examples, the work has no American counterpart.

Curie, Eve. *Madame Curie.* (Tr. by Vincent Sheean.) Doubleday, 1949. xi+412 pp. illus. $5.95; Pocket Books 75017, 1959, paper 75¢. 48–40660. (JH–SH)
A very good biography of the discoverer of radium, written by her daughter, which is well documented and maintains the reader's interest at a high intensity.

Davis, Kenneth S. *The Cautionary Scientists: Priestley, Lavoisier, and the Founding of Modern Chemistry.* Putnam's, 1966. 256 pp. illus. $5.75. 66–15580. (SH–C)
Devoting alternate chapters to each of his subjects, the author narrates the work of these two scientists. In addition to developing the historical importance and the philosophical significance of Priestley and Lavoisier, the author shows their intimate involvement in contemporary affairs. Bibliography and index. Good collateral reading for high school or introductory college chemistry, and will also interest the nonspecialist.

Farber, Eduard. *The Evolution of Chemistry: A History of Its Ideas, Methods, and Materials.* (2nd ed.). Ronald, 1969. ix+349 pp. illus. $10.00. 52–9465. (SH–C)
History of chemistry is a part of all history. Every beginning student of chemistry should be assigned this book as collateral reading. The author is not only a historian of chemistry, but was a well-known research chemist.

Farber, Eduard (ed.) *Great Chemists.* Interscience, 1961. xxvi+1742 pp. illus. $39.00. 60–16809. (C)

An impressive anthology of excerpts from the biographies of more than 100 chemists, ranging over 3,000 years from the chemical technology in Mesopotamia to 1937 when the last chemist who is the subject of a biography died. A scholarly work without peer in the English language. Recommended for larger libraries and advanced readers.

Farber, Eduard (ed.). *Milestones of Modern Chemistry: Original Reports of Discoveries.* Basic, 1966. ix+237 pp. illus. $5.95. 66–23492. (SH–C)

A collection of original writings in chronological order that portray the major events in chemistry since 1859. It is absorbing to the reader because each breakthrough is described vividly in the words of its discoverer. Each selection is prefaced by a brief discussion by the editor which places it in historical perspective and explains its significance. A good bibliography and an index.

Farber, Eduard. *Nobel Prize Winners in Chemistry (1901-1961)* (rev. ed.). Abelard-Schuman, 1963. vii+341 pp. illus. $6.50. 62–17263. (SH)

Brief biographies introduce each Prize winner and then follows a description of the prize-winning work and a brief evaluation of its significance. A bibliography indicates the sources of the published account of the prize-winning work.

Findlay, Alexander. *A Hundred Years of Chemistry* (3rd ed., rev. by Trevor I. Williams). Humanities, 1965. 335 pp. $6.50. 65–3170. (SH–C)

A valuable history of the foundations of modern chemistry as it developed in the 18th and 19th centuries. The beginning and development of inorganic chemistry and radioactivity, organic chemistry (including naturally-occurring compounds), physical chemistry, and industrial chemistry are traced expertly. Ideal background reading for sophisticated secondary school students and all college undergraduates.

Gibbs, F. W. *Joseph Priestley: Revolutions of the Eighteenth Century.* Doubleday, 1967. xii+258 pp. illus. $6.00. 67–14120. (SH–C)

Gives a rather detailed chronological account of Priestley's life that emphasizes his interactions with society that grew out of his scientific work, his religious views, and his political activities.

Ihde, Aaron J. *The Development of Modern Chemistry.* Harper & Row, 1964. 851 pp. illus. $14.50. 64–15152. (SH–C)

A well-documented coverage of the history of chemistry which is divided into four parts: (1) "The foundations of chemistry," (2) "The period of fundamental theories," (3) "The Growth of specialization," and (4) "The century of the electron." Excellent in almost every respect because it devotes most of its pagination to newer developments in chemistry and has unique appeal to students as collateral reading. The appendices are on discovery of elements and Nobel laureates; the bibliography is arranged by chapters.

Kendall, James. *Great Discoveries by Young Chemists.* Crowell, 1954. xii+231 pp. illus. $5.00. 54–10375. (JH)

Many important discoveries were made by young men. A young chemist of today probably knows much more about the science at the same age as Davy, Faraday, Perkin, Cooper, and others were, when they made the discoveries for which they became famous.

Kuslan, Louis, and A. Harris Stone. *Liebig, the Master Chemist.* (History of Science Series.) Prentice-Hall, 1969. 84 pp. illus. $3.95. 69–14808. (JH–SH)
As a student, Leibig discovered that understanding chemistry was easier through experiment and demonstration than by the lecture technique, and used this knowledge with great success in his own teaching. The evolution of chemistry as a science is related in this biography.

McKie, Douglas. *Antoine Lavoisier, Scientist, Economist, Social Reformer.* Abelard-Schuman, 1952. 440 pp. illus. $6.00; Macmillan (Collier), paper 95¢. 52–13428. (JH)
The reforms Lavoisier instituted through his new doctrine and new nomenclature and other accomplishments paved the way for modern chemistry; hence, his biography is worth reading.

Multhauf, Robert P. *The Origins of Chemistry.* (Watts History of Science Library.) Watts, 1967. 412 pp. illus. $7.95. 67–24564. (C)
Examines the early background of chemistry which laid the groundwork for rapid period of growth resulting from the work of Lavoisier and his contemporaries. An interesting contribution to the history of science and chemistry.

Palmer, William G. *A History of the Concept of Valency to 1930.* Cambridge, 1965. vii+178 pp. illus. $8.00. 65–14348. (SH–C)
A fascinating account of the origins and history of the theories of valency. Pertinent biographical notes are included. This authoritative book is an ideal example of good history of science.

Schofield, Robert E. (ed.). *A Scientific Autobiography of Joseph Priestley (1733–1804).* M.I.T. Press, 1966. xiv+415 pp. $13.50. 67–14099. (SH–C)
A collection of letters by Priestley and his correspondents during the interval 1762–1767. Schofield fills the gaps between the letters with interesting biography and critical commentary.

541.2 CHEMICAL STRUCTURE OF MATTER

Asimov, Isaac. *Building Blocks of the Universe* (rev. ed.). Abelard-Schuman, 1961. 280 pp. illus. $3.50. 61–8933. (JH)
Asimov has written for the young student and layman a discussion of 102 of the basic elements of nature and the atomic laboratory. The structure as well as everyday forms and uses of the most common elements (such as oxygen, hydrogen, and carbon) are studied in detail.

Asimov, Isaac. *The Search for the Elements.* Basic, 1962. 158 pp. illus. $4.50. 62–15833. (JH)
Outlines the history and development of chemistry—the enduring quest for better understanding of those most basic particles which make up all matter. Includes the discoveries and theories of the Greek philosophers, ancient and

medieval alchemists, and Renaissance founders of the science, as well as the modern research which has not only investigated the remaining natural elements, but synthesized others. A good companion to *Building Blocks of the Universe*.

Campbell, J. Arthur. *Why Do Chemical Reactions Occur?* (Foundations of Modern Chemistry Series.) Prentice-Hall, 1965. viii+117 pp. illus. $4.95, paper $1.50. 65–12337. (SH–C)
A good introduction to the theory of chemical reactions. Principles are generally taught using a specific example which is thoroughly described and explained. Sample problems for each chapter. Diagrams, tables, an index, and a suggested reading list.

Friend, J. Newton. *Man and the Chemical Elements* (2nd ed.). Scribner's, 1961. ix+354 pp. illus. $6.00. 61–16758. (JH–SH)
Taking the entire list of elements as his theme, the author provides sidelights of discovery, development and use that every student should read as one of his early literary excursions into the history of science.

Grunwald, Ernest, and Russell H. Johnson. *Atoms, Molecules and Chemical Change* (2nd ed.). Prentice-Hall, 1965. xii+252 pp. illus. $9.95. 60–8778. (SH–C)
An easy-to-read general introduction to physical chemistry for the person with a limited background in mathematics who wishes to understand the structure, composition, and behavior of matter. Good collateral study material for secondary students.

Pauling, Linus. *The Nature of the Chemical Bond and the Structure of Molecules and Crystals: An Introduction to Modern Structural Chemistry* (3rd ed.). Cornell, 1960. xx+644 pp. illus. $11.50. 60–16025. (C)
The "standard" work on this important subject. It should be available for teachers using the "CHEM Bond" curriculum and some of their best students can use parts of it for advanced study and reference.

Pauling, Linus, and Roger Haywood. *The Architecture of Molecules*. Freeman, 1964. x+116 pp. illus. $10.00. 64–7755. (JH–SH–C)
An eminent chemist and gifted architect have collaborated in a fascinating volume, designed to provide insight in the inherent beauty of molecular architecture.

Seaborg, Glenn T. *Man-Made Transuranium Elements*. Prentice-Hall, 1963. 120 pp. illus. $4.50, paper $1.50. 63–15410. (C)
A detailed survey of the discovery, properties and applications of the recently discovered elements heavier than uranium by the Nobel Prize-winner whose work helped to pioneer this major extension to the field of chemistry.

Seaborg, Glenn T. *The Transuranium Elements*. Yale, 1958. 328 pp. illus. $7.00. 58–11258. (SH–C)
Based on the Silliman Lectures delivered by the author in Yale in 1957, which have been expanded for this book to include more detail. The sections are: (1) "The plutonium story;" (2) "Chemical properties of the actinide elements;" (3) "Nuclear properties of the transuranium elements;" (4) "Future synthetic elements;" also bibliography.

Seaborg, Glenn T., and Evans G. Valens. *Elements of the Universe*. Dutton, 1958. 253 pp. illus. $4.95, paper (D171) $2.15. 58–9576. (JH)
Using the periodic table as a unifying theme, early beliefs and subsequent developemnts over the centuries are contrasted with recent discoveries in chemistry, including that of element 102.

Weeks, Mary Elvira. *Discovery of the Elements* (7th ed.). (Rev. and new material added by Henry M. Leicester; illus. collected by F. B. Dains.) Journal of Chem. Ed., 1968. x+896 pp. $12.50. 56–6382. (SH)
"The material blessings that man enjoys today have resulted largely from his ever-increasing knowledge of about one hundred simple substances, the chemical elements, most of which were entirely unknown to ancient civilizations."

Wohlrabe, Raymond A. *Exploring Giant Molecules*. World, 1969. 95 pp. illus. $3.75. 68–26979. (JH–SH)
This is an interesting attempt to introduce a complete concept to the un-initiated. The problems include the determination of composition, structure, and the synthesis of sugars, starches, and proteins. Advances in polymer science are treated, with stress on their applications to man.

541.3   PHYSICAL CHEMISTRY

Asimov, Isaac. *The Noble Gases*. Basic, 1966. x+171 pp. $4.50. 66–13510. (JH–SH)
Describes first the discovery and identification of the rare gases. There follows a discussion of atomic structure in modern, quantum chemical language. Concludes with three typically well-done chapters on the newly-found chemistry of the formerly inert gases.

Granet, Irving. *Elementary Applied Thermodynamics*. Wiley, 1965. 288 pp. $8.95. 65–16410. (C)
An introduction to thermodynamics suitable as a first course for students in colleges or technical institutes with the inclusion of instructive and appropriate problem material taken from professional engineering examinations of New York. References, upon which the textual material is based, are at the ends of chapters and may be used for supplementary study.

Hamm, Donald I. *Chemistry: An Introduction to Matter and Energy*. Appleton-Century-Crofts, 1965. xxiii+1000 pp. illus. $8.95. 65–14499. (C)
Emphasis is on energy in chemical reactions. The stage is set by a careful background in physics for the historical evolution of atomic theory. Enough background in spectral phenomena is given to lay a sound base for the beginning of atomic orbital theory and the necessity of non-classical mechanics for small particles. The text is also an excellent reference for classes presenting a purely descriptive first course. There are numerous useful tables throughout, as well as appendices on units, acidity constants, solubility products, stability constants, and the use of pH, pOH, and pK.

Mason, Brian. *Principles of Geochemistry* (3rd ed.). Wiley, 1958. vii+310 pp. illus. $9.95. 58–6080. (C)

Designed for both the chemistry and geology major, this advanced text concentrates on the distribution, migration, and relative as well as absolute abundances of the elements and isotopes which occur in geologic materials and processes. Equations for some chemical reactions in rock and mineral formation, and thermodynamics are also studied. Index and extensive bibliographies make it a useful reference.

541.38. RADIO CHEMISTRY

Hahn, Otto. *Otto Hahn: A Scientific Autobiography.* (Introduction by Glenn T. Seaborg.) Scribner's, 1966. xxiv 296 pp. illus. $7.95. 66–25149. (C)
Hahn (1878–1968) details his scientific achievements in the years 1904–1945 when he was laying the foundations of radiochemistry with its more recent development into nuclear chemistry. His scientific training is described along with reminiscences of many of his famous colleagues. The account of his 30-year collaboration with Lise Meitner culminating in their epoch-making discovery of nuclear fission in 1938 is particularly significant. Full technical details are given, and three of Hahn's most significant papers are translated and reprinted in full.

542 CHEMISTRY—LABORATORIES AND EQUIPMENT

Steere, Norman V. (ed.). *CRC Handbook of Laboratory Safety.* Chemical Rubber Co., 1967. xii+568 pp. illus. $19.00. 67–29478. (SH–C)
The prevention or control of accidents, injuries, fires, and losses in all laboratories where there are chemical hazards is the purpose of this handbook, a compilation of writings by 40 contributors. Topics included are protective equipment, ventilation, water supply, hazards, legal liabilities, animal care and handling, and injuries.

543 ANALYTIC CHEMISTRY

American Public Health Association, Inc. *Standard Methods for the Examination of Water and Wastewater Including Bottom Sediments and Sludges* (12th ed.). American Public Health Assn., 1965. xxi+626 pp. $15.00. 55–1979. (C)
The official manual used by limnologists, public health officers, sanitary engineers, and others who have need for the analysis of water and water supplies. An indispensable reference for all college and public libraries. A basic knowledge of physics and chemistry is needed by the reader.

Moore, Walter J. *Physical Chemistry* (3rd ed.). Prentice-Hall, 1962. xiii+844+xxv pp. illus. $13.95. 62–10559. (C)
A basic text and reference work in physical chemistry that is of considerable importance because of the interdisciplinary involvements of physical chemistry. For the reader with the required mathematical background, this book will "open new doors."

544　QUALITATIVE ANALYSIS

Layde, Durward C., and Daryle H. Busch. *Introduction to Qualitative Analysis* (2nd ed.). Allyn and Bacon, 1968. ix+267 pp. illus. $6.95. 68–14573. (C)
Theory, laboratory, and appendix make up the three major parts of this textbook that teaches the fundamentals necessary to understand qualitative analysis. These three major sections are followed by an index and note pages. The chemical terminology and treatment of theory are adequate.

544.6　SPECTROSCOPY

Conley, Robert T. *Infrared Spectroscopy.* Allyn and Bacon, 1966. ix+293 pp. illus. $9.50. 66–12106. (SH–C)
A good orientation and introductory text to elementary spectroscopy. Lends itself also to selected reading and study for reference, so that each level of interest from advanced secondary school through college undergraduates will profit. Additional readings are suggested and problems are included.

544.92　CHROMATOGRAPHY

Bobbitt, James M. *Thin-Layer Chromatography.* Reinhold, 1963. 208 pp. illus. $9.50. 63–21843. (C)
An exposition of thin-layer chromatography which was developed to supply the need for a rapid method of separating small amounts of compounds. This newer technique should be considered as supplementing previously developed methods.

Mikes, O. (ed.). *Laboratory Handbook of Chromatographic Methods.* Van Nostrand, 1966. 434 pp. illus. $15.00. 66–21548. (C)
Beginners in chromatography as well as experienced investigators will find this a satisfactory handbook. Includes enough theory for the investigator interested in application of chromatography as a working tool. As stated in the Preface, no attempt has been made to present an exhaustive survey. A good guide to general chromatographic methodology.

Stock, Ralph, and C. B. F. Rice. *Chromatographic Methods* (2nd ed.). Reinhold, 1967. 206 pp. illus. $6.75, paper $3.50. 63–24898. (SH–C)
A summary account of techniques in current use, emphasizing paper and gas chromatography because they are the most widely used, although thin-layer chromatography has become popular recently. Model experiments are described.

545　QUANTITATIVE ANALYSIS

Pierce, Willis C., Edward L. Haenisch, and Donald T. Sawyer. *Quantitative Analysis* (4th ed.). Wiley, 1958. 497 pp. illus. $6.50. 58–7905. (C)
A well-known and highly recommended traditional textbook which is very lucid as to methodology, and hence has considerable value for secondary students working on individual projects. Includes instructions on colorimetric

analysis, potentiometric titrations, and electrodeposition. Miscellaneous useful appendices.

547 ORGANIC CHEMISTRY

Asimov, Isaac. *The World of Carbon.* Abelard-Schuman, 1958. 178 pp. illus. $3.50; Macmillan (Collier AS229), paper 95¢. 58–6063. (JH)
An account for the layman of approximately one-half of the world of organic chemistry—that part of chemistry concerned with those substances that are "vegetable" and "animal."

Asimov, Issac. *The World of Nitrogen.* Abelard-Schuman, 1958. 160 pp. illus. $3.50; Macmillan (Collier AS220), paper 95¢. 58–11066. (JH)
Nitrogen plus carbon are the distinguishing elements involved in the chemical composition of living things. Having read Asimov's book on carbon, the reader should proceed immediately to this consideration of nitrogen.

Campaigne, Ernest. *Elementary Organic Chemistry.* Prentice-Hall, 1961. viii+ 312 pp. illus. $9.95. 61–16607. (SH)
An outstanding and well-organized text designed for a background course for nonscience majors, which therefore can be used by good secondary school students who have had or are taking a first chemistry course.

Catch, John R. *Carbon-14 Compounds.* Butterworth, 1961. vii+128 pp. illus. $5.75. 62–5863. (C)
A guide to the use of Carbon-14 compounds and to the literature. Recommended for collateral or background reading because of the insight it provides into experimental techniques as well as of organic chemistry itself. The bibliographies are extensive.

Conrow, Kenneth, and Richard N. McDonald. *Deductive Organic Chemistry: A Short Course.* Addison-Wesley, 1966. 405 pp. illus. $7.95. 66–10504. (C)
The authors say in their introduction: " . . . a deductive approach is adopted in this text. . . . A continuing use of the chemical principles learned in modern freshman courses. . . . An emphasis of the principles serves as an efficient route to the knowledge of the descriptive aspects of organic chemistry." Each chapter ends with a generous supply of thought-requiring questions. The cross-references are excellent as is the index. Good outlines and summaries are quickly available inside the front and back covers. The better students will find this text invaluable.

Depuy, Charles H., and Kenneth L. Rinehart, Jr. *Introduction to Organic Chemistry.* Wiley, 1967. xii+392 pp. illus. $8.95. 67–12561. (C)
Intended as a college text, this introduction to modern organic chemistry emphasizes organic structures rather than methods of synthesis. The text is logically organized and content is systematically developed. Old-school chemists will find this refreshing.

Fieser, Louis F., and Mary Fieser. *Introduction to Organic Chemistry.* Heath, 1957. 613 pp. illus. $9.25. 57–6310. (C)
Part I develops the essential principles of organic chemistry, each chapter concluding with a summary, an analysis, problems, and references. Part II

illustrates the ways in which the reactions and principles of organic chemistry have been embodied in specific research projects.

Geissman, Theodore A. *Principles of Organic Chemistry* (2nd ed.). Freeman, 1962. x+854 pp. illus. $9.75. 62–10138. (C)

Somewhat more difficult than other organic texts, but unique because it stresses and gives insights into the theoretical aspects of organic chemistry. Descriptions of reactions therefore are presented to illustrate fundamental principles.

Gutsche, C. David. *The Chemistry of Carbonyl Compounds.* (Foundations of Modern Organic Chemistry Series.) Prentice-Hall, 1967. ix+141 pp. illus. $5.95, paper $2.95. 66–29093. (C)

One of a series designed to augment a basic text, to keep teacher and student abreast of the rapid developments in organic chemistry. Serves as supplementary material in an undergraduate organic chemistry course. Each chapter ends with a series of problems and a list of selected references. Index.

Smith, L. Oliver, Jr., and Stanley J. Cristol. *Organic Chemistry.* Reinhold, 1966. xv+966 pp. illus. $13.50. 66–18631. (C)

Organized around types of reactions rather than functional groups, this represents a fresh and challenging approach to organic chemistry. Includes an essential introductory unit on principles of chemistry. A good text for first-year college organic students.

548 CRYSTALLOGRAPHY

Bennett, Allan, Donald Hamilton, Alexei Maradudin, Robert Miller, and Joe Murphy. *Crystals: Perfect and Imperfect.* (A Westinghouse Search Book.) Walker, 1965. 237 pp. illus. $5.95. 65–15989. (SH–C)

Discussion of crystal symmetries, structures, forms and habits, imperfections, effects of symmetry on crystal properties, the crystallography of deformation of single crystals, and methods of growing crystals and determining their structures. Appendices give a list of State and Federal agencies that may be able to supply information interesting to rock collectors and a list of references that spans a wide range of subject matter. The book is profusely, often ingeniously, illustrated. If this book is to be used for collateral reading an instructor should be handy to help with the occasional jumps in level of treatment.

Burke, John G. *Origins of the Science of Crystals.* U. of California, 1966. 198 pp. illus. $6.50. 66–13584. (C)

The history of crystallography is ably recorded: early history, classification, Hauy's theory of crystal structure, crystals in chemistry and optics, and the concept of crystal symmetry. Experience in organic chemistry is helpful.

De Jong, Wieger F. *General Crystallography: A Brief Compendium.* Freeman, 1959. 281 pp. illus. $7.00. 58–5969. (C)

A good complementary work for those of Gilman, and of Holden and Singer, since it emphasizes the geometric, structural, chemical and physical aspects of crystallography. A mathematical background at least including trigonometry is essential for comprehension.

Gilman, John Joseph (ed.). *The Art and Science of Growing Crystals.* Wiley, 1963. x+493 pp. illus. $23.00. 63–11432. (SH–C)
An authoritative and exhaustive treatment of all aspects of crystals, crystallography, and crystal growing, consisting of 23 separate articles by outstanding authorities. Full appreciation of the book requires some background in mathematics, physics and chemistry, yet the serious amateur will find it an excellent reference.

Holden, Alan, and Phylis Singer. *Crystals and Crystal Growing.* (Science Study Series S7.) Doubleday, 1960. 320 pp. illus. $1.45 (paper). 60–5932. (JH)
An outstanding member of the Science Study Series, concerning the beautiful and fascinating world of crystals and of the crystallographer's art. Because of the great interest of young people in crystal growing every library needs this book.

Wells, Alexander F. *The Third Dimension in Chemistry.* Oxford, 1956. x+143 pp. illus. $5.60. 56–13674. (C)
The aim of this book is to help the student appreciate the three-dimensional arrangement of atoms in crystals as an integral part of structural chemistry. A good companion for De Jong's book. Excellent photographs.

549 MINERALOGY (see also 552 PETROLOGY)

Bateman, Alan M. *Formation of Mineral Deposits.* Wiley, 1951. x+371 pp. illus. $8.50. 51–13033. (SH)
Tells where mineral substances occur in nature and how they were formed, related in terms that do not presuppose previous training in geology and mineralogy; good glossary.

Berry, Leonard G., and Brian Mason. *Mineralogy: Concepts, Descriptions, Determinations.* Freeman, 1959. x+630 pp. illus. $9.25. 59–7841. (S)
Emphasizes general principles and the significance of mineralogical data in interpreting geological phenomena, especially in the fields of petrology and economic geology.

Dana, James D. *Dana's Manual of Mineralogy* (17th ed.). (Rev. by Cornelius S. Hurlbut, Jr.) Wiley, 1959. xi+609 pp. illus. $11.95. 59–11820. (C)
For over a century this manual has been widely used by students, amateurs, and professional workers. Frequent revisions have incorporated new knowledge and introduced quantitative data.

Deer, W. A., R. A. Howie, and J. Zussman. *An Introduction to the Rock-Forming Minerals.* Wiley, 1966. x+528 pp. illus. $11.95. 66–7468. (C)
An excellent, thorough presentation on rock-forming minerals, written clearly and concisely for advanced undergraduate and graduate students. As an adjunct to laboratory work in mineralogy and petrology, it is excellent.

Dennen, William H. *Principles of Mineralogy* (rev. ed.). Ronald, 1960. v+453 pp. illus. $8.50. 60–16452. (C)
An introductory textbook of mineralogy that emphasizes fundamental geometrical, chemical and physical relationships and hence is more than a guide to mineral recognition and identification.

Desautels, Paul E. *The Mineral Kingdom.* (Photos. by Lee Boltin.) Grosset & Dunlap, 1968. 252 pp. 29.8 cm. $14.95. 67–10532. (SH–C)

The curator of gems and minerals at the Smithsonian Institution shares the lore, history, science, technology, and beauty of the mineral world with outstanding monochrome and color photographs. It is intended for general readers and amateur "rock hounds" and can serve as a first reference for students.

English, George L., and David E. Jensen. *Getting Acquainted with Minerals.* McGraw-Hill, 1958. x+362 pp. illus. $7.95. 58–7417. (SH)

A complete guide for the amateur collector, including information on mineral resources, equipment, preparation, identification and classification. Also has information on crystals, and on fluorescent and radioactive minerals.

Fisher, P. J. *The Science of Gems.* Scribner's, 1966. 189 pp. illus. $7.95. 66–22665. (JH–SH)

Chemistry, physics, mathematics, biology, earth science, history, economics, and art are interwoven into a fascinating story of gems. There are chapters devoted to diamonds; rubies and sapphires and other important gem stones; and such organic gems as pearls, amber and coral. Additional chapters deal with gem-faceting and the testing of gems for value and quality. Good collateral reading for secondary school students. Attractively illustrated and contains a helpful glossary and usable index.

Gleason, Sterling. *Ultraviolet Guide to Minerals.* Van Nostrand, 1960. xii+244 pp. illus. $8.50. 60–16925. (SH)

Fluorescence is fascinating to all "rock-hounds," and this book reduces identification to looking up the glowing color in the field charts and comparing the description with the specimen. Other important data on mineral resources are included.

Hawkes, Herbert E., and J. S. Webb. *Geochemistry in Mineral Exploration.* Harper & Row, 1962. xiv+415 pp. illus. $14.25. 62–8889. (C)

A synthesis of the theory and technique of geochemistry in mineral exploration representing the authors' personal experiences in various areas of North America, Africa, and Southeast Asia. Extensive bibliography.

Hurlbut, Cornelius S. *Minerals and Man.* Random, 1968. 304 pp. illus. 28x22 cm. $15.00. 68–28329. (SH–C)

Addressed to novice, student, connoisseur, specialist, or practically anyone who wants to know more about minerals and their role in the technology and economics of the modern world, this text by a well-known specialist is accompanied by a spectacular and accurate set of colored, and black-and-white illustrations. An appendix contains the major physical and chemical properties of all minerals mentioned in the book. Not a textbook or manual of mineralogy, this is a very beautiful and enjoyable book to read that has good references value for amateur mineralogists.

McDivvit, James F. *Minerals and Men.* Johns Hopkins, 1965. viii+158 pp. illus. $1.95. (paper). 64–17636. (JH–SH–C)

This well-written book deals with mineral supplies and the factors which influence industrial and governmental policy decisions. The book has three parts: the place of minerals in modern society, commodity studies, and summary.

Pearl, Richard M. *How to Know the Minerals and Rocks.* McGraw-Hill, 1955. 192 pp. illus. $6.95; paper $1.95; also NAL (Signet T2647) 95¢. 53–5783. (JH)

A guide to more than 125 of the most important minerals and rocks, including ores, native metals, meteorites and gems. Includes identification data.

Sinkankas, John, *Mineralogy for Amateurs.* Van Nostrand, 1964. 585 pp. illus. $12.50. (SH)

A book on mineralogy and petrology that fills the gap between very elementary manuals and the college textbooks that require a background of mathematics, physics and chemistry uncommon among amateurs. Part I is an authoritative exposition of the subsciences of mineralogy and crystallography. Part II is a descriptive catalog based upon actual examination of specimens in the leading institutional and private collections. Appendices contain identification tables and a bibliography.

Smith, George F. Herbert. *Gemstones* (13th ed.). Pitman, 1958. 560 pp. illus. $15.00. 59–16106. (SH)

A semi-technical, reliable, and interesting reference work which includes information on precious, semi-precious and synthetic gems. Includes data on crystallography, chemical composition, and physical and optical properties.

U. S. Bureau of Mines. *Mineral Facts and Problems.* (2nd ed.). U. S. Bureau of Mines (Bull. 585), 1960. 1016 pp. $6.00. (Purchase from U. S. Govt. Printing Office.) 56–60859. (C)

A detailed inventory of the mineral resources of the United States and of foreign countries that are the principal producers, with accompanying data on strategic considerations, costs, government policies, research and industrial problems. A useful reference for all large libraries.

Zim, Herbert S., and Paul R. Shaffer. *Rocks and Minerals.* Golden Press, 1957. 160 pp. illus. $3.95; LB $2.99; paper $1.00. 57–3710. (JH)

An elementary and popular guide with colored illustrations and miscellaneous data of interest to amateurs.

550   EARTH SCIENCES   (see also 525 PLANET EARTH)

American Geological Institute. *Glossary of Geology and Related Sciences, with Supplement* (2nd ed.). AGI, 1960. xii+325 pp; supplement 72 pp. $7.50. 60–60083. (SH–C)

An authoritative and complete glossary of current, traditional, obsolete, and foreign terms from each of the earth sciences. Includes definitions, synonyms, variant usages and spellings, and preferred forms.

Branley, Franklyn M. *The Earth: Planet Number Three.* (Illus. by Helmut K. Wimmer.) Crowell, 1966. 151 pp. illus. $4.50; LB $4.40. 66–12668. (JH–SH)

The composition of the earth and its atmosphere, its origins and age, its motions, shape, the force of gravity, and a discussion of geomagnetism are the major topics of this readable account. An Appendix, "Some facts about the earth," contains a summary of the mathematical data. Contains historical material and descriptions of the process of scientific inquiry. Reading list and an index.

Dunbar, Carl O. *The Earth*. (World Natural History Series.) World, 1966.
xiii+252 pp. illus. $12.50. 66–25428. (SH–C)
The author has attempted "to present a broad view of the earth for the intelligent reading public," with a minimum of technical jargon. Topics include
the physical properties, age and cosmic setting of the earth; the chemical and
physical properties of the lithosphere, hydrosphere and atmosphere; the origin
and evolution of the earth; and ancient climates, earth dynamics and paleontology. A well-balanced and coherent synthesis of the geological sciences. The
illustrations and bibliography are excellent.

Namowitz, Samuel N., and Donald B. Stone. *Earth Science: The World We
Live In* (3rd ed.). Van Nostrand, 1965. x+597 pp. illus. $5.96. 64–23969.
(JH–SH)
Material is arranged logically so that successive chapters give the reader
a coherent picture of the interrelatedness of the major divisions of the earth
sciences. Two-color diagrams and line sketches encourage clear and rapid
comprehension. May be used as a text. A useful adjunct for the materials of
the ESCP.

National Academy of Sciences. *Geology and Earth Sciences Sourcebook*. Holt,
Rinehart and Winston, 1962. xv+496 pp. illus. $3.84 (paper). 62–10549.
(JH–SH)
An invaluable handbook of suggested questions, problems, activities, and references for the earth sciences teacher at all levels. Teaching aids and "unsolved problems" help stimulate interest in each topic, and the appendix lists
Federal and State geological survey centers, suppliers of teaching aids, and
additional reference books and films. Especially useful to students for the
bibliographies, experimental procedures, and summaries of basic facts and
concepts.

Raikes, Robert. *Water, Weather and Prehistory*. (Introd. by Sir Mortimer
Wheeler.) Humanities, 1967. xvi+208 pp. illus. $8.50. 67–23022. (C)
An attempt to set in proper perspective the relative roles of hydrology and
climate in relation to human prehistory, this work is addressed to all interested in archeology. Useful for reference and for collateral reading.

Wolfe, C. Wroe, Louis J. Battan, Richard H. Fleming, Gerald S. Hawkins, and
Helen Skornik. *Earth and Space Science*. Heath, 1966. ix+630 pp. illus.
$6.20. 66–1231. (SH)
A well-written and comprehensive book on earth and space science which
is divided into: (1) physical geology, (2) astronomy, (3) meteorology, (4)
oceanography, and (5) historical geology. At the end of each chapter are
essential vocabulary questions and activities. For senior high or junior high
students. Useful as an earth-science background course for prospective elementary teachers and for in-service teacher training.

551 PHYSICAL GEOLOGY

Chapman, Sydney. *IGY: Year of Discovery, The Story of the International
Geophysical Year*. U. of Michigan, 1959. 111 pp. illus. $4.95. 59–9733. (JH–
SH)

The president of the central committee of scientists that directed IGY tells of the discoveries and progress made in the study of the earth and sun from June 1957 to January 1959. The history of international scientific cooperation is also outlined, and the development of the IGY from its predecessors, the two International Polar Years, is briefly described.

Day, F. H. *The Chemical Elements in Nature.* Reinhold, 1964. 372 pp. illus. $9.50. (C)
The relative abundance, economic sources, and natural activity of the chemical elements are discussed briefly but factually. Tells of the general occurrence and importance of the elements in geologic and biologic materials and processes, then discusses the elements individually and in families; includes information on world production and deposits. Relatively little chemical and geological background is required.

Hunt, Charles B. *Physiography of the United States.* Freeman, 1967. 480 pp. illus. $7.50. 66–24952. (C)
A comprehensive description of the natural features and resources of the United States for college students who may have no other course in earth science, as well as for other interested readers. Emphasizes the relations of landforms to geologic structures and processes, resulting in an excellent summary of United States geology along with discussion of the physiography. Climate, surficial deposits and soils, water and mineral resources, vegetation, agriculture and pertinent historical matters are also included. Abundant maps, diagrams, sketches, and tables are included.

Lobeck, Armin K. *Things Maps Don't Tell Us: An Adventure into Map Interpretation.* Macmillan, 1956. x+159 pp. illus. $5.95. 56–10626. (SH)
An unusual and stimulating presentation of geography and geology by a famous map maker, designed to encourage "reading between the lines" (or under them). Seventy-two examples of distinctive coast lines, islands, rivers, lakes, and cities are developed on opposing pages. On the left side the "problem" is presented, with a flat map of the area and a discussion of its noticeable features. The right-hand page then gives "the explanation," telling what forces, materials, or structures account for the particular shapes and sizes under discussion.

Mather, Kirtley F. *The Earth Beneath Us.* Random, 1964. 320 pp. illus. $15.00. 64–17101. (JH–SH–C)
This is the admirable fulfillment of Dr. Mather's "long-nurtured ambition to produce a pictorial geology for the layman." Dignified, very readable prose. The carefully chosen illustrations, some in color, are magnificent. Includes a geological time table, a time table of the great ice age, a brief but good bibliography, and an index.

Milne, Lorus J., Margery Milne, and The Editors of LIFE. *The Mountains.* Time, Inc., 1962. 192 pp. illus. $6.60. 62–13534. (JH)
A colorful exposition of mountain climates, plant and animal life, formation, and climbing. Illustrated with many large and beautiful photographs, it includes discussions of the problems and dangers man faces on and around mountains and volcanoes; gives brief descriptions of some famous disasters

and the life of various mountain peoples. The limited text is broad and selective, but covers many basic geologic principles.

Ogburn, Charlton, Jr. *The Forging of Our Continent.* (The Smithsonian Library.) American Heritage (dist. by Van Nostrand), 1968. 160 pp. illus. $4.95; LB $4.98. 68–22959. (JH–SH)

Proceeds logically from oldest to youngest areas, including the Precambrian shield, the Appalachians, the Rocky Mountains, and glaciated regions. Within this framework, major geologic processes are discussed in connection with the geographic area where each is best displayed. Appendixed are an annotated list of national parks and monuments, the geologic time scale, a glossary, suggestions for further reading, and an index.

Putnam, William C. *Geology.* Oxford, 1964. 480 pp. illus. $10.95 (ext ed. $8.00). 64–11237. (SH)

Presents a well-rounded introduction to geology, covering the basic land forms, their component rocks and minerals, and the natural forces of construction and destruction. Brief surveys of historical and petroleum geology are also included, and each chapter is accompanied by an extensive list of suggested references.

Schultz, John R., and Arthur B. Cleaves. *Geology in Engineering.* Wiley, 1955. ix+592 pp. illus. $11.50. 55–7317. (C)

"It has been said that civil engineers are to a greater or less extent engaged in the practice of applied geology whether they know it or not." On this assumption the author offers this book as a background in the earth sciences for the engineer. The discussions of the elements, structures, and dynamics of geology give special attention to engineering applications and problems.

Shelton, John S. *Geology Illustrated.* (Illus. by Hal Shelton.) Freeman, 1966. xii+434 pp. $10.00. 66–16380. (SH–C)

Illustrates some of the main principles of geology. The excellent photographs are drawn mainly from the North American West and Southwest. Often the details in the photographs are explained by line drawings. The work is basically divided into six major parts: (1) materials, (2) structure, (3) sculpture, (4) time, (5) case histories, (6) implications. A brief list of references, conversion charts, and an index. An excellent reference for students and for nonspecialist adults.

Spar, Jerome. *Earth, Sea, and Air: A Survey of the Geophysical Sciences.* Addison-Wesley, 1962. vii+152 pp. illus. $3.75; paper $2.50. 62–15533. (SH)

The author divides geophysics into four areas, the earth as a planet, the solid earth, the hydrosphere, and the atmosphere, devoting a section to each. The book is a condensed and factual introduction for non-majors or the general reader and requires little scientific knowledge. A series of six laboratory experiments covering time, gravity, geomagnetism, ocean currents, waves, and weather observation is appended.

Sullivan, Walter. *Assault on the Unknown: The International Geophysical Year.* McGraw-Hill, 1961. xiv+460 pp. illus. $8.50. 60–53222.

The *New York Times* correspondent who covered IGY describes the development and findings of the program, with proper attention to the achievements of foreign as well as American scientists. Each of the twelve major areas of research is discussed, but there is more emphasis on Antarctica and the first artificial satellites. Also includes a fairly detailed account of Project *Argus,* the series of upper atmospheric atomic tests carried out by the U. S. Government and observed by IGY scientists.

Takeuchi, H., S. Uyeda, and H. Kanamori. *Debate About the Earth: Approach to Geophysics through Analysis of Continental Drift.* (Tr. by Keiko Kanamori.) (Illus. by James K. Levorsen.) Freeman, Cooper, 1967. 253 pp. $4.50. 67–21261. (SH–C)
Examines the evolution of the theory of continental drift from the skepticism with which the original ideas of the young German meteorologist, Alfred Wegener, were greeted at the beginning of this century to the widespread acceptance of the theory by geophysicists and geologists of the 1960's.

Thorarinsson, Sigurdur. *Surtsey: The New Island in the North Atlantic.* (Tr. by Sölvi Eysteinsson.) Viking, 1967. 54 pp. illus. $6.00. 67–10220. (JH–SH–C)
Superb illustrations and a well-written text vividly describe the growth of a new volcanic island off the coast of Iceland. This topical narrative has broad appeal both to the young student and to the scholar, who will find a helpful reference section.

Thornbury, William D. *Principles of Geomorphology.* Wiley, 1954. ix+618 pp. illus. $11.50. 54–7553. (C)
An introduction to the structure, formation, and evolution of land forms which presents the history and development as well as the basic principles of the science. Conflicting theories and the men who proposed them are discussed and contrasted, accompanied by the evidence available then and now and the currently accepted or favored hypotheses. Some background in geology is required.

Walker, John. *Lectures on Geology: Including Hydrography, Mineralogy, and Meteorology, With an Introduction to Biology.* (Ed. by Harold W. Scott.) U. of Chicago, 1966. xlvi+280 pp. illus. $8.50. 65–24986. (SH–C)
An excellent starting point for the reader who wishes to consider the early historical background of various natural sciences. The subject matter considered is much more inclusive than geology. Extensive consideration is given to meteorology, particularly development and uses of the barometer. The lectures cover many phases of the earth. Any student of the natural sciences should find many portions of this book interesting reading.

Wilson, J. Tuzo. *IGY: The Year of the New Moons.* Knopf, 1961. xxi+350+ix pp. illus. $6.75. 61–14195. (SH)
This nontechnical description of the IGY project combines both the scientific and human aspects of the science. The former president of the International Union of Geodesy and Geophysics relates his travels to many of the worldwide research sites. There is less emphasis on the development and overall organization of the project than in Sullivan's book, but they both are authoritative and factual accounts.

551.07 GEOLOGY—STUDY AND TEACHING

Bates, Robert L., and Walter C. Sweet. *Geology: An Introduction.* Heath, 1966. x+367 pp. illus. $7.50. 66–10074. (SH–C)
Designed for college students desiring to obtain a nontechnical introduction to

geology in one semester. Emphasis is on the cultural values to be derived from a limited acquaintance with geologic processes and materials. The book has two parts: physical geology and historical geology. The geologic processes such as the work of streams, glaciers, ground water, vulcanism and deformation are briefly but adequately handled.

Croneis, Carey, and William C. Krumbein. *Down to Earth: An Introduction to Geology.* U. of Chicago, 1936. xvi+501 pp. illus. $8.50, paper (Phoenix PSS501). $2.95. 36–10420. (SH)
A complete introduction to the basics of geology, written in a clear and informal style and accompanied by many diagrams, sketches, and photographs. Although slightly dated on some topics, this is a useful and very readable reference.

Eardley, A. J. *General College Geology.* Harper & Row, 1965. xviii+499 pp. illus. $10.50. 65–10165. (SH–C)
Covers both the physical and historical-biological fields of geology. Emphasizes the cultural values to be derived from earth science. Geological materials and processes are treated from a descriptive point of view that requires little previous knowledge of chemistry and physics. Descriptions are supplemented by clear illustrations and excellent photographs. Parts of this book are suited equally well to the senior high school.

Earth Science Curriculum Project. *Investigating the Earth.* Houghton Mifflin, 1967. xiv+594 pp. illus. $8.00. (JH–SH)
Written by a large group of scientists and teachers, this classroom-tested presentation of the earth sciences provides an up-to-date treatment of those topics traditionally included in introductory courses. In addition, such topics as the nature of scientific research, its assumptions and techniques, circulation of the ocean and atmosphere, astronomy, and planetary sciences are treated.

Gilluly, James, et al. *Principles of Geology* (3rd ed.). Freeman, 1968. 689 pp. illus. 68–14228. (C)
An excellent introductory textbook including the facts, evidence and theories which have led to the conclusions presented. The techniques of mapping, experimentation, and chemical analysis used in modern geology are also presented, accompanied by frequent illustrations and chapter review questions. Various useful appendices. A good reference for advanced secondary school students.

Holmes, Arthur. *Principles of Physical Geology* (2nd ed.). Ronald, 1965. xv+ 1288 pp. illus. $12.00. 65–21811. (SH–C)
This new and fully revised edition represents the most extensive treatment of introductory physical geology in the English language. Examples and illustrations are abundant and wide ranging. Its readability, objectivity, and diversity make it a requirement for every geology or earth science library.

Leet, L. Don, and Sheldon Judson. *Physical Geology* (3rd ed.). Prentice-Hall, 1965. 406 pp. illus. $9.95. 65–10094. (SH–C)
This considerably revised edition is suitable for a one-semester physical geology course and valuable to secondary school students for collateral reading. New chapters concern mountains and the deformation of the earth's crust,

earth magnetism, continental drift, and the earth as a planet. Relatively nonmathematical, and features an attractive format and excellent illustrations, maps, and diagrams. Brief bibliographies at the chapter ends, and a glossary of terms.

Longwell, Chester R., and Richard F. Flint. *Introduction to Physical Geology* (2nd ed). Wiley, 1962. 504 pp. illus. $9.95. 62–8779. (SH–C)
Another good text in the fundamentals of physical geology, which has roughly the same scope as Leet and Judson, with similar appendices and chapter summaries in sentence form. An extensive glossary is incorporated, and each term, as well as the page on which it is defined (listed in the general index) is printed in bold type, affording quick and accurate reference to information in context.

Read, H. H., and Janet Watson. *Beginning Geology.* (2nd ed.). St. Martin's, 1968. 246 pp. illus. $12.50. 66–18762. (SH–C)
A concise, comprehensive and exceedingly well-written book about basic geologic investigations, including earth history. The authors have woven an unusually large number of significant geological terms into the text, well defined and very well indexed for reference. English examples are used because the authors are English, the islands have a rather complete geologic record, and early British geologists profoundly affected geological investigations.

Spencer, Edgar Winston. *Basic Concepts of Physical Geology.* Crowell, 1962. xv+480 pp. illus. $9.95. 62–7082. (C)
This basic text puts more emphasis on physical geology as a science than either Longwell or Leet and Judson (see preceding annotations). Fundamental concepts are often presented as theory, and opposing hypotheses are fully discussed, with reference to their authors and current acceptance. It lacks a glossary, chapter summaries, and well-integrated illustrations, but is well organized, with frequent itemization and outlining of key facts and concepts.

Stokes, William Lee, and Sheldon Judson. *Introduction to Geology: Physical and Historical.* Prentice-Hall, 1968. xii+530 pp. illus. $9.50. 67–14773. (SH–C)
Combines the material of their two previous texts: *Physical Geology* (Leet and Judson) and *Essentials of Earth History* (Stokes). Divided equally between physical and historical geology. An excellent general geology text.

Strahler, Arthur N. *The Earth Sciences.* Harper & Row, 1963. xii+679 pp. illus. $11.50. 63–11290. (C)
Believing in the necessary unity of the earth sciences, the author presents a comprehensive analysis of geophysics, oceanography, meteorology, and geology. With the detail and breadth of several books, this single, well-integrated text is an excellent introductory cultural course. A useful reference for students as well as for teachers and specialists in any of the related fields.

Wolfe, C. Wroe, Louis J. Battan, Richard H. Fleming, Gerald S. Hawkins, and Helen Skornik. *Earth and Space Science.* Heath, 1966. ix+630 pp. illus. $6.20. 66–1231. (SH)
A comprehensive work with units on physical geology, astronomy, meteorology, oceanography, and historical geology. Includes glossary, questions, and

activities. Useful as textbook or reference; good resource book for elementary science teachers.

Zumberge, James H. *Elements of Geology* (2nd ed.). Wiley, 1963. viii+342 pp. illus. $7.95. 63–17488. (SH)

The what, why, and when of geology are presented under the two traditional subdivisions of the science. Part One, "The Dynamic Earth," explains the materials and language of physical geology and features the natural forces of sculpture and change. Part Two unfolds "The Geological Story," following an introduction to the methods and theory of historical geography.

551.2–551.3 VOLCANOES, GLACIERS, AVALANCHES, EARTH-QUAKES

Bullard, Fred M. *Volcanoes in History, in Theory, in Eruption*. U. of Texas, 1961. xvi+441 pp. illus. $7.50. 61–10043. (SH–C)

This clear and well-illustrated study of the development, principles, and applications of volcanology is comprehensive yet requires no previous knowledge of geology. The various types of volcanic landforms, eruptions, and materials are explained by extensive reference to well-known examples, and some of the theory of volcanic formation and eruption cycles is included. Complete bibliography and glossary.

Dyson, James L. *The World of Ice*. Knopf, 1962. xvii+292+xiii pp. illus. $6.95. 62–8682. (SH)

A lively discussion of the formation, movement, and effects of glaciers and ice caps, including chapters on the role of ice masses in climatology and the problems of permanently frozen soil. Emphasis on the present and future glacial cycles makes this winner of the 1962 Phi Beta Kappa Science Award an intriguing introduction to a science too often considered remote and historical.

Fraser, Colin. *The Avalanche Enigma*. Rand McNally, 1966. xvi+301 pp. illus. $6.95. 66–23794. (SH–C)

Covers the history and physics of avalanches, snow structure, rescue methods, and control and protection against avalanches. An excellent and permanent reference, this is the only modern book on avalanches written in English. Although slightly repetitive and restricted, it is well illustrated and has a good bibliography.

Rittmann, A. *Volcanoes and Their Activity* (2nd ed.). (Tr. by E. A. Vincent.) Wiley, 1962. xiv+305 pp. illus. $11.95. 62–10523. (C)

The author presents his own views and experiences in volcanology along with his descriptions of basic concepts and principles of volcanic origins, activity, products, and geomorphic effects. An advanced and detailed theoretical analysis, most useful to those with a good background in other sciences as well as geology.

Roberts, Elliott. *Our Quaking Earth*. Little, Brown, 1963. vi+247 pp. illus. $4.50. 63–14238. (JH)

Surveys famous earthquake disasters and various myths and superstitions concerning these phenomena. Then the author explains their actual causes, distribution, and the various scientific and nonscientific attempts to explain and predict their occurrence. The principles and applications of seismology are also clearly presented, including some career material.

Tazieff, Haroun. *When the Earth Trembles*. (Tr. by Patrick O'Brien.) Harcourt, Brace, 1964. x+245 pp. illus. $4.95. 64–11542. (JH–SH)
The first section is an account of the author's six-week journey through areas affected by the 1960 Chilean earthquake. The second section deals with the geography and geophysics of earthquakes, including accounts of the most destructive ones. Finally there is a discussion of the methods and instruments of seismology.

Wilcoxson, Kent H. *Chains of Fire: The Story of Volcanoes*. Chilton, 1966. 235 pp. illus. $6.95. 66–25046. (JH–SH–C)
Primarily a nontechnical description of volcanic eruptions in which most of the chapters are each devoted to a specific volcano or volcanic area. Interspersed are four chapters devoted to theory and discussions of volcanism in general. Intentionally omits the majority of the world's volcanoes and concentrates only on the few dozen that have had spectacular or well-documented eruptive histories. A very good job of presenting numerous and detailed eyewitness accounts of eruptions.

551.44 SPELEOLOGY

Mohr, Charles E., and Howard N. Sloane (eds.). *Celebrated American Caves*. Rutgers, 1955. xii+339 pp. illus. $6.00. 55–12228. (SH)
An excellent selection of authoritative, often firsthand accounts, of the exploration and history of many famous caves in the Western Hemisphere. Some also contain descriptions and explanations of geologic structures, materials, and processes. The beautiful formations, strange wildlife, and perils for human intruders in these natural wonders are part of the story.

Mohr, Charles E., and Thomas L. Poulson. *The Life of the Cave*. McGraw-Hill, 1966. 232 pp. illus. $3.95. 66–24465. (JH–SH)
Surveys cave biology-ecology in North America with beautiful and diverse photographs. Emphasis is on the cave environment, trophic patterns, behavior patterns, paleoecology, circadian rhythms, and conservation. Intended for young readers.

Siffre, Michel. *Beyond Time*. (Tr. and ed. by Herma Briffault.) McGraw-Hill, 1964. 228 pp. illus. $5.95. 63–22554. (SH)
A young French geologist describes the unusual experiment in which he spent two months alone camping on a subterranean glacier in a cave almost 400 feet beneath the French-Italian Alps. He made geologic observations and for the entire duration lived completely without a clock, "losing" 25 days in his attempted estimations of time. Largely based on his daily journal, this is a personal and often weird insight into the physiological and psychological effects of deep caves.

551.46 OCEANOGRAPHY

Bardach, John. *Harvest of the Sea*. Harper & Row, 1968. viii+301 pp. illus. $6.95. 67–22536. (SH–C)

Concerned with the uses and potentials of the sea and with the consideration of conservation and legal ramifications. Effects of tides and currents on organisms are covered in some detail, and there is a thorough explanation of agricultural techniques, from planting of seed organisms to various methods of harvesting.

Bascom, Willard. *A Hole in the Bottom of the Sea, The Story of the Mohole Project.* Doubleday, 1961. 352 pp. illus. $5.95. 61–7638. (SH)

Despite recent developments, this authoritative account is valuable for the complete, firsthand description of the organization, methods, and aims of the Mohole Project, and also for background material in basic principles of geology, oceanography, engineering, and geophysics. In a style varying from humorous to technical the author traces the history of the "Mohole idea" from its beginnings in theory and speculation, and explains the scientific and technological developments which suggested the project and made it feasible.

Bascom, Willard. *Waves and Beaches: The Dynamics of the Ocean Surface.* (Science Study Series.) Doubleday, 1964. xii+267 pp. $1.45 (paper). 64–11735. (SH)

The story of "the struggle for supremacy between sea and land" is told authoritatively and enthusiastically. Drawing frequently from his own experiences, professionally and privately, the author explains the theory of wave motion, the types of waves and swells, their origins and effects, and the problems they create for engineers and coastal inhabitants. Various types of beaches and coastlines are also analyzed, and many of the methods and instruments of oceanographic research are introduced where relevant.

Bergaust, Erik, and William O. Foss. *Oceanographers in Action.* Putnam's 1968. 96 pp. illus. LB $3.29. 68–15038. (JH–SH)

This overview of the diversity of oceanographic studies is followed by guidance to the courses potential oceanographers should pursue in high school and a listing of required college courses. Summer employment opportunities for college students and a brief glossary of terms ends this work. All vocational counselors and school and public libraries should own this book.

Boyd, Waldo T. *Your Career in Oceanology.* Messner, 1968. 219 pp. illus. LB $3.64. 68–25099. (JH–SH)

An interesting and complete exposure for those totally unacquainted with the marine sciences. It can be useful to individuals seeking sound career guidance. It will, of course, need updating in four or five years, and possibly sooner. Tables in the rear of the book list colleges, universities, government offices, and companies associated with the field. Indexed.

Carson, Rachel L. *The Sea Around Us* (rev. ed.). Oxford, 1961. 237 pp. illus. $5.75; (Large typed ed., Watts, $6.95); NAL (Signet P2361), paper 60¢. 61–6295. (JH)

This second edition of a popular classic has been brought up to date with an appendix of 16 short discussions which are keyed to the original text and summarize relevant developments and discoveries which have been made in the 10 years following its original publication. It offers a stimulating and rewarding combination of professional knowledge and literary perfection which can be enjoyed by all.

Carter, Samuel III. *Kingdom of the Tides: The Gigantic Forces that Have Shaped Our Past and Can Determine Our Future.* Hawthorn, 1966. 160 pp. illus. $4.95. 66–15251. (SH)
This is the first popular book dealing exclusively with tides. The reason for tidal study, its history, types and causes of tides, the biological effects of tides, and the effect of tides on man, are all examined.

Cotter, Charles H. *The Physical Geography of the Oceans.* American Elsevier, 1966. 317 pp. illus. $7.00. 66–13009. (SH–C)
Clearly written and nonmathematical introduction to oceanography. Deals primarily with the geology of the ocean basins and the ocean circulation, but also includes limited discussions of oceanic weather and climate, marine biology, tides, instrumentation, and the history of oceanography. The bibliography is poor—almost exclusively British books—but the index is useful.

Cowen, Robert C. *Frontiers of the Sea: The Story of Oceanographic Exploration.* Doubleday, 1960. 307 pp. illus. $5.95. 60–8859. (SH)
Traces the history and recent developments of oceanography, including the theory and applications of the science. Descriptions of methods and instruments emphasize the modern science itself. The origin, biology, and dynamics of the seas are covered, and the problems and promise which they offer are clearly presented.

Cox, Donald W. *Explorers of the Deep.* (Illus. by Jack Woodson.) Hammond, 1968. 95 pp. $3.50. 68–27451. (JH–SH)
Sketches of the work and achievements of some early workers in the field as well as some who are presently active: Franklin, Wilkes, Maury, Alexander Agassiz, Holland, Beebe, Rickover, W. R. Anderson, the Picards, Carson, Ewing, Cousteau, Link, Bond, and Carpenter. The final chapter is the vision of what tomorrow's work may be in the world of the ocean as man seeks to utilize its resources. Aside from its educational and career-guidance value, it is an interesting book nonspecialist adults will enjoy for recreational reading.

Cromie, William J. *Exploring the Secrets of the Sea.* Prentice-Hall, 1962. xiii+ 300 pp. illus. $6.95. 62–16312. (JH)
The history, plant and animal life, movements, and future possibilities of the oceans are presented in a series of exciting accounts of the voyages and discoveries of many famous scientists and explorers, from Darwin to IGY and beyond. Although not as detailed as Cowen, this is a more attractive and readable survey for the general reader.

Defant, Albert. *Ebb and Flow: The Tides of Earth, Air and Water.* (Tr. by A. J. Pomerans.) U. of Michigan, 1958. 121 pp. illus. $4.00; paper (Ann Arbor AA5506) $1.95. 58–62520. (SH)
The tides—what they are, how they operate, how they are caused—are explained in this story of the forces that move our whole world—the oceans, the air, and the earth.

Dietrich, Günther. *General Oceanography: An Introduction.* (Tr. by Feodor Ostapoff.) Interscience, 1963. xv+588 pp. illus. $21.00. 63–17346. (C)
Provides an introduction into all branches of the subject and points out to the reader the diversified aspects which are receiving considerable attention in

modern research. Intended for the general reader with appropriate scientific background as well as for the beginning student. Extensive bibliography of professional literature.

Dugan, James (ed.). *Jacques-Ives Cousteau's World Without Sun*. Harper & Row, 1965. 205 pp. illus. $10.00. 65–14690. (JH–SH–C)
Conshelf Two (Continental Shelf Station No. 2) in the Red Sea is the locale of this magnificent pictorial account of man's first undersea colony. A 10-page solidly written preface by Captain Cousteau sets the stage for the magnificent "picture show" that follows. The 102 colored photographs and the 140 in black and white, together with the descriptive legends in a quasi-diary form provide for the reader the exciting experience of first-hand participation in the exploration at the bottom of the sea.

Eibl-Eibesfeldt, Irenaüs. *Land of a Thousand Atolls*. (Tr. by Gwynne Vevers.) World, 1966. 195 pp. illus. 66–29853. (JH–SH)
An effective example of popular science writing constituting an excellent introduction to tropical islands, to tropical marine biology and coral reefs, and to observational, natural history-type science.

Fairbridge, Rhodes W. (ed.). *The Encyclopedia of Oceanography*. Reinhold, 1966. xiii+1021 pp. illus. $25.00. 66–26059. (JH–SH–C)
Offers a one-volume source of information on oceanography that emphasizes the geological aspects of the marine environment. Despite the magnitude of the subject matter, the author writes with clarity. It can be a handy reference for all interested in oceanography.

Freuchen, Peter, and David Loth. *Peter Freuchen's Book of the Seven Seas*. Messner, 1957. 512 pp. illus. $8.95. 57–11723. (SH)
This varied and highly entertaining narrative combines, in the author's words, "both the science and the dreams—the facts and the fancies which make the Seven Seas endlessly fascinating." In addition to his lively descriptions of oceanic origins, dynamics, and biology, Freuchen outlines the history of man's adventures and conquests of the sea, including the development of ships, great naval battles, and famous scientific and exploratory voyages.

Gaber, Norman H. *Your Future in Oceanography*. Richards Rosen, 1967. 143 pp. illus. LB $3.78. 67–10288. (SH–C)
Complete and authoritative, this is a guidebook to careers in oceanography. Various appendices list general information and sources for further guidance.

Interagency Committee On Oceanography (Comp.). *Opportunities in Oceanography*. (ICO Pamphlet No. 8; 3rd ed.) Smithsonian Press, 1967. 32 pp. illus. $1.00 (paper). (JH–SH–C)
This inexpensive pamphlet is designed to answer the myriad of questions on career fields in oceanography. The excellent bibliography should stimulate collateral reading. A necessary guide for school libraries, academic counselors, and interested students.

Kovalik, Vladimir, and Nada Kovalik. *The Ocean World*. Holiday House, 1966. 191 pp. illus. $4.50. 66–3323. (SH)
A good introduction to oceanography which can be used as collateral reading in high school earth science courses. There are nontechnical and interesting

discussions of the physical and biological aspects of the oceans. Includes "suggested reading," a list of colleges and universities offering degrees in marine science, a compilation of some oceanographic institutions and a roster of some private industries which have an interest in oceanography. Indexed.

Marx, Wesley. *The Frail Ocean.* Coward-McCann, 1967. viii+248 pp. illus. $5.95. 67–15286. (SH–C)

A very readable, factual, and startling portrayal of the great disaster man and his enterprise are causing to the oceans and their resources. Comparable in scope and message to Rachel Carson's *Silent Spring.*

Miller, Robert C. *The Sea.* Random, 1966. 316 pp. illus. $15.00. 66–21844. (SH–C)

Written in a lively nontechnical style, this book on oceanography, marine biology, and man's use of the sea is definitely not a text, and will have only limited use as a reference book. However, its many illustrations—well executed drawings and striking color photographs—make a handsome book.

Neumann, Gerhard, and Willard J. Pierson, Jr. *Principles of Physical Oceanography.* Prentice-Hall, 1966. xii+545 pp. illus. $26.00. 66–12784. (C)

Strikes a happy balance between theoretical consideration and a basic popularized approach to the wide range of subject matter of physical oceanography. It is an intensely readable book and many portions should appeal to the casual scientific reader. Throughout, the classic views of physical oceanographic processes are updated by discussions of recent developments. An exhaustive and up-to-date bibliography.

Pell, Claiborne, and Harold L. Goodwin. *Challenge of the Seven Seas.* Morrow, 1966. xi+306 pp. illus. $6.95. 66–24965. (SH–C)

A practical guide to the sea. Chapters are developed within the narrative history of a man whose mission is to encourage the use of the sea. Although exciting to the imagination and generally accurate, the book has some characteristics of a public document. An especially good book for one wishing to see how the Federal Government operates in oceanography.

Shepard, Francis P. *The Earth Beneath the Sea.* (rev. ed.). Johns Hopkins, 1967. xi+242 pp. illus. $6.95. 67–12522. (SH–C)

A traditional description of the ocean bottom and the processes affecting it are presented in this revision of the author's *Submarine Geology.* Useful for collateral reading in elementary courses in geology and oceanography.

Spilhaus, Athelstan. *The Ocean Laboratory.* (Introd. by F. G. Walton Smith.) Creative, 1967. 112 pp. illus. $5.95. 66–24868. (JH)

A good beginner's introduction to the sea: its origin and structure, the terrain of the ocean floor, the zones of the sea and their relation to climate, the life in the sea, modern methods of oceanographic research, and the resources of the sea and man's dependence on them. Complete with excellent drawings and photographs, a glossary of oceanographic terms, and an index.

Stewart, Harris B., Jr. *Deep Challenge.* Van Nostrand, 1966. 202 pp. $5.95. 66–27523. (JH–SH)

Provides a relatively complete answer to the question, "What does an oceanographer do?" The roles of marine geologists and geophysicists, the physical oceanographer, the role of marine biology, and that of the marine chemist are described.

Stommel, Henry. *The Gulf Stream: A Physical and Dynamical Description* (2nd ed.). U. of California, 1965. xiii+248 pp. illus. $6.00. 64–23710. (SH–C) Brings together the scientific information on the forces that combine to produce the Gulf Stream. This second edition of a successful earlier work includes information collected in the 1960's concerning meanders and undercurrents in relation to the other forces bearing on a description of the dynamic stream. Useful for reference by serious students.

Sverdrup, Harald U., Martin W. Johnson, and Richard H. Fleming. *The Oceans: Their Physics, Chemistry, and General Biology.* Prentice-Hall, 1942. x+1087 pp. illus. $22.00 (text ed. $16.50). 43–51021. (C) Despite its age, this monumental work has not been superseded as the most comprehensive and authoritative basic textbook. Since oceanography integrates those aspects of geology, physics, chemistry, biology, and mathematics involved in the study of the sea, this book is for the student who has mastered fundamentals in the various disciplines. Although primarily intended for advanced college students it merits a place among science reference works in all major secondary school, college, and public libraries. Extensive bibliographies.

Terrell, Mark. *The Principles of Diving.* Barnes & Noble, 1967 [c. 1965]. 240 pp. illus. $8.50. 67–13135. (SH–C) A complete and authoritative manual on diving. Begins with a discussion of the seas as an environment and proceeds to examine diving techniques and methods, applications and uses of diving, and future trends and possibilities in diving.

Tricker, R. A. R. *Bores, Breakers, Waves, and Wakes: An Introduction to the Study of Waves on Water.* American Elsevier, 1964. xiii+250 pp. illus. $6.50. 65–13164. (SH–C) A lucid explanation of wave motion in water requiring only a knowledge of simple algebra. Illustrated by numerous diagrams; a number of photographs of phenomena are included that have been cleverly taken by methods described in detail. Recommended as an adjunct to the study of wave motion. Many experiments are described that can readily be repeated in the classroom to illustrate wave phenomena.

Troebst, Cord-Christian. *Conquest of the Sea.* (Tr. from the German by Brian C. Price and Elsbeth Price.) Harper & Row, 1962. vii+269 pp. illus. $6.95. 61–10219. (SH) "There is no doubt," says Troebst, "that the future of all life on our planet will one day be decided in the very ocean where it once began." His book is an urgent appeal for more interest and support in our marine research and defense, as well as a vivid account of the undersea arms race, deep-sea explorations, submarine solutions to space, population, food and fuel problems, and the possible threats to our successful conquest of the sea.

Walford, Lionel A. *Living Resources of the Sea: Opportunities for Research and Expansion.* Ronald, 1958. xv+321 pp. illus. $7.50. 58–5862. (C) An appraisal of what we know and what we need to know about the seas and their resources in order to assist in feeding an ever-increasing world popula-tion, and to provide other economic resources for industry and commerce. The initial discussion, aided by 23 maps, gives a broad orientation in the geographical distribution, ecology, natural history, and abundance of marine resources. The concluding portions are concerned with past, present and future conservation problems, utilization, and potential.

551.48 LIMNOLOGY

Arnov, Boris, Jr. *Secrets of Inland Waters.* Little, Brown, 1965. 143 pp. illus. $4.50. 65–11780. (JH–SH)
Identifies and describes physical, chemical and biological characteristics and important processes such as the freezing of water, photosynthesis, the hy-drologic cycle and the fresh-water habitat. Most of the essentials of the world's moving and standing fresh-water bodies are detailed. A stimulating, well-written contribution to the literature on the water resource.

Bardach, John. *Downstream: A Natural History of the River from its Source to the Sea.* Grosset & Dunlap, 1966. ix+278 pp. illus. $2.95. (paper) (JH–SH–C)
Readers of diverse interests and training will recognize with pleasure familiar scenes and gain an enrichment of their understanding of the ecology of streams. Discusses historic and new problems concerning inland waters: flood control, pollution, conservation, and uses and abuses of the waters. Concludes with some advice and a prediction of problems that will yet arise. Though most of the book is concerned with North American waterways, many aspects of foreign rivers are mentioned. Bibliography for each chapter.

Ruttner, Franz. *Fundamentals of Limnology* (3rd ed.). (Tr. by D. G. Frey and F. E. J. Fry.) U. of Toronto, 1963. xvi+295 pp. illus. $6.95. A53–6149. (C)
A well-organized text covering the major physical, chemical, and biological aspects of lakes, bogs, and streams. In "Water as an Environment" Ruttner explains the properties of lake water and its substances in solution, water movements, and the role of insolation. The second part, "Biotic Communities" is the larger of the two and includes discussions of plankton and other micro-scopic fresh-water organisms. A glossary and bibliography are appended.

Wisler, Chester O., and Ernest F. Brater. *Hydrology.* Wiley, 1959. xiv+408 pp. illus. $10.95. 59–14981. (C)
A comprehensive introduction to both the descriptive and analytic concepts used in the study of surface and ground water. The modern methods of measuring and predicting precipitation, infiltration, and evaporation are presented, including the unit hydrograph method—a new means of determining the magnitude of floods of a given probability to be expected on individual streams.

551.5 METEOROLOGY

Atkinson, Bruce W. *The Weather Business.* (Doubleday Science Series). Double-day, 1969. 192 pp. illus. $5.95; $2.45 (paper).

Discussion ranges from early statistical forecasting to computer solution of the primitive equations and from storm "modification" by a fusillade of arrows to silver iodide seeding of hurricanes. This should be required reading for introductory meteorology courses.

Barrett, E. C. *Viewing Weather from Space*. Praeger, 1967. xii+140 pp. illus. $6.00. 67–23367. (SH–C)
Presents an interesting and informative picture of the utilization of weather satellites for increasing knowledge of the structure and behavior of the atmosphere. Deals with an historical perspective, weather satellite systems, and analysis and utilization of these satellite data. Also gives some insight into possible future developments.

Battan, Louis J. *Harvesting the Clouds: Advances in Weather Modification*. Doubleday, 1969. xii+148 pp. illus. $4.95; paper (Anchor) $1.95. (JH–SH)
This is a valid plea for more research in weather modification. Considers the question of why man-made changes in the weather may be desirable, the history of weather control, the unintentional effects of changing the ecology of the earth, and the prospects for the future.

Battan, Louis J. *Radar Observes the Weather*. (Science Study Series). Doubleday, 1962. 158 pp. illus. paper (Anchor) $1.25. 62–9505. (SH)
A specialist explains the operation of radar and its applications to meteorological observation and prediction. The many contributions which it has made toward better understanding of the nature and formation of precipitation and storms are discussed along with its practical use in aviation, forecasting, and weather statistics. The methods of interpretation and inference used in evaluating radar information are also presented, including illustrations of typical screen patterns.

Berry, Frederick A., and Sidney R. Frank. *Your Future in Meteorology*. Rosen, 1962. 155 pp. $2.95; LB $2.79. 62–11570. (JH–SH)
Describes some of the various positions and types of work in this general field, the major organizations which employ meteorologists, the advantages and drawbacks of the vocation, and offers suggestions for education. Brief sketches of duties and qualifications for specific positions assist in self-evaluation, and this little book contains a wealth of other information for the interested or aspiring weatherman.

Cantzlaar, George L. *Your Guide to the Weather*. Barnes & Noble, 1964. xiv+242 pp. illus. $4.50, paper $1.50. 64–14817. (SH)
An orderly and factual introduction to the science of meteorology, which requires no previous knowledge and covers in considerable detail many aspects of atmospheric physics and dynamics, global and regional weather patterns, and current methods of observing, measuring, and forecasting. The history and development of man's efforts to understand and control the weather are also outlined. A useful guide to the understanding of public weather reports, maps, and forecasts.

Day, John A. *The Science of Weather*. Addison-Wesley, 1966. x+214 pp. illus. $6.95. 66–22573. (SH–C)
A course in elementary meteorology with only occasional use of simple al-

gebra. Nevertheless, numerical data are given frequently for temperature, pressure, height, electric charge in thunderstorms, dimensions in microns of cloud droplets, radar pulse length, etc. Concise summaries precede each chapter. Pictures and diagrams are numerous. Basic physics is cleverly worked into the book. Further, the text is related to the real world of science, international affairs, and daily life. Easily grasped by a good high school science student and nonscience students in college.

Donn, William L. *Meteorology* (3rd ed.). McGraw-Hill, 1965. xv+484 pp. illus. $8.95. 65–17492. (SH–C)
The emphasis on maritime aspects of the subject is the one aspect which distinguishes this book from other conventional elementary treatments. The Preface states "more than half of the original text has been rewritten" yet both the organization and the actual wording are so nearly the same that libraries having the second edition will gain little by obtaining the third.

Hubert, Lester F., and Paul E. Lehr. *Weather Satellites*. Blaisdell, 1967. vii+ 120 pp. illus. $2.95 (paper). 67–10547. (SH–C)
The impact of the space age on the science of meteorology and on its increasingly important and supplemental role in meteorological measurements made from the ground is described by two competent meteorologists. The diagrams and photographs are particularly good.

Knight, David C. *The Science Book of Meteorology*. Watts, 1965. 215 pp. illus. $4.95. 64–17779. (JH–SH)
One of the better elementary descriptive introductions to the field of meteorology. Most of the basic elements of the subject such as the structure of the atmosphere, clouds and precipitation, weather systems and forecasting, as well as some recent information about satellite meteorology and instrumentation are included. Contains many helpful diagrams and pictures.

Laird, Charles, and Ruth Laird. *Weathercasting*. Prentice-Hall, 1955. vii+163 pp. illus. $4.95. 55–8468. (JH)
Presents a complete "do-it-yourself" program in weather forecasting and study. The symbols and techniques used in weather reporting and mapping are fully explained and numerous charts and tables aid the amateur in his calculations and predictions. Detailed instructions for building and using home-made instruments permit the interested beginner to carry on advanced and accurate studies economically.

Landsberg, Helmut E. *Weather and Health: An Introduction to Biometeorology*. Doubleday, 1969. x+148 pp. illus. $4.95. 68–27126. (SH–C)
The science dealing with the relations between atmospheric and living processes is called biometeorology. This book gives warning to man to abstain from further destruction of his environment, and it emphasizes the necessity of living in harmony with environment for optimal health. This is more than a description of a new science, it is an important commentary on an issue that deserves more thoughtful discussion.

Middleton, W. E. Knowles. *The History of the Barometer*. Johns Hopkins, 1964. xx+489 pp. illus. $12.50. 64–10942. (SH)
Beginning with Toricelli's Experiment, the author traces meticulously the

development of the present-day widely used meteorological instrument. The work is a product of the author's extensive travels and examination of a great amount of published and manuscript material. Other instruments such as aneroid barometers, barographs, wheel barometers, and many others are mentioned in their historical context. Valuable for all libraries.

Middleton, W. E. Knowles. *Invention of the Meteorological Instruments*. Johns Hopkins, 1969. xiv+362 pp. illus. $12.00. 68–31640. (C)
Barometer, thermometer, hygrometer, rain gauge, atmometer, windvane, anemometer, sunshine recorder, meterograph, measuring upper winds, and the height and motion of clouds, upper air soundings without telemetry, telemetorography, and the radiosonde are treated in this book. Its reference value to students of meteorology, astronomy, and other disciplines who are interested in the development and uses of scientific instruments is obvious. Index lists the year of birth and death of all historical individuals named in the book.

Pothecary, I. J. W. *The Atmosphere in Action*. St. Martin's, 1965. 111 pp. illus. $3.95. 65–27028. (JH–SH)
A quick and accurate overview of meteorology, supplying good nonmathematical explanations of basic physical concepts. The book obviously is intended for British readers, but this fact does not impair its value to readers elsewhere. Recommended for secondary school teachers, as supplementary reading for high school and lower-division college students, and for interested laymen.

Reiter, Elmar R. *Jet Streams: How Do They Affect Our Weather?* (Science Study Series.) Doubleday, 1967. x+189 pp. illus. $5.95, paper $1.25. (67–12894. (SH–C)
This work will interest budding meteorologists. The author sees almost all atmospheric events as the consequence of jet streams. Trigonometry, vector arithmetic, physics and a great deal of meteorology is covered.

van Straten, Florence W. *Weather or Not*. Dodd, Mead, 1966. xiii+237 pp. illus. $5.00. 66–20447. (SH–C)
A highly readable and enjoyable account of modern meteorology. Developments of the science since World War II are presented in a personal, quasi-autobiographical fashion starting with the author's activities as a Navy Wave weather officer evaluating the effects and use of weather in naval engagements and ending with her recent research activities as a civilian scientist in weather modification. More suited for general informational reading than as a reference source.

551.55 ATMOSPHERIC FORMATIONS AND DISTURBANCES

Battan, Louis J. *The Nature of Violent Storms*. (Science Study Series.) Doubleday (Anchor S19), 1961. 158 pp. illus. $1.25 (paper). 61–7639. (SH)
The causes and effects of thunderstorms, tornadoes, hurricanes, and cyclonic air masses are explained and contrasted, including a discussion of the atmospheric and thermodynamic forces which direct and maintain all such disturbances. Opposing theories and current hypotheses are presented to

emphasize the developing nature of the science. The present limitations in our knowledge and research methods are stressed throughout.

Blumenstock, David I. *The Ocean of Air*. Rutgers, 1959. xii+457 pp. illus. $7.50. 59–7509. (C)

A comprehensive study of atmospheric phenomena for the lay and learned alike, with emphasis on the weather's effect on man and vice versa. The author first describes the patterns of air flow, precipitation, and temperature found in many parts of the world, explaining not only their natural causes and principles, but also their effects on climate, vegetation, and civilization. The science of observation, prediction, and control is then taken up, showing its development as well as its present knowledge and techniques. The final section attacks specific problems such as nuclear air pollution. Considers also physiological and historical importance of weather and climate.

Edinger, James G. *Watching for the Wind: The Seen and Unseen Influences on Local Weather*. (Science Study Series.) Doubleday, 1967. 148 pp. illus. $4.50, paper $1.25. 66–24874. (SH)

Describes prevailing large-scale air currents and shows their effects on the weather, and how mountains, and other land forms, bodies of water and local breezes affect major air currents and help to determine temperatures and other features of local weather.

Fisher, Robert Moore. *How about the Weather?* (rev. ed.). Harper & Row, 1958. xviii+172 pp. illus. $4.95. 58–8869. (SH)

An informal and entertaining guide to understanding and predicting the weather from maps, household instruments, and amateur observations. The author examines the history and importance of meteorology and covers many atmospheric phenomena such as clouds, fronts, air masses, winds, and fog, explaining their use in forecasting as well as their origin and function.

Ohring, George. *Weather on the Planets: What We Know about their Atmospheres*. (Science Study Series.) Doubleday (Anchor), 1966. x+144 pp. illus. $1.25 (paper). 66–17455 (JH–SH)

Describes electromagnetic radiation, how it is transmitted through space and planetary atmospheres, and how it is measured. With this background the author goes on to describe the planets. He devotes separate chapters to the atmospheres of Mars and Venus. Another chapter contains a description of what is known about the atmospheres of the other planets, and the final chapter outlines problems for the future.

Schonland, Basil. *The Flight of Thunderbolts* (2nd ed.). Oxford, 1964. 182 pp. illus. $5.50. 65–1110. (SH–C)

The knowledge of lightning gained by man since the dawn of history is summarized. Some of the topics are early observations and superstitions, Franklin's experiments, the use of lightning rods and protection against lightning, the forms of lightning and their effects, luminous and electrical processes, the indirect and distant effects of thunderstorms, etc. Technically accurate but nonmathematical. Many illustrative drawings; indexed.

Sutton, O. G. *The Challenge of the Atmosphere*. Harper & Row, 1961. xiii+227 pp. illus. $5.95; text ed. $4.50. 61–6442.

A highly condensed and factual presentation of the theoretical science of meteorology, emphasizing its analytic and deductive nature, and considering applications such as forecasting only in a brief fashion. The advanced yet largely nonmathematical text explains the hypothetical and accepted principles of atmospheric movement, storms, moisture balance and climatic conditions. The limitations of knowledge are stressed throughout, and the author reveals both weather and climate as complex and worldwide solar-driven processes.

Viemeister, Peter E. *The Lightning Book: The Nature of Lightning and How to Protect Yourself from It.* Doubleday, 1961. 316 pp. illus. $4.50. 61–8908. (SH)
Following an introduction to lightning in history and mythology, the discussion deals with man's early and modern attempts to study it, the development and electrification of thunderstorms, the nature of bolts and their targets, and the safety precautions which can minimize personal and property hazards on land, sea, and air. Many well-chosen examples of lightning destruction are included. Bibliography by chapters.

551.56 CLIMATOLOGY

Blair, Thomas A., and Robert C. Fite. *Weather Elements* (5th ed.). Prentice-Hall, 1965. xiv+414 pp. illus. $11.35 (text ed. $8.50). 57–5047. (C)
The fundamentals of weather and climate are presented clearly and non-technically, with an appendix of selected references, conversion tables, and practice exercises. The techniques and instruments of modern meteorological observation, analysis, and research are explained; also the principles of atmospheric circulation, precipitation, temperature, and pressure. Describes the applications of meteorology to forecasting, aviation, and climatology.

Critchfield, Howard J. *General Climatology* (2nd ed.). Prentice-Hall, 1966. xiii+465 pp. illus. $9.95 (text ed.). 60–5131. (C)
Following a unit on the fundamentals of weather processes and forces, the climates of the world are classified and described, and the effects of climate on soil, vegetation, water cycles, and human life are taken up. Some of the research methods and theory of climatology are also covered, and many of the problems caused by adverse weather and climate conditions are discussed. Bibliography and tables included.

Silverberg, Robert. *The Challenge of Climate: Man and His Environment.* Meredith, 1969. 326 pp. $5.95. 69–19052. (SH–C)
Changes in climate throughout Earth's history, especially the Pleistocene, recent and future eras have been discussed competently and comprehensively, with emphasis on the effects on evolution of man and his civilization. The book offers an admirable summary of current knowledge and hypotheses on Earth's history, atmospheric circulation, climatic changes, human evolution, and of man's adaptation to climate.

Sutcliffe, R. C. *Weather and Climate.* Norton, 1966. 206 pp. illus. $5.95. 66–11653. (SH–C)
An excellent and interesting introduction to the atmosphere. Topics range

from an introduction to the structure of the atmosphere through a presentation on clouds and weather systems to a discussion of forecasting, weather control, and climate.

Trewartha, Glenn T. *The Earth's Problem Climates.* U. of Wisconsin, 1961. 334 pp. illus. $7.50. 61–5187. (C)
In almost all latitudes and climate zones there are areas which do not conform to the surrounding pattern. Describes many of these "problem" climates and points out the local terrain, geographic location, or oceanic and atmospheric currents which could account for their specific peculiarities. Many less unusual regions are also discussed, and the causes of climatic differences in general are examined. Considerable background in meteorology and climatology is required.

551.57 HYDROMETEOROLOGY

Battan, Louis J. *Cloud Physics and Cloud Seeding.* (Science Study Series.) Doubleday (Anchor), 1962. xii+144 pp. illus. $1.25 (paper). 62–14693. (SH)
Explains the things we know or need to know about the natural processes of atmospheric condensation and precipitation in order to artificially squeeze more water from the clouds to meet our ever-increasing needs. The physics of condensation nuclei, cloud formation, and the production of different forms of precipitation are explained, as well as the actual attempts made so far to stimulate these processes chemically.

Bell, Corydon. *The Wonder of Snow.* Hill and Wang, 1957. xvi+269 pp. illus. $5.00. 57–5837. (JH)
A broad and entertaining combination of adventure and science which describes many winter hardships, explorations, and recreations as well as the principles of snow crystallization, natural and artificial precipitation, glaciology, and snow hydrology. The causes of seasonal and climate differences are also covered, including some of the popular theories on the relationship between climate and cultural advancement. A brief glossary of meteorological and physical terms is appended.

Mason, Basil John. *Clouds, Rain & Rainmaking.* Cambridge, 1962. 145 pp. illus. $5.50, paper $1.95. 62–52729. (C)
Traces the history of artificial rain-making attempts and discusses the phenomenon of thundercloud electrification, including theoretical and quantitative aspects as well as fundamental principles. Advanced experiments and demonstrations in each chapter encourage further observation and study. Although similar to *Cloud Physics and Cloud Seeding* by Battan, this book requires considerably more background in physics and mathematics.

Middleton, W. E. Knowles. *A History of the Theories of Rain and Other Forms of Precipitation.* (Watts History of Science Library.) Watts, 1966. viii+ 223 pp. $5.95; LB $4.46. 66–15982. (SH–C)
The formation and dissolution of clouds, the generation of rain and hail, and finally the formation of dew and hoar frost, are discussed. Covers the his-

torical spectrum from the time of ancient Greece up to about 1914. Therefore, modern theories of precipitation are not dealt with. Nevertheless, anyone interested in man's intellectual struggles to understand his atmospheric
environment should read this book.

551.7 HISTORICAL GEOLOGY

Bailey, Edward B. *Charles Lyell*. Doubleday, 1962. x+214 pp. illus. $3.50. 62–
18878. (C)
This short biography by the director of the British Geological Survey covers
not only Lyell's important contributions to modern geology, but also his
great influence on the sciences of archeology and anthropology. Some
geological background is required to read it.

Bailey, Edward B. *James Hutton—the Founder of Modern Geology*. American
Elsevier, 1967. xii+161 pp. $9.00. 66–28574. (C)
James Hutton was a keen observer of geologic features and processes whose
major contribution was the concept that the earth's surface is continuously
being worn down by erosion, the detritus carried to the sea to make new rock
formations. This biography is a good contribution to the history of the development of geology.

Bartlett, Richard A. *Great Surveys of the American West*. U. of Oklahoma,
1962. xxiii+408 pp. illus. $7.95. 62–16475. (SH–C)
Between the end of the Civil War and the establishment of the U. S. Geological Survey in 1879 the vast uncharted lands between the Rockies and the
Sierra Nevadas were explored, surveyed, and mapped by four great expeditions. This narrative on the surveys led by Hayden, King, Powell, and
Wheeler describes the hardships, adventures, and discoveries that transformed
vast areas of myth and mystery to a reality of promise and potentiality.

Clark, Thomas H., and Colin W. Stearn. *The Geological Evolution of North
America: A Regional Approach to Historical Geology*. (2nd ed.). Ronald,
1968. vi+434 pp. illus. $9.50. 60–6154. (C)
The basic formulations, principles and theories of historical geology are presented through the study of the major structural units of the North American continent. Examines possible solar and cosmic origins and the fundamental theories of folding and sedimentation, outlining their effect on the
total development of North America. Next, the great eastern and western
American ranges and nearby plains are taken up, covering their structure,
composition, and history, and the geological evolution of the Arctic islands
and the Canadian Shield is discussed. The final unit is devoted to the appearance and development of life, and the appendix summarizes biological classification and natural history.

Clayton, Keith. *The Crust of the Earth: The Story of Geology*. (Nature and
Science Library.) Nat. Hist. Press, 1967. 154 pp. illus. LB $4.95. 66–21172.
(JH)
Introduces the founders of geology, demonstrates the evolution of concepts,
contrasts old beliefs and present-day ideas. Emphasizes the dynamic nature

of geology and presents the science as a dynamic discipline with ample opportunities for further study and revelations.

Cloos, Hans. *Conversation with the Earth.* Knopf, 1953. xi+413+xiii pp. $5.95. 52–12186. (SH)

A noted German geologist describes his travels and discoveries in the U.S.A., Germany, Africa, and the East Indies. The often poetic language communicates the author's love and respect for the whole of nature as well as his own specialty. His descriptions of local stratigraphy and the eternal forces of geologic change are accurate and factual.

Dunbar, Carl Owen. *Historical Geology* (2nd ed.). Wiley, 1960. xi+500 pp. illus. $9.50. 60–5598. (SH–C)

The geologic and biologic development of the earth is covered chronologically, with explanations of the methods of dating and the different kinds of fossil evidence. The geology is almost entirely North American but there is some history of Europe and other countries and the biologic scope is world-wide. Somewhat easier than Clark and Stearn, although both require some background in physical geology.

Dunbar, Carl Owen, and John Rodgers. *Principles of Stratigraphy.* Wiley, 1957. xii+356 pp. illus. $10.00. 57–8883. (C)

The composition, formation, and historical relationships of layered sedimentary and metamorphic rocks are examined in detail, with emphasis on original sedimentation processes and little or no tectonic analysis or description. Frequent reference is made to modern deposits which parallel and help to explain the past. The techniques and principles used by stratigraphers to "read" history in the rocks are included in addition to the routine descriptive and correlational methods. Bibliography.

Fenton, Carroll Lane. *Earth's Adventures: The Story of Geology for Young People.* John Day, 1942. 207 pp. illus. $5.00. 43–934. (JH)

The natural forces and processes which have shaped the earth from its beginning are identified in their present-day manifestations and are simply but accurately described as the history of our planet is unfolded.

Gamow, George. *A Planet Called Earth.* Viking, 1963. x+257 pp. illus. $5.75. 63–8853. (SH)

The earth's history, life, features, and future, are written in the popular style which has distinguished the author. The moon and the planets are also discussed, and the probable origin and evolution of life is traced. Gamow's own views are included. Well developed for the general reader.

Gould, Charles N. *Covered Wagon Geologist.* U. of Oklahoma, 1959. xiii+282 pp. illus. $4.00. 59–7954. (SH)

A famous geologist and explorer of Oklahoma and the Southwest describes his life, from the inspiration which led him to a career in geology to his retirement. A light and adventuresome insight into a rewarding vocation.

Hurley, Patrick M. *How Old is the Earth?* (Science Study Series.) Doubleday (Anchor S5), 1959. 160 pp. illus. $1.25 (paper). 59–11599. (SH)

Following a discussion of natural radioactivity and its use in dating the earth and explaining its structure and change, the author reviews the terrestrial chronology as it now stands and explores the probable origin of the solar system.

Kummel, Bernhard. *History of the Earth: An Introduction to Historical Geology.* Freeman, 1961. xii+610 pp. illus. $9.75. 61–6783. (C)
The physical and biological history of the earth is discussed on a worldwide scope not found in other historical geology texts. Deals with each of the major eras of the earth's past, covering its structural and biological developments. Also included are excellent units on the Pleistocene Epoch and on the "Gondwana" formations and fossils which have led to propositions that the Southern Hemispheric land masses were once one, or at least connected. The appendix contains a remarkable survey of plant and animal classification and structure, plus 45 pages of stratigraphic correlation charts.

Lurie, Edward. *Louis Agassiz: A Life in Science.* U. of Chicago, 1960. xiv+ 449 pp. illus. $7.95, paper (Phoenix P248; abridged ed.) $2.95. 60–11623. (SH–C)
A thorough and honest biography of a famous scientist and teacher. His accomplishments on both sides of the Atlantic as the originator of the Ice Age Theory and famed naturalist are presented along with his obstinate rejection of Darwin's theories and over-confident extensions of his glacial hypotheses. Written objectively and yet with understanding, this is an unusually personal account of a great "life in science." No other Agassiz biography is comparable.

Moore, Raymond Cecil. *Introduction to Historical Geology* (2nd ed.). McGraw-Hill, 1958. ix+656 pp. illus. $9.95. 58–8043. (SH–C)
A systematic stratigraphical study of the major geological divisions in the earth's history follows an explanation of the materials and methods of historical geology. The evolution of plant and animal forms is traced concurrently and fully illustrated. Review questions, glossary, and appendices of fossil biology and geological symbols are included.

Moore, Ruth. *The Earth We Live On: The Story of Geological Discovery.* Knopf, 1956. xiv+416+x pp. illus. $6.95. 56–8924. (SH)
Traces man's attempts to explain the origin and changes of the earth from ancient mythology to geophysics. The theories and accomplishments of Hutton, Agassiz, Lyell, Powell, and many others are presented along with pertinent biographical material.

Powell, John Wesley. *The Exploration of the Colorado River.* U. of Chicago, 1957. xxi+138 pp. illus. $3.95; Doubleday (Anchor N11), paper $1.25 57–6988. (JH)
The classic adventures of the first expedition to travel the entire length of the Colorado River and the Grand Canyon, written by the leader of the party and taken from his journal. First published in 1875 and long out of print, this edition is an exciting glimpse of the vast wilderness of a century ago, and an authoritative description of the regional geology.

Shimer, John A. *This Sculptured Earth: The Landscape of America.* Columbia U., 1959. xii+255 pp. illus. $10.00. 59–10628. (JH)
An exciting and educational explanation of the geologic basis of "scenery."

Taking many of the famous landmarks and natural wonders to be found while vacationing in the United States, the author describes how they were formed and have evolved. Accompanied by an illustrated glossary and many handsome photographs, this entertaining book can stimulate deeper appreciation of scenic beauty as well as develop interest in the earth sciences.

Spencer, Edgar Winston. *Basic Concepts of Historical Geology.* Crowell, 1962. xvi+504 pp. $9.95. 62–7081. (C)

A broad introduction to evolution, geologic history, and the methods of stratigraphic analysis and geochronology. Contains more material up to the Paleozoic Era than some texts, with sections on cosmic, solar, oceanic, and atmospheric origins as well as discussions of the world's Precambrian shields, deposits, and early forms of life. Each subsequent historical period is then taken up, covering its structural changes, distribution of materials, and biological advancements. Although written to be used as a one-year combined course with the author's *Physical Geology,* can be used separately and is useful for its paleogeography and appended study maps.

552 PETROLOGY (see also 549 MINERALOGY)

Fenton, Carroll Lane, and Mildred A. Fenton. *The Rock Book.* Doubleday, 1940. xiv+357 pp. illus. $9.95. 40–30728. (JH)

An introduction to the properties, occurrence, and importance of common rocks and minerals. Identifying characteristics, geologic formation, and regional distribution are given, and the historic significance and present uses of various ores and minerals are discussed. Contains suggestions for collecting, preparing, and displaying specimens, and includes references.

Kraus, Edward H., and Chester B. Slawson. *Gems and Gem Materials* (5th ed.). McGraw-Hill, 1947. ix+332 pp. illus. $8.95. 48–379. (SH)

Precious and semiprecious gems and crystals are treated thoroughly, explaining how they are mined, cut, classified, and artificially produced as well as the properties which identify them and contribute to their characteristic beauty. The principles of crystal structure and its importance in gem mineralogy are also presented and fully illustrated. Tables and photographs facilitate identification.

Loomis, Frederick Brewster. *Field Book of Common Rocks and Minerals* (rev. ed.). Putnam's, 1948. xviii+352 pp. illus. $3.95. 50–6033. (JH–SH)

A useful guide to observation, collection, and study which requires little background or equipment. The minerals are grouped according to their chemical composition, appearance, specific gravity, and many other distinguishing characteristics. Uses, natural combinations, and important national deposits are discussed. The rocks are then taken up and grouped by origin, giving their composition, visible features, and natural occurrence.

Pearl, Richard M. *American Gem Trials.* McGraw-Hill, 1964. x+173 pp. illus. $5.50. 63–22431. (JH)

This compact guide to gem identification and principal localities in each of the 50 States also covers many historical aspects of the subject. The author explains first the appearance, composition, distribution, and geologic occurrence of many common species and subspecies. Some chapters deal with a single mineral, elaborating on the earlier description and briefly examining the location, discovery, and history of its major deposits.

Pough, Frederick H. *A Field Guide to Rocks and Minerals* (3rd ed.). Houghton Mifflin, 1953. xv+349 pp. illus. $4.95. 52–9593. (SH)
More advanced than Loomis, with much greater emphasis on minerals, this book can also be used by beginners while easily accommodating the needs of the serious collector. Explains many specific procedures for distinguishing between very similar materials and discusses some related topics such as crystal structure, uranium ores, and collecting techniques. Illustrated with many line drawings and photographs, often in color.

Riley, Charles M. *Our Mineral Resources.* Wiley, 1959. x+338 pp. illus. $7.95. 59–11807. (C)
Important metallic and non-metallic minerals are fully examined in this nontechnical study explaining their formation and concentration, principal world deposits, economic importance, and means of mining and refining. A lengthy discussion of the known processes of ore deposition introduces the metals unit, and the section on non-metallic resources covers ground water, gems, petroleum, building materials, etc. Some background in chemistry and geology is assumed.

Sinkankas, John. *Gemstones and Minerals: How and Where to Find Them.* Van Nostrand, 1961. xvi+387 pp. illus. $8.95. 61–4196. (SH)
This advanced and practical guidebook to rock and mineral study is written for the serious collector. Tells how to plan a field expedition, the use of various tools and methods, and the preparation, display, and storage of collections. Explains how and where many valuable deposits form, how to recognize them for surface features, and how to extract specimens. Extensive references are suggested and pictures and diagrams illustrate everything from gold-panning to dynamiting.

Sinkankas, John. *Gemstones of North America.* Van Nostrand, 1959. xv+675 pp. illus. $15.00. 59–13853. (SH)
Following a brief introduction to the properties of gems and crystals the text takes up each of the members of the various families and classes of gems, describing physical and chemical properties, natural occurrence, and history and special features in the principal localities where it can be found. Some organic gemstones such as pearls, amber, and anthracite coal are mentioned. Appendices include bibliography, glossary, and aids to collecting and identification.

Spock, L. E. *Guide to the Study of Rocks: A Systematic Guide to Their Identification, Physical Characteristics, and Origins* (2nd ed.). Harper & Row, 1962. xvi+298 pp. illus. $8.75; $6.95 (text ed.). 61–7429. (C)
The origin, properties, and composition of igneous, sedimentary, and metamorphic rocks are studied. Brief introductions to mineral identification and the physical characteristics of rocks are included, and each chapter contains experiments, exercises, and some references. Tips on handling specimens, field study, writing reports, and making elementary computations are appended.

554 GEOMORPHOLOGY

Billings, Marland P. *Structural Geology*. Prentice-Hall, 1954. xiv+514 pp. illus. $10.95 (text ed.). 54–8507. (C)

A comprehensive and detailed text in the principles of crustal folding, faulting, intrusion, and extrusion. A series of 13 lengthy exercises and problems is appended, including a folder of practice maps. The terminology is not difficult, but the diagrams and illustrations quickly present advanced descriptive and analytic methods and symbols. Readers need some broad general scientific background.

Fraser, Ronald. *The Habitable Earth*. Basic, 1964. 155 pp. illus. $4.50. 65–17083. (SH–C)

An expert, meaty, but overly concise summary of recent discoveries in certain geographical areas, such as the earth's magnetism, the earth's interior, mountain building and continental growth, continental drift, and atmospheric and oceanic circulation. Probably this is the best of several fairly successful popular works stressing solid earth geophysics. Will appeal mainly to the layman or student with some knowledge of geology.

Harland, W. B. *The Earth: Rocks, Minerals, and Fossils*. Watts, 1960. 255 pp. illus. $4.95. 60–10810. (JH)

The history, composition, and structure of the earth are presented in a clear and well-organized text. Includes suggestions for home study and explanations of the development and applications of the science of geology. A useful combination glossary and index is appended.

Jacobs, John A., et al. *Physics and Geology*. McGraw-Hill, 1959. xii+424 pp. illus. $12.50. 59–8572. (C)

Deals with the descriptive and theoretical applications of physics to the study of the earth's structure, composition, history, and natural forces, including also some geochemistry and astronomy. Can be used to teach physics to geology majors or vice versa. Written by three geophysics professors, with little condescension in language or mathematics.

Wyckoff, Jerome. *Rock, Time, and Landforms*. Harper & Row, 1966. 372 pp. illus. $8.95. 66–10662. (SH–C)

Natural history enthusiasts, travelers, students, and almost everyone else will enjoy this magnificently-illustrated account of the processes—erosion, volcanism, earthquakes, glaciers, avalances, floods—that have produced and are constantly changing the configurations of Planet Earth. Emphasis is strongly on North American features. A glossary of technical terms, a selected bibliography, and an adequate index.

560 PALEONTOLOGY

Ager, Derek V. *Principles of Paleoecology*. McGraw-Hill, 1963. 383 pp. illus. $12.50. 62–21239. (C)

Fossils are the remains of animals and plants that once lived and breathed, fed, reproduced, and died as they waged a continuous battle with their environment as do living animals today. This book is an account of the ecology of various fossil groups and assemblages distilled from available evidence.

Andrews, Henry N., Jr. *Ancient Plants and The World They Lived In.* Cornell U., 1947. 279 pp. illus. $7.50. 47–5386. (SH)
Although the subject is paleobotany, since many people interested in geology think of fossil plants as "rock specimens", basic elements of botany are included so that the relationships between living and fossil plants may be discerned. An excellent introduction.

Andrews, Henry N., Jr. *Studies in Paleobotany.* Wiley, 1961. xii+487 pp. illus. $9.95. 61–6768. (C)
The author has prepared a story of fossil plants that is suitable for advanced botany students as well as for geology students who have little botanical background. Some information on living plants has been included so as to be helpful to the non-botanist. A chapter on palynology by Charles J. Felix is included, also a statement of techniques for studying fossil plants.

Baity, Elizabeth Chesley. *America Before Man* (rev. ed.). (Illus. by C. B. Falls). Viking, 1964. 224 pp. $4.50. (JH)
Based primarily on a study of background literature, Mrs. Baity has written an interesting, chronologically arranged, and popular paleontological history of North America which contains interesting original diagrams and drawings as well as photographs and reproductions of paintings from other works.

Beerbower, James R. *Search for the Past: An Introduction to Paleontology.* Prentice-Hall, 1960. xiii+562 pp. illus. $13.50. 60–5040. (C)
Paleontology is presented in this basic textbook as a dynamic, many-sided discipline involving morphology, genetics, ecology, and natural history. The reader is taught to follow the fascinating and revealing "detective work" of paleontological research.

Cohen, Daniel. *The Age of Giant Mammals.* Dodd, Mead, 1969. 160 pp. illus. $4.00. 73–81627. (SH–C)
The combination of interesting material and good presentation make this a worthwhile book for student, naturalist and intelligent layman. Dwells on the history of discoveries and interpretations of mammal bones and the legends and myths that have evolved. Theories on dinosaurs and the history and evolution of mammals, the geological evolution, and finally present day mammals are discussed.

Colbert, Edwin H. *The Age of Reptiles.* Norton, 1965. 228 pp. illus. $8.50, $2.25 (paper). 65–13325. (JH–SH–C)

Colbert, Edwin H. *Dinosaurs: Their Discovery and Their World.* Dutton, 1961. 300 pp. illus. $7.95. 61–9765. (JH–SH–C)

Colbert, Edwin H. *Men and Dinosaurs: The Search in Field and Laboratory.* Dutton, 1968. xviii+283 pp. illus. $8.95. 68–12457. (JH–SH–C)
Colbert, Curator of Paleontology at the American Museum of Natural History and a student of dinosaurs, is highly qualified to appraise the history of and research on dinosaurs. All three volumes are a unique blend of history and science, from which laymen and specialists will benefit. *Dinosaurs: Their Discovery and Their World* discusses the most important characters and their significance as found in each group of dinosaurs, together with basic principles and techniques of paleontology. The story of tetrapods at the end

of the Paleozoic and during all of the Mezozoic, when primitive reptiles developed from primitive types to the climax of reptilian evolution, is presented in *The Age of Reptiles*. The most recent book, *Men and Dinosaurs,* is the story of the paleontologists in the field and in the laboratory who have discovered and studied dinosaurs, and have developed the science of paleontology. Each volume has numerous black-and-white photographs, line drawings, extensive bibliographies, an index, and appropriate appendices.

Colbert, Edwin H. *The Dinosaur Book: The Ruling Reptiles and Their Relatives* (2nd ed.). McGraw-Hill, 1951. 156 pp. illus. $5.95. 51–11524. (JH)
A popular guide book on fossil reptiles and amphibians that gives special attention to the dinosaurs. Written to supplement displays in the American Museum of Natural History, but not aimed particularly at the museum visitor.

Ericson, David B., and Goesta Woolin. *The Deep and the Past*. Knopf, 1964. xiv+292+ix pp. $6.95. 64–17698. (SH–C)
A fascinating story of how seagoing geologists, who study long cores of sediments drawn up from the ocean bottom, have recently worked out the geologic time scale of the Pleistocene. The authors assign the various types of prehistoric men to their proper places in Ice Age history. There are also brief but suggestive chapters on the causes of the Ice Age and on future developments in sea-bottom research.

Fenton, Carroll Lane. *Tales Told by Fossils*. Doubleday, 1966. 182 pp. illus. $4.95. 66–11731. (JH–SH)
Highlights of the story of life on the earth from the first simple forms to the dawn of civilization. After depicting the thrill of fossil collecting and describing the various forms in which fossils occur and what these forms reveal, the author establishes a simple but understandable time scale of earth history. The step-by-step development of life as revealed by fossil remains is then carried from the first faint traces in the Precambrian to the almost modern forms. Concludes with a listing of museums where fossils can be viewed and books where additional information can be obtained. Good glossary and index with pronunciations.

Fenton, Carroll Lane, and Mildred Adams Fenton. *The Fossil Book: A Record of Prehistoric Life*. Doubleday, 1959. xiii+482 pp. illus. $15.00. 58–13278. (JH)
Since the examples and literature in the field are voluminous, the authors have used illustrations of North American species whenever possible in this survey of the evidence concerning the earliest forms of plants and animals and progressing up to the discussion of recently-extinct species. Many useful appendices enhance the value of this reference and general reading book.

Huxley, Thomas Henry. *On a Piece of Chalk*. (Edited, introduction and notes by Loren Eiseley.) (Illus. by Rudolf Freund.) Scribner's, 1967. 90 pp. $4.95. 67–14100. (SH–C)
A delightful exposition on the role of minute animals in the abyssal seas in forming the chalk beds of Europe and North Africa. Provides a fundamental lesson in elementary geology and paleontology.

Petersen, Kai. *Prehistoric Life on Earth*. (Ed. by Georg Zappler.) Dutton, 1961. 163 pp. illus. $4.95. 59–11505. (JH)

Adapted from the Danish original, this refreshing investigation of the development of life on earth is valuable both as advanced reading and as a reference. In addition to precise and extensive coverage of animal evolution, it contains information on the plant environment, on classification, heredity and selection, and on the contributions of the men who have greatly influenced our understanding of evolution. Print is small and set in double columns; the illustrations are beautiful.

Ransom, Jay Ellis. *Fossils in America.* Harper & Row, 1964. xii+402 pp. illus. $8.95. 63–16546. (SH)
The collection, preparation, and identification of fossils is a hobby enjoyed by many. This descriptive guide for the beginning student and amateur provides a rudimentary background in paleontology, and gives directions. It also includes a list of collecting localities in the United States, a glossary and a list of libraries and museums with paleontology collections.

Raymond, Percy E. *Prehistoric Life.* Harvard U., 1947. ix+324 pp. illus. $8.00. 48–1065. (SH)
The history of life through the 500-600 million years of which fossils provide the only record is told in a nontechnical fashion that will interest the average reader. A straightforward scholarly presentation that is basic, despite its age.

Romer, Alfred S. *Vertebrate Paleontology* (3rd ed.). U. of Chicago, 1966. viii+468 pp. illus. $12.50. 66–13886. (C)
A masterful and foundational work on vertebrate paleontology that has been enhanced in value by this latest revision. Essential to all college and large public libraries. Contains many new illustrations in addition to complete textual revision.

Scheele, William E. *The First Mammals.* World, 1955. 128 pp. illus. $5.95. 55–8251. (JH–SH)
In word and picture Scheele describes the major steps in the development of the early mammals and shows, by means of good drawings, their probable appearance when alive. Distribution maps, evolutionary chart, and bibliography.

Scheele, William E. *Prehistoric Animals.* World, 1954. 125 pp. illus. $5.95. 54–5351. (JH–SH)
A summary of important facts with excellent drawings, including simulated living animals and skeletal features, concerning representative prehistoric fishes, amphibians, reptiles and birds. Presents compact information on where and how the animals live, what they looked like, and the place they occupied in the evolutionary scale. A good companion for the author's *The First Mammals.*

Seeley, Harry G. *Dragons of the Air: An Account of Extinct Flying Reptiles.* (Introd. by Edwin H. Colbert.) Dover, 1967. xxi+239 pp. illus. $1.75 (paper). 67–15003. (C)
First published in 1901. A new Foreword by Edwin H. Colbert provides a biographic sketch, and pays tribute to his desire to communicate Seeley's scientific work to the general public. Interesting to read and important to the history of geology and biology.

Simpson, George Gaylord. *Life of the Past: An Introduction to Paleontology.*
Yale U., 1953. xi+198 pp. illus. $6.00, paper $1.75. 52–12078. (JH–SH)
An outstanding specialist has written a nontechnical discussion, for the
younger student and general reader, of the scope and significance of paleontology. It covers principles and the interpretation of the history of life
gained from fossil evidence.

572 ANTHROPOLOGY—ETHNOLOGY (See also 390 CULTURAL ANTHROPOLOGY)

Beals, Ralph L., and Harry Hoijer. *An Introduction to Anthropology* (3rd ed.).
Macmillan, 1965. xxi+711+x pp. illus. $8.95. 59–5397. (SH)
The 22 chapters of this text cover such topics as "The nature and scope
of anthropology," "Race, evolution, and genetics," "Language," "Space, time,
and culture," "Education and the formation of the personality," and "Accul-
turation and applied anthropology." Well-illustrated and with maps, charts,
and an ethnographic bibliography.

Berrill, Norman J. *Inherit the Earth: Man on an Aging Planet.* Dodd, Mead,
1966. 218 pp. $5:00. 66–21753. (JH–SH)
Discusses significant problems that have developed by recent studies of
earth and man. Subjects range from the genetic code to planet formation.
Provocative and informative for intelligent readers of any age.

Brace, C. Loring. *The Stages of Human Evolution: Human and Cultural Origins.*
(Foundations of Modern Anthropology Series.) Prentice-Hall, 1967. xi+
116 pp. illus. $4.50, paper $1.75. 67–22426. (SH–C)
An unusually lucid account of man's continuing development. Man is defined in terms of specific ecological interactions. Includes such modern
concepts of population genetics as genetic drift and the "founder effect."
Highly recommended as supplementary reading.

Carrington, Richard. *A Million Years of Man.* World, 1963. xvi+336 pp. illus.
$7.50; NAL (Mentor MT576, 1964), paper 75¢; also Meridian, 1965. $2.95.
62–15712. (SH)
The human species should not be considered in isolation, but in relation to
all living things and the physical universe as well. With this theme in mind,
the author covers the geological evolution of the earth, the origins of life,
evolutionary principles, the development of *Homo sapiens,* the development
of civilizations, and closes with a discussion of the individual in relation to
these phenomena.

Coon, Carleton S. *The Story of Man* (2nd ed.). Knopf, 1962. xxii+438+xiv
pp. illus. $7.95. 62–11055. (SH)
An anthropological history of man that is as interesting as it is informative.
Traces the development of man from the near-ape through the hunting and
gathering bands, the domestication of plants and animals, the rise of technology,
to the triple choice facing man today: "Either the world will be destroyed,
or nature will regain its balance at man's expense, or man will restore
nature without loss of his cultural heritage when he learns to unify the cultures of the world, just as his ancestors made man a single species."

Freed, Stanley A., and Ruth S. Freed. *Man from the Beginning*. Creative, 1967. 144 pp. illus. $4.95. 65–28357. (JH)
The methods, goals and content of anthropology are explained simply and succinctly, with the aid of numerous illustrations. Chapters deal with biological evolution, the evolution and characteristics of the primates, modern human races, methods of dating archeological remains, and the phenomenon of culture. The text is well written and accurate, and there are numerous well-chosen illustrations. Index.

Fried, Morton H. *Readings in Anthropology* (2 vols.). Crowell, 1959. 1080 pp. in all. (Paper) $4.95; $5.95. 59–6441. (C)
A collection of 73 diverse, thought-provoking articles by authors expert in their fields. Volume 1 covers physical anthropology, linguistics, and archeology; Volume 2 deals extensively with cultural anthropology. While some of the writing is fairly difficult, all of it is good. The selection and organization of the articles and the layout of the set makes it a valuable reference work, and well worth binding for libraries.

Hall, Edward T. *The Hidden Dimension*. Doubleday, 1966. xii+201 pp. illus. $4.95. 66–11173. (SH–C)
After establishing his concept of "culture as communication," Dr. Hall examines distance regulation, crowding, and social behavior of animals. Then perception of space is explored followed by a discussion of intimate, personal, social, and public distance in man. Discusses the Germans, English (and Americans), French, Japanese and Arabs. The final sections deal with space concepts in cities and culture, and the future of man. Bibliography and an index. Excellent assigned reading for secondary school and college social science students.

Hoebel, E. Adamson. *Man in the Primitive World* (2nd ed.). McGraw-Hill, 1958. xvi+678 pp. illus. $11.25 (text ed. $7.95). 57–7233. (C)
The first quarter of this general introduction to anthropology covers physical anthropology and archeology, including evolution, ancient man, fossils, the beginnings of culture, the dawn of civilization, and human races. The rest deals with cultural anthropology—subsistence, crafts, marriage, kinship, social control, property, religion—with a final section on the dynamics of culture. Comprehensive glossary and bibliography.

Holmes, Lowell D. *Anthropology: An Introduction*. Ronald, 1965. ix+384 pp. illus. $5.50. 65–27071. (SH–C)
Ethnology and social anthropology are introduced through this clearly-written, well-arranged general text. The anthropological viewpoints expressed are neither out of date nor radically new. Although oriented to beginning students, it presents compactly a substantial number of essential elements of anthropology. Many suggested readings from popular as well as professional journals; lists of relevant paperbound books follow each chapter.

Service, Elman R. *Profiles in Ethnology*. Harper & Row, 1963. xxix+509 pp. illus. $7.50. 63–9050. (C)
Precise, well-written accounts of 21 nonindustrialized cultures, arranged according to five levels of complexity: bands, tribes, chiefdoms, primitive states, and modern folk societies. Each level is represented by several geographically

and historically diverse cultures. A sound, readable introduction to ethnology and a valuable reference work for every library.

Silverberg, Robert. *To the Rock of Darius: The Story of Henry Rawlinson.* Holt, Rinehart and Winston, 1967. 218 pp. illus. $3.95; LB $3.59. AC 66–10116. (JH–SH)
The major theme is young Henry Rawlinson's decipherment of cuneiform writing. A minor theme running through the book is the role of the British East India Company in the Orient—its policies, wars, and methods of dealing with Asiatic peoples. There are good descriptions of nineteen-century oriental culture and of the ancient cultures of the Assyrians, Babylonians and Sumerians. Presents the intricate problem of cuneiform decipherment most skillfully and plainly. Illustrations are excellent, good bibliography.

Vlahos, Olivia. *Human Beginnings.* (Illus. by Kyuzo Tsugami.) Viking, 1966. 255 pp. LB $5.63. AC 66–10435. (JH–SH)
The findings from paleontology, physical anthropology, prehistory, and culture anthropology are reported to provide understanding of the evolution of man and the origins and development of human culture.

572.7 ANTHROPOLOGY—PRIMITIVE RACES

Breuil, Henri, and Raymond Lantier. *The Men of the Old Stone Age (Palaeolithic & Mesolithic)* (2nd ed.). (Tr. by B. B. Rafter.) St. Martin's, 1965. 272 pp. illus. $6.00. 65–10190. (JH–SH–C)
The first edition was based on 20 lectures given at the University of Lisbon in 1942 by Henri Breuil. In this second edition the text has been revised and amplified, including paragraphs dealing with hunting and abodes of men and their gods, and two chapters on funeral customs of fossil man and religious practices of leptolithic man.

Driver, Harold E. *Indians of North America.* U. of Chicago, 1961. xviii+668 pp. illus. $10.95, paper $5.00. 61–6504. (C)
Homo sapiens migrated across the Bering Strait to the New World between 11,000 and 40,000 years ago. This detailed book describes the peoples that have inhabited the region from the Arctic to Panama, using the "culture area" concept, which divides North America into 17 areas whose tribes share many customs and cultural institutions. An extensive map section and large bibliography make it a valuable reference work.

Matthiessen, Peter. *Under the Mountain Wall: A Chronicle of Two Seasons in the Stone Age.* Viking, 1962. xvi+256+xxxii pp. $7.50. 62–16796. (SH)
An account of the Kurelu of New Guinea, one of the last tribes in the world to come in contact with white culture, who were studied by the Harvard-Peabody Expedition in 1961 only months before the arrival of white missionaries and traders in their secluded valley.

Rapport, Samuel, and Helen Wright (eds.). *Anthropology.* New York U., 1967. xix+332 pp. illus. $4.95. 66–12601. (SH–C)
Selections from the writings of well-known anthropologists provide the content for chapters on "The evolution of man," "The study of man's culture," "Aspects of culture," and "Some primitive cultures." All selections are highly readable, making the volume a painless introduction to some of the

aims and content of anthropology, especially the ethnology of primitive peoples.

Service, Elman R. *The Hunters*. (Foundations of Modern Anthropology Series.) Prentice-Hall, 1966. x+118 pp. illus. $4.50, paper $1.75. 66–10095. (SH–C) A highly readable introduction to the anthropology of hunting peoples that summarizes specific information on each of nine societies. Fills a gap in the treatment of prehorticultural society.

Vlahos, Olivia. *African Beginnings*. (Illus. by George Ford.) Viking, 1967. 286 pp. LB $6.43. 67–24859. (SH–C) The historical and ethnographic accounts of a number of African peoples are discussed. Geography, physical type, language, social structure, political organization, subsistence patterns, local art and legends, and particular tribal characteristics are reported. Of interest to those seeking the historical cultural background of rising African nationalism.

572.8 ANTHROPOLOGY—SPECIFIC RACES

Cohoe, William. *A Cheyenne Sketchbook*. (Commentary by E. Adamson Hoebel and Karen Daniels Peterson.) U. of Oklahoma, 1964. xv+96 pp. illus. $5.95. 64–21708. (JH–SH–C) Cohoe, a Plains Indian imprisoned in a St. Augustine fort for five years, made sketches of the life he and his people led prior to his surrender. Twelve colorful drawings and accompanying comments by an anthropologist depict the buffalo and turkey hunts, the Sun, Wolf and Osage dancers, and a war dance. Information on customs, clothing and Indian lore. Bibliography and index.

DuBois, Cora. *The People of Alor*. Harvard, 1960. xxvi+654 pp. illus. $10.00; Harper & Row (Torchbooks TB 1042–1043, 2 vols.), paper vol. 1, $2.45, vol. 2, $2.75. 60–16359. (C) Anthropological and psychological methods are combined in this important study of the relationship between culture and personality in a primitive society. The four parts deal with the social and geographic setting; personal development, institutions, and religion, all with psychological emphasis; eight autobiographies of natives with psychological comments; and an interpretation of various psychological tests given the Alorese. A difficult work that requires knowledge of psychology as well as anthropology.

Fejes, Claire. *People of the Noatak*. Knopf, 1966. xii+368 pp. illus. $6.95. 66–11344. (SH–C) The author lived for months at a time in primitive Eskimo villages where old-fashioned hunting life is still practiced. The descriptions and the events of these visits make up the contents of this book. Besides a map, it is illustrated with numerous representative black-and-white reproductions of the author's drawings and paintings. The text includes many quotations from the literature of the Arctic and the acknowledgments of their sources constitute a rich bibliography.

Freuchen, Peter. *Peter Freuchen's Book of the Eskimos*. World, 1961. 441 pp. illus. $7.50; $0.95 (paper). 61–5815. (SH) The people who live at the top of the world in Greenland, northern Canada, and Alaska, are tough and hardy, but have a friendliness and spirit all their own. Freuchen, a Danish explorer, describes their tribal social structure, beliefs, and customs with an earthy enthusiasm and firsthand knowledge that reflect the years he lived with them.

Grant, Campbell. *Rock Art of the American Indian.* Crowell, 1967. xiv+178 pp. illus. $12.95. 67-12402. (SH-C)

Photographs and drawings are combined with a reliable and readable text, portraying the art left on rocks by North American Indians. A brief review of the peopling of the New World and the crystallization of regional cultural differences helps to place the art in context. Methods of production, problems of and clues to dating, and techniques of recording are enumerated and described. Rich bibliography and good index.

Kluckhohn, Clyde. *Navaho Witchcraft.* Beacon, 1962. xxii+254 pp. $4.95, paper $1.95. 62-13533. (C)

More specialized and difficult than *The Navaho,* this classic work in human behavior combines depth psychology, cultural history, and linguistics to unravel the complexities of Navaho witchcraft beliefs. The careful reader will get valuable insight into interviewing techniques, data collection and write-up, and the general method of anthropological field work.

Kluckhohn, Clyde, and Dorothea Leighton. *The Navaho.* Howard, 1964. xx+ 258 pp. $5.00 (text ed. $3.75); Doubleday (Anchor N28), paper $1.45. 62-6779. (SH)

A depth study of the largest Indian tribe in the United States by a cultural anthropologist and a psychiatrist. All aspects of Navaho culture are covered —history, kinship and relationships with one another, their "world-view" and the supernatural, language—with special emphasis on the conflicts between Navaho and white cultures.

Kroeber, Theodora. *Ishi in Two Worlds: A Biography of the Last Wild Indian in North America.* U. of California, 1961. 257 pp. illus. $5.95, paper $1.95. 61-7530. (SH)

Ishi's tribe, the Yahi of California, is described in the first half of this narrative, while the second covers the last five years of his life, from the time he came wandering out of the hills to his death in San Francisco in 1916. The author, wife of anthropologist A. L. Kroeber who worked closely with Ishi, describes the gruesome history of the last years of the tribe, and the problems that faced Ishi when he tried to adapt to our society, a stone age man in an industrial world.

Malinowski, Bronislaw. *Argonauts of the Western Pacific.* Dutton, 1922. xxxi+ 527 pp. illus. $10.00, paper (Everyman D74) $2.45. 22-16057. (C)

First published in 1922, this monograph on the Trobriand Islanders continues to be a milestone in ethnographic literature. Malinowski describes the society and culture of the Trobrianders in relation to the *Kula,* a complex economic trading system that is a major force in their culture. Also includes an interesting discussion of ethnographic field methods and goals.

Newcomb, Franc Johnson. *Navaho Neighbors.* U. of Oklahoma, 1966. ix+236 pp. illus. $5.95. 66-13432. (JH-SH-C)

A sensitive picture of Navaho life by one who lived with the people, learned their customs and traditions, and gained their confidence. There are descriptions of many ceremonies, including a healing rite, and the moccasin game. A nonprofessional but valuable study of the everyday life of the Navahos. Captures the Navaho way of life and thought-pattern to a degree seldom accomplished even by professional ethnologists.

Read, Kenneth E. *The High Valley.* Scribner's, 1965. xvii+266 pp. illus. $6.95. 65–20581. (SH–C)
A social anthropologist's account of two years spent among the Gahuku in the Central Highlands of New Guinea. While the book centers on Gahuku social organization, related through native personalities well known to the author, it is also frankly autobiographical. This beautifully-written book contains a glossary, an index, excellent black-and-white photographs of the Gahuku, and two maps.

Steward, Julian, and Louis C. Faron. *Native Peoples of South America.* McGraw-Hill, 1959. xi+481 pp. illus. $9.95. 58–10010. (C)
A summary and interpretation of the information contained in a six-volume set published by the Smithsonian Institution, *Handbook of South American Indians.* Describes the history and general trends of South American cultures and the tribes found in different areas with their distinctive features.

Thomas, Elizabeth Marshall. *The Harmless People.* Knopf, 1965. ix+266 pp. $4.95. 59–5437. (SH–C)
This first-person account of the lives and world of the elusive African Bushmen who eke out a precarious primitive existence on the great Kalahari desert of central Africa presents vivid living vignettes of Bushman life as it was experienced by the author during three visits to the Kalahari. Written as a nontechnical literary essay, yet it is an accurate informal ethnography of professional importance. Admirable collateral reading for both high school and college courses dealing with social and cultural development.

Turnbull, Colin M. *Wayward Servants: The Two Worlds of the African Pygmies.* Nat. Hist. Press, 1965. xiv+390 pp. illus. $7.95. 65–17265. (C)
A thorough and well-written ethnological study of a single band of Mbuti pygmies, hunters and gatherers in the Ituri Forest of the Congo who share the area with Negro cultivators living in the same region. The author studied a single band of Mbuti throughout an entire calendar year. Various appendices include maps, legends, glossary, and bibliography.

Underhill, Ruth M. *Red Man's America.* U. of Chicago, 1953. x+400 pp. $7.50. 53–10535. (SH–C)
A well-organized survey of the Indian tribes of the United States for the general reader. Covers the migration of peoples to the New World, the tribes of various geographical areas, and the treatment and protective measures of the government toward Indians today.

Wissler, Clark. *Indians of the United States.* (Rev. by Lucy Wales Kluckhohn.) Doubleday, 1966. 336 pp. illus. $5.95, paper $1.95. 66–12215. (SH)
The original edition was published in 1940. Revisions in the present version by Lucy Wales Kluckhohn add bibliographic detail and update those sections where the most substantial advances have been made in recent years. The primary objective is to describe the pattern of life of the Indian of the

historic period. The presentation of descriptive detail is organized in terms of major linguistic stocks. A warm and sympathetic introduction to the American Indian.

573 PHYSICAL ANTHROPOLOGY

Clark, Grahame. *The Stone Age Hunters*. McGraw-Hill, 1967. 143 pp. illus. $5.50. 67–15599. (C)
Traces man's physical and cultural development from the earliest Pleistocene to Neolithic times, with a section on recent hunters and gatherers. Highly recommended for use in introductory college courses in anthropology and archeology.

Coon, Carleton S. *The Origin of Races*. Knopf, 1962. xli+724+xix pp. illus. $10.00 (text ed. $7.50). 62–14761. (C)
Coon presents his well-researched but somewhat controversial theories concerning the development of the five modern races of man. He proposes that races have existed for a much longer time than is commonly supposed and that their development was separate but parallel, the more common theory being that they emerged after man reached the *Homo sapiens* level. He supports his thesis with a vast quantity of evidence and considerable personal experience, and despite the debatable aspect, his book is a milestone in the field.

Garn, Stanley M. *Human Races* (2nd ed.). Thomas, 1965. xiv+137 pp. illus. $6.50. (SH)
A concise, rigorous statement of modern findings in human raciation. Explains the concepts of geographical, local, and micro-races and the relation of race to physiology, biochemistry, blood groups, natural selection, disease, behavior, and intelligence. Concludes with a taxonomy for man—a listing of the 9 geographical races and over 30 local race groups with their descriptions.

Hulse, Frederick S. *The Human Species: An Introduction to Physical Anthropology*. Random, 1963. xxii+504 pp. illus. $11.50 (text ed. $7.95). 63–8269. (SH)
A new and thorough introductory text in physical anthropology and the related fields of paleontology, biochemistry and genetics. Discusses evolution in general, human evolution, human races and their origin and distribution, basic genetics, natural history, and laboratory analysis.

McKern, Thomas W. *Readings in Physical Anthropology*. Prentice-Hall, 1966. viii+199 pp. illus. $4.75 (paper). 66–13727. (SH–C)
Designed as a survey text for an introductory college course in physical anthropology, also to supplement biology and psychology courses. The book can be used for resource material by advanced high school students. The readings demonstrate how the evidence and tools of physical anthropology are used and focus on the classical areas of physical anthropology, represented by the work of the leading scholars. Covers such topics as race, future man, living primates, man's place in nature, cultural prehistory, human genetics, human evolution, and fossil man. Only a panoramic vista is given, with few specific details.

Montagu, Ashley. *Man's Most Dangerous Myth: The Fallacy of Race* (4th ed.). World, 1964. 499 pp. $7.50, paper $2.65. 64–12067. (C)

Designed to "expose the most dangerous myth of our age, the myth of 'race,' by demonstrating the falsities of which it is compounded," this is a forcefully written, well-documented book that contains a 60-page bibliography and several useful appendices, one of which gives the UNESCO Statement on Race.

Rodnick, David. *An Introduction to Man and His Development.* Appleton-Century-Crofts, 1966. xiv+443 pp. illus. $3.95 (paper). 66–12109. (SH–C)

Attempts to summarize the concepts and data of the social sciences (specifically history, anthropology, sociology, economics, and political sciences) and thus provide a coherent interpretive account of man from his Paleolithic beginnings to the present. Three main parts: (1) a survey of culture growth from the Paleolithic through the rise of "Primary Civilization," (2) the rise of "Secondary Civilization," and (3) a survey of the relations between past and present cultural forces in Africa, India, China, Japan, and the Soviet Union. A concluding section appraises the prospects for "Tertiary civilization" in the decades to come. A remarkable and stimulating book.

Silverberg, Robert. *Man Before Adam.* Macrae Smith, 1964. 253 pp. illus. $4.50. 64–14871. (JH–SH)

A popular historical presentation of the collection of data which developed the science of human paleontology and replaced the biblical interpretation of the origin of man. A thorough job of reporting fossil discoveries and current interpretations of them. The reader is neither misled by a confusion of facts with theories, nor bewildered by too much technical detail. The historical approach makes it interesting to all groups of readers. Few books available to the general public gives as current and complete a discussion of fossil men.

Weiner, Joseph S. *The Piltdown Forgery.* Oxford, 1955. xii+214 pp. illus. $2.00. 55–14367. (SH)

The story of one of the greatest deceptions ever perpetrated on science, written by a scientist whose investigations helped expose it. He tells why it was believed, the difficulties of interpretation that led to suspicion, the scientific proof of its falsity, and the probable persons responsible for it.

574 BIOLOGY IN GENERAL

Andrews, Roy Chapman. *This Amazing Planet.* Putnam's, 1940. ix+231 pp. $4.50. 40–34045. (JH)

Contains 80 miscellaneous items of biological information and natural history notes gleaned from his varied experiences as scientist, field investigator, and museum collector. Many young people have been inspired by Andrews' books to seek careers in the biological sciences.

Blackwelder, Richard E. *Taxonomy: A Text and Reference Book.* Wiley, 1967. xiv+698 pp. $19.95. 67–13520. (SH)

Designed as a reference book to make the simple rules of taxonomy more readily available. An important work of reference for beginners, collectors, and specialists.

Bonner, John Tyler. *The Ideas of Biology.* Harper & Row, 1962. xi+180 pp. illus. $5.95, paper (Torchbooks TB570) $1.60. 62–9884. (SH)
A book that explains the themes of biology, in a discursive and somewhat popular fashion, supposing that the reader knows the fundamental "facts" of elementary biology. Valuable because it stimulates thinking.

Bonner, John Tyler. *Size and Cycle: An Essay on the Structure of Biology.* (Illus. by Patricia Collins.) Princeton, 1965. viii+219 pp. $7.50. 65–14306. (SH–C)
The author attempts to bring the entire science of biology into a new set of concepts. The view taken here is that the life cycle is the central unit in biology, rather than just the adult organism. The science of biology—genetics and biochemistry and development—are thus brought into one view. This simple approach leads to different views of understanding. Many examples are presented, from the smallest to the largest organism. The explanations and comparisons are supplemented by excellent figures and plates.

Carlson, Elof Axel (ed.). *Modern Biology: Its Conceptual Foundations.* Braziller, 1967. viii+327 pp. illus. $7.50. 67–12476. (SH–C)
Selected noteworthy papers have been collected that made significant contributions to the development of biological concepts or the elucidation of biological processes. The papers range in time of original publication from Hooke in 1695 to Nirenberg and Matthaei in 1963. The editor provided an introductory chapter, one-paragraph commentary on each paper, and a glossary. Papers are included on two major biological units, five major concepts, and four major processes. The units are the cell and the gene. The concepts are (1) the cell doctrine, (2) heredity, (3) development, (4) molecular biology, and (5) evolution. The processes are (1) mitosis, (2) meiosis, (3) genetic decoding, and (4) metabolism.

Frisch, Karl von. *Biology: The Science of Life.* (Tr. by Jane M. Oppenheimer; Foreword by Paul Weiss.) Harper & Row, 1964. xv+516 pp. illus. $12.95. 64–10219. (SH–C)
A collateral reading and reference work, emphasizing life in all its aspects and the ecological relationships of all organisms, that no public, secondary school, or college library should lack. The author says it is "a book in which the student reads more according to his own impulse than his teacher requires, and which will perhaps be his companion beyond school because it deals with things that concern us all, the understanding of which is good and is valuable." The writing is magnificent. Excellent illustrations and a scholarly bibliography.

Frisch, Karl von. *Man and the Living World.* Harcourt, Brace, 1963. 304 pp. illus. $7.50. 62–16734. (SH)
Introduces fundamentals of vital processes beginning with the structure of organisms from the single cell to the highly specialized vertebrates. He demonstrates form and function of sense organs in animal and man; deals with migrations, social behavior, and other concepts. Highly recommended cultural reading.

Gamow, George, and Martynas Ycas. *Mr. Tompkins Inside Himself: Adventures in the New Biology.* Viking, 1967. xiv+274 pp. illus. $6.95. 67–20297. (SH–C)

Presents advances in biological knowledge and research in a fictional style through the adventures of a fictional bank teller—Mr. Tompkins. A high school background in biology and chemistry is required. Designed for the nonspecialist reader and student.

Goldstein, Philip. *Triumphs of Biology.* Doubleday, 1965. v+304 pp. illus. $4.95. 65–12365. (SH–C)
A narrative of six triumphs of biology, each beginning with a chance observation that develops with a chronological pattern of plateaus of achievement to success. (1) Leeuenhoek in 1703 discovered hydra, on which Trembley and others performed classic experiments. (2) Reports in 1910 of sickle-cell anemia was followed by heredity and biochemical studies. (3) An English chemist in 1794 said pink is blue. Color blindness was later tied to heredity and functioning physiology. (4) Nineteenth-century plant hybridizing lagged until Austria's Mendel used pure strains. Following him Wilson, Correns and deVries, hydridizing independently, formulated today's gene theory. (5) Certain water flea eggs require no fertilization. (6) Genetic molecular biology developed rapidly from the 1899–1908 investigation with black urine.

Jaques, Harry E. *Living Things: How to Know Them* (rev. ed.). Wm. C. Brown, 1946. iv+172 pp. illus. $3.25, paper $2.50. 39–8451. (JH)
An introductory field guide to plant and animal life for the younger hobbyist, or the older amateur, which provides suggestions for study projects, and pictured keys to the phyla, classes, and orders of plants and animals.

Keosian, John. *The Origin of Life.* (2nd ed.). (Selected Topics in Modern Biology Series.) Reinhold, 1968. ix+118 pp. illus. $2.25 (paper). 64–20141. (C)
Opens with a brief historical survey of past theories. Discusses the hypothesis of Oparin in light of the subsequent research. Speculations as to the possibility of the existence of extraterrestrial life are presented. The historical approach and presentation of unanswered questions make fascinating reading for both the student and the lay reader.

Krutch, Joseph Wood. *The Great Chain of Life.* Houghton Mifflin, 1956. xiv+227 pp. illus. $4.50. 56–13104. (JH)
In a wide-ranging discussion covering the whole spectrum of life from protozoa to giant mammals, the author unfolds the richness and dignity of man's animal inheritance. Good required reading to stimulate rational thinking.

Langdon-Davies, John. *Seeds of Life: The Story of Sex in Nature from Amoeba to Man.* Devin Adair, 1955. xix+172 pp. $3.95; NAL (Signet Ks 345), paper 35¢. 55–7742. (JH)
Tells of the many ways in which living things reproduce themselves and examples used are algae, disease germs, snails, onions, orchids, termites, earthworms, robins, and man himself. An elementary discussion of the mechanisms of inheritance is included.

Marteka, Vincent. *Bionics.* (Illus. by John Kaufmann.) Lippincott, 1965. 157 pp. $4.25. 65–13422. (JH–SH)
Bionics is the application of mechanisms possessed by living organisms to the design of man-made machines or devices. Biological clocks, the unerring

navigation of birds, the periodic migrations of certain insects, fishes, and mammals; the moth's power dive from sonar probes of a bat; how the blind "see" objects through sound echoes; actual communication between men and animals—these are some of the topics investigated. An interesting and useful synthesis that will introduce students and nontechnical adult readers to bionics. Bibliography.

Palmer, E. Laurence. *Fieldbook of Natural History*. McGraw-Hill, 1949. x+ 664 pp. illus. $11.95 (text ed. $8.50). 49–8179. (SH)

A well-known and useful one-volume compendium of natural history suitable for the naturalist and teacher, and for serious students, dealing briefly with the universe, the solar system, the earth, rocks, and minerals, and primarily with plants and animals. It needs taxonomic revision, but there is no substitute.

Platt, Robert B., and George K. Reid. *Bioscience*. (Reinhold Book in the Biological Sciences.) (Illus. by Charles W. Schmidt.) Reinhold, 1967. xvi+ 528 pp. $9.95. 67–18708. (C)

Uses the "levels of organization" framework to present a version of the modern biology. Ecology, the chemical nature of living matter, cellular processes, and organ systems are covered in detail, while taxonomy and anatomy are left for the laboratory or independent research.

574.03 BIOLOGY—ENCYCLOPEDIC WORKS

Crow, W. B. *A Synopsis of Biology* (2nd ed.). Williams & Wilkins, 1964. xx+1056 pp. illus. $11.00. (C)

A summary, in telegraphic style, of the whole of biology. Divided into sections on morphology, physiology, ecology, genetics, systemics, applications, and biotechnique. Elementary biological background needed.

Delaunay, Albert (Scientific editor), and Hans Erni (Artistic editor). *Encyclopedia of the Life Sciences*. Vol. 1, *The Living Organism;* Vol. 2, *The Animal World;* Vol. 3, *The World of Plants;* Vol. 4, *The World of Microbes;* Vol. 5, *The Human Machine: Mechanisms;* Vol. 6, *The Human Machine: Disorders;* Vol. 7, *The Human Machine: Adjustments;* Vol. 8, *Man of Tomorrow*. Doubleday, 1965. 160 pp. ea., $9.95 ea. Vol. 1, 64–17276; Vol. 2, 64–17277; Vol. 3, 65–10501; Vol. 4, 65–10502; Vol. 5, 65–11367; Vol. 6, 65–11781; Vol. 7, 65–11782; Vol. 8, 65–11783. (SH–C)

English translations of the handsome eight-volume French set, *La Vie et l'Homme: Encyclopedie des Sciences Biologiques*, first published in 1961. Each volume was written by a panel of experts. The exquisite, large illustrations—mostly photographs in color—make these books an exciting reading experience for laymen and for scientists. The English translation is accurate and in good style to hold the interest of the reader. Excellent indexing in each volume makes them suitable for reference.

Gray, Peter. *The Dictionary of the Biological Sciences*. Reinhold, 1967. xx+ 602 pp. $14.75. 67–24690. (SH–C)

This is a valuable and scholarly tool for the educated layman, college student, and professional worker, and is recommended for acquisition by the libraries that they use. It is also a useful reference in larger secondary school libraries for use by faculty and talented students. Contains over 40,000 entries and has great advantages for scholars because of a unique arrangement of related terms.

Gray, Peter (ed.). *The Encyclopedia of the Biological Sciences* (2nd ed.). Reinhold, 1970. xxv+1027 pp. illus. $24.95. 77–81348. (SH–C)
Contains over 800 articles on biological subjects and the terminology and scientific thoroughness are above the level of comparable articles in standard encyclopedias. Essential for all college, public, and large high school libraries.

Henderson, Isabelle F., and W. D. Henderson. *A Dictionary of Biological Terms* (8th ed., by J. H. Kenneth.) Van Nostrand, 1963. xv+640 pp. $12.50. 61–383 Rev. (SH–C)
The 1st (1920) to 7th (1960) editions were entitled *A Dictionary of Scientific Terms*. Includes terms in biology, botany, and zoology, as well as in the related disciplines of anatomy, physiology, cytology, genetics, embryology, and other cognates. A well-known dictionary compiled and printed in England.

574.07 BIOLOGY—STUDY AND TEACHING

Alexander, Gordon. *General Biology* (2nd ed.). Crowell, 1962. xiv+904 pp. illus. $7.50. 62–10276. (SH–C)
Because of the broad scope and detailed treatment of biology in terms of concepts, principles, morphology, and classification surveys of plants and animals, this college text is a valuable reference and resource book for secondary students and college undergraduates.

Beaver, William C., and G. B. Noland. *General Biology* (7th ed.). Mosby, 1962. 765 pp. illus. $8.95. 62–8327. (SH–C)
A very comprehensive college textbook, almost encyclopedic in scope, with excellent bibliography and good treatment of man as a typical mammalian vertebrate. Special sections on ecology and evolution, and a good glossary. Can be used by superior secondary students.

Berrill, Norman J. *Biology in Action: A Beginning College Textbook*. Dodd, Mead, 1966. xvi+878 pp. illus. $9.95. 66–19412. (C)
Intended for students who have had one of the BSCS high school courses, and useful for those students still in high school as a reference. Organization centers around the three major areas of cell biology, organismal biology, and communities in space and time. Traditional and modern biology are integrated effectively.

Biological Sciences Curriculum Study. *Biological Science: An Inquiry into Life* (2nd ed.). (Arnold Grobman, Director.) Harcourt, Brace, 1968. xix+840 pp. illus. $7.96. 63–2510. (SH)
The "Yellow Version" BSCS biology textbook for high schools (supplemented by a teachers' guide, laboratory manual, and other materials listed in the book and priced separately). Treats the subject of biology by conceptual themes; cells, microorganisms, plant life, animal life, genetics, evolution, and ecology. The historical development of the concepts is interwoven as appropriate. Needed by all high school libraries, whether it is or is not being used as a text in the biology course.

Biological Sciences Curriculum Study. *Biological Science: Interaction of Experiments and Ideas*. Prentice-Hall, 1965. xvi+429 pp. illus. $7.50 (text ed. $5.96). 65–6181. (SH–C)

Intended for an advanced secondary school biology course exclusively for students who have had a first course in biology. The objectives of the BSCS "second level" course are broader than those of most advanced high school offerings in biology. Brings together a series of experiments and activities with selected journal reprints and historical perspectives. Extensive sections on experimental design, statistical analysis, and literature resources; based largely on a composite of BSCS Laboratory Blocks.

Biological Sciences Curriculum Study. *Biological Science: Molecules to Man* (2nd ed.). (Arnold Grobman, Director.) Houghton Mifflin, 1968. xx+872 pp. illus. $8.80. 63-24630. (SH)
The "Blue Version" BSCS biology textbook for high schools (supplemented by a teachers' guide and other materials listed in the textbook and priced separately). The major emphasis is upon the recent advances in molecular biological resarch through physiology, biochemistry, and evolution. Many people have some understanding of the *products* of science; few know anything of its *process*. To provide an understanding of the nature of science as *process* which often leads to *product* is the dynamic objective of this course. The sections of the text are: (1) *Biology, the introduction of facts and ideas;* (2) *The evolution of the cell;* (3) *The evolving organism;* (4) *Multicellular organisms: new individuals;* (5) *Genetic continuity;* (6) *Energy utilization;* (7) *Integrative systems;* (8) *Higher levels of organization.* Laboratory investigations occupy 130 pages in the back of the book.

Biological Sciences Curriculum Study. *Biology Teachers' Handbook.* (Joseph J. Schwab, Supervisor.) Wiley, 1963. xvii+585 pp. $7.00. 63-18628. (SH)
This manual and resource book explains the BSCS approach to the teaching of biology as a process of scientific inquiry. It compares and explains the content of the "Blue, Green, and Yellow Versions", provides a summary of the principles and concepts of the physical sciences, of statistical methods, and of the principles of biochemistry that are needed by high school biology teachers. It includes information on visual aids, laboratory equipment, preparation of laboratory solutions and reagents, lists sources of laboratory supplies, contains a general bibliography, a list of career information sources, and other information. This is an indispensable manual for all teachers of biology, irrespective of whether they are using BSCS texts. It should be owned by every high school library as a resource book for talented students working on individual projects.

Biological Sciences Curriculum Study. *High School Biology.* (Arnold Grobman, Director.) Rand McNally, 1963. xv+749 pp. illus. $8.76. 63-6945. (SH)
The "Green Version" BSCS biology textbook for high schools (supplemented by a teachers' guide, laboratory manual, and other materials listed in the textbook and priced separately). All three versions are based on the assumption that, for the majority of students, the course will be their only formal instruction in biology. The emphasis in this version is on the ecological and behavioral aspects. The major sections are: (1) *The world of life: the biosphere;* (2) *Diversity among living things;* (3) *Patterns in the biosphere;* (4) *Within the living organism—the cell, plant, and animal morphology, reproduction, and heredity;* (5) *Adaptation;* (6) *Man and the biosphere.* Each of the three BSCS texts is good collateral reading material for students using another BSCS text, or a traditional text. High school libraries should own all three.

Biological Sciences Curriculum Study. *Research Problems in Biology: Investigations for Students* (4 series). Doubleday (Anchor), 1963, 1965. Series 1 —232 pp.; Series 2—240 pp.; Series 3—210 pp.; Series 4—244 pp. $1.25 ea. (paper). 63–13080. (SH–C)
Investigations for high school students who wish to pursue research that borders on the frontiers of current biological knowledge. Successful work on many of the research problems may extend over several months or years; these are not projects intended for quick "science fair" exhibits. The intent is to provide better-than-average students with a suggested research area which will give them some insight into the work of a research biologist.

Curtis, Helena. *Biology*. Worth (dist. by Nat. Hist. Press). 1968. xvii+854 pp. illus. $10.75 (trade ed. $15.00). 68–18187. (C)
A text for college students taking their first college science course. An animal-oriented book of six sections: "Cells," "Organisms," "Genetics," "Development," "Behavior," and "Population Biology and Ecology." Appendices include classification of plants and animals and a glossary. A good reference for advanced high school students and teachers.

Elliott, Alfred M., and Charles Ray, Jr. *Biology* (2nd ed.). Appleton-Century-Crofts, 1960. ix+723 pp. illus. $8.95. 60–6282. (C)
Intended for the college student pursuing an introductory course either looking toward major work in science, or for cultural purposes. Because of its appealing narrative style and the very good illustrations, it could be a useful reference work for secondary students. The central theme of the book is evolution, the guiding principles in any panoramic view of biology.

Farris, Edmond J. (ed.). *The Care and Breeding of Laboratory Animals*. Wiley, 1950. xvi+515 pp. illus. $15.95. 50–10593. (SH–C)
A manual based upon the practical experiences of the experts who have contributed the material concerning particular species. Intended for the research worker, technician, or the instructor who wishes to undertake the care and breeding of laboratory animals. It will also be useful to commercial breeders and dealers. Includes information on monkeys, rats, mice, guinea pigs, hamsters, rabbits, cats, dogs, fowl, amphibia, fishes, reptiles, and other species.

Graubard, Mark. *The Foundations of Life Science*. Van Nostrand, 1958. xii+627 pp. illus. $8.95. 58–7097. (SH–C)
Intended as a text for an orientation course in fundamental biological principles for nonscience majors. Because of the cultural approach it is interesting collateral reading and references material for younger students.

Hardin, Garrett. *Biology, Its Principles and Implications* (2nd ed.). Freeman, 1966. xi+682 pp. illus. $9.50. 61–14908. (SH–C)
The most strongly recommended (by referees) college biology textbook which is suitable also for reference by secondary students and especially those studying BSCS texts. Contains glossary, bibliography and useful appendices.

Johnson, Willis H., Richard A. Laubengayer, Louis E. Delanney, and Thomas A Cole. *Biology* (3rd ed.). Holt, Rhinehart and Winston, 1966. x+788 pp. illus. $10.95. 66–16951. (SH–C)
All chapters have been revised and updated and nine are new or completely

rewritten. It is well balanced and the emphasis is neither molecular nor classical. The major parts of the text deal with: cells; classification structure and function of flowering plants; structure, function, and development of vertebrates; representative plants and plant phylogeny; representative animals and animal phylogeny; and dynamics of organisms and species. Illustrations are well selected. For advanced high school or beginning college students the work will provide a sound and broad introduction.

Keeton, William T. *Biological Science*. (Illus. by Paula DiSanto Bensadoun.) Norton, 1967. xiii+955 pp. $9.50. 67–10720. (C)

For use in a college course, both for future majors and those who will not pursue biology further. Coverage is complete and broader than that of many similar books. The total perspective achieved is good. Good features include the illustrations, emphasis on the tentative nature of hypotheses, many references to outstanding biologists, pertinent comments on current problems of society in which biology has a role, and very full and useful lists of references at the end of each of the 22 chapters.

Kimball, John W. *Biology* (2nd ed.). Addison-Wesley, 1968. 776 pp. illus. $9.50. 68–15801. (SH–C)

Intended for mature preparatory school students or college undergraduates, the test is organized along functional lines ranging from the molecular and cellular organization of living things to features of their metabolism, responsiveness, reproduction, evolution, and ecology. Includes but is not limited to the topics of the *Advanced placement program syllabus*. References, exercises, and glossary included.

Knudsen, Jens W. *Biological Techniques: Collecting, Preserving, and Illustrating Plants and Animals*. Harper & Row, 1966. xi+525 pp. illus. $13.00. 66–10839. (SH–C)

Methods and materials for the collection and preservation of biological materials including the entire range from bacteria to flowering plants and mammals. The language is direct and simple. Readers are cautioned on safe and legal use of poisons, electrical and mechanical devices, and guns used in obtaining and preparing specimens. Disposal of bacterial cultures to guard against release of pathogens is also described.

Mason, John M., and Ruth T. Peters. *Life Science: A Modern Course*. Van Nostrand, 1965. ix+390 pp. illus. $4.96. 64–23970. (JH)

A beginning elementary introduction to biology, which might serve as a good preface to the descriptive biology required by students prior to modern programs such as BSCS including some early understandings in chemistry related to the life sciences. Develops content logically from cell, protoplasm, energy understandings through classification, man as a biological entity to ecology, conservation, and a very limited introduction to evolution.

Miller, David F., and Glenn W. Blaydes. *Methods and Materials for Teaching the Biological Sciences* (2nd ed.). McGraw-Hill, 1962. x+453 pp. illus. $8.50. 61–10137. (SH)

Intended as a source book for secondary school teachers, but valuable also for their students who are working on individual research projects and special assignments.

Milne, Lorus J., and Margery Milne. *The Biotic World and Man* (3rd ed.). Prentice-Hall, 1964. v+665 pp. illus. $9.75. 65–10174. (SH–C)

A number of chapters describe the interrelationship between structure and function of the organ systems of higher animals, emphasizing the chordates and man in particular. Many interesting and important details of comparative anatomy are included, as well as some physiological data not usually found in an introductory text. An appendix presents much of the chemical background and terminology, including structural organic formulae, necessary for a complete understanding of these important chemical pathways and reactions. A discussion of the taxonomy, structure, and life cycles of the multicellular algae, the bryophytes, and the vascular plants. The major characteristics of living organisms are covered in a series of chapters which include many well-chosen specific details and illustrations from both plants and animals. Elaborately illustrated.

Otto, James H., and Albert Towle. *Modern Biology.* viii+792 pp. $6.96.

Otto, James H., Albert Towle, and Elizabeth H. Crider. *Teacher's Guide to the Modern Biology Program.* xvi+326 pp. $3.60.

Otto, James H., Albert Towle, and Elizabeth H. Crider. *Biology Investigations: Laboratory Manual to Accompany Modern Biology.* vi+346 pp. $2.64 (paper). Holt, Rinehart and Winston, 1966. illus. (SH)
Modern Biology is the latest revision of the well-known series of texts in high school biology initiated by the late Truman J. Moon. The style is simple and oriented to the average student. In general, the content is inclusive and the eight units are somewhat interchangeable so that a teacher need not be bound to a specific sequence. The *Teacher's Guide* is an excellent adjunct to the text. It lists all of the apparatus, material, supplies and specimens needed. The guide should enable even the beginning teacher to have confidence in handling the material. The laboratory manual is adequate, but not as impressive as the textbook.

Rapport, Samuel, and Helen Wright (eds.). *Biology.* New York U., 1967. xiv+268 pp. illus. $4.95. 67–10286. (SH–C)
This text can be used as collateral reading or for reference and classroom use. Each of the five sections: the concept of biology, the cell, the origin of life, evolution and genetics, and the ways of living things, are written by different authors.

Silvan, James. *Raising Laboratory Animals: A Handbook for Biological and Behavioral Research.* Nat. Hist. Press, 1966. viii+225 pp. illus. $4.95, paper $1.45. 66–17442. (JH–SH–C)
Elementary, secondary school, and college teachers with little experience raising animals will find this an excellent guide and reference. The illustrations and construction directions are clear. The information about laboratory animals is accurate and documented. Also appropriate for high school students.

Simpson, George Gaylord, and William S. Beck. *Life: An Introduction to Biology* (2nd ed.). Harcourt, Brace, 1965. xviii+869 pp. illus. $10.50. 65–14384. (SH–C)
A college text which uses a "principles approach." Systematics and ecology are emphasized. All of the main phyla are described with emphasis on vertebrates. Man's role in the ecosystem, the community organization of

organisms, and paleontology are covered. Oriented towards students who have more than a passing interest in biology. The illustrations and diagrams are exceptional. An excellent reference and collateral reading book for high school and public libraries.

Simpson, George Gaylord, and William S. Beck. *Life: An Introduction to Biology.* (Shorter Edition). NY: Harcourt, Brace, 1969. xiii+546 pp. illus. $8.95. 69–14399. (C)

This shorter edition gives the first year biology student an overall view by approaching the subject from a unified standpoint. For example, in dealing with transport, animal circulatory systems as well as those of lower invertebrates and plants are discussed. The emphasis is on transport rather than the individual organisms.

Smith, Ella Thea, and Thomas G. Lawrence. *Exploring Biology: The Science of Living Things* (6th ed.). Harcourt, Brace, 1966. 766 pp. illus. $6.40. (SH)

A well-organized, good introduction to biology for secondary schools. Can be used as the text in a course designed to acquaint students with the total subject matter of biology. Will arouse the curiosity of students and excite them to further study, either by collateral reading or in subsequent courses offered at the senior high school level. A teacher's manual, tests, and a student workbook are available.

Taylor, William T., and Richard J. Weber. *General Biology* (2nd ed.). Van Nostrand, 1968. x+945 pp. illus. $9.95. 61–4574. (SH–C)

"We hope to bring the student onto the threshold of biological knowledge where he will see the taxonomic aspects . . . the morphology . . . and finally the functional and molecular aspects. . . ." A college text that secondary students can use for collateral reading; good bibliography.

Van Norman, Richard W. *Experimental Biology.* Prentice-Hall, 1963. xi+243 pp. illus. $8.50 (text ed.). 63–7313. (SH)

A valuable manual for teachers and students who are beginning their own experimental biological research. Chapters on research in general, using biological literature, selection of experimental organisms, and on methods, techniques and instruments used in experimental biology.

Villee, Claude A. *Biology* (5th ed.). Saunders, 1967. xxiii+730 pp. illus. $8.25. 67–11774. (C)

A widely used text which is also an invaluable reference work and rich in bibliographic suggestions. It begins with general discussions of the scientific method and major achievements, then proceeds to cell structure and functions, plants, animals (treated in evolutionary order), reproduction, heredity, and organic evolution.

Weisz, Paul B. *Elements of Biology* (2nd ed.). McGraw-Hill, 1965. 486 pp. illus. $8.25. 64–21077. (SH–C)

This is a simplified and rewritten version of *The Science of Biology*, Weisz's noted larger volume. It omits much of the biochemical background of the larger text and assumes no science background. The emphasis is on principles of biology. It is lucidly written and can well serve as an adjunct text in advanced high school classes. Elegantly simple diagrams supplement the text, and suggested collateral readings at the end of each chapter of readily available articles encourage students to delve deeper.

Weisz, Paul B. *The Science of Biology* (2nd ed.). McGraw-Hill, 1963. xv+ 796 pp. illus. $8.95. 58–59679. (C)

Based on the maturity and dynamism of modern biology, this college text uses the biochemical approach to the study of properties and organization of living creatures, their physiology, reproduction, adaptation and evolution.

Whaley, W. Gordon, Osmond P. Breland, C. Heimsch, A. Phelps, A. R. Schrank, and O. Wyss. *Principles of Biology* (3rd ed.). Harper & Row, 1964. xviii+ 776 pp. illus. $9.95. 64–11187. (C)

The new edition of a widely used text places major emphasis on development and function and integrates those fundamentals of physics and chemistry essential to an understanding of molecular biology; hence, it is useful as background for more advanced reading or study in biophysics and biochemistry.

Witherspoon, James D., and Rebecca H. Witherspoon. *The Living Laboratory: 200 Experiments for Amateur Biologists*. Doubleday, 1960. 256 pp. illus. $4.50. 60–12968. (JH–SH)

Any school that sponsors science fairs or affords opportunity for individual investigation needs this important guide so that students may obtain suggestions and may undertake work with live materials in accordance with approved techniques and humane procedures.

574.09   BIOLOGY—HISTORY

Gardner, Eldon J. *History of Biology* (2nd ed.). Burgess, 1965. iii+376 pp. illus. $6.00. 60–14970. (JH–SH)

A particularly enjoyable feature of this book is the subdivision of individual biographies so that one can see the contributions of a number of individuals to the development of an idea. The author has wisely included some rather good biographies as references following each vignette. The twentieth century has been neglected for the most part; the chapter on "Biology in the twentieth century" could almost be titled "The major areas of the unsolved problems of biology." References and index.

Locy, William A. *Biology and Its Makers* (3rd ed.). Holt, Rinehart and Winston, 1935. xxvi+477 pp. illus. $8.50. 8–21045. (SH–C)

Since publication of the first edition in 1908 this history has been required reading for many college majors, and is now a "classic" that all libraries should own. Its bibliography is a good key to older historical literature.

Singer, Charles. *A History of Biology* (rev. ed.). Abelard-Schuman, 1959. 580 pp. illus. $6.00. 50–70001. (SH)

Biology in retrospect from Hippocrates to Mendel, which shows how medicine, anatomy, philosophy, geology, chemistry, and psychology each has contributed to the whole of biology. A standard history of biology and a good companion for Locy.

Taylor, Gordon Rattray. *The Science of Life: A Picture History of Biology*. McGraw-Hill, 1963. 368 pp. illus. $9.95. 62–22414. (SH)

A profusely illustrated history that spans the whole period of man's study of living things up to the publication date; contains a chronological table. Good for reference by beginning students and for cultural reading by everyone.

574.1 GENERAL PHYSIOLOGY

Hall, Thomas S. *Ideas of Life and Matter: Studies in the History of General Physiology, 600 B.C.–1900 A.D.* Vol. 1: *From Pre-Socratic Times to the Enlightenment;* Vol. 2: *From the Enlightenment to the End of the Nineteenth Century.* U. of Chicago, 1969. xii+419 pp.; vii+399 pp. illus. $20.00 set. 69–16999. (SH–C)
Hall sees the history of physiology as an attempt to find a solution to the "life-matter" problem. He views the changing answers to the questions "what is life?" and "what is matter?" in historic context and attempts to trace the interaction of the answers from the pre-Socratic era to the beginnings of the twentieth century.

Michelmore, Susan. *Sexual Reproduction.* Nat. Hist. Press, 1965. 229 pp. illus. $4.50, paper $1.25. 65–10688. (SH–C)
A fascinating, easy-to-read story. Many examples are used to illustrate each phase of sexual reproduction, but many of the phenomena—examples and concepts—are given only superficial treatment. Diagrams are few but clear. Most of the text deals chiefly with animals and very little mention, except for lower forms, is made of plants. Indexed, but no bibliography.

Telfer, William H., and Donald Kennedy. *The Biology of Organisms.* Wiley, 1965. xi+374 pp. illus. $7.95. 65–19472. (SH–C)
Deals with the cell and the organism and the classification of organisms, the biology of microorganisms, multicellular organisms, growth, metabolism, and endocrine and nervous controls. This integrated consideration of embryology and metabolism provides insights into the modern findings of biophysics and biochemistry. Students need a good background in general biology, physics, and chemistry. A brief but significant list of additional references. Good reference and collateral study material.

574.19 BIOPHYSICS AND BIOCHEMISTRY

Ackerman, Eugene. *Biophysical Science.* Prentice-Hall, 1962. xiv+626 pp. illus. $15.95 (text ed. $12.95). 62–11880. (C)
A more detailed and rigorous presentation than Casey's, dealing with the major topics of special sensory systems, nerves and muscles, physical microbiology, molecular biology, thermodynamics and transport systems, and specialized instrumentation. The mathematical level is that of elementary calculus.

Baker, Jeffrey J. W., and Garland E. Allen. *Matter, Energy, and Life: An Introduction for Biology Students.* (Principles of Biology Series.) Addison-Wesley, 1965. xi+180 pp. illus. $3.95, paper $2.95. 64–20832. (JH–SH–C)
Chemical concepts are presented in comprehensible terms so that the reader may appreciate the intra-doctrine principles of science. Includes discussions of theories of chemical bonding, energy conversions, molecular versus organismic structuralization, and molecular activity. Includes adequate amino acid tables, reading lists, good illustrations and an index.

Baldwin, Ernest. *The Nature of Biochemistry.* Cambridge, 1962. xiii+111 pp. illus. $3.50, paper $1.65. 62–52370. (C)
A sound and sophisticated presentation of what biochemistry is about, to

be read by those with some background in biology and elementary organic chemistry. A superb cultural experience for high school teachers and college undergraduates.

Bennett, Thomas Peter, and Earl Frieden. *Modern Topics in Biochemistry: Structure and Function of Biological Molecules.* Macmillan, 1966. vi+186 pp. illus. $2.95 (paper). 66–17380. (SH–C)
A guide book for beginners to modern molecular biochemistry. Abundant and excellent diagrams. It is exceptionally well written: the clarity bears marks of genius in simplicity and elegance of presentation. A superb book easily read by anyone with a high school knowledge of chemistry. Can serve as a quick review even for professionals.

Borek, Ernest. *The Atoms Within US.* Columbia, 1961. 272 pp. illus. $6.00, paper $1.95. 61–6159. (SH)
Traces a path through enzymes, vitamins, metabolism, isotope techniques, the architecture of proteins, RNA and DNA, to present—via the historical route—an overview of modern molecular biology. This book brings up to date the author's earlier and successful excursion into biochemistry, *Man, the Chemical Machine* (out of print).

Chambers, Robert W., and Alma Smith Payne. *From Cell to Test Tube.* Scribner's 1960. 216 pp. illus. $3.50 paper $1.45. 60–6412. (JH)
An introduction for the younger student and interested layman to the chemistry of living organisms, or how these complicated chemical machines function.

Conn, Eric E., and P. K. Stumpf. *Outlines of Biochemistry* (2nd ed.). Wiley, 1963. 391 pp. illus. $9.95. 63–8054. (C)
An introductory textbook of biochemistry for those with a basic elementary knowledge of biology, physics, chemistry, and mathematics. For larger secondary schools with many talented students, and for their teachers, it is a valuable technical reference.

Downes, Helen R. *The Chemistry of Living Cells* (2nd ed.). Harper & Row, 1962. 645 pp. illus. $10.75. 62–7431. (C)
An up-to-date text for students with a background in biology and college chemistry. The first part is historical and explains the properties of solutions and the structure of living forms; the second deals with the organic constituents of cells; the last part with studies of metabolism. Includes references and a good index.

Frankel, Edward. *DNA—Ladder of Life.* McGraw-Hill, 1964. 127 pp. illus. $2.95. 64–24599. (JH–SH)
An elementary introduction to biochemistry, discusses the composition of DNA and tells how proteins are broken down by enzymes and reassembled by RNA "templates" produced by DNA in the cell nucleus. It concludes with a discussion of the role of DNA in biology and medicine. Brief reading list and index.

Hutchins, Carleen M. *Life's Key-DNA: A Biological Adventure into the Unknown.* Coward-McCann, 1961. 64 pp. illus. LB $2.86. 61–13411. (JH)
Theories and discoveries of the last two decades have made deoxyribonucleic acid a very exciting topic in biochemistry, since it probably provides the basis for a better understanding of the nature and processes of heredity

and evolution as well as of malignancies and other metabolic diseases. This
may appear as a child's book, but its contents are for mature young people
and adults.

McElroy, William D. *Cell Physiology and Biochemistry* (2nd ed.). (Foundations of Modern Biology Series.) Prentice-Hall, 1964. viii+120 pp. illus.
$3.95, paper $1.95. 64–12159. (C)
Explains the mechanism of the cell and how it functions with up-to-date treatment of DNA and RNA structure. Also presents information on the chemistry of photosynthesis, control of protein synthesis, and control of cell function.

Mahler, Henry R., and Eugene H. Cordes. *Basic Biological Chemistry*. Harper
& Row, 1968. x+527 pp. illus. $10.75. 68–12927. (C)
An excellent introduction to the fundamentals of biochemistry that emphasizes metabolism rather than physical chemistry. References for each
chapter. Recent work on current topics such as RNA structure is included.

Mallette, M. Frank, Paul M. Althouse, and Carol O. Claggett. *Biochemistry of
Plants and Animals: An Introduction*. Wiley, 1960. 552 pp. illus. $9.95.
60–11726. (C)
General biochemistry, the biochemistry of plants, and the biochemistry of
animals, constitute the major divisions of this text. Basic training in organic
and inorganic chemistry is presumed, and the book is intended primarily for
those who intend to major in the agricultural sciences.

Oparin, Aleksandr I. *The Origin of Life on the Earth* (3rd ed.). (Tr. from
the Russian by Ann Sunge.) Academic, 1957. xviii+495 pp. illus. $7.50;
Dover (2nd ed.; tr. by S. Margulis), paper $1.75. 58–1444. (C)
Presents a classical discourse on the chemical basis of ontogeny, and in
each revision the author has rethought his basic arguments as dictated by
new research findings.

Stern, Herbert, and David L. Nanney. *The Biology of Cells*. Wiley, 1965.
xii+548 pp. illus. $8.95. 65–19473. (SH–C)
Develops for biological science the basic generalizations of physics, chemistry and physiology for which those studying biology at the university
level now usually are prepared. Deals with the cell doctrine and the gene
concept, and the search for physico-chemical mechanisms by considering
properties of aqueous solutions, impact of organic chemistry and study of
chemical change. Considers cellular substance in detail, molecular aspects,
and regulation of cell behavior. Introduces a revolutionary turn to the study
of modern biology.

Sullivan, Navin. *The Message of the Genes*. (Science and Discovery Series.)
Basic, 1967. vi+198 pp. illus. $5.95. 67–17395. (SH–C)
Anything that concerns man's growing understanding of the biochemistry
of life should be understood not only by the scientist, but by the layman too.
This uncomplicated but accurate survey of molecular biology is presented with
the latter in mind. The reader is introduced to the notion of a gene, informed of its chemical structure and how it functions, and in the last chapter
he is included in a bit of speculation on future prospects of "biological engineering," i.e. manipulation of genes to improve the human race. Useful as
supplementary reading in a college biology course. Thoughtful high school

students will find the simpler concepts stimulating, though certain portions of the book might prove too difficult.

Williams, Roger J., and Edwin M. Lansford, Jr. (eds.). *The Encyclopedia of Biochemistry*. Reinhold, 1967. xvii+876 pp. illus. $25.00. 67–15466. (C)
Contains valuable information, but caution must be exercised when using it as a reference: some major topics are well presented, whereas others are sketchy. Each article cites reference material. A vast amount of worthwhile material with few errors, in a compendium that will supply readers who wish to supplement standard biochemistry texts with concise references to specific topics. Useful in college, university and research libraries, and to students and professional workers.

574.5 BIOECOLOGY

Allen, Durward L. *The Life of Prairies and Plains*. (Our Living World of Nature Series.) McGraw-Hill, 1967. 232 pp. illus. $4.95. 67–15849. (JH–SH)
This excellent work on desert ecology is well illustrated with a variety of color photographs, drawings, sketches, and diagrams.

Amos, William H. *The Life of the Pond*. (Our Living World of Nature Series.) McGraw-Hill, 1967. 232 pp. illus. $4.95. 67–16306. (JH–SH–C)
A vivid portrayal of the life of a pond from its formation, growth, and natural destruction and its myriad of inhabitants. Good color photographs. A seasonal approach is presented in one chapter, the animals and plant species in another, followed by an examination of little-known pond inhabitants. Glossary, bibliography, and index included.

Bates, Marston. *The Forest and the Sea*. Random, 1960. vii+277 pp. $4.95, paper (Vintage V292) $1.65. 60–5564. (SH)
An examination of the life of forests, lakes, grasslands, deserts, coral reefs and the open seas, not for the purpose of describing life in nature, per se, but as an attempt to demonstrate how man must learn to live harmoniously and in cooperation with the other occupants of his own biological environment.

Bonner, John Tyler. *Cells and Societies*. Princeton U., 1955. 234 pp. illus. $4.50. 55–5002. (JH)
Vivid descriptions of the ways in which different forms of life carry out their respective life cycles—reproduction, feeding, social coordination—make up this study. Examples are diverse, including mammals, social insects, and even protozoa.

Buchsbaum, Ralph, and Mildred Buchsbaum. *Basic Ecology*. Boxwood, 1957. viii+192 pp. illus. $3.50, paper $2.35. 57–28684. (JH)
Living plants and animals cannot be studied thoroughly without a consideration of their natural environment, and the various forces, interactions and associations inherent therein. Plans for the utilization, conservation and restoration of natural living resources must be based upon a consideration of the ecological complex. Here is a good, simplified presentation.

Carson, Rachel. *Silent Spring*. Houghton Mifflin, 1952. x+368 pp. illus. $5.95; Fawcett (Crest T681, 1964), paper 75¢. 60–5148. (SH–C)

A foremost biologist and writer spent over four years obtaining data from all over the world, but principally in North America, demonstrating the effect of chemical pesticides upon the flora and fauna in addition to those species that were intended to be brought under control. Although highly controversial, the book has stimulated serious concern for our resources and has initiated comprehensive testing and evaluation studies. Very necessary for all libraries and valuable to all readers.

Collins, W. B. *The Perpetual Forest.* Lippincott, 1959. 288 pp. $4.50. 59–7773. (SH)
Twenty years of work and study in the tropical forest have provided the author with an opportunity to observe and to study its peculiar and endless cycles of life, death and decay. The day-to-day experiences of a man dedicated to the preservation of the forests have resulted in excellent word-pictures. No illustrations.

Dasmann, Raymond F. *A Different Kind of Country.* Macmillan, 1968. viii+276 pp. illus. $6.95. 68–11193. (C)
Presents an understanding of man's relationships to his environment amid the physical and political forces that are shaping a constantly changing world. A plea for the "preservation of natural diversity" in which the author suggests management programs that could bring highly desirable results in a relatively short time.

Eckert, Allan W. *Wild Season.* (Illus. by Karl E. Karalus.) Little, Brown, 1967. 244 pp. $4.95. 67–14449. (JH–SH–C)
Written with meticulous care for such biological principles as the energy cycle, instinctive behavior as defined by the ethologists, and a somewhat modernized "survival of the fittest." In his diary-like account of the month of May in and around Oak Lake (near the Illinois-Wisconsin border), his descriptions of the constant series of predator-prey encounters make interesting but not necessarily pleasant reading. Descriptions of the appealing aspects of a natural environment and the animal's behavior in it are well done.

Engel, Fritz-Martin. *Life Around Us: The Strange Planet Earth and the Stranger Creatures that Live on It.* Crowell, 1966. 206 pp. illus. $6.95. 65–13816. (JH–SH)
Extraordinary and as broad as ecology. Short declarative sentences and incisive presentation. Principles are stated, examples predominate. These provide a comprehensive survey of the world's plants and animals, large and small, fitted into the intricate pattern of their interrelationships and environments.

Farb, Peter. *Living Earth.* Harper & Row, 1959. ix+178 pp. illus. $4.95; LB $4.43. 59–6330. (SH)
An ecological study of those animals that live in the soil in the three main types of environment—the forest, the grassland, and the desert—each with its characteristic living communities.

Farb, Peter, and The Editors of *LIFE. Ecology.* (Life Nature Library.) Time, Inc., 1963. 192 pp. illus. $6.60. 63–22074. (JH–SH)
Prince Bernhard of the Netherlands, in the introduction, points out that it is the duty of the ecologists to open the eyes of all mankind to the delicate balance of nature and to the relationships between the living organism and

its surroundings, for preservation of our natural resources should be the concern of all men.

Grossman, Shelly. *The Struggle for Life in the Animal World.* Grosset & Dunlap, 1967. 128 pp. illus. $4.95. 67–23802. (JH–SH)

This is the story of the North American biomes, their plant and animal life, and the complex pattern of relationships, with emphasis on predator-prey. Excellent factual presentation and photographs add greatly to the text.

Hirsch, S. Carl. *The Living Community: A Venture into Ecology.* (Illus. by William Steinel.) Viking, 1966. 128 pp. LB $3.56. 66–14415. (JH)

Three basic biological concepts of themes are brought into focus for the young student: ecology, evolution and conservation. Stimulates serious thought and opens doors for further studies and debate. Because of its universal themes and their importance to man and his communities, this book is recommended for *all* young students; not exclusively those interested in biology or the other sciences.

Kormondy, Edward J. (ed.). *Readings in Ecology.* Prentice-Hall, 1965. xiv+ 219 pp. illus. $4.50 (paper). 65–26924. (SH–C)

A wide range of papers that have contributed significantly to the currently-held, broad concepts of ecology. Specifically, the field of ecology is treated by studies separated into the physical and chemical environment, plant and animal communities, population dynamics and concepts of the ecosystem. Each paper, which may be complete or a pertinent portion of a larger work, is preceded by editorial comments regarding the contents and, in many cases, their importance in developing new ideas. Good for collateral reading in a course in ecology for college juniors and seniors. Will prove of considerable value in advanced biology courses in high school.

Krutch, Joseph Wood. *The Voice of the Desert: A Naturalist's Interpretation.* Sloane (dist. by Morrow), 1955. 223 pp. illus. $5.00. 55–8485. (JH)

This book examines the plants and animals which the author has observed in a desert of Arizona, and the style is the philosophical approach of an informed enthusiast.

MacArthur, Robert, and Joseph Connell. *The Biology of Populations.* Wiley, 1966. xv+200 pp. illus. $6.95. 66–21070. (C)

Fills a void in background understanding for courses in ecology. Examines environmental patterns, population evolution, and population biology. The text is complimented with illustrative tables and figures.

McCormick, Jack. *The Life of the Forest.* (Our Living World of Nature Series.) McGraw-Hill, 1966. 232 pp. illus. $4.95. 66–14241. (JH–SH)

An excellent account of plant and animal communities in the forests of North America. Treats the ecologic forest system by season and geographic location. While designed for younger readers, even the specialist will enjoy the photographs.

Odum, Eugene P., and Howard T. Odum. *Fundamentals of Ecology* (2nd ed.). Saunders, 1959. 546 pp. illus. $7.50. 58–12125. (C)

Biologists and the public alike are realizing that ecological research is vital to the solution of mankind's environmental problems. Part I deals with basic principles and concepts. Part II emphasizes descriptive phases. Part

III deals with practical applications. Readers should have had basic training in biology and chemistry.

Platt, Robert B., and John F. Griffiths. *Environmental Measurement and Interpretation.* Reinhold, 1964. xii+235 pp. illus. $10.00. 64–20139. (SH–C)
The first part is an excellent presentation of the philosophy and reasoning of research—making it especially useful for those just entering some research endeavor. The remainder is a complete but elementary presentation of methods of measuring environmental variables. Useful as a supplementary text for undergraduate courses in ecology, agriculture and horticulture.

Pruitt, William O., Jr. *Animals of the North.* (Illus. by William D. Berry.) Harper & Row, 1967. 173 pp. $5.95. 66–13923. (JH–SH–C)
Combines qualities rarely found into an extraordinary book. The evocative, narrative style gives the reader a vivid picture of plant and animal life in the northland, the changes through seasons and over years, and the impact of man. Moves easily, and with fine detail, into meteorology, population dynamics, ecology, soil conservation, and the behavior of snow masses. This is an adult book, but its narrative quality should attract many younger readers.

Raskin, Edith. *The Pyramid of Living Things.* (Illus. by Joseph Cellini.) McGraw-Hill, 1967. 192 pp. $4.00. 67–22961. (JH–SH)
Considers the flora and fauna of major areas from the Arctic tundra to the tropical rain forests and savanna. Shows that in each biome, whether in the water or on the land, highly productive green plants, lichens, mosses or microscopic plants and animals form the base of a huge triangle of living things. Discusses the decrease in productivity of the complex animals in the food chains of the different regions, ultimately culminating in man at the apex of the triangle.

Sears, Paul B. *Lands Beyond the Forest.* (Illus. by Stanley Wyatt.) Prentice-Hall, 1969. xii+206 pp. $7.95. 68–8126. (SH–C)
More than just a study of plant and animal communities, Sears has produced a study of ecology, anthropology, evolution, history, and economics. No aspect of the land has been omitted. The book considers the grasslands, the wetlands, the scrub forest and the savannah, the desert, the development of continents, the changes in the land, and the people on the land. Index is excellent.

Sears, Paul B. *The Living Landscape.* Basic, 1966. 199 pp. illus. $4.95. 66–23379. (SH–C)
Covers the sum and parts of ecology, including current topics of general interest. While logically unified, students may use this as a supplemental reference with work in the biological and earth sciences.

Schmidt-Nielsen, Knut. *Desert Animals: Physiological Problems of Heat and Water.* Oxford, 1964. xiv+277 pp. illus. $7.50. 64–2298. (JH–SH–C)
Explores all aspects of the problem of desert existence, including adaptations made by camels, donkeys, sheep, squirrels, birds, reptiles and man himself. The physiological functions of heat stress, water loss and temperature regulations are thoroughly discussed in relation to body size, color and the animal concerned. Also, some myths are resolved. Highly readable.

Shelford, Victor E. *The Ecology of North America.* U. of Illinois, 1963. xxii+
610 pp. illus. $10.00. 63–7255. (C)
The author is known to thousands as the "Dean of Ecology" for he, more
than any other man, has raised the study of floras and faunas from mere
descriptive natural history to an orderly scientific specialty that is fundamental to all studies of life in its native habitats. This book is the only
available comprehensive discussion of the ecology of North America from
1500–1600—during the period that preceded European colonization and
the consequent utilization and exploitation of the natural resources.

Wallace, Bruce, and Adrian M. Srb. *Adaptation* (2nd ed.). (Foundations
of Modern Biology Series.) Prentice-Hall, 1964. 113 pp. illus. $3.95, paper
$1.95. (SH–C)
Traces the phenomenon of adaptation of living things to their changing environments and describes the evolution of natural populations. Ecology, gene
pools, formation of species, and communication also are discussed.

574.8 HISTOLOGY AND CYTOLOGY

Afzelius, Björn. *Anatomy of the Cell.* (Tr. by Birgit H. Satir.) U. of Chicago,
1966. 127 pp. illus. $6.50. 66–13860. (SH–C)
Up-to-date, illustrated account of the anatomy of the cell; excellent collateral
reading for beginning students in biology. The book begins with a description
of cellular ultrastructures and their functional enzymes. Well-written chapters on mitochondria, cell membranes, nucleus, ribosomes, microbodies, and
lysosomes. Structure and function are emphasized throughout. Good organization and adequate references.

Austin, Colin R. *Fertilization.* (Foundations of Developmental Biology Series.)
Prentice-Hall, 1965. 145 pp. illus. $4.95, paper $2.95. 65–17794. (SH–C)
A general coverage of fertilization, and of the various cytological, physiological and behavioral mechanisms concerned with the union of the gametes.
Emphasizes the principles and mechanisms of fertilization common to all
levels. Each chapter begins with a fundamental concept, illustrated by detailed examples in a variety of organisms. A great volume of specific data
on the variations make this book a valuable comparative study. Remarkable
for its clarity and ease of comprehension, supplemented with numerous
illustrations and classified and partly annotated. Bibliography.

Chambers, Robert, and Edward L. Chambers. *Explorations into the Nature of
the Living Cell.* Harvard U., 1961. xxiv+352 pp. illus. $8.00. 61–8845. (C)
Presents the lifetime research of the senior author and his co-workers in
relation to advances in cellular biology. Micromanipulative methods have
enabled him to make excellent detailed studies of individual living cells. An
unusually good bibliography is appended, enhancing the book's value for
advanced students.

Cohn, Norman S. *Elements of Cytology.* Harcourt, Brace, 1964. xvi+368 pp.
illus. $9.75. 64–15777. (C)
Modern cytologists are no longer content with identification and descriptions
of cells and of their components and activities, but consider these elements
meaningful only if they are related to the heredity, physiology and develop-

ment of the whole organism. This introduction is developed on that broad comprehensive base. Similar to the text by Wilson and Morrison, but perhaps less suited for high school student reference, and preferable for more mature readers.

Giese, Arthur C. *Cell Physiology* (2nd ed.). Saunders, 1962. xxi+592 pp. illus. $11.50. 62–11601. (C)

Students with a background in the basics of biology, chemistry, physics, and some mathematics will find this a useful text and reference work. Its major divisions are "Functional organization of the cell," "The cell environment," "Exchange of materials across the cell membrane," "Conversions of energy and matter," "Irritability and contractibility," and "History of cell physiology."

Hughes, Arthur. *A History of Cytology*. Abelard-Schuman, 1959. x+158 pp. illus. $5.00. 59–5337. (SH)

The origins of cytology are traced from the beginnings of microscopical investigations in the 17th century through the step-by-step progress that was made in the subsequent 200 years, culminating in a description of present-day types of research with highly refined and specialized equipment.

Scientific American, Readings from. *From Cell to Organism*. (Introd. by Donald Kennedy.) Freeman, 1967. vi+256 pp. illus. $9.00, paper $4.45. 66–30156. (SH–C)

These 24 reprinted articles lead the reader from the simplest intercellular communication to the social behavior of complex organisms. They are woven together by brief comments of Donald Kennedy who introduces the common theme (e.g. exchange and transport, sensory receptors), shows how each article develops the theme, and, when appropriate, brings the subject up to date. The articles are written with obvious authority and style. Each topic is well illustrated with a few photographs and with drawings in which one color is used effectively.

Swanson, Carl P. *The Cell* (3rd ed.). (Foundations of Modern Biology Series) Prentice-Hall, 1969. xii+150 pp. illus. 32.5 x 19.5 cm. $5.95; $2.50 (paper). 69–10438. (SH–C)

The previous editions (1960 and 1964) have been widely used to provide for in-depth study of cellular structure, function, and behavior which is fundamental to an understanding of modern biology. The text has been thoroughly revised and updated to reflect the recent advances in knowledge of the cell and its components. Libraries that own one or both of the previous editions should replace them.

Weiner, Dora B. *Raspail: Scientist and Reformer*. Columbia, 1968. xiv+336 pp. illus. $11.00. 68–19761. (JH–SH–C)

This is a carefully documented and generally well-written biography of the founder of histochemistry and chemical microscopy. The reader is given a vivid picture of a brilliant young scientist whose arrogance, politics, and compassion for the underprivileged became his worst enemy of his scientific pursuits.

Wilson, George B., and John H. Morrison. *Cytology* (2nd ed.). Reinhold, 1966. 297 pp. illus. $8.95. 61–17442. (SH–C)

The authors have provided an excellent "map" of the field with the expectation that the student will fill in the details by liberal collateral reading from among the list of references that concludes each chapter. Although intended as an introductory college text, the high school student can readily read and understand chapters or other parts selected and assigned by the teachers for collateral study.

574.9 NATURAL HISTORY

Baker, Laura Nelson. *A Tree Called Moses.* (Illus. by Penelope Naylor.) Atheneum, 1966. x+69 pp. LB $3.41. AC 66–10408. (JH–SH)
Fascinating narrative, full of dramatic interest, partly imagined as to minor details and conversational aspects, but the basic facts, place names, and persons are genuine. One may learn much basic natural history of the flora and some of the fauna, the ecology of the redwoods, the menace of storms and fire. The drawings are attractive and the author's list of sources suggests additional reading.

Bartlett, Jen, and Des Bartlett. *Nature's Paradise, Africa.* Houghton Mifflin, 1967. 360 pp. illus. 32.5 cm. $19.95. 66–19834. (JH–SH–C)
Tells how mammals, birds, reptiles, fishes, and insects of Africa depend on their habitat and on each other. Organized by regional sections on the coral reef, thornbush country, open bush and woodland, plains, seasonal changes, lakes, forests, and high mountains.

Brooks, Maurice G. *The Appalachians.* (The Naturalist's America Series.) (Illus. by Lois and Louis Darling.) Houghton Mifflin, 1965. xvii+346 pp. $6.95. 64–12870. (SH–C)
Beautiful drawings, and outstanding photographs in black and white and color, combined with excellent natural history writing provide an informative tour of this mountain range. Presents all aspects of geology, botany, forestry, zoology, and finally the peoples with their culture and crafts. Enjoyable and informative to read for pleasure, assigned collateral reading by students in the natural and social sciences, and—aided by the index—a reference work on Appalachia.

Brooks, Maurice G. *The Life of the Mountains.* (Our Living World of Nature Series.) McGraw-Hill, 1967. 232 pp. illus. $3.95. 67–16307. (JH–SH)
Escorts the reader on a trip to many of the wonderful sights in the mountain ranges of North America: shows the trees, flowers, insects, birds, and mammals found at various altitudes. Includes a great deal of physical geology and ecology.

Brown, Leslie. *Africa: A Natural History.* (The Continents We Live On Series.) Random, 1965. 299 pp. illus. (part col.) 31 cms. $20.00. 65–19636. (JH–SH–C)
A vivid descriptive natural history of most of Africa arranged largely by geographic regions. The climate, surface geology, and ecology of each area are mentioned with discussions of the fauna. Tends to emphasize the larger mammals and only incidentally considers other groups of vertebrates and the invertebrates. Repeatedly mentions the beneficial and the harmful effects of man upon the living resources of Africa. Good recreational and background reading in natural history and zoogeography. Indexed, no bibliography.

Cruickshank, Helen Gere (ed.). *John and William Bartram's America.* Devin Adair, 1957. xxii+418 pp. illus. $6.00. 57–8862. (JH)
The Bartrams, contemporaries of Benjamin Franklin, established the first botanical garden in the Western Hemisphere (in Philadelphia). William Bartram made the first natural history exploration of many places in the Southeastern United States. The book consists of a biological sketch of father and son, and excerpts from their journals and correspondence.

Darwin, Charles. *The Voyage of the Beagle.* Dutton, 1906. 469 pp. $2.25. (Various paperback editions of other publishers.) 36–37576. (JH–SH)
A history of Darwin's famous voyage and of his observations in natural history and geology which has furnished delightful reading and enlightenment for more than a century.

Farb, Peter. *Face of North America: The Natural History of A Continent.* Harper & Row, 1963. xv+316 pp. illus. $7.95. 62–14598. (SH)
Marston Bates thinks that everyone who tours North America should carry this book in his car, for the appendix lists national parks and forests, suggests places to visit and outlines special natural history areas of each State and Canadian Province. The author is a well-known science writer who provides here a solid background of the continent's geological and natural history. Good reading for armchair tourists and tourists-in-fact.

Hay, John, and Peter Farb. *The Atlantic Shore: Human and Natural History from Long Island to Labrador.* (Illus. by Edward and Marcia Norman.) Harper & Row, 1966. 246 pp. $6.95. 66–13919. (SH–C)
A clear perspective of the general biology and the complex ecology of the organisms living along the North Atlantic shore, the geological and meteorological forces that worked together to create this environment and the influences of man in shaping this stretch of land. An appendix provides a useful abbreviated guide to the natural areas of the North Atlantic Coast and a list of suggested readings. Indexed.

Headstrom, Richard. *Nature in Miniature.* Knopf, 1968. xviii+412+xiii pp. illus. $10.00. 67–11122. (JH–SH)
Deals with the semi-microscopic forms in the plant and animal kingdoms, and the necessary apparatus and procedures needed for investigation. A useful reference with a good index, glossary, and a selected bibliography. Good nature writing.

Hey, Douglas. *Wildlife Heritage of South Africa.* Oxford, 1966. 246 pp. illus. $11.30. (JH–SH)
Illustrated descriptive account of the South African flora and fauna which contains good natural history information on the condition of the species, and recommendations for their preservation. Of great interest to students and laymen who want to learn about the unusual species of wildlife in South Africa. Recommended as a supplement to book collections that have adequate material on North American and European flora and fauna.

Hillcourt, William. *Field Book of Nature Activities and Conservation.* Putnam's 1961. 432 pp. illus. $4.95. 61–10331. (JH)
A general hobbyist's guide for the amateur or young student involving the collection and study of rocks, minerals, fossils, plants and animals, with suggestions as to equipment and projects to be undertaken by individuals or

groups. The user will need other references for classification and identification of his collections and these are suggested.

Hylander, Clarence J. *Wildlife Communities: From the Tundra to the Tropics in North America.* Houghton Mifflin, 1966. 342 pp. illus. $5.00. 66–19940. (JH–SH)

This lucid outline of basic ecology is presented in a logical and easily read fashion. Part I takes the reader on "An ecological adventure" in a typical North American forest and then expands on the limiting factors, physical and living. Part II describes many examples of representative flora and fauna in each biome occurring in North America. Well illustrated by maps, diagrams, drawings and photographs. The concluding chapter is an extensive (but not complete) list of wildlife sanctuaries with a description of their biota in North America.

Jaeger, Edmund C. *North American Deserts.* Stanford, 1957. vii+308 pp. illus. $5.95. 57–9307. (SH)

A retired zoology professor, widely traveled, presents a comprehensive narrative study of all of the North American deserts, five in number, extending from central Mexico almost to the Canadian border. Numerous photographs reveal their natural beauty and their characteristic flora and fauna. Bibliography.

Kieran, John. *An Introduction to Nature: Birds—Wild Flowers—Trees.* (Illus. by Don Eckelberry, Tabea Hofmann, Michael H. Bevans.) Doubleday, 1966. 223 pp. $7.50. 66–10864. (JH–SH)

John Kieran has written a new introduction for this combined reprint of his three older books: *An Introduction to Birds* (1965); *An Introduction to Wild Flowers* (1948, 1952); and *An Introduction to Trees* (1954). The 300 colored illustrations and descriptions are equally divided among the birds, flowers, and trees.

Kieran, John. *A Natural History of New York City.* Houghton Mifflin, 1959. xi+428 pp. illus. $6.25. 59–7703. (JH)

Even in the largest metropolis in the Western Hemisphere, the opportunity for studying and enjoying natural history has not been obliterated. Since boyhood, Kieran has looked for and made notes on all forms of natural wildlife in the five boroughs of the city and now he shares these with his readers. Those who do not live in New York will enjoy this book too, for it will encourage amateurs to become observant in other urban localities.

Leopold, Aldo. *A Sand County Almanac, with Other Essays on Conservation from Round River.* (Illus. by Charles W. Schwartz.) Oxford, 1966. xv+ 269 pp. $6.50. 66–28871. (SH–C)

In this republication all of *The Sand County Almanac* is included, together with eight essays from *Round River*, rearranged in sequence. A clear presentation of the only sound basis for conservation of natural resources— a love and respect for the land and resources themselves. A highly recommended acquisition for all libraries, and worthwhile reading for all students and laymen.

Matthiessen, Peter. *Wildlife in America.* Viking, 1959. 304 pp. illus. $6.00, paper (Compass C148) $1.95. 59–11635. (JH)

The writer has prepared a comprehensive story of American wildlife from the time of the first colonists to the approximate present. It presents a progressive tale of the decimation, and sometimes obliteration, of important species. The book is well documented and all Americans interested in the natural living resources should read it.

Milne, Lorus J., and Margery Milne. *The Valley: Meadow, Grove, and Stream.* Harper & Row, 1963. xii+178 pp. illus. $5.95. 62–14597. (JH)
Guides the reader through meadows and woodlands, along riffles and ponds, to observe the plants and animals in an environment common to most glaciated valleys in the temperate zone. The style is somewhat "precious" but the tale will interest almost everybody.

Neill, Wilfred T. *The Geography of Life.* Columbia, 1969. xvi+480 pp. illus. $12.95. 68–8877. (C)
A discussion of the scope of biogeography and the sources of biogeographic data, this book covers both botanical and zoological material. The most thought provoking chapters are on the animals and plants together in the changing environment, principally since the Jurassic. A lively, fast moving book to read many times. Appended list of literature and a good index make it a useful reference work.

Peterson, Roger Tory, and James Fisher. *Wild America.* Houghton Mifflin, 1955. xii+434 pp. illus. $6.00; Sentinel 35, paper $2.85. 55–876. (SH)
The authors covered much of the North American Continent in 100 days, following partially the trail made by Audubon 100 years before, beginning in Newfoundland and ending in the Pribilof Islands in the Bering Sea. This is the absorbing story of their travels and observations.

Peterson, Russell. *Another View of the City.* McGraw-Hill, 1967. xiii+220 pp. illus. $6.50. 67–15039. (SH–C)
A sensitive personal view of nature is in the tradition of Thoreau, Blatchley, Krutch, and others, describing a year in the natural history of a yet-to-be-urbanized coastal backwater in New Jersey. The author's prose style makes pleasurable reading. Provides us with a view of the city from the sort of vantage point that few of us will ever again have an opportunity to occupy. Students and laymen will find the book thoroughly absorbing. Provides anecdotal fodder for endless debate on the precepts and goals of conservation.

Pimentel, Richard A. *Natural History.* Reinhold, 1963. xii+436 pp. illus. $10.00. 63–21379. (JH–SH)
A general work that introduces the reader to the full content of natural history—the substrate as well as the forms of plant and animals life thereon and therein. A straightforward professional book for students and novices of all ages. Technical terms are italicized in the text and defined in the glossary. Useful in all collections.

Roedelberger, Franz A., and Vera I. Groschoff. *African Wildlife.* (Tr. by Nieter O'Leary and Pamela Paulet; introd. by Edwin Way Teale.) Viking, 1965. 224 pp. illus. $8.95. 65–11630. (JH–SH–C)
Magnificent monochrome and colored action photographs combined with a descriptive narrative have resulted in this panorama of typical insects, fishes, amphibians, birds and mammals which represent the African fauna. Back-

ground and information on natural history, ecology, folklore, and conservation. It is a good book for all libraries and personal collections.

Rublowsky, John. *Nature in the City.* Basic, 1967. viii+152 pp. illus. $4.95.
67–13776. (JH–SH)

For the city-stranded student, this is an excellent guide to urban plant and
animal life. Considerable emphasis is placed on bird life, however there is
good coverage of all animals dependent on man and his activities.

Silverberg, Robert. *The Auk, the Dodo and the Oryx: Vanished and Vanishing
Creatures.* (Illus. by Jacques Hnizdovsky.) Crowell, 1967. 247 pp. $3.95.
AC 67–10476. (JH–SH)

Discusses trends of species diminution, and contrasts natural extinction with
extermination by acts of man. A most informative book for anyone interested
in natural history and wildlife conservation.

Sterling, Dorothy. *The Outer Lands.* (Illus. by Winifred Lubell.) Nat. Hist.
Press, 1967. 192 pp. $4.50. 67–11253. (JH–SH)

A guide to the natural history of the New England coast. The subject
matter is limited to the ecological communities of the shore and coast and
is therefore of most interest in New England schools and libraries.

Teale, Edwin Way. *Autumn Across America.* Dodd, Mead, 1956. xviii+386
pp. illus. $6.95. 56–10059. (JH–SH)

A famous nature writer takes his readers on a leisurely westward journey
during which they poke into out-of-the-way places, encounter interesting
people, and see the signs of coming winter in the migrations, seasonal
reactions, and other manifestations of various invertebrate and vertebrate
forms of life.

Teale, Edwin Way. *Journey into Summer.* Dodd, Mead, 1960. xviii+366 pp.
illus. $6.95. 60–11923. (JH–SH)

While reading this book one can feel the warm laziness of a summer day as
the last of the spring flowers are around, the clouds sweep by on a gentle
wind, and you experience the silence of the woodlands broken only by the
scurryings of the insects, birds and other forms of life. You will visit 26
States beginning at the White and Green Mountains to the western Rockies,
and your path will be a meandering one.

Teale, Edwin Way. *North with the Spring.* Dodd, Mead, 1951. xviii+358 pp.
illus. $6.95. 51–13966. (JH–SH)

Spring has a different meaning for almost everyone, but it involves a sense of
awakening and of excitement. Spring moves northward in eastern America
at the rate of about 15 miles a day. Beginning at the Florida Everglades
the author takes you on a trip covering 17,000 miles in 23 States in an
elapsed time of 130 days. The story of this adventure is a composite natural
history study of Spring.

Teale, Edwin Way. *Wandering Through Winter.* Dodd, Mead, 1965. xx+370 pp.
illus. $6.95. 65–23773. (JH–SH)

Rich in observations about nature and full of examples relating the history of
science to modern studies. Geographic, climatological, physical, and biological
observations hold the reader's interest. As collateral reading and as a reference the book should help to interest young persons in scientific work.

574.92 AQUATIC BIOLOGY (See also 551.46 OCEANOGRAPHY)

Carson, Rachel L. *The Edge of the Sea.* Houghton Mifflin, 1955. viii+276 pp.
  illus. $5.50; NAL (Signet P2360), paper 60¢. 54–10759. (JH)
  Miss Carson gives here a vivid description of the variety of life along the
  seashore in three types of environments—the rocky shore, the sand beach,
  and the coral coast. Bob Hines provides the illustrations for this exemplary
  piece of natural history writing which is a good companion for *The Sea
  Around Us* (Miss Carson's first best-seller which is listed under "Oceanography").

Duddington, C. L. *Flora of the Sea.* Crowell, 1966. 207 pp. illus. $6.95. 67–
  11665. (JH–SH)
  A very readable, lively, and for the most part nontechnical résumé. A well-
  rounded and understanding view of many aspects of the biology and economic
  uses of these organisms from diatoms to giant kelp. The book is an excellent source of this kind of material, ordinarily widely scattered and often
  hard to find. An abbreviated glossary and an index, no bibliography. A
  desirable reference for school and public libraries and for the layman naturalist.

Fraser, James. *Nature Adrift: The Story of Marine Plankton.* Dufour Editions,
  1962. v+178 pp. illus. $8.95. 62–20435. (SH)
  The tremendous popular appeal of the science of oceanography brings into
  focus the important study of the small plants and animals that provide food
  for countless larger forms of life. This layman's account explains collection
  and study methods, the principal planktonic groups, geographic and seasonal
  distribution, food chains, and the potential use of plankton as human food.

Hardy, Alister C. *The Open Sea* (2 vols. bound in 1). 1: *The World of Plankton; 2: Fish and Fisheries.* Houghton Mifflin, 1956, 1959. xv+335 pp; xvi+
  322 pp. illus. $15.00. A57–3220. (SH–C)
  Professor Hardy served as Chief Zoologist aboard the research vessels
  R. R. S. *Discovery I* and R. R. S. *Discovery II* of the British Ministry of
  Agriculture and Fisheries when he made the firsthand observations upon
  which these books are based. The color paintings are also the work of the
  author. These books are interesting to read and valuable for collateral
  study.

Idyll, Clarence P. *Abyss: The Deep Sea and the Creatures that Live in It.*
  Crowell, 1964. xviii +396 pp. illus. $6.95. 64–13911. (SH)
  The abyssal deeps of the oceans—the 80% of the oceans' cubic volume of
  which we know least—contains some of the most curious and fascinating
  forms of vertebrate and invertebrate life. In addition to describing the life
  at great depths, the author discusses the origin of the seas, their chemico-
  physical characteristics as an environment for life, and other basic facts. A
  book for the armchair traveler who desires relaxation and knowledge.

Klots, Elsie B. *The New Field Book of Freshwater Life.* (Illus. by Suzan
  Noguchi Swain.) Putnam's 1966. 398 pp. $4.95. 66–15583. (JH–SH–C)
  Clearly written and beautifully illustrated. Provides a clear account of all
  those fresh-water organisms, both vertebrate and invertebrate, that are likely

to be encountered by an inquiring collector in the fresh waters of the United States. The work is unusual in providing an introduction to the ecology of waters, and has an excellent glossary of ecological terms. Brief chapter on collecting and preserving. A 16-page key at the end will be invaluable to the collector in the identification of insect nymphs. The index is detailed and thorough.

MacGinitie, G. E., and Nettie MacGinitie. *Natural History of Marine Animals* (2nd ed.). McGraw-Hill, 1968. xii+523 pp. illus. $11.00. 67–24441. (SH–C)

Since 1949 this general descriptive account of marine life has been widely used. Initial chapters cover general topics of food, comparison of land and ocean fauna, animal relationships, and topics of morphology and physiology. Then chapters are devoted to the various phyla. Appendices list chemical elements in sea water, and notes on recent research related to chapter materials. Bibliography and index.

Miner, Roy Waldo. *Field Book of Seashore Life*. Putnam's, 1950. xv+888 pp. illus. $8.00. (JH–SH)

A pocket guide for amateurs to common invertebrates in the shallow waters of the Atlantic Coast from Labrador to the Cape Hatteras region of North Carolina, extending from the upper tide limit to the edge of the Continental Shelf. Arranged by phyla from the protozoa to the protochordates, with line drawings accompanying brief descriptive notes. A few color plates are interspersed.

Pennak, Robert W. *Fresh-Water Invertebrates of the United States*. Ronald, 1953. ix+769 pp. illus. $15.00. 52–12522. (C)

An indispensable manual for aquatic zoologists, limnologists, fishery biologists, and entomologists. Natural history, ecology, and taxonomy have been emphasized. In general it includes only the free-living, fresh-water invertebrates that occur in the United States. Cestodes, trematodes, parasitic nematodes, etc., are omitted. Some chapters contain keys to species, others only to genera. References at end of each chapter.

Phillips, Craig. *The Captive Sea*. Chilton, 1964. xii+284 pp. illus. $6.50. 64–11425. (SH–C)

Describes the behind-the-scenes activities of the operator of a marine aquarium, and fascinating peculiar characteristics of diverse sea creatures. Particularly suitable for young adults interested in aquaria.

Reid, George K. *Pond Life: A Guide to Common Plants and Animals of North American Ponds and Lakes*. (Illus. by Sally D. Kaicher and Tom Dolan.) Golden Press, 1967. 160 pp. $3.95. 67–16477. (JH–SH)

An aid in studying the total ecology of lakes and ponds and in identifying the plants and animals encountered there. Prefaced by general discussions of characteristics of lake and ponds, habitats, food webs, and instructions on collecting and preserving specimens. A sound fieldbook for the novice.

Russell, Frederick S., and Charles M. Yonge. *The Seas: Our Knowledge of Life in the Sea and How it is Gained* (rev. ed). Warne, 1963. xiii+376 pp. illus. $7.95. 63–10574. (SH)

A comprehensive account of all the principles, methods, and concepts of biological oceanography, together with descriptions of many representatives of the ocean's flora and fauna. Good assigned collateral reading for students and interesting reading for anyone.

Yonge, Charles M. *The Sea Shore*. Atheneum 45, 1963. xvi+311 pp. illus. $1.95. (paper). 49–6598. (SH)

A naturalist's delight is the best designation that can be given to this study of the flora and fauna along the shores of Britain. Since similar, but not identical, species occur in North American coastal waters, the book is very useful for students. The colored and black-and-white illustrations are excellent.

575 EVOLUTION

Bibby, Cyril. *T. H. Huxley: Scientist, Humanist and Educator*. Horizon, 1960. xxi+330 pp. illus. $5.00. 60–8165. (SH)

Huxley was the contemporary of Darwin and an outstanding protagonist of his theory of evolution. This biography, based on a careful study of original materials, presents Huxley as a scientific humanist and educator—an outstanding nineteenth-century thinker.

Broms, Allan. *Thus Life Began*. Doubleday, 1968. ix+326 pp. illus. $5.95. 67–15354. (SH–C)

Traces the science and history of living things from the early theories of spontaneous generation and vitalism to modern discussions of cellular biology. Scientific and historic documentations are good and not beyond the comprehension of advanced high school students and interested laymen.

Carlquist, Sherwin. *Island Life*. Nat. Hist. Press, 1965. viii+451 pp. illus. $10.95. 65–19897. (SH–C)

A book on evolution containing a unique compilation and collation of old and new information on the physical and biological characteristics of the isolated land masses. Beautifully documents the processes of evolution which are set in motion when species become separated from the competition and habitat of the mainland. Discusses most of the island groups and explains in lucid detail why certain animal and plant species are found on each and how all ecological niches have been occupied. Serves as a necessary adjunct to or even as a text for any course in biology in which evolution forms a portion of the subject material.

Colbert, H. Edwin. *Evolution of the Vertebrates* (2nd ed.). Wiley, 1969. xvi+535 pp. illus. $12.95. 674–84960. (SH–C)

Colbert successfully acquaints the reader with major events and trends in the evolution of background animals by referring to fossilized evidence. Principles of evolution are mentioned where applicable and an attempt is made to reconstruct the mode of life of the fossils by analyzing their anatomy and reviewing the geological and climatic circumstances of the relevant times. Summary classification of the chordates and references are included.

Darwin, Charles. *On the Origin of Species: A Facsimile of the First Edition with an Introduction by Ernst Mayr*. Atheneum, 1967. xxxvii+502 pp. $3.45 (paper). 63–17196. (C)

Since, as Mayr indicated, Darwin softened his original statements and withdrew some of his original claims in subsequent editions, and because the sixth edition is the one most generally available, it is necessary to turn to the first edition to determine actually what Darwin's views were. For those interested in such a study this work is a boon.

Darwin, Charles. *The Origin of Species and The Descent of Man.* (Modern Library ed. G27.) Random, 1936. xvi+1000 pp. illus. $3.95. 36–27228. (C)
A complete and unabridged text of both works in a single volume, with a composite index.

Darwin, Francis (ed.). *The Life and Letters of Charles Darwin* (2 vols.). Basic, 1959. 1120 pp. illus. $10.00 set. 59–16177. (SH)
The editor is the son of Charles Darwin and has selected letters that illustrate the personal character, activities, and accomplishments of his father. The Foreword is by George Gaylord Simpson.

De Beer, Gavin. *Charles Darwin: Evolution by Natural Selection.* Doubleday, 1964. xi+290 pp. illus. $4.95, paper $1.45. 64–10233. (SH)
Sir Gavin has conducted research in Darwiniana and on evolution in various scientific disciplines. It therefore is fortunate that he had wrapped up the essence of his varied experiences into a popular, accurate, and highly informative book. If any student is in doubt as to whether he should read the book, let him decide after reading the preface.

Dobzhansky, Theodosius. *Evolution, Genetics, and Man.* Wiley, 1955. ix+398 pp. illus. $7.50. $2.45 (Science Eds., paper). 55–10868. (SH–C)
A foremost geneticist has prepared this outstanding guide to the study of biological evolution which is also solid collateral reading for courses in general biology, zoology, botany, and beginning anthropology.

Dobzhansky, Theodosius. *Mankind Evolving: The Evolution of the Human Species.* Yale, 1962. xiii+381 pp. illus. $8.50, paper $2.45. 62–8243. (C)
Evolution of the human species has two aspects: the scientific and the cultural which are elucidated in the author's effort to portray what has been learned about man. Genetics, natural selection, adaptation, and sociology are among the facets considered.

Dodson, Edward O. *Evolution: Process and Product* (rev. ed.). Reinhold, 1960. xvi+352 pp. illus. $8.50. 60–11083. (SH)
Beginning with an exposition of evolution by natural selection, the author proceeds to consider evidences of evolution, first in terms of the various biological disciplines, and secondly in a progressive phylogenetic order. An invaluable text and reference for students and interested nonspecialists.

Elliott, H. Chandler. *The Shape of Intelligence: The Evolution of the Human Brain.* Scribner's, 1969. xiv+303 pp. illus. $12.50. 68–17353. (SH–C)
The evolution of the nervous system is traced from the lowly level of single-celled organisms to the highest level, the human brain. Emphasis is on those attributes that enable the organism to cope with the environment. This is not a technical text, but rather a popular account of factual material.

Glass, H. Bentley, Owesei Temkin, and William L. Straus, Jr. (eds.). *Forerunners of Darwin: 1745–1859.* Johns Hopkins, 1959. 471 pp. illus. $7.50. 59–9978. (C)

Issued on the centennial of the publication of Darwin's major work, this anthology makes available to the historian of science the background of knowledge and scientific evidence available to Darwin as a basis for his own work.

Grant, Verne. *The Origin of Adaptations.* Columbia, 1963. x+606 pp. illus. $15.00 63–11695. (SH–C)

"The purpose of this book is to set forth the casual theory of evolution as applied to diploid sexual organisms. . . . I have attempted to provide a general framework in which we can organize our present knowledge concerning the evolution process in higher plants and animals."—Preface. A college text or collateral reading book; useful for bright secondary school students.

Gregory, William K. *Our Face from Fish to Man.* Hafner, 1963. xi+295 pp. illus. $8.00. 63–14246. (SH)

This reprint of the 1929 original is still fundamentally good, though some important later findings have shed new light on certain evidence, for example the Piltdown man which has since been proved false. Nonetheless, this is a generally accurate and fascinating account of the evolution of the face, tracing its development from early fish, reptiles, rodents, and mammals to the first men, the present-day face, and some possibilities for the future.

Hardy, Alister. *The Living Stream: Evolution and Man.* Harper & Row, 1965. 292 pp. illus. $6.95. 67–15975. (C)

Hardy applies all the knowledge and philosophical insights gained during 40 years of scientific work to the problems of how man evolved. His ideas are supported and illustrated by examples from the works of many famous scientists and philosophers. Few books are available that present the evidence and theories of evolution as completely, vividly, and creatively as this one, both for the professional scientist and the literate adult.

Hellman, Hal. *The Right Size.* (Science Survey Series.) (Illus. by Sam Salant.) Putnam's, 1968. 126 pp. LB $3.49. 68–15054. (JH–SH)

The author explains why some creatures have survived and others are extinct in an accurate and popular discussion of the size of individuals and their morphological proportions and anatomical structure. A remarkably good and interesting introduction to the processes of organic evolution.

Hooton, Earnest Albert. *Up From The Ape* (rev. ed.). Macmillan, 1946. xxii+788 pp. illus. $9.50. 31–15189. (SH–C)

Lectures to Harvard undergraduates over a period of many years provided the initial inspiration for this well-known discussion of the evolution of vertebrates which has found favor with students and lay readers for many years.

Hotton, Nicholas, III. *The Evidence of Evolution.* (The Smithsonian Library.) American Heritage (dist. by Van Nostrand), 1968. 160 pp. illus. $4.95; LB $4.98. 68–24491. (JH–SH)

Portrays the course of evolutionary change and the evidence of that change disclosed through fossil records. A section of color illustrations from the murals of the hall of North American Mammals follows. A timetable of evolution, a simplified tree of life, brief notes on theorists and evolution, excerpts from the *Voyage of the Beagle,* a descriptive glossary, bibliography, and index complete the book.

Howells, William W. *Mankind in the Making: The Story of Human Evolution* (rev. ed.). Doubleday, 1967. 384 pp. illus. $5.95. 67–10973. (SH–C)
Since the publication of the first edition in 1959, this book has been acclaimed as an outstanding, courageous, readable, and scientifically authoritative introduction to human evolution and physical anthropology. The revised edition has incorporated new information. Librarians should replace the older edition, and if the old one isn't on the shelves they should buy this one. Fundamental collateral reading.

Huxley, Julian, and H. Bernard D. Kettlewell. *Charles Darwin and His World*. Viking, 1965. 144 pp. illus. $6.50. 65–10184. (SH–C)
This brief biographical sketch of Darwin, based on his original notebooks and published works as well as the recollections of his granddaughter, Lady Barlow, possesses the authenticity and unadorned factual literary style characteristic of two eminent biologists who have the necessary depth of background to produce a scholarly work of this nature.

Mayr, Ernst. *Animal Species and Evolution*. Harvard, 1963. xiv+797 pp. illus. $11.95. 63–9552. (C)
A comprehensive exposition and critical evaluation of man's current knowledge of the morphological characteristics and biological properties of animal species, of the factors involved in speciation, and of the relation of all of this knowledge of our understanding of the process of evolution. The chapter on "Man as a biological species" and the bibliography are especially valuable.

Moody, Paul Amos. *Introduction to Evolution* (2nd ed.). Harper & Row, 1962. xi+553 pp. illus. $8.50. 62–7428. (C)
An introduction to organic evolution for those with an elementary background in biology. Presents both the physical and the biological approach. For able secondary students and their teachers. An alternative to Dodson's *Evolution: Process and Product* (Reinhold, 1960).

Ravielli, Anthony. *From Fins to Hands: An Adventure in Evolution*. Viking, 1968. 47 pp. illus. LB $2.96. 68–27574. (JH)
Attractive two-color illustrations by the author complement his lucid text that traces the evolution of paired forelimbs and hands from the lobe-finned fishes, stage-by-stage to the versatile arms and hands of modern man. An excellent overview of one example of organic evolution.

Romer, Alfred S. *The Procession of Life*. World, 1969. 323 pp. illus. $12.50. 68–15191. (SH–C)
After a brief description of the evidence for and mechanisms of evolution, this presents the evolution of animal life, from Protozoa to Primates. Balance is maintained between discussion of fossils and living organisms, and much attention is paid to functional morphology.

Ross, Herbert H. *A Synthesis of Evolutionary Theory*. Prentice-Hall, 1962. xiii+387 pp. illus. $10.60 (text ed. $8.25). 62–8859. (C)
The integration of evolutionary theory and ecology is the most novel aspect of this important treatment of the historical and contemporary background of evolutionary theory in earth science and biological science. At least a high school background in biology and the physical sciences is needed for comprehension of this multidisciplinary exposition.

Ross, Herbert H. *Understanding Evolution.* Prentice-Hall, 1966. ix+175 pp. illus. $4.95. 66–28112. (C)

A skilled synthesis on the definition and operation of evolution recommended for anyone interested in the subject. Particularly suited for students seeking a readable book to clarify evolutionary concepts.

Stebbins, G. Ledyard. *Processes of Organic Evolution.* (Concepts of Modern Biology Series.) Prentice-Hall, 1966. xii+191 pp. illus. $2.95 (paper). 66–16917. (SH–C)

A concise and well-organized account of current concepts of evolution. Describes the processes and uses extensive examples from his own work and the biological literature to support them. The accepted facts of evolutionary history are described, and the reader is given a real feeling of evolution as a dynamic process which can be seen in action and which has many important implications for the future. Each chapter has a summary and questions for thought and discussion. List of additional readings and an index.

575.1 GENETICS

Asimov, Isaac. *The Genetic Code.* Grossman, 1963. xiv+187 pp. illus. $3.95; NAL (Signet P2250), paper 75¢. 63–12156. (SH)

A famous biochemist and popular science writer explains the biochemistry of the cell and the process of heredity, and in particular describes the role in the complex mechanisms of heredity played by DNA, which was discovered in 1944. This and other revelations of recent research have provided new data valuable in the all-important human search for an explanation of the basic mystery of life.

Auerbach, Charlotte. *Genetics in the Atomic Age* (rev. ed). Oxford, 1965. vii+111 pp. illus. $3.00. 65–6920. (JH–SH–C)

Provides nonbiologists with the essential facts on the relationship between radiation and mutations with a minimum of technical verbiage. On the way the nonbiologist will absorb some principles of genetics which may be helpful to him in other contexts. Useful as collateral reading in high-school biology, and at higher academic levels as background material for political science.

Auerbach, Charlotte. *The Science of Genetics.* Harper & Row, 1961. x+273 pp. illus. $5.95, paper (Torchbooks Tb568) $1.95. 61–6429. (SH)

This straightforward presentation of the science of genetics for students and laymen takes into account the classic principles and the fruits of modern research and discovery. An excellent approach to a practical knowledge of heredity and eugenics which should become part of the secondary school preparation for all students.

Beadle, George, and Muriel Beadle. *The Language of Life.* Doubleday, 1966. x+242 pp. illus. $5.95. 66–12195. (SH–C)

A beautifully written and effectively illustrated book that permits the layman to understand and appreciate the significance of genetics. Treats the essentials of chemistry, evolution theory, Mendelian genetics, cell biology, biochemical genetics and the structure of genetically important material, along with the transmission of genetic information. There is a particularly good treatment of the questions left to be answered. Highly recommended for use in a wide range of liberal arts curricula, particularly as outside reading for courses in the social sciences.

Bonner, David M. *Heredity* (2nd ed.). (Foundations of Modern Biology Series.) Prentice-Hall, 1964. xiv+112 pp. illus. $3.95, paper $1.95. (SH)
A concentrated and informative consideration of the basic principles and "mechanisms" of heredity and an exposition of how modern genetic research involves the coordinated efforts of biologists, chemists, and physicists.

Borek, Ernest. *The Code of Life*. Columbia, 1965. xi+226 pp. illus. $5.95. 65–10944. (JH–SH–C)
All chapters are informative and vocabulary-building for the field of molecular biology—undoubtedly the most exciting area being investigated by scientists today. Excellent accounts of the function of the gene and the function of nucleic acids. The remaining chapters are about the biochemical experiments unveiling the link between structure and function of nucleic acids.

Brewbaker, James L. *Agricultural Genetics*. (Foundations of Modern Genetics Series.) Prentice-Hall, 1964. xiv+156 pp. illus. $4.95, paper $3.75. 54–16038. (C)
The author's primary objective is to explain those principles of genetics that should be understood and applied in agriculture and animal husbandry. Recognizing variations in a population, breeding systems and the control of variation, mutations, parasitism and symbiosis, and genetic advance through selection are the major topics discussed. The reader should know elementary genetics and statistical methods.

Burns, Marca, and Margaret N. Fraser. *Genetics of the Dog: The Basis of Successful Breeding*. Lippincott, 1966. viii+230 pp. illus. $9.00. 67–1778. (JH–SH)
A clear and literate summary of existing information on the genetics and breeding of dogs, treating fully the inheritance of color, morphology, and behavior. Recommended for all interested in dogs.

Carson, Hampton L. *Heredity and Human Life*. Columbia, 1963. xvi+218 pp. illus. $6.50, paper $1.95. 63–9808. (SH–C)
An introduction to human genetics for the intelligent layman is the first objective of this book and the second is to discuss impartially available evidence opposed to "racism." This is a thoughtful presentation that all high school seniors and college freshmen should read.

Dobzhansky, Theodosius. *Heredity and the Nature of Man*. Harcourt, Brace, 1964. 179 pp. illus. $4.75. 64–22666. (SH–C)
A distinguished scientist challenges the reader living in today's complex society to become "better acquainted" with himself through an improved understanding of genetics, the science of heredity. By treating the humanistic aspects of genetics, the author makes a timely social contribution when race considerations are in the forefront of society. Presented in an interesting fashion is the origin of genetics and the historical and scientific significance of DNA and RNA. Excellent for collateral reading and reference.

Ehrlich, Paul R., and Richard W. Holm. *The Process of Evolution.* McGraw-Hill, 1963. xvi+347 pp. illus. $9.50. 63–15891. (C)
Contains a basic exposition of genetics and cytology to provide a foundation for their discussion of the process of evolution involving a consideration of population genetics, natural and artificial selection, variation within and between species, and other topics.

Engel, Leonard. *The New Genetics.* Doubleday, 1967. xv+220 pp. illus. $5.95. 63–12986. (SH–C)
Presents the origins and development of the principal concepts of molecular genetics in a clear and nontechnical format. High school students who have taken a modern biology course will understand this excellent text. Useful for collateral reading.

Fast, Julius. *Blueprint for Life: The Story of Modern Genetics.* St. Martin's, 1964. ix+206 pp. illus. $5.00. 64–13484. (JH)
An elementary introduction to the principles of heredity for the beginning student or for the adult with little or no biological background. An acceptable complement or alternate for Philip Goldstein's book, *Genetics is Easy* (Lantern, 1961).

Gardner, Eldon J. *Principles of Genetics* (3rd ed.). Wiley, 1968. ix+518 pp. illus. $9.50. 67–30462. (SH–C)
A widely used text and reference, this new edition has been completely revised and rewritten with many new examples, illustrations, and problems. The book is comprised of three divisions: basic genetics, nature and functions of genetic material, and population genetics and evolution.

Goldstein, Philip. *Genetics is Easy* (2nd ed.). Lantern, 1961. 238 pp. illus. $6.00; Viking (Compass C151), paper $1.45. 55–8304. (JH)
A brief but lucid presentation of genetics. The first edition was a "sellout" because of its simplicity and authoritativeness. The second edition includes revisions that incorporate later discoveries and research.

Herskowitz, Irwin H. *Genetics* (2nd ed.). Little, Brown, 1965. x+554+135 pp. illus. $10.00. 65–17335. (SH)
The basic principles of genetics are faithfully explained and with these the author has skillfully interwoven the results of modern genetic research. A supplement contains a letter of Gregor Mendel and the Nobel Prize Lectures of six famous geneticists. An outstanding college text and resource book for secondary school teachers, as well as a reference for their outstanding students. Indexed.

Jinks, John L. *Extrachromosomal Inheritance.* (Foundations of Modern Genetics Series.) Prentice-Hall, 1964. xiv+177 pp. illus. $4.95, paper $3.50. 64–16032. (C)
Readers with an elementary knowledge of cell structure will be interested in this lucid discussion of evidence for the transmission of traits by means other than chromosomes. The evidence, largely ignored in standard genetics textbooks, has been accumulating since 1909.

Kendrew, John C. *The Thread of Life: An Introduction to Molecular Biology.* Harvard, 1966. 112 pp. illus. $4.00. 66–6434. (JH–SH)
Professor Kendrew communicates the joy and delight that comes from being

one of the first to dig through a complicated intricate problem and see the light of reason that comes from unraveling it. He uses the principles and facts of molecular biology to illustrate the superficial complexity of these problems and the simple principles which underlie them. Stimulates one to go on and learn more, for it conveys the enthusiasm of a great scientist for the process of life.

Lessing, Lawrence, and The Editors of FORTUNE. *DNA: At the Core of Life Itself.* (Illus. by Max Gschwind). Macmillan, 1967, 85 pp. $3.95. 67–22155. (SH–C)
Discusses the nature of DNA and its function, how it controls synthesis in the cell and development of the organism, and the nature of the link between DNA and brain function. Written in a lively and comprehensive style.

Levin, Louis. *Biology of the Gene.* Mosby, 1969. xii+334 pp. illus. $9.50. (C)
This well written text is suitable for the one semester undergraduate course in genetics. DNA, RNA, the genetic code, and protein synthesis are interestingly and informatively covered. Careful consideration is given to the role of genes in development, in the carrying out of metabolic processes (both normal and aberrant), in helping to produce various types of behavior, and in the formation of populations, races and species.

McKusick, Victor A. *Human Genetics.* (Foundations of Modern Genetics Series.) Prentice-Hall, 1964. xii+148 pp. illus. $4.95, paper $3.75. 64–16044. (C)
Begins with an historical introduction followed by a discussion of the various approaches to genetic analysis in man: cytological studies, family analysis, biochemical analysis, and studies of homologies in closely related species. Also includes chapters on genes in populations, genes in evolution, genes and disease, and genes in society. A basic knowledge of genetics is presumed.

Moody, Paul Amos. *Genetics of Man.* Norton, 1967. viii+444 pp. illus. $7.50. 67–10610. (C)
Introduces human genetics in clear, precise, and readable fashion. Logical concept development enables the reader to follow with little background in genetics. Problems and a good glossary add to the value of the book.

Moore, Ruth. *The Coil of Life: The Story of the Great Discoveries in the Life Sciences.* Knopf, 1961. xv+418+viii pp. illus. $6.95. 60–14469. (SH)
"The discovery of the role of DNA and its drawing together of all the life sciences has at long last made it possible for the layman to approach the subject of life, and to aspire to an understanding of the amazing knowledge science is gaining of it. It was this that emboldened me to undertake this book."—Preface.—This is the story of the men and the ideas which step-by-step amassed our knowledge of the workings of life itself.

Papazian, Haig P. *Modern Genetics.* (Advancement of Science Series.) Norton, 1967. xx+350 pp. illus. $8.50. 66–18629. (SH–C)
Devoted to laws governing the inheritance of characteristics, the mechanism of inheritance at the subcellular level, and the chemical nature and action of genes. An excellent supplement to an undergraduate biology course.

Ravin, Arnold W. *The Evolution of Genetics*. Academic, 1965. x+216 pp. illus. $6.00, paper $2.95. 65–18434. (SH–C)

Although this book is not a history of the science of heredity, it surveys adequately past accomplishments to show relevance between past and present research and a connection of current findings to future research. Clearly demonstrated is the intertwining and interdependence of the fields of genetics, biochemistry, physiology, cytology, development, systematics, ecology and evolution. Generous list of references, well-developed index. Excellent collateral reading and reference.

Scheinfeld, Amram. *Your Heredity and Environment*. Lippincott, 1965. xxiv+ 830 pp. illus. (part col.). $14.95. 64–14468. (SH–C)

No single book, it is estimated, has communicated to more students and nonspecialist adults the essential facts of human embryology, genetics, heredity, and eugenics, than the author's *New You and Heredity* (1939, 1950) which this completely rewritten and enlarged text replaces. New sections deal with human biochemistry and the nucleic acids. Added features are a glossary, "Wayward Gene Tables," inheritance forecast tables, and a selected list of periodicals. The bibliography has been completely revised, and is now keyed to the various chapters of the book. Well indexed.

Singleton, W. Ralph. *Elementary Genetics* (2nd ed.). Van Nostrand, 1967. xiii+482 pp. illus. $9.50. 62–6088. (SH)

Examines the principles of genetics as they have been developed, modified and amplified in the years since the rediscovery of Mendel's law (1900). It leads the student directly into research techniques, for solving problems provides a part of the learning process. Good for nonscience majors in college or college preparatory students.

Sinnott, Edmund W., L. C. Dunn, and Theodosius Dobzhansky. *Principles of Genetics* (5th ed.). McGraw-Hill, 1958. xiv+459 pp. illus. $9.95. 57–13342. (SH–C)

One of the most widely used textbooks which has been kept abreast of modern developments through periodic revisions. A superb resource book for high school teachers, and good reference material for their students.

Stahl, Franklin W. *The Mechanics of Inheritance*. (Foundations of Modern Genetics Series.) Prentice-Hall, 1964. xiii+171 pp. illus. $4.95, paper $3.75. 64–16042. (C)

The elucidation of the chemical structure of the genetic material, particularly DNA, has led to a tremendous increase in research and in the total knowledge of the process of inheritance. Students and others with a knowledge of basic genetic principles will find here a complete story of modern principles and concepts.

Stern, Curt and Eva R. Sherwood (eds.). *The Origin of Genetics: A Mendel Source Book*. Freeman, 1966. xii+179 pp. illus. $4.50; paper $2.25. 66–27948. (C)

Six original papers by Mendel (1865, 1869), DeVries (1900), Correns (1900), R. A. Fisher (1936), and S. Wright (1966), trace the origin and development of classical genetics. Ten letters written by Mendel to Nägeli in 1866–1873 show in interesting detail Mendel's problems, reasoning, and experimental tests of his theories of plant inheritance. Reading these excellent new translations of Mendel's papers is an exciting experience.

Sturtevant, Alfred H. *A History of Genetics*. (Modern Perspectives in Biology.) Harper & Row, 1965. viii+165 pp. $5.50. 65–25993. (SH–C)

A very illuminating and critical chronological history of genetics to 1950. Since the author was a principal participant in the making of the history of genetics, the book is characterized by his extraordinary perception and evaluation of events. A classic reference for exceptional senior high and beginning college students.

Wallace, Bruce. *Chromosomes, Giant Molecules, and Evolution*. (Illus. by Frances Ann McKittrick.) Norton, 1966. xi+171 pp. $5.00, paper $1.95. 65–20237. (SH–C)

Giant chromosomes and the story they tell are carefully and thoroughly presented. The testimony stemming from recent knowledge concerning human hemoglobins and other protein polymorphisms is recounted in an interesting and convincing way. The 39 appropriate and well-made illustrations, mostly line drawings, are gathered at the end of the book. Glossary and index.

Wallace, Bruce, and Theodosius Dobzhansky. *Radiation, Genes, and Man*. Holt, Rinehart and Winston, 1959. xii+205 pp. illus. $6.95. 59–14275. (SH)

A discussion of the genetic problems raised by atomic radiation, including a summary of the known effects on the mutation of genes. For lay readers who know the basic principles of heredity.

Watson, James D. *The Double Helix: A Personal Account of the Discovery of the Structure of DNA*. Atheneum, 1968. xvi+226 pp. illus. $5.95. 68–16217. (C)

A highly personal account of how the discovery of DNA's helical structure came about. Interweaves the facts on the matter—names, dates, places, and events—with Watson's own candid opinions and impressions.

Winchester, A. M. *Genetics: A Survey of the Principles of Heredity* (3rd ed.). Houghton Mifflin, 1966. vii+504 pp. illus. $8.95. 66–1347. (C)

Introduces the important ideas of genetics in an updated edition of an old standard textbook. The chemical basis of heredity, the genetic code, DNA duplication and synthesis, chemical control of enzyme production by virus transfer in bacteria, and the biochemical patterns of gene actions are discussed and explained in an elementary and nonprofessional manner.

576 MICROBIOLOGY

Allen, John M. (ed.). *Molecular Organization and Biological Function*. (Modern Perspectives in Biology Series.) Harper & Row, 1967. x+243 pp. illus. $9.95, paper $5.00. 67–10112. (C)

This collection of articles on molecular organization and cellular substructure is noteworthy because each contributor took pains to avoid the tedious compilation of dusty facts without interpretation. The chapters are well written, current, and stimulating. All assume considerable prior knowledge by the reader both in cellular ultrastructure and in biochemistry. Not intended to serve as a basic text but as a supplemental one in which readers in respective areas summarize and interpret current information and current concepts in their fields. Beautiful electron micrographs.

Andrewes, Christopher H. *The Natural History of Viruses.* (The World Naturalist. Norton, 1967. xiii+237 pp. illus. $9.50. (C)
Although emphasis is on viruses infecting man and animals, plant viruses are adequately covered in this excellent book on the role of viruses in nature. Covers a host of topics in virology: transmission, ecology, classification, latency, evolutionary role, and cancer and congenital viruses.

Asimov, Isaac. *The Wellsprings of Life.* Abelard-Schuman, 1960. 238 pp. illus. $3.75; NAL (Signet P2066), paper 75¢. 60–9914. (SH)
Through a discussion of "Life and the species," "Life and the cell," and "Life and the molecule"—the three major wellsprings—we learn how man evolved on earth and explored the mystery of life's origin and continuity.

Carpenter, Philip L. *Microbiology.* Saunders, 1967. 476 pp. illus. $7.50. 67–11767. (C)
Deals with a survey of microorganisms, the biology of bacteria, ecological relationships and roles of microorganisms, and interactions of pathogenic microorganisms with their hosts. A good general college text.

Clark, Paul F. *Pioneer Microbiologists of America.* U. of Wisconsin, 1961. xiv+369 pp. illus. $6.00. 60–11441. (SH–C)
A history of microbiology in America in terms of biographical notes of the scientists and narrative accounts of their research and accomplishments. Includes an extensive bibliography keyed to the text.

Curtis, Helena. *The Viruses.* Nat. Hist. Press, 1965. x+228 pp. illus. $4.95, paper $1.95. 65–17266. (SH–C)
It is obviously impossible to cover all of virology with a limited space. Enough is given, however, to provide a good background of earlier work as well as of recent progress. Informative and interesting reading for high school or college students as well as adults in general, including a few professional virologists who may enjoy a chance to survey their subject in broad outlines.

Dubos, René J. *Louis Pasteur: Free Lance of Science.* Little, Brown, 1950. xii+418 pp. illus. $6.00. 50–5543. (SH)
An excellent biography of Pasteur written by an eminent scientist and hence a reliable account of Pasteur's many scientific accomplishments, including the structure of the chemical molecule, the process of fermentation, the role of microorganisms in disease and technology, the theory and practice of immunization, and the policy of public hygiene.

Dubos, René J. *The Unseen World.* Rockefeller U., 1962. 112 pp. illus. $4.75. 62–19210. (JH)
With the aid of the most recent electron micrographs, an authority on microbiology describes the form, structure, and activity of a variety of cells and viruses, and shows how microbes interact with other forms of life. He outlines the development of the germ theory of disease, shows how microbes aid as well as hinder man, and concludes with a chapter on science as a way of life.

Foster, Edwin M., F. Eugene Nelson, et al. *Dairy Microbiology.* Prentice-Hall, 1957. xvi+492 pp. illus. $9.75. 57–6457. (C)
A valuable reference for libraries in agricultural communities, since it emphasizes practical information and methodology for those interested in the

sanitation and quality of dairy products. The reader should have a biological and general chemistry background.

Fraser, Dean. *Viruses and Molecular Biology.* (Current Concepts in Biology.) Macmillan, 1967. iii+124 pp. illus. $2.25 (paper). 67–13144. (C)
Using the virus as a model, the author deftly leads the reader into an account of current concepts in molecular biology. The reader is introduced to biochemical genetics and protein synthesis in an informative and painless manner. The illustrations are very well done and relate closely to the text. A colorful, coherent account of the fundamental biology of viruses.

Frobisher, Martin. *Fundamentals of Microbiology* (7th ed.). Saunders, 1962. xviii+610 pp. illus. $8.00. 62–8806. (C)
A basic college textbook that will be useful as a reference and for advanced study by high school students who have had courses in biology and chemistry. In addition to the biology of microorganisms, taxonomy, etc., chapters on techniques, therapy, industrial problems and other practical applications are included.

Mann, John. *Louis Pasteur: Founder of Bacteriology.* Scribner's, 1964. 160 pp. $3.50; LB $3.31. 64–22753. (JH–SH)
His major accomplishments are covered: his work with crystals which led to development of the field of stereochemistry; his studies of fermentations; his successful efforts to disprove the theory of spontaneous generation; control of diseases in silkworms; development of immunization procedures for control of chicken cholera, anthrax in sheep, and rabies in man. Brief bibliography, index, glossary, and a calendar listing events in Pasteur's life and concurrent happenings in the world.

Rosebury, Theodor. *Life of Man.* Viking, 1969. xvi+239 pp. $6.95. 69–18804. (SH–C)
A scientific and social commentary on the dynamic interaction between microbes and man that is entertaining and illuminating for scientist and layman alike. The author reveals how man acquired his own set of microbes from the world-at-large and what a Gulliver half the size of a flea might see on the surface of man.

578 MICROSCOPY

Anderson, M. D. *Through the Microscope: Man Looks at an Unseen World.* Nat. Hist. Press, 1965. 156 pp. illus. $4.95. 65–12163. (JH–SH)
This absolutely first-class book surveys microscopy clearly from its earliest history to the most recent electron microscope. The balance between applications of these instruments to daily life and to theoretical biology is maintained throughout. The writing is clear and crisp. The illustrations, many in color, are interesting and descriptive of the text.

Bradbury, S. *The Evolution of the Microscope.* Pergamon, 1967. x+357 pp. illus. $12.50. 67–18485. (SH–C)
A grand sweep of the history of the microscope is experienced over the past 40 years. A valuable reference work on microscopes that should be used for any basic laboratory course in which they are used.

Corrington, Julian D. *Exploring with Your Microscope.* McGraw-Hill, 1957. 229 pp. illus. $6.50. 57–7998. (JH)

The purpose of this introductory manual is to acquaint the reader with the construction and use of microscopes and to afford an insight into the use of microscopy in various fields of science and applied technology.

Cosslett, V. E. *Modern Microscopy or Seeing the Very Small.* Cornell, 1966. 160 pp. illus. $6.00. 66–23407. (SH–C)

In addition to bringing informed laymen, teachers, and students up to date on microscopy, this work also anticipates future trends in research possibilities and microscope applications in advancing technology. The electron microscope and its relatives are well covered.

Gray, Peter. *Handbook of Basic Microtechniques* (3rd ed.). McGraw-Hill, 1964. xii+302 pp. illus. $8.95. 63–22430. (SH–C)

An introductory guide for students of microbiology, botany, zoology, premedicine, and medical technology. The first part explains the microscope, its use, and principles of photomicrography; the second is devoted to the preparation of slides; the third part explains specific examples of slide making; and the fourth the techniques for demonstrating a wide variety of anatomical, histological, and cytological details in animals and plants. Can be used by advanced secondary students.

Humason, Gretchen L. *Animal Tissue Techniques.* Freeman, 1962. xv+468 pp. illus. $9.00. 61–17383. (C)

A more advanced manual than Gray, and intended for zoology and premedical students and for working technicians. Outlines procedures applicable to both normal and pathological tissues in zoological and medical study. Extensive bibliography.

Jacker, Corinne. *Window on the Unknown: A History of the Microscope.* (Illus. by Mary Linn.) Scribner's, 1966. 188 pp. $3.95. 66–24490. (JH–SH–C)

The role of the microscope and microscopy in the history of science is examined in this unique chronology. Primary sources are listed in the bibliography, the glossary is good, and the illustrations are helpful.

Jones, Ruth McClung. *Basic Microscopic Technics.* U. of Chicago, 1966. xiv+334 pp. illus. $6.50. 66–20579. (SH–C)

Detailed microscopy techniques are accurately covered. Valuable as a text, reference and work manual. An appendix containing reagents, supplies, stains, formulas, and equipment supplements the text.

Lindeman, Edward. *Water Animals for Your Microscope.* (Illus. by Christine Sapieha.) Crowell-Collier, 1967. 119 pp. illus. $4.50; LB $3.74. 67–10577. (JH–SH)

An introduction to the study of small and microscopic animals commonly found in ponds, streams, and along the seashore has been written in a very simple and straightforward manner. It can stimulate an intelligent student to learn a great deal about water animals, largely by engaging in various projects. The author suggests equipment needed and provides a simple but accurate explanation of principles involved.

Needham, George Herbert. *The Microscope: A Practical Guide.* Thomas, 1968. xi+115 pp. illus. $6.50. 68–20792. (SH–C)

This will make a good additional reference or text for a short course in microscopy. There is much practical advice, and the book should be good for people in chemistry and the biological sciences.

Sass, John E. *Botanical Microtechnique* (3rd ed.). Iowa State U., 1958. xi+ 228 pp. illus. $5.95. 58–13416. (C)
A manual for teachers, investigators, technicians, and advanced students that teaches the rudiments of preparing smears, sections, whole mounts, etc., for classroom use or in research.

Schneider, Leo. *You and Your Cells*. Harcourt, Brace, 1964. 157 pp. illus. $3.75. 64–11496. (JH–SH)
Begins with the amoeba and draws parallels with the cells and cell components of higher animals, progressing to cell differentation—epithelial, muscle, bone, fat, nerve, blood and reprodutive. Then the reader is introduced to the biochemistry of the cell, energy, food cycles, and enzymes. Finally the systems of the human body are described and interrelated as one functional organism.

Smith, Alice Lorraine. *Principles of Microbiology* (6th ed.). Mosby, 1969. xi+669 pp. illus. $9.75. 71–84012. SBN 8016–4679–0. (C)
The entire scope of microbiology is covered here in a very clear and concise manner. The book is divided into three sections: the introductory section properly defines the subject and provides insight into the tools of study and basic cellular morphology biologic activities and classification; the second provides a panoramic view of the various types of microbial organisms; and the final section deals with human disease and its identification and control.

Smith, Kenneth M. *Viruses*. Cambridge, 1962. 134 pp. illus. $4.25; paper $1.95. 62–4466. (C)
Written to explain to the student and the layman certain recurring questions concerning viruses such as: What do they look like? How do they multiply? How can they be purified and crystallized? Do they cause cancer?

Stainier, Roger Y., et al. *The Microbial World* (2nd ed.). Prentice-Hall, 1963. xiii+753 pp. illus. $12.95 (text ed. $10.95). 63–8887. (C)
A general introduction to the biology of microorganisms, with special emphasis on the properties of bacteria. Biological background information is included to enable those with limited training to utilize the book for reference. All aspects of the subject are treated so that it meets the needs of students, medical technicians, public health workers, and others.

Stanley, Wendell M. and Evans G. Valens. *Viruses and the Nature of Life*. Dutton, 1961. 224 pp. illus. $4.95; paper $1.95. 61–5876. (JH–SH)
The senior author received the Nobel Prize in Chemistry in 1946 for this work in virology, and the junior author produced a series of TV films on viruses at the Virus Laboratory at the University of California directed by Dr. Stanley. This book explains what scientists know and what they do not know about viruses. Excellent assigned reading for superior junior high and all senior high biology students.

Weidel, Wolfhard. *Virus*. U. of Michigan, 1959. 159 pp. illus. $4.50; (Ann Arbor Bks. AA5509) paper $1.95. 59–7295. (SH)
The author holds that "virus" must be considered as a "concept" rather than

a "thing" at our present state of knowledge. This exposition for student and layman is a compact discussion of the biology, reproduction, biochemistry and other information about viruses, and includes information on prevention and cure of virus disease.

581 BOTANY

Beaty, Janice J. *Plants in His Pack.* Pantheon, 1964. 182 pp. illus. LB $3.89. 64–18324. (JH)
A clearly written and carefully researched biography of Edward Palmer, a naturalist and early scientific collector (1830–1911). The author weaves into the biography the necessary history and descriptions to give an exciting picture of Palmer's life, including accounts of his Civil War experience as a "doctor," his travels throughout the early West, and records of the plant lore of American Indians.

Benson, Lyman. *Plant Taxonomy: Methods and Principles.* Ronald, 1962. ix+ 494 pp. illus. $12.95. 62–11646. (C)
This text is more than a taxonomic guide; it lays a foundation for the application of taxonomic methods to the study of botany by a review of the various sources of data that are fundamental to taxonomy; field and herbarium studies, morphology, paleobotany, and biogeography. Then the principles of botanic classification, the international code, keys, definitive description, and documentation are explained. Fundamental training for botanical research is the objective.

Coulter, Merle C., and Howard J. Dittmer. *The Story of the Plant Kingdom* (3rd ed.). U. of Chicago, 1964. ix+467 pp. illus. $5.95. 64–10093. (SH)
A condensed introduction to the plant kingdom in terms of morphology and evolution, incorporating references to recent work in genetics, nucleic acids, photosynthesis, and radioisotopes. Glossary.

Dickinson, Alice. *Carl Linnaeus: Pioneer of Modern Botany.* (Immortals of Science.) Watts, 1967. ix+209 pp. illus. $3.95. 67–18897. (JH–SH)
An interesting portrayal of Linnaeus that informs the reader of the scope and diversity of this pioneering botanist and physician, who extended himself as a dynamic and powerful leader of students and author of numerous publications. Concludes with a chronology of the major events in his life.

Dupree, A. Hunter. *Asa Gray.* Harvard U., 1959. x+505 pp. illus. $7.50. 59–12967. (SH)
A definitive biography of the leading American botanist of the 19th century who was the center of a network of botanical explorations; he stimulated field collectors, identified and classified specimens, sold collections, and published scientific results. He knew and corresponded with Darwin and acted as Darwin's sponsor in America.

Graustein, Jeannette E. *Thomas Nuttall Naturalist: Explorations in America 1808–1841.* Harvard U., 1967. 481 pp. $11.95. 67–13253. (SH–C)
This is the first definitive biography of one of America's greatest botanists. Portrays Nuttall as a diversified and devoted naturalist with many interests outside of botany.

Jaques, Harry E. *Plant Families: How to Know Them.* Wm. C. Brown, 1949. 177 pp. illus. $3.25, paper $2.50. 42–3823. (JH)
A guide for the identification of the families of most of the members of the Plant Kingdom, in which drawings are used as details of the keys.

Klein, Richard M. and Deana T. Klein. *Discovering Plants.* (A Nature and Science Book of Experiments.) Nat. His. Press, 1968. 124 pp. illus. $4.50. 68–14178. (JH–SH)
Botanical experimentation is usually very simple and instructional for young students, and this book presents that experimentation at its best. The emphasis is on activities or functions of plants, including plant movements, growth, water uptake and loss, nutrition and metabolism. Suggestions for additional reading occur regularly throughout the book. Of considerable interest to junior high school teachers as well as students, and certain of the experiments could be modified for use in senior high school courses in botany or biology.

Kraft, Ken, and Pat Kraft. *Luther Burbank: The Wizard and the Man.* Meredith, 1967. xvi+270 pp. illus. $7.95. 67–12638. (JH–SH–C)
This fascinating biography of the famous American plant breeder is both informative and inspiring. A thorough research on the life and time of this unusual man who played a crucial role in horticultural history.

Milne, Lorus J., and Margrery Milne. *Plant Life.* Prentice-Hall, 1959. xiii+283 pp. illus. $8.95. 59–9558. (JH)
An account of methods of botanical study for the student and layman, with considerable attention to the leading figures in the plant sciences, and the economic aspects thereof. The two concluding chapters, "Basic interpretations of life," and "Looking toward the future," are thought-provoking.

Novak, F. A. The Pictorial Encyclopedia of Plants and Flowers. (J. G. Barton, Ed.; H. W. Rickett, Consultant.) Crown, 1966. 589 pp. illus. $10.00. 66–18549. (JH–SH–C)
The beautiful illustrations, consisting of over 1,000 black-and-white and 49 colored photographs are organized by families and are usually accompanied by brief notes about form and distribution. By no means a textbook, this is a useful reference which should be in all school libraries.

Reisgl, Herbert (ed.). *The World of Flowers.* Viking, 1965. 240 pp. illus. $12.50. 65–10265. (JH–SH).
An anthology beginning with an essay on the aesthetical value of flowers. Then follow essays by different authors on the unique flowers and plant life of many countries. There is a detailed listing by continents and countries of botanical gardens, parks, and floral regions, which will be valuable to travelers and museum goers.

Steere, William C. (ed.). *Fifty Years of Botany: Golden Jubilee Volume of the Botanical Society of America.* McGraw-Hill, 1958. xiii+638 pp. illus. $12.50. 57–14685. (C)
A series of invited papers that, collectively, provide material for a history of botanical research in America during the first half of the present century. Useful as a general reference, with the assistance of the index.

581.07 BOTANY—STUDY AND TEACHING

Cronquist, Arthur. *Introductory Botany.* Harper & Row, 1961. ix+902 pp. illus. $11.75. 61–5461. (SH–C)
An alternative to some of the other college botany texts—this one uses the evolutionary arrangement, prefaced by introductory chapters dealing with cells, and botanical principles. Concludes with keys to divisions, classes and orders, and a glossary.

Devlin, Robert M. *Plant Physiology.* Reinhold, 1966. xi+564 pp. illus. $11.00. 66–25438. (SH–C)
Carefully organized and very readable. As a textbook, it is a refreshing addition to the works available and should suit its primary objective better than the rest. It necessarily condenses much and eliminates more that occupied chapters in earlier textbooks, yet without leaving the impression that the coverage has been skimped. It adds much that is new and current. Further, it is well documented. A useful sourcebook and reference manual for the biological department of any school library and will be an indispensibly up-to-date (or as up to date as any published book can be) guide to the literature on the workings of plants.

Fuller, Harry J., and Oswald Tippo. *College Botany* (rev. ed.). Holt, Rinehart and Winston, 1954. xiv+993 pp. illus. $12.95. 54–6603. (C)
An almost encyclopedic and phylogenetic approach to the study of botany, with excellent sections on morphology, physiology, and ecology. Not the type of book that goes out of date quickly; hence a valuable reference.

Greulach, Victor A., and J. Edison Adams. *Plants: An Introduction to Modern Botany* (2nd ed.). Wiley, 1967. xvi+636 pp. illus. $8.95. 66–26744. (SH–C)
Those who desire to study botany as potential science majors and those who wish an acquaintance with the subject as part of a liberal education will find this up-to-date account will meet their needs for a survey of physiology, morphology, cytology, ecology, and genetics. Bibliography at the end of each chapter.

Robbins, Wilfred W., T. Elliot Weier, and C. Ralph Stocking. *Botany: An Introduction to Plant Science* (3rd ed.). Wiley, 1964. 614 pp. illus. $9.95. 64–20081. (SH–C)
Noted for its outstanding photographs and camera lucida drawings, this college textbook includes both traditional material for a broad introductory course and newer concepts that have resulted from recent research in gene activity, photosynthesis, respiration, and molecular biology. Good reference for advanced high school students. Glossary.

Sinnott, Edmund W., and Katherine S. Wilson. *Botany: Principles and Problems* (6th ed.). McGraw-Hill, 1963. 515 pp. illus. $9.50. 62–12094. (C)
A well-known introductory textbook, useful for reference and advanced work in secondary schools. Special features are the list of questions for discussion and the scientific papers listed at the end of each chapter which encourage individual thinking and research.

Weisz, Paul B., and Melvin S. Fuller. *The Science of Botany*. McGraw-Hill, 1962. xi+562 pp. illus. $9.95. 61–18052. (SH–C)
Designed for college botany courses and useful as a reference for advanced secondary school students, this text is an analytical and dynamic approach, with an appropriate experimental outlook. The major sections are: "nature of science;" "the living world;" "the world of plants;" "metabolism;" "self-perpetuation, reproduction and adaptation, including heredity and evolution." Appendices include a glossary and outline of plant classification.

Wilson, Carl L., and Walter E. Loomis. *Botany* (4th ed.). (Illus. by Hannah T. Croasdale.) Holt, Rinehart and Winston, 1967. xii+626 pp. $10.95. 67–11746. (C)
This college textbook balances a discussion of the structure and function of the flowering plants and a survey of the plant kingdom. This new edition should further promote its status as one of the most popular botany textbooks.

581.1 PLANT PHYSIOLOGY

Cook, Stanton A. *Reproduction, Heredity, and Sexuality*. (Fundamentals of Botany Series.) Wadsworth, 1964. vi+117 pp. illus. $2.50 (paper). 64–21771. (SH–C)
An exposition of the many means and mechanisms used in the biological processes by living organisms in order to remain suited to their immediate environment while retaining the ability to change genetically through mutation, segregation, and recombination. The book is packed with information, often telegraphic in style.

Galston, Arthur W. *The Life of the Green Plant* (2nd ed.). (Foundations of Modern Biology Series.) Prentice-Hall, 1964. xii+116 pp. illus. $3.95, paper $1.95. 61–8697. (SH–C)
The green plant in the economy of nature is the initial discussion subject of this book, which then moves on to consider the green plant cell, plant nutrition, growth and differentiation, and morphogenesis. Excellent collateral reading.

Meeuse, Bastiaan J. D. *The Story of Pollination*. Ronald, 1961. x+243 pp. illus. $7.50. 61–15612. (JH)
No other book for the layman presents half so well the fascinating and important process of pollination. The presentation is aided by superb drawings in color and black and white. Good bibliography.

Meyer, Bernard S., Donald B. Anderson, and Richard H. Böhning. *Introduction to Plant Physiology*. Van Nostrand, 1960. v+541 pp. illus. $8.75. 60–9034. (C)
Designed as a concise introduction to plant physiology, yet comprehensive in the treatment of subject matter. The discussion is coordinated from chapter to chapter to emphasize the interrelationships among the various plant processes. Basic instruction in botany and chemistry are readers' prerequisites.

Rosenberg, Jerome L. *Photosynthesis: The Basic Process of Food-Making in Green Plants* (Vol. 21, Holt Library of Science). Holt, Rinehart and Winston, 1965. 127 pp. illus. $2.50, paper $1.60. 64–23015. (SH–C)
Current knowledge relative to the photosynthetic process, without extensive use of chemical, physical and mathematical terminology, is presented for the beginning student. Ten short chapters present basic ideas relative to the role of photosynthesis, the photosynthetic apparatus, its structure and how it works within the living cell and when separated from it, an analysis of the different steps in the process and the various factors which tend to limit some of these steps. Charts and micrographs are well chosen. Short bibliography and adequate glossary.

Stewart, William D. P. *Nitrogen Fixation in Plants.* Athlone (dist. by Oxford U.), 1966. 168 pp. illus. $4.50. 66–2150. (SH–C)
One of the first fairly comprehensive monographs on the physiology and chemistry of nitrogen fixation in plants with 612 literature citations, well illustrated, and an excellent index. This classical contribution to knowledge of the life processes in plants should prove valuable to students in botany, soils, microbiology and all fields in agriculture. Also an excellent reference for advanced courses in soil microbiology.

Van Overbeek, Johannes. *The Lore of Living Plants.* (Vistas of Science Series.) (Activities by Harry K. Wong.) McGraw-Hill, 1954. 160 pp. illus. $2.50. 63–22650. (JH)
A very elementary introduction, with suggested individual activities, to plant physiology. Includes some historical background.

Walker, John Charles. *Plant Pathology* (2nd ed.). McGraw-Hill, 1957. xi+ 707 pp. illus. $13.50. 56–10335. (C)
An introduction, for students who have had basic botanical training, to the foundations of plant pathology. It contains an historical review of the subject as a basis for understanding recent trends, discusses specific disease groups, considers environmental factors, host-parasite relationships, and control methods. Reference list at end of each chapter.

581.4 PLANT MORPHOLOGY

Bold, Harold C. *Morphology of Plants* (2nd ed.). Harper & Row, 1967. xxix+ 541 pp. illus. $12.75. 67–10790. (SH–C)
Offers a discussion of the morphology and reproduction of the more important plant types. Designed for a year's course when paralleled with an integrated study of the types, but anyone interested in the structure of plants will find this valuable.

Esau, Katherine. *Plant Anatomy* (2nd ed.). Wiley, 1965. xx+767 pp. illus. $14.95. 65–12713. (SH–C)
The most thorough coverage of plant anatomy that has ever appeared in this country. It is up to date on the results of research on the ultrastructure of the cell wall and the protoplast. Covers all the primary and secondary tissue systems in plants and is well illustrated. Admirably suited for the first course in plant anatomy at the college level, it is likewise an excellent

reference book for high school botany and of value as a reference to students in plant physiology, cytology and wood technology. Lists of carefully selected additional sources after each chapter are valuable assets.

581.5 PLANT ECOLOGY

Anderson, Edgar. *Plants, Man, and Life.* U. of California, 1967. ix+251 pp. illus. $6.00; $1.95 (paper). (SH–C)
Emphasizes the necessity and pleasure every biologist-scholar may find in the natural history, taxonomy, and ecology of the organisms with which he works and those with which he lives.

Corner, E. J. H. *The Life of Plants.* (World Natural History Series.) World, 1964. xiv+315 pp. illus. $12.50; NAL (Mentor MW 796), 1966. $1.50 (paper). 63–14794. (SH–C)
An outstanding natural history book by a leading British specialist which presents a unified treatment of the evolution and ecology of microscopic and macroscopic plants. Excellent line drawings and photographs, glossary, bibliography, and index.

Daubenmire, Rexford F. *Plants and Environment: A Textbook of Plant Autecology* (2nd ed.). Wiley, 1959. xi+422 pp. illus. $8.50. 59–6762. (C)
A knowledge of plant ecology is needed whenever plant behavior is to be studied in relation to the production of cultivated crops or the management of vegetation, for production and management techniques must comprehend the natural or cultural environment. The reader should have some background in biology and the earth sciences.

Oosting, Henry J. *The Study of Plant Communities* (2nd ed.). Freeman, 1956. xiii+440 pp. illus. $6.50. 56–11029. (C)
The vegetation of North America serves as the primary source of illustrative material for this introduction to the ecology of plant communities. References are primarily to American literature.

Watts, May Theilgaard. *Reading the Landscape: An Adventure in Ecology.* Macmillan, 1957. x+230 pp. $5.95. 57–6359. (JH)
An interesting introduction to plant ecology written in the manner of a detective story, in which the reader is shown how to discern, detect, and evaluate the evidence recorded in the landscape in determining the history of the plant life of a region. The lessons are taught by analyzing various typical areas.

581.6 ECONOMIC BOTANY

Baker, Herbert G. *Plants and Civilization.* (Fundamentals of Botany Series.) Wadsworth, 1965. vi+183 pp. illus. $2.50 (text ed.). 65–11580. (SH–C)
An up-to-date account of the most important plants used by man, historical in approach, but incorporating many recently-established facts. Coverage of economic botany is complete. Attention is given to the putative origin of the principal crop plants and, where it is known, the genetic and archeological evidence is stressed.

Fernald, Merritt L., and Alfred C. Kinsey. *Edible Wild Plants of Eastern North America* (rev. ed.). Harper & Row, 1958. 452 pp. illus. $6.95. 58–7977. (SH)

A history of the use of all of the edible flowering plants and ferns, plus the important mushrooms, seaweeds and lichens east of the Great Plains and north of the Florida Peninsula. Most useful to those who have had an introductory course.

Gray, William D. *The Relation of Fungi to Human Affairs.* Holt, Rinehart and Winston, 1959. xiii+510 pp. illus. $11.95. 59–8695. (SH–C)
Begins with an introduction to the morphology, classification, identification, and cultivation of the fungi; then in two major parts—beneficial activities and harmful activities—considers the relations of fungi in business, industry, public health, human food and nutrition, etc. The detailed analytic index makes it a valuable reference.

Hill, Albert F. *Economic Botany: A Textbook of Useful Plants and Plant Products* (2nd ed.). McGraw-Hill, 1952. xii+560 pp. illus. $12.50. 51–12617. (C)
A standard reference work providing information on the economically useful plants of the world. Intended for the reader with an elementary botanical background to provide him with an appreciation of the importance of plants and plant products in industry, commerce, medicine, etc.

King, Lawrence J. *Weeds of the World: Biology and Control.* (Plant Science Monographs.) Wiley, 1966. xxxii+526 pp. illus. $18.00. (SH–C)
Ecology, physiology, reproduction, genetics, chemical and non-chemical control are competently and succinctly described, discussed and documented in this outstanding work. College and senior high students and their teachers will find the book interesting and informative. Scientists and other professionals, as well as farmers, will find it useful. There are 80 figures, and 16 tables, and appendices on properties and uses of herbicides, data on recently introduced herbicides, and chemical weed control for vegetable crops.

Kingsbury, John M. *Poisonous Plants of the United States and Canada* (3rd ed.). Prentice-Hall, 1964. xiii+626 pp. illus. Trade ed. $15.00 (text ed. $9.75). 64–14394. (C)
Replaces and updates Walter C. Muenscher's book of the same title (Macmillan, 1951) and represents a thorough survey of the literature on poisonous plants and the toxicology of plant poisoning in man and other animals. In all, about 700 species are discussed. Data include description, habitat and distribution, poisonous principle, toxicity, symtoms, etc.

Montgomery, F. H. *Weeds of the Northern United States and Canada.* Warne, 1965. xxviii+226 pp. illus. $3.95. 65–9406. (SH–C)
Describes 365 weeds of northern United States and southern Canada, most illustrated with line drawings. Simplified and useful keys to families and species are provided at strategic points. Common names as well as scientific names are used and indexed. The descriptions are brief but clear and the illustrations are adequate.

Schery, Robert W. *Plants for Man.* Prentice-Hall, 1952. viii+564 pp. illus. $10.95. 52–7748. (SH–C)
Man's dependence on plants for food, fibre, shelter, drugs, etc., is the subject of this comprehensive work which is well organized and authoritative. Its major sections deal with forest products and fibres, plant-cell exudates and extractions, and plants and plant parts used primarily for food and beverages.

581.9   PLANT GEOGRAPHY

Fernald, Merritt Lyndon. *Gray's Manual of Botany* (8th ed.). American, 1950.
lxiv+1632 pp. illus. $15.00. 50–9007. (C)
The standard handbook of the flowering plants and ferns of the Central and
Northeastern United States and adjacent Canada. The original edition published in 1848 was done by Gray who carried the work through the 5th
edition in 1867. Others have prepared the 6th and later editions. An indispensable reference for all college and public libraries, and for large
high school collections.

Goodspeed, T. Harper. *Plant Hunters in the Andes* (2nd ed.). U. of California,
1961. ix+378 pp. illus. $7.50. 61–7533. (SH)
Describes the six botanical expeditions to the Andes directed by the author
and his associates who were seeking South American relatives of tobacco
and other previously unknown plants of potential scientific and agricultural
importance. These expeditions covered thousands of miles in Columbia,
Peru, Bolivia, Chili, Argentina, and Uruguay. Fascinating reading for student and layman.

Hylander, Clarence J. *The World of Plant Life* (2nd ed.). Macmillan, 1956.
xv+653 pp. illus. $12.95. 56–7311. (SH)
Written wholly for the layman, and from this point of view, Dr. Hylander
conducts the reader on a phylogenetic tour of the plant kingdom. This is
not only a general reading book; it also contains aids in the identification of
plants and information on plant evolution, morphology, and physiology.

Polunin, Oleg. *Flowers of Europe, a Field Guide.* Oxford, 1969. 662 pp. illus.
$15.00. (SH–C)
Contains information on about 2,800 species of an estimated 17,000 species
found in Europe today. Of the species included, over 1,000 are excellently
illustrated by color photographs and an additional 280 by line drawings.
The important agricultural, industrial, medicinal, or herbal uses of a species
is included.

Polunin, Nicholas. *Introduction to Plant Geography and Some Related Sciences.*
McGraw-Hill, 1960. xix+640 pp. illus. $12.00. 60–50391. (C)
An authoritative and comprehensive text and reference book that covers the
fundamentals of plant geography and then delves into physiology, dispersal
and migration, evolutionary development, types of distribution, environmental
factors, crops, and finally a series of comprehensive chapters on various
regional vegetational types.

582.13   FLOWERING PLANTS

Cuthbert, Mabel Jaques. *How to Know the Fall Flowers.* (Ed. by Harry E.
Jaques.) Wm. C. Brown, 1948. 199 pp. illus. $3.25; paper $2.50. 49–959.
(JH)
A pictured-key to the more common fall flowering plants with instructions
on keeping a herbarium for the amateur.

Cuthbert, Mable Jaques. *How to Know the Spring Flowers* (rev. ed.). (Ed. by Harry E. Jaques.) Wm. C. Brown, 1949. vi+194 pp. illus. $3.25; paper $2.50. A51–32. (JH)
The counterpart of the preceding title, dealing with spring flowers.

Hausman, Ethel Hinckley. *Beginner's Guide to Wild Flowers.* Putnam's, 1955. viii+376 pp. illus. $4.95. 48–3074. (JH)
Will assist in the identification of every wild flower east of the Mississippi and, except for rare or restricted species, is useful as far west as the Rocky Mountains. Individual drawings show the flower, leaf, and stem of every species. Descriptive notes include scientific names, period of bloom, color, size, and geographic range.

Hawkes, Alex D. *Orchids: Their Botany and Culture.* Harper & Row, 1960. xii+297 pp. illus. $7.95. 60–10431. (JH–SH)
The present-day interest in amateur orchid culture has disproved the old fallacy that orchids were intended only for aristocrats. This excellent introduction to orchidaceous plants contains botanical background, detailed information on culture, descriptive notes on many species and varieties, a phylogenetic list and a glossary.

House, Homer D. *Wild Flowers.* Macmillan, 1961. 362 pp. illus. (mostly col.) $17.95. 34–28489. (SH–C)
Originally published in 1934 and reissued with new color plates in 1961, this illustrated list of wild flowers has become a "classic" in the field. The descriptive notes, keyed to the plates, contain information on geographic distribution.

Lemmon, Robert S., and Charles C. Johnson. *Wildflowers of North America in Full Color.* Doubleday, 1961. vii+280 pp. illus. $9.95. 61–5862. (JH)
A descriptive and illustrated introduction to typical wildflowers, organized by geographical regions.

Mathews, F. Schuyler. *Field Book of American Wild Flowers* (rev. by Norman Taylor). Putnam's, 1955. xxix+601 pp. illus. $5.95. 55–5778. (SH)
A well-known, complete, convenient and authoritative guide to the wild flowers of eastern and central North America. Arrangement is by families, identification being based chiefly on leaf and flower character. Includes both common and scientific names and mentions general characteristics and habitats.

Porter, C. L. *Taxonomy of Flowering Plants* (2nd ed.). Freeman, 1967. 472 pp. illus. $7.75. 66–19914. (SH–C)
Although only slightly altered from the first edition, this serves as a successful textbook for specialists and as profitable reference for interested amateurs. Well illustrated with adequate references. Coverage is given to only the flowering plants.

Price, Molly. *The Iris Book.* (Photos by author; drawings by Allianora Rosse.) Van Nostrand, 1966. xviii+204 pp. $7.95. 66–16905. (SH–C)
Written for amateur naturalists and gardeners, covers thoroughly the horticulture of irises with discussions and illustrations of the "tried and true" varieties and appropriate companion plants. Includes basic botany of irises,

and gives adequate guidance for hybridizers. Appendix lists reliable varieties of garden plants. Lists of awards, iris societies, and glossary. More suitable for public libraries and college libraries than for schools.

Reisigl, Herbert (ed.). *The World of Flowers.* Viking, 1965. 240 pp. illus. $12.50. 65–10265. (JH–SH–C)

The anthology begins with an essay on the aesthetic value of flowers, followed by a brief description of the life of a botanist on an expedition. Then the various authors, through their essays, escort the reader on a pictorial journey to see the world's rare flowers. The 98 pages of photographs, many in color, portray some of the most beautiful and unusual flowers. An authoritative compilation that will interest many students and adult readers.

Synge, Patrick M. *The Complete Guide to Bulbs.* Dutton, 1962. 320 pp. illus. (mostly col.) $6.95. 62–528929. (JH–SH)

An alphabetically-arranged descriptive guide to bulbs and corms which imparts good botanical information as well as practical instructions and notes on cultivation.

582.15   WOODY PLANTS

Boom, B. K., and H. Kleijn. *The Glory of the Tree.* (Illus. by G. D. Swanenburg de Veye.) Doubleday, 1966. 128 pp. $12.95. 66–21005. (SH–C)

A magnificently illustrated story of trees, mainly concerned with the beauty of the trees, their history, folklore and economic importance. The selection of trees is restricted to Western and Southern Europe and various parts of the United States. The description of each tree includes some of the identifiable characteristics, the land of origin, history of importation, sages, legends, features of wood, and usage in medicine. There are 200 beautiful color photographs, and numerous line drawings of leaves. The trees are arranged according to families and genera.

Brockman, C. Frank. *Trees of North America: A Field Guide to the Native and Introduced Species North of Mexico.* (Illus. by Rebecca Merrilees.) Golden Press, 1968. 280 pp. $5.95, paper $2.95. 68–23532. (JH–SH)

This field guide facilitates identification of 594 of the total of about 865 species of trees native to North America north of Mexico in addition to many important introduced species that have become naturalized and some that are grown commercially. Trees are defined as "woody plants at least 15 feet tall." Descriptions include identification details, distribution maps, dimensions, and colored illustrations. Bibliography and index.

Grimm, William Carey. *The Book of Trees.* Stackpole, 1962. xviii+487 pp. illus. $7.95. 62–14047. (JH–SH)

Includes descriptions of virtually every species of tree from Northern Canada to the Gulf Coast. Drawings in the text facilitate identification.

Grimm, William Carey. *Recognizing Native Shrubs.* Stackpole, 1966. 319 pp. illus. $7.95. 66–12781. (JH–SH–C)

Over 400 native shrubs and woody vines of the Eastern United States from Canada to the Gulf of Mexico are described and illustrated with adequate line drawings in this useful work, provided with keys to various groups, families, and some species. An index includes common and scientific names.

The terminology is carefully selected for simplicity and usefulness. The glossary defines a sufficient number of terms, and some terms are illustrated. Instructions on identifying shrubs and vines are given.

Hough, Romeyn B. *Handbook of the Trees of the Northern States and Canada East of the Rocky Mountains.* Macmillan, 1947. x+470 pp. illus. $10.00. 7–31197. (SH)
A well-known handbook for the identification of trees of the region of North America that lies north of the northern boundaries of North Carolina, Tennessee, Arkansas and Oklahoma and east of the Rocky Mountains, and extending southward in the Appalachian Region to Alabama and Georgia. Although an older book and set in small type, it is authoritative. One feature of importance is the distribution map for each major species.

Mathews, F. Schuyler. *Field Book of American Trees and Shrubs.* Putnam's, 1915. xvii+537 pp. illus. $4.95. 15–5896. (C)
Since its publication in 1915, the guide has been one of the most popular aids to the identification of American trees and shrubs. While some of the taxonomy is outmoded, it is still a compact and useful work for beginner and layman.

Menninger, Edwin A. *Fantastic Trees.* Viking, 1967. xii+304 pp. illus. $8.95. 66–19165. (SH–C)
Presents a fascinating story and excellent photographs of abnormal growth and flowering habits in trees. Abnormalities discussed include: trees with dissimilar trunks, wild roots, parasitic roots, strange leaves, fruits and nuts, obesity in trees, sex switches, odoriferous trees, and those twisted by heredity. Insightful, informative, and useful as a reference and as casual reading.

Peattie, Donald Culross. *A Natural History of Trees of Eastern and Central North America.* Houghton Mifflin, 1950. xv+606 pp. illus. $8.50. 50–10354. (SH)
One of America's best-known and gifted nature writers is responsible for this descriptive account of the tree flora of Eastern and Central North America, which includes a compact key, a glossary and an index of scientific names.

Peattie, Donald Culross. *A Natural History of Western Trees.* Houghton Mifflin, 1953. xiv+751 pp. illus. $8.50. 52–5263. (SH)
A companion to the preceding volume, this one is concerned with Western North America. Both volumes should be owned by every library.

Petrides, George A. *A Field Guide to Trees and Shrubs.* Houghton Mifflin, 1958. xxix+431 pp. illus. $4.95. 57–10783. (SH)
Roger Tory Peterson collaborated in providing the illustrations for this work which provides field marks of all trees, shrubs, and woody vines that grow wild in the Northeastern and North-Central United States and in Southeastern and South-Central Canada.

Platt, Rutherford. *Discover American Trees.* (Illus. by Margaret Cosgrove; photos by the author.) Dodd, Mead, 1968. 256 pp. $4.50. 68–27440. (JH–SH)
A revised edition of *American Trees* (1952) which is a descriptive pocket guide with text figures illustrating leaves, buds or other identifying characteristics and a central signature of photographs of trees in actual land-

scape situations. An appendix contains illustrated keys of leaves, seeds, berries and fruits, twigs, flowers, bark smell, taste, and cones. A good handbook for amateurs.

Preston, Richard J., Jr. *North American Trees* (Exclusive of Mexico and Tropical United States) (rev. ed.). Iowa State U., 1961. xxxii+395 pp. illus. $4.50. 60–16604. (SH–C)
Includes all species of trees within the area covered, except 162 species of hawthorn and 20 usually shrubby willows which can be identified only by specialists, and commonly planted exotic species. Includes 135 genera and 568 species. Drawings show maps and give concise descriptions. Excellent taxonomic keys.

Silverberg, Robert. *Vanishing Giants: The Story of the Sequioas.* Simon & Schuster, 1969. 160 pp. illus. $4.50; LB $4.29. (JH)
This is probably the best book on the subject of the great trees written for young people. The threat of their imminent destruction makes the book even more desirable; it vividly portrays the rapacious practices of the past. There are many excellent photos, and an index tells where the great trees may be seen.

584   MONOCOTYLEDONS

Corner, E. J. H. *The Natural History of Palms.* U. of California, 1966. 393 pp. illus. $12.95. 66–25698. (C)
An outstanding authority on the tropics treats the natural history of palms, with emphasis on structure, morphology and evolutionary trends; it is neither taxonomic nor systematic. Emphasizes the range of variation in structural features in different kinds of palms; also has information on evolution and geographic distribution. Good for general reading despite the rather heavy botanical terminology. Glossary, bibliography, and index.

McClure, F. A. *The Bamboos: A Fresh Perspective.* Harvard, 1966. xv+347 pp. illus. $10.00. 66–10126. (SH–C)
The best reference work that exists on bamboos. It is not a text, but will be useful to the occasional student who wishes to learn something about bamboos, either superficially or in depth. Although some of the chapters are moderately technical, others are relatively nontechnical and reflect the author's long experience with bamboos in the Orient and in tropical America.

Pohl, Richard W. *How to Know the Grasses* (2nd ed.). (Ed. by Harry E. Jacques.) Wm. C. Brown, 1968. 192 pp. illus. $4.00, paper $3.25. 54–1268. (SH)
A pictured-key manual that will aid in collecting, studying, and identifying the common American grasses.

586   SEEDLESS PLANTS

Ahmadjian, Vernon. *The Lichen Symbiosis.* Blaisdell, 1967. viii+152 pp. illus. $5.75. 66–21101. (SH–C)
This is a must for any student specializing in mycology, microbiology, or basic biology. Contains exciting expositions on historical background, physiology, isolation techniques, synthesis experiments, etc. There is no other source for the information summarized in this illustrated study of lichens.

Alexopoulos, Constantine John *Introductory Mycology* (2nd ed.). Wiley, 1962. xvi+613 pp. illus. $12.95. 62–18348. (C)
Intended to follow a basic botany course, this mycology textbook is basically morphological and taxonomic, but significant physiological and genetic knowledge is interwoven whenever the facts can be discussed at an introductory level. References appended to each chapter and extensive final glossary, subject and author indexes.

Bonner, John T. *The Cellular Slime Molds* (2nd ed.). Princeton, 1967. x+205 pp. illus. $7.50. 66–22732. (C)
A thorough introduction to the biology of cellular slime molds. Emphasis is placed on experimental aspects and problems, and practical guidance on laboratory methods is included. An indispensable basic work.

Christensen, Clyde M. *The Molds and Man: An Introduction to the Fungi* (3rd ed.). U. of Minnesota, 1965. viii+284 pp. illus. $5.50. 65–17718. (SH–C)
Since the first edition was published in 1951, Dr. Christensen's book has won great popularity as a layman's introduction to the biology of the fungi. Also devotes a chapter to the industrial uses of fungi in the preparation of food and pharmaceutical products. For students and biology teachers there is an excellent chapter on experiments with fungi, sources of culture materials and laboratory equipment, and references. The final chapter is devoted to a summary classification of fungi.

Hutchins, Ross E. *Plant Without Leaves: Lichens, Fungi, Mosses, Liverworts, Slime-Molds, Algae, Horsetails.* Dodd, Mead, 1966. 152 pp. illus. LB $3.46. 66–20449. (JH)
An excellent series of black-and-white close-up photographs bring to life the vast array of small, often inconspicuous plants covering the earth's surface. The accompanying narrative, written for those with little botanical background, provides pertinent information on geographic distribution, reproductive methods, and economic uses.

Kavaler, Lucy. *Mushrooms, Molds, and Miracles.* John Day, 1965. 318 pp. $6.50; NAL (Signet T2978), paper 75¢. 65–13747. (SH–C)
"In this book I have tried to give some idea of the tremendous number, diversity and ubiquitous nature of these organisms [fungi] and to explain their tremendous importance to mankind. . . ." Thus the author explains the purpose of her book. She has diligently researched her subject as evidenced by the extensive bibliography. A "reading book" for laymen as well as a good basic collateral text for students.

Krieger, Louis C. C. *The Mushroom Handbook.* Dover, 1967. vii+560 pp. illus. $3.50 (paper). 67–28792. (SH–C)
First published in 1936, and now reprinted with a new preface and an appendix covering nomenclatural changes since the first printing, this field guide will fill a very real need. It gives much information about the varied aspects of the biology and importance of fungi.

Phaff, H. J., M. W. Miller, and E. M. Mrak. *The Life of Yeasts: Their Nature, Activity, Ecology, and Relation to Mankind.* Harvard, 1966. vii+186 pp. illus. $5.50. 66–14452. (C)
Contains a great deal of information not to be found elsewhere. Written for the nonspecialist, but some knowledge of biology and organic chemistry

is assumed. Topics include morphology, cytology, reproduction, genetics, distribution and propagation in nature, and metabolism.

Prescott, Gerald W. *How to Know the Fresh-Water Algae.* (Ed. by Harry E. Jaques.) Wm. C. Brown, 1954. xii+211 pp. illus. $3.50, paper $3.00. 55–3214. (SH)

A key illustrated by drawings for the identification of the more common fresh-water algae genera.

Sculthorpe, C. D. *The Biology of Aquatic Vascular Plants.* St. Martin's, 1967. xviii+610 pp. illus. $20.00. 67–15291. (C)

Thoroughly reviews the research literature and provides a comprehensive bibliography on aquatic vascular plants, and area ill-treated in botany texts. These plants are described at the structural, physiological, and ecological level. Of considerable reference value for teachers of botany and biology.

Shuttleworth, Floyd S., and Herbert S. Zim. *Non-Flowering Plants.* (Illus. by Dorothea Barlowe, et al.) (A Golden Nature Guide.) Golden Press, 1967. 160 pp. $3.95; $1.00, (paper). AC 67–16476. (JH–SH)

Successfully introduces the uninitiated reader to the diversity of non-flowering plants. An excellent addition to libraries. All amateur nature lovers with an interest in the lower plants will find this intriguing.

Smith, Alexander H. *The Mushroom Hunter's Field Guide* (rev. ed.). U. of Michigan, 1963. 264 pp. illus. $6.95. 63–14007. (JH–SH–C)

The foremost handbook and guide to the collection and identification of mushrooms. Descriptions are primarily nontechnical and are facilitated with more than 200 monochromatic photographs and 89 color plates that are realistic and accurate.

Sterling, Dorothy. *The Story of Mosses, Ferns, and Mushrooms.* Doubleday, 1955. 159 pp. illus. $3.25. 55–7012. (JH)

An interesting and lively text, plus good photographs, that introduces the reader to three important groups of seedless plants.

Thomas, William Sturges. *Field Book of Common Mushrooms, with a Key to Identification of the Gilled Mushrooms and Directions for Cooking Those That are Edible* (3rd ed.). Putnam's, 1948. 369 pp. illus. $5.00. (SH)

A well-known guide with detailed identification keys, a color identification chart, and other useful information for beginning student, field botanist, collector, etc.

Waksman, Selman A. *The Actinomycetes: A Summary of Current Knowledge.* Ronald, 1967. vi+280 pp. illus. $12.00. 67–14487. (C)

Certainly the most authoritative volume on all aspects of the topic, by the outstanding expert—who developed the field. Although frankly technical, this book will be helpful to serious advanced students who want the best and the most timely publication on the actinomycetes.

587   FERNS

Cobb, Boughton. *A Field Guide to the Ferns and Their Related Families of Northeastern and Central North America, with a Section on Species also found in the British Isles and Europe.* (Illus. by Laura L. Foster.) (Peterson Field Guide Series.) Houghton Mifflin, 1956. xviii+281 pp. $4.95. 55–10024. (SH–C)

Like other guides in the series, it is based on visual identification with descriptive characters on one page and drawings on the opposite page. General information on morphology and natural history.

Durand, Herbert. *Field Book of Common Ferns, for Identifying Fifty Conspicuous Species of Eastern America, with Directions for their Culture* (rev. ed.). Putnam's 1949. 223 pp. illus. $3.95. (JH–SH) There are black-and-white photographs of some species, but line drawings with descriptive material on a facing page is the general format of this useful field guide that has appeal to field naturalists as well as gardeners.

Wherry, Edgar T. *The Fern Guide: Northeastern and Midland United States and Adjacent Canada.* (Illus. by J. C. W. Chen.) (Doubleday, Nature Guide Series.) Doubleday, 1961. 318 pp. $5.50. (JH–SH)
Covers some 135 species of ferns in the area with descriptions and drawings on facing pages. Information on habitat and culture, technical and common names included.

Wherry, Edgar T. *The Southern Fern Guide: Southeastern and South-Midland United States.* (Illus. by J. C. W. Chen and K. C. Y. Chen.) (Doubleday Nature Guide Series.) Doubleday, 1964. 349 pp. $4.95. 64–19307. (JH–SH)
An account of 185 species of Southern ferns, with illustrations and descriptions. Indexed to both common and technical names. It complements the author's *The Fern Guide* (1961).

589.3 ALGAE

Chapman, Valentine J. *The Algae.* St. Martin's, 1961. viii+472 pp. illus. $8.00. 62–2640. (C)
Algae comprise an interesting group of plant life ranging from microscopic forms up to huge seaweeds. The author has prepared a survey of them, arranged by groups in evolutionary order, and followed by chapters on ecology, physiology, reproduction, geographical distribution, and utilization of algae. Suitable for readers with a basic knowledge of botany.

Dawson, E. Yale. *How to Know the Seaweeds.* (Ed. by Harry E. Jacques.) Wm. C. Brown, 1956. 197 pp. illus. $3.25, paper $2.50. 56–14426. (SH)
Identifying the common forms of marine algae of the Atlantic and the Pacific Coasts of North America is facilitated by this manual in which drawings portray the salient characteristics.

Prescott, Gerald W. *The Algae: A Review.* Houghton Mifflin, 1968. xi+436 pp. illus. $7.95. (C)
Based on 40 years of teaching and research, this compact yet comprehensive work on algae (plants that have no true roots, stems, leaves or leaflike organs) summarizes their phylogeny, taxonomy, morphology, physiology, and economic aspects. An excellent introductory work for students and a good summary for others. Bibliography, glossary and index.

590.744 ZOOLOGICAL GARDENS

Fisher, James. *Zoos of the World: The Story of Animals in Captivity.* (Nature and Science Library.) Nat. Hist. Press, 1967. 253 pp. illus. $5.95. 67–14047. (JH–SH–C)

Traces the history of zoos from ancient to modern times, and, in more detail, discusses the development of modern zoos. Emphasizes the theme of the zoo as an educational institution involved in the conservation of wildlife.

Kirchshofer, Rosl (ed.). *The World of Zoos.* (Tr. by Hilda Morris.) Viking, 1968. 327 pp. illus. $12.95. 68–15015. (JH–SH–C)

A collection of ten essays by various authors, each a recognized authority dealing with the theory and philosophy of modern zoo-keeping, aquarium management, animal trade, veterinary medicine, etc. The second section is a survey of zoological gardens and public aquariums throughout the world with descriptive details and other information for prospective visitors. Outstanding illustrations, some in color.

591 ZOOLOGY

Burton, Maurice. *Living Fossils.* Vanguard, 1954. xiv+282 pp. illus. $5.00. 54–11515. (SH)

A study of those species which are living representatives of groups that are largely extinct and are characteristic of geological ages long past. Among these are the coelacanth, the opposum, the duck-bill platypus, peripatus (the worm-insect), the king crab, and many others.

Burton, Maurice, Léon Bertin, et al. *The Larousse Encyclopedia of Animal Life.* (Foreword by Robert Cushman Murphy.) McGraw-Hill, 1967. 640 pp. illus. 29.5 cm. $22.50. 67–16267. (JH–SH–C)

A handsomely printed, well-illustrated volume, with entries arranged in taxonomic sequence. An immense amount of information is offered in a highly organized and readable form. A valuable reference.

Clark, James L. *In the Steps of the Great American Museum Collector, Carl Ethan Akeley.* Evans (dist. by Lippincott), 1968. 127 pp. illus. $3.95. 67–28198. (JH)

Akeley is widely known to zoologists as an outstanding naturalist, big-game collector, and as a sculptor and taxidermist who developed new techniques that resulted in lifelike exhibits for museums. This is the story of his African expeditions, of the collection and preparation of specimens, and other details of his life and accomplishments.

Dasmann, Raymond F. *Wildlife Biology.* Wiley, 1964. 231 pp. illus. $5.95. 64–25894. (SH–C)

Discusses the history of wildlife conservation, the esthetical and ethical, as well as the practical values, communities, the principles and techniques of wildlife management, and methods of studying wildlife. The dynamics of wildlife populations and their regulation are presented in the concluding chapters. May be used effectively as a college text or for collateral reading by the general public and high school student.

Dembeck, Hermann. *Animals and Men.* (Tr. from the German by Richard and Clara Winston.) Nat. Hist. Press, 1965. x+390 pp. illus. $7.50. 65–10416. (JH–SH–C)

A well-organized account of the important role that vertebrate animals have played in the advance of human welfare, economics, culture and enjoyment. Materials are drawn from a wide variety of sources. Animals are considered in their roles as prey, as sources of human food and clothing, as servants, and as companions of man. Most of the numerous illustrations are excellent.

George, Wilma. *Biologist-Philosopher: A Study of the Life and Writings of Alfred Russell Wallace.* Abelard-Schuman, 1964. xiv+320 pp. illus. $6.00. 64–12738. (SH)

Wallace formulated his theory of natural selection quite independently of Darwin. He also did voluminous work in zoogeography and is considered to be the father of that branch of zoology. He also worked and wrote in other fields, sometimes with distinction, sometimes not. The author is not concerned with the details of Wallace's personal life, but with presenting a history of his scientific ideas and accomplishments.

Hegner, Robert W. *Parade of the Animal Kingdom.* Macmillan, 1935. vi+675 pp. illus. $8.95. 35–27342. (JH)

A very comprehensive nontechnical account of all forms of animal life, from protozoa to man. Each phylum or major class is treated thoroughly, with many photographs and drawings. Includes information on appearance, structure, habits, defense, diet, reproduction, and effect on man. Despite its age, it continues to hold the interest of persons of all ages.

Kinkead, Eugene. *Spider, Egg, and Microcosm.* Knopf, 1955. vii+244 pp. $4.95. 55–9287. (SH)

"The egg! The spider! The protozoan! Promise of life, web of life, life invisible to the naked eye. One may fairly inquire what purpose it serves to delve briefly into these miraculous designs and into the private lives of three scientific men who enjoy an almost total preoccupation with a single object in nature. I can only answer . . . that to go with Petrunkevich on the track of spider, to gaze with Romanoff at the avian egg, to magnify with Vishniac the shape and beauty of the microcosm, is the same sort of experience as taking a walk in the spring woods."

Ley, Willy. *Dawn of Zoology.* Prentice-Hall, 1968. viii+280 pp. illus. $7.95. 68–13648. (SH)

Traces man in the biotic world as hunter, thinker, collector, allegorizer, cleric, reformer, systematizer, and digger. This is the story of the attempt of various ages of man to understand and interpret the animal world. Woodcuts and other illustrations from old works greatly add to the flavor.

Ley, Willy. *Exotic Zoology* (rev. ed.). Viking, 1959. xii+468 pp. illus. $5.95. 59–8356. (SH)

Ley has investigated each of the topics from the standpoint of its historical development in the knowledge of man and its place, if any, in the evolutionary scale of life. It is a synthesis of some of his earlier books which provide the basic material for individual parts, and to these he has added new material. This is fascinating reading for everyone who wants to know what is zoological fact, what is myth, and what is lore.

Simpson, George Gaylord. *Principles of Animal Taxonomy.* Columbia, 1961. xii+247 pp. illus. $7.50. 60–13939. (C)

From the author's long experience in the taxonomy of fossils and of living animals, he has written a detailed introduction to the fundamental principles of animal taxonomy which should be understood by all serious students of the biological sciences.

591.03   ZOOLOGICAL ENCYCLOPEDIAS

Pennak, Robert W. *Collegiate Dictionary of Zoology.* Ronald, 1964. vi+583 pp.
$8.50. 64–13331. (SH–C)
A dictionary of zoological terms and proper names, many of which are not
listed in standard dictionaries. The entries are cross-indexed, tying together
synonyms, popular, and scientific names. The appendix consists of a taxonomic outline of the animal kingdom.

591.07   ZOOLOGY—STUDY AND TEACHING

Elliott, Alfred M. *Zoology* (4th ed.). Appleton-Century-Crofts, 1968. x+799
pp. illus. $9.95. 68–13000. (SH–C)
A college textbook in zoology, organized on the central theme of organic
evolution but cognizant of new observations at the cellular and molecular
level which fortify the evolutionary concept. In essence the book is an
historical account of the earth and its animal life. This is a useful supplemental resource for senior high students.

Goodnight, Clarence J., Marie L. Goodnight, and Peter Gray. *General Zoology.*
Reinhold, 1964. xi+564 pp. illus. $9.50. 64–20138. (SH–C)
A substantial text embodying principles of molecular biology which begins
with basic discussions of the nature of living things and the physical state
of matter, proceeds to discuss the origin of life and the classification of
animals, then takes up each major group of vertebrates giving their characteristics, passes on to a consideration of the various systems, and ends with
discussions of genetics, evolutions and ecology. Bibliography and glossary.
An excellent reference for high schools.

Guthrie, Esther L. *Home Book of Animal Care.* Harper & Row, 1966. xiv+
302 pp. illus. $5.95. 66–11474. (JH–SH)
A handbook on the care of almost any animal a child might bring home.
Part I, arranged by taxonomic groups of vertebrates, provides information
on the natural history and care of exotic pets, with accurate, interesting, and
concise descriptions of habitat, range, reproduction, housing, and food. Scientific and common names of animals are used. Part II deals with different
ecological types of balanced terrariums and aquariums. Part III tells how
to prepare special diets for pets. Part IV gives plans for cages. Suggested
reading list.

Hickman, Cleveland P. *Integrated Principles of Zoology* (3rd ed.). Mosby, 1961.
972 pp. illus. $8.50. 61–6383. (C)
An almost encyclopedic textbook divided into six major parts: (1) functional organization; (2) morphology and physiology by phyla; (3) résumé
of organ systems; (4) evolution, heredity, and embryology; (5) ecology
and adaptation; (6) chronological history. Glossary.

Milne, Lorus, and Margery Milne. *The Nature of Animals.* (Illus. by Thomas
R. Funderburk.) Lippincott, 1969. 225 pp. $5.95; LB $5.82. 69–11999.
(SH–C)
An excellent job that brings together many facts of the animal kingdom
to show the many ways in which all animals are alike and in what respects

they differ. Chapters are devoted to how animals live, reproduce, inherit their characteristics, and the place of animals in the balance of nature. Aquatic animals are treated separately from land animals before the discussions of the major attributes of animal life.

Moment, Gairdner B. *General Zoology* (2nd ed.). Houghton Mifflin, 1967. 717 pp. illus. $9.50. 67–13056. (SH–C)
Retains the excellent quality of the first edition. The text is organized into seven parts: basic concepts, primitive phyla, stream of life, protosomes, deuterostomes, organ systems, and animals and their world. Highly recommended for elementary college and advanced high school students.

Pettit, Lincoln Coles. *Introductory Zoology*. Mosby, 1962. 619 pp. illus. $7.50. 62–7485. (SH–C)
A modern treatment of classical zoology which is as readable as a tradebook. It deals with zoology as a science; surveys the animal kingdom in phylogenetic order, devoting a section to the human organism; discusses heredity, populations, ecology and evolution; and finally deals with zoology and human destiny. Historical résumé and glossary.

Storer, Tracy I., and Robert L. Usinger. *General Zoology* (4th ed.). McGraw-Hill, 1965. vi+741 pp. illus. $8.95. 64–7739. (SH–C)
This well-known college textbook, first published in 1943, has been kept up to date by thorough revisions to incorporate the growing body of knowledge resulting from ongoing research. Many illustrations are new. References are listed at the end of each chapter; a comprehensive glossary.

Villee, Claude A., Warren F. Walker, Jr., and Frederick E. Smith. *General Zoology* (3rd ed.). Saunders, 1968. xxi+844 pp. illus. $9.75. 68–13956. (C)
Designed for an introductory college course and valuable as a collateral reading and reference book. The first part explains the general concepts including a review of scientific method and history, discussion of the physical and chemical basis of life, cells, physiology, and reproduction. The second part is taxonomic in organization with a chapter for each phylum. Glossary and index, references for each chapter.

591.1. ANIMAL PHYSIOLOGY

Dröscher, Vitus B. *The Mysterious Senses of Animals*. Dutton, 1965. 255 pp. illus. $5.95. 64–21859. (SH–C)
Considers topics about which new knowledge has been secured recently: the avoidance behavior of nocturnal moths to the ultrasonic sounds of bat predators, the communication of dolphins, the heat-ray eye of rattlesnakes, the social life of the prairie dog town, and the use of the stars by migrating birds at night. An appendix provides an excellent bibliographical introduction to some of the most exciting new areas of investigation.

Frisch, Karl von. *A Biologist Remembers*. (Tr. by Lisbeth Combrich.) Pergamon, 1967. ix+199 pp. illus. $6.00. 67–16653. (SH–C)
A fascinating autobiography by a famous zoologist traces the history of his research on communication in bees and the behavior of fishes. Strongly recommended for anyone interested in the sciences.

Galambos, Robert. *Nerves and Muscles*. (Science Study Series.) Doubleday, 1962. 158 pp. illus. $1.25 (paper). 62–10797. (SH)
"In order to uncover the secrets of nerve and muscle action, biologists have borrowed most of their tools from the physics laboratory, improved or modified them, and put them to work on living tissues. Consequently, cathode-ray oscilloscopes along with vacuum tubes and transistorized amplifiers are today as much the trade-mark of a biological laboratory as the stethoscope used to be of the country doctor."—Introduction.

Klein, H. Arthur. *Bioluminescence*. Lippincott, 1965. 184 pp. illus. $4.25. 64–19043. (JH–SH–C)
The strange creatures that luminesce, inhabiting the world's oceans usually at great depths, are the subject of this book. The manner in which this light is produced, the groups of organisms that possess it, and the function of luminescence are discussed. Scientific terminology is not shunned, rather it is presented in a manner easily understood by the layman.

Mason, George F. *Animal Vision*. Morrow, 1968. 95 pp. illus. $2.95. 68–14231. (JH)
Shows how the overall behavior of the animal is related to the position of the eye, the type of pupil, the relative number of rods and cones, and the ability to see at low levels of illumination. Examines how different species have adapted its vision to its needs and environment.

Milne, Lorus J., and Margery Milne. *The Senses of Animals and Men*. Atheneum, 1962. x+305 pp. illus. $6.95. 62–9411. (JH)
Students of biology and those who have an amateur's fascination for nature will enjoy this nontechnical introduction to the basic five senses, as well as responses to stimuli such as hot and cold, wetness and dryness, direction, hunger and thirst, etc. Points out how studies of the sense organs of animals have led to technological developments.

Scheer, Bradley T. *Animal Physiology*. Wiley, 1963. xii+409 pp. illus. $10.95. 63–12289. (C)
"In this book, I have tried to make a synthesis, not alone of human or cellular or comparative physiology, but of the physiology of animals. The synthesis begins at the level of molecules and of energy and moves upward through a scale of increasing complexity . . . into activities of organisms."
—Preface. For those who know fundamentals of zoology and chemistry.

Wells, Robert. *Bionics: Nature's Ways for Man's Machines*. Dodd, Mead, 1966. 159 pp. illus. LB $3.23. 66–20514. (SH)
Bionics is an interdisciplinary study involving biology and engineering that investigates biological function for the purpose of machine design. Although lacking in strict biological orientation, the young reader will probably be inspired to learn biology after reading this book.

Wendt, Herbert. *The Sex Life of the Animals*. (Tr. by Richard and Clara Winston.) Simon & Schuster, 1965. 383 pp. illus. $7.95. 65–11168. (SH–C)
Sex in this interesting, thorough, and delightfully written book is considered from an evolutionary point of view beginning with the emergence of sex in the lowest forms of life. Whereas sex in man is not covered directly, constant comparisons are made with man, particularly with his interpretation

of events and behavior as noted in lower animals. The large mammals are included and the material concerning them is valuable for many young readers. Worthwhile collateral reading and reference book for both high school students and college undergraduates; good resource material on sex education for teachers and parents.

Wooldridge, Dean E. *The Machinery of the Brain.* McGraw-Hill, 1963. xii+ 252 pp. illus. $5.95, paper $1.95. 63–13940. (C)
Aside from Chapter 1 and 3 which deal with the electrical property of nerves and peripheral data processing in the nervous system, the reader needs no background in physics and very little in biology. The book calls attention to similarities between electronic computers and the brain. The author's main purpose is to indicate the potential for future research through collaboration of physical and biological scientists.

591.33 EMBRYOLOGY

Barth, Lester George. *Embryology* (rev. ed.) Holt, Rinehart and Winston, 1953. xi+516 pp. illus. $11.50. 49–9187. (C)
A comprehensive account dealing with ovulation, fertilization, gastrulation, differentiation, and discussion of the embryology of amphibians, the chick, and the pig. An interesting feature are drawings of cross-sections of chick and pig embryos from various regions and at different stages, with explanations on the back. These drawings, three to a page, are on stiff paper and perforated so that the student may use them for comparison in microscopic studies.

Berrill, Norman J. *Growth, Development and Pattern.* Freeman, 1961. 555 pp. illus. $12.00. 61–8356. (C)
A basic work that provides a background in cells and cell aggregations, growth and form in metazoa, morphogenesis in plants, and concludes with sections on vertebrate morphogenesis. A scholarly treatment for collateral reading.

Huettner, Alfred F. *Fundamentals of Comparative Embryology of the Vertebrates* (rev. ed.). Macmillan, 1949. xvii+309 pp. illus. $6.95. 49–9357. (SH–C)
A study of comparative embryology from a morphological point of view that has proved to be valuable reference and collateral reading for secondary students and their teachers, although intended for an introductory college course.

Needham, Joseph. *A History of Embryology.* Abelard-Schuman, 1959. 304 pp. illus. $7.50. 59–6081. (C)
Needham's now classical work, covering the approximate period 1450-1900, is the only basic history of comparative embryology. Fascinating illustrations from historical works; excellent bibliography.

Oppenheimer, Jane M. *Essays in the History of Embryology and Biology.* M.I.T. Press, 1967. ix+374 pp. $12.50. 67–14098. (C)
This collection of essays on the origins of basic ideas in development of embryology will be admired by established biologists, but will be appreciated by students only after they have acquired more than an introductory knowledge of the field. None of these essays sounds dated or currently unimpor-

tant, though the earliest was written in 1940 and the most recent in 1965. An important book as a historical record. It will be preserved by students, but it will be hard reading for any but the most advanced.

Patten, Bradley M. *Early Embryology of the Chick* (4th ed.). McGraw-Hill, 1957. 244 pp. illus. $6.25. 62–51763. (C)

Patten, Bradley M. *Embryology of the Pig* (3rd ed.). McGraw-Hill, 1948. 352 pp. illus. $6.95. 48–10874. (C)
These well-known introductory embryology texts are still valuable for both classroom and reference use.

Rugh, Roberts. *Vertebrate Embryology: The Dynamics of Development.* Harcourt, Brace, 1964. ix+600 pp. illus. $10.50. 64–12975. (C)
The dynamic nature of embryology is emphasized in this text by presenting the normal sequence of events that transforms an apparently structureless egg into an individual having all of the structures, systems, functions, and bodily processes characterisic of a vertebrate species. The arrangement by complete and separate development histories— frog, chick, mouse, pig, and man— gives the student a concise concept of each organism as it develops. The reader should have completed a substantial basic course in general biology or zoology.

Sussman, Maurice. *Animal Growth and Development.* (Foundations of Modern Biology Series.) Prentice-Hall, 1964. 114 pp. illus. $3.95, paper $1.95. (SH–C)
This condensed treatment includes not only fundamentals of embryology, but includes a discussion of developmental phenomena based on present-day concepts of cell physiology and genetics.

Witschi, Emil. *Development of Vertebrates.* Saunders, 1956. xvi+588 pp. illus. $8.50. 56–5835. (C)
Written primarily for premedical students and zoology majors who have completed a course in general biology. The illustrations, both photographs and line drawings, are very good. Includes complete treatment of oögenesis, ontogenesis, and uses types for discussion of development by stages in amphibians, fishes, birds, and mammals. Good bibliography. A good reference for high school and public libraries.

591.4 ANIMAL MORPHOLOGY

Goin, Coleman J., and Olive B. Goin. *Comparative Vertebrate Anatomy.* Barnes & Noble, 1965. xiii+242 pp. illus. $1.75 (paper). 65–17010. (SH–C)
This clearly-written condensation of the basic facts and concepts of comparative vertebrate anatomy provides a concise course review for college students and a useful introduction for anyone. Each system of the vertebrate body is covered from structural, functional, and comparative points of view. An outline summary of precursor, unique, and critical traits of each class is appended. Good glossary and index.

Gray, Sir James. *Animal Locomotion.* Norton, 1968. xi+479 pp. illus. $15.00. (C)
Based on original research and a thorough survey of primary sources, Sir James explains essential features of animal locomotion with examples from

the vertebrates, arthropods, segmented worms, and molluscs. The number of mechanical principles is small and explained through a restatement in biological terms of Newton's three laws of motion. Mathematical computations, graphs, sketches and photos amplify the text. Valuable reference for biologists, medical physiologists and students; also of interest to physicists and engineers. Bibliography and index.

Hanson, Earl D. *Animal Diversity* (2nd ed.). (Foundations of Modern Biology Series.) Prentice-Hall, 1964. x+118 pp. illus. $3.95, paper $1.95. 64–12158. (SH–C)
A résumé of the principles and factors that define and characterize the diversity of animals is presented concisely in terms of systematics, phylogeny, evolution, paleontology, and ecological zoogeography.

Hyman, Libbie H. *Comparative Vertebrate Anatomy* (2nd ed.). U. of Chicago, 1942. xx+544 pp. illus. $5.50. 42–21814. (C)
A traditional and substantial textbook in comparative anatomy of vertebrates designed for college students who have had a fundamental zoology course. It defines the anatomical characteristics of vertebrates, describes the major stages of embryology, and then considers each of the systems in a phylogenetic and comparative fashion. Glossary and bibliography.

Romer, Alfred S. *The Vertebrate Body* (4th ed.). Sanders, 1970. viii+601 pp. illus. $10.25. 75–92143. (C)
Romer's treatment of comparative anatomy is written from the standpoint of paleontological and morphologic history, with due attention to embryology which he holds to be crucial in the consideration of homology. In breadth, depth, and wealth of illustrations it is one of the best texts and general reference works currently available. It has a glossary of scientific terminology and a bibliography arranged by subjects in which the most useful works are listed first.

Thompson, D'Arcy W. *On Growth and Form.* (Abridged edition by John Tyler Bonner, ed.) Cambridge U., 1961. xiv+346 pp. illus. $5.95. (SH–C)
Professor Bonner has prepared this condensation of the author's 2-volume original (2nd ed., 1942), by eliminating out-of-date materials and some of the more difficult mathematical analyses. In a classical style, the author describes mathematical relationships between living things and inanimate objects. This is an important and exciting work that serious students should be encouraged to read and digest.

Weichert, Charles K. *Elements of Chordate Anatomy* (3rd ed.). McGraw-Hill, 1967. 472 pp. illus. $8.95. 67–10881. (C)
Deals with classification and early development of the chordates and then devotes a separate chapter to each organ system. Very useful for the serious student of anatomy.

591.5 ANIMAL ECOLOGY

Bates, Marston. *Animal Worlds.* Random, 1963. 316 pp. illus. $15.00. 63–14144. (JH–SH)
Millions of animals live in various "worlds" in a state of association, sometimes interdependence, sometimes as hunter and hunted. These "worlds" are

generally called seas, deserts, mountains, prairies, and shores. They are described with the aid of black-and-white and colored photographs.

Breland, Osmond P. *Animal Life and Lore.* Harper & Row, 1964. x+388 pp. illus. $6.95; Avon, paper 95¢. 63–17708. (JH–SH)
Includes materials from two previous books, *Animal Facts and Fallacies* and *Animal Life and Lore,* with added topics and details to provide a series of informal notes and essays on natural history and ecology. Running speed, jumping ability, speed of flight, and other interesting details are compared. Mainly devoted to vertebrates.

Caras, Roger A. *Last Chance on Earth.* (Illus. by Charles Fracé.) Chilton, 1966. xii+207 pp. $12.95. 66–28805. (JH–SH)
Forty species from the list of animals considered endangered have been selected for a two- or three-page essay about each; accurate summaries of information drawn from respected sources. Because the book is a plea for public support, emphasis is on the dangers to wildlife; less is said about the efforts being made to save threatened species. Bibliography and a list of conservation organizations. Handsome printing and monochrome illustrations.

Cloudsley-Thompson, J. L. *The Zoology of Tropical Africa.* (World Naturalist Series.) Norton, 1969. 354 pp. illus. $12.50. (SH–C)
Various groups of the fauna are delt with according to general habitat: savanna; scrub and desert; mountains; rivers, lakes, and fresh water swamps; and mangrove swamps and coral reefs. Following sections discuss populations and migrations, the various rhythms, and ecological adaptations, and man as an ecological factor. Bibliography is a rich reference source, principally for periodical literature.

Eckert, Allan W. *Bayou Backwaters.* (Marlin Perkins' Wild Kingdom Series.) (Illus. by Joseph Cellini.) Doubleday, 1968. 155 pp. $4.95. 68–11758. (JH)
Presents the drama of life in a Louisiana bayou as seen through the eyes of the animals who live there. Conflict, climax, tragedy, all the qualities that make for good story-telling, are presented in an extremely well-written style.

Errington, Paul L. *Of Predation and Life.* Iowa State U., 1967. xii+277 pp. illus. $6.95. 67–20153. (SH–C)
Packed with valuable and fascinating accounts of predations, predominately among the vertebrates, this work points out the difficulty in distinguishing between the predatory and antagonistic intentions of many predators and how man can readily misinterpret these acts.

Etkin, William (ed.). *Social Behavior and Organization Among Vertebrates.* U. of Chicago, 1964. xii+307 pp. illus. $7.50. 64–13947. (C)
A layman's account of some of the outstanding research that has been undertaken by zoologists and experimental psychologists to analyze the social behavior of animals from an evolutionary viewpoint. The book consists of 10 chapters on various related topics by the editor and five other contributors. Valuable collateral reading for students of biology and psychology.

Gronefeld, Gerhard. *Understanding Animals.* (Tr. by Gwynne Vevers and Winwood Reade.) Viking, 1965. 319 pp. illus. $7.95. 65–19275. (JH–SH)
Intended for a nonprofessional audience, an excellent series of animal photographs, some in color, supplemented by a very readable text. Descriptions of animal activities include a mixture of objective reporting and subjective

interpretation. Introduces several current theoretical concepts in animal behavior. May be used as supplementary reading for an introduction to animal behavior studies in secondary schools. Laymen and students of behavior will find this sensitive, well-written text, a welcome change from the usual popular books about animals.

Grzimek, Bernard, and Michael Grzimek. *Serengeti Shall Not Die*. (Tr. by E. L. and D. Rewald.) Dutton, 1961. 344 pp. illus. $11.95. 61–6001. (JH)
The Director of the Frankfort Zoo, who is a veterinarian, accompanied by his son, explored, and made a census of the wild herds of the Serengeti Park which is the last great concentration of wild animals in the world. Aside from its population of resident animals, the area is a migration route, and a breeding and pasturing area for itinerant species.

Jones, Arthur W. *Introduction to Parasitology*. (Illus. by Allan D. Jones.) Addison-Wesley, 1967. xiii+458 pp. $10.75. 67–12830. (C)
Parasite coverage is arranged according to place in the animal kingdom, and this volume compares most favorably with other parasitology textbooks. Intended for a semester course in the senior year of the biology curriculum, this text interrelates parasitology with many other subjects.

Kendeigh, Samuel Charles. *Animal Ecology*. Prentice-Hall, 1961. x+468 pp. illus. $12.95. 61–12332. (SH–C)
Ecology, "the science of the relation of the animal to its organic and its inorganic environment," is introduced through this comprehensive summary and analysis based on hundreds of references. It discusses and describes all of the various types of local habitats, the ecological processes that occur therein, and, finally, discusses various biomes.

Laycock, George. *The Alien Animals: The Story of Imported Wildlife*. Nat. Hist. Press, 1966. 240 pp. illus. $4.95. 66–15772. (JH–SH–C)
Concerns efforts to introduce exotic birds, mammals, and fishes into the United States. An extensive and carefully used bibliography. The result is a highly readable book which is also a handy secondary reference. Makes a strong case that our knowledge of ecology is as yet too limited to predict the multiple effects of animal introductions.

Milne, Lorus J., and Margery Milne. *The Balance of Nature*. (Illus. by Olaus J. Murie.) Knopf, 1960. vii+329+vii pp. illus. $5.95. 60–13433. (SH)
Basically this book is a study of the effects of human activities on the precarious balance of nature. There is a dynamic equilibrium among associated species, sometimes very vivid and sometimes obscure to human observers. Understanding these complexes in animal societies is an obligation of intelligent laymen.

Milne, Lorus J., and Margery J. Milne. *The Mating Instinct*. (Illus. by Olaus J. Murie.) Little, Brown, 1954. 243 pp. illus. $4.75. 54–5134. (JH)
The sexual recognition and behavior in the animal kingdom—the colorful and instinctive activities of animals at mating time—based on actual observations of two well-known naturalists have been described in this book. Illustrations produced by a third famous field zoologist.

Milne, Lorus J., and Margery J. Milne. *Paths Across the Earth*. Harper & Row, 1958. xi+216 pp. illus. $5.95. 58–5431. (JH)
Individual movements of animals in search of food, shelter, to avoid enemies,

etc., and group or population movements, such as seasonal or other mass migrations, are discussed for the layman in this elementary natural history book.

Milne, Lorus J., and Margery Milne. *Patterns of Survival.* (Illus. by Stanley Wyatt.) Prentice-Hall, 1967. xii+339 pp. $7.95. 67–22801. (SH–C)
A successful attempt to illustrate how structural, functional, and behavioral adaptations among animals interact with ecological factors to determine survival and population level. Excellent collateral reading.

Morgan, Ann H. *Field Book of Animals in Winter.* Putnam's, 1939. xv+527 pp. illus. $5.00. 39–27745. (JH)
The activities of animals in winter—both the active and the inactive—are discussed in this handbook that conveys sound principles, details of natural history, migrations, hibernation, and reproduction. Following the general discussions there are chapters devoted to each major zoological group.

Murie, Olaus J. *A Field Guide to Animal Tracks.* Houghton Mifflin, 1954. xxii+374 pp. illus. $4.95. 54–9602. (JH)
This field guide to animals "you didn't see" enables you to identify their movements in deserted fields, over the snow, along the river bank, along a dusty trail. It includes a great many North American mammals, birds, and insects. Drawings were made by the author in the field.

Owen, D. F. *Animal Ecology in Tropical Africa.* Freeman, 1966. viii+122 pp. illus. $5.00. (SH–C)
Not a rehash of temperate zone ecology, but stresses the situation in tropical Africa. An excellent nontechnical introduction. Covers the biological scene now and in the past; the number, abundance and diversity of species; populations; the seasons and other periodic events; ecological genetics of populations; the ecology of man in tropical Africa; and ecological research.

Pedersen, Alwin. *Polar Animals.* (Tr. by Gwynne Vevers.) Taplinger, 1967 [c. 1966]. 188 pp. illus. $5.50. 66–20234. (SH–C)
The geographic scope is northeast Greenland. Deals with the principal mammals of the region, notably the musk-ox, wolf, hares, fox, polar bear, lemming, ermine, and aquatic mammals, and with a selection of birds. It is a detailed factual account of the behavior of these animals, their diets, mating, gestation periods, birth, and development, interspersed with many illustrative anecdotes. The 70-odd photographs are excellent.

Portmann, Adolf. *Animals as Social Beings.* (Tr. by Oliver Coburn.) Viking, 1961. 249 pp. illus. $6.00. 61–11425. (SH)
Describes the social interaction, relationships and behavior of various species of higher invertebrates in an interesting fashion for the edification of the nonprofessional reader.

Portmann, Adolf. *Animal Camouflage.* U. of Michigan, 1959. 111 pp. illus. $4.50, paper $1.95. 59–5066. (JH)
Camouflage is instinctive to animals and the methods are as varied and ingenious as many human techniques of self-preservation. Camouflage and vision, shape and camouflage, protective coloration, and the significance of camouflage are the major divisions of the book.

Schaller, George B. *The Deer and the Tiger: A Study of Wildlife in India.* U. of Chicago, 1967. 370 pp. illus. $10.00. 66–23697. (C)

A well-planned field study, technically well-documented, that is a complete appraisal of the animals previously found in India and of those found in India today. Can be used as collateral reading in ecology, mammalogy, and animal behavior.

Scott, John Paul. *Animal Behavior.* U. of Chicago, 1958. xi+281 pp. illus. $5.00; Doubleday, paper $1.45. 57–6989. (SH)
An intriguing introduction to animal psychology. The text is organized according to the groups of factors affecting behavior which operate at each level of biological organization. Treats heredity, instinct, behavior, social organization, environment, communication among animals, and other topics with interest and precision.

Selsam, Millicent E. *How Animals Tell Time.* (Illus. by John Kaufmann.) Morrow, 1967. 94 pp. $3.25; LB $3.14. AC 67–10311. (JH)
Demonstrates the many rhythms in animal existence, geared to seasons, tides, and sunrises. Also describes experiments which isolated some of the specific environmental factors controlling the biological cycles. The illustrations are attractive.

Street, Philip. *Vanishing Animals: Preserving Nature's Rarities.* Dutton, 1963. 232 pp. illus. $4.95. 63–8609. (SH)
The problem of preserving wildlife in the face of expanding human populations, resource utilization, changing patterns in land use and development, etc., is a world-wide concern to many zoologists and laymen. The author states the general problem, discusses the techniques of controlled exploitation and conservation, and then discusses examples of critical species.

Terres, John K. (Editor). *Living World Book* (Series). Lippincott, 1962–69. illus. $5.95. (JH–SH)
Austing, G. Ronald. *The World of the Red-Tailed Hawk,* 1964. 128 pp.
Austing, G. Ronald, and John B. Holt. *The World of the Great Horned Owl,* 1966. 160 pp.
Costello, David G. *The World of the Ant,* 1968. 160 pp.
Costello, David G. *The World of the Porcupine,* 1966. 160 pp.
Keefe, James F. *The World of the Opossum,* 1967. 144 pp.
Porter, George. *The World of the Frog and Toad,* 1967. 160 pp.
Rue, Leonard Lee, III. *The World of the Beaver,* 1964. 160 pp.
Rue, Leonard Lee, III. *The World of the Raccoon,* 1964. 152 pp.
Rue, Leonard Lee, III. *The World of the White-Tailed Deer,* 1962. 160 pp.
Rutter, Russell J., and Douglas H. Pimlott. *The World of the Wolf,* 1967. 160 pp.
Schoonmaker, W. J. *The World of the Grizzly Bear,* 1968. 190 pp.
Schoonmaker, W. J. *The World of the Woodchuck,* 1966. 160 pp.
Van Wormer, Joe. *The World of the Black Bear,* 1966. 160 pp.
Van Wormer, Joe. *The World of the Bobcat,* 1964. 128 pp.
Van Wormer, Joe. *The World of the Canada Goose,* 1968. 192 pp.
Van Wormer, Joe. *The World of the Coyote,* 1964. 152 pp.
Van Wormer, Joe. *The World of the Pronghorn,* 1968. 191 pp.
Although each one is brief, this series provides an introduction to animal ecology and life history. Each volume follows the same general pattern. The reader is first introduced to the animal, then the species is examined seasonally. During each season the development from very young to adult is

traced. A final chapter examines the relation of the animal to man. Most of the volumes contain a listing of subspecies, excellent bibliographies, and adequate indexes. Perhaps the strong point of each volume is the outstanding photography, excellent in the portrayal of the natural history and ecology of the species. This series is a substantial addition to any library.

591.9 ZOOGEOGRAPHY

Wallace, Alfred Russell. *The Geographical Distribution of Animals* (2 vols.) Hafner, 1962. $24.00. 62–15789. (C)
One of the great classical works of zoology, originally published in 1876, which is fundamental to the study of zoogeography. For mature students.

592 INVERTEBRATES IN GENERAL

Barnes, Robert D. *Invertebrate Zoology* (2nd ed.). Saunders, 1968. xiii+632 pp. illus. $10.50. 62–11596. (C)
A complete, advanced and authoritative general text, resource book and reference work that is intended for students who have had a good foundation course in biology or zoology. Professional zoologists and others need this basic book in their professional libraries. Covers phylogeny, embryology, morphology, and physiology of the representative invertebrate groups. Bibliographies for each chapter.

Buchsbaum, Ralph. *Animals Without Backbones* (rev. ed.). U. of Chicago, 1948. xii+405 pp. illus. $9.00 (text ed. $7.00). 48–9508. (JH–SH)
Over 550 photographs from actual life and 327 drawings illustrate this outstanding introduction to invertebrates. The text is written for high school students and college freshmen, but younger children and nonspecialist adults can easily read and understand it. Since invertebrates comprise 95 percent of the animal kingdom, this book is indispensable to all libraries.

Buchsbaum, Ralph, and Lorus J. Milne. *The Lower Animals: Living Invertebrates of the World.* (Mildred Buchsbaum and Margery Milne, collaborators.) Doubleday, 1960. 303 pp. illus. $12.50. 60–10650. (JH–SH)
A pictorial account and descriptive natural history of outstanding representatives of the invertebrate world, ranging from microscopic radiolarians to giant squids, from spiders on high mountains to sea cucumbers in the ocean deeps.

Hegner, Robert W., and Joseph G. Engeman. *Invertebrate Zoology* (2nd ed.). Macmillan, 1968. xviii+619 pp. illus. $10.95. 68–10280. (C)
The usefulness of the first edition (1933) is proclaimed by the fact it was reprinted 25 times. The revision updates the text and incorporates new knowledge and revised concepts, but retains the classic type approach which provides a basis for understanding adaptation and variation. Includes parasites and insects. For those who have had a basic biology or zoology course.

Hickman, Cleveland P. *Biology of the Invertebrates.* Mosby, 1967. x+673 pp. illus. $10.75. 67–10718. (C)
Scholarly and easily read. The formate, in contrast to the sequential presentation of type forms seen in other publications, should give the student a

wider appreciation of the various invertebrate groups as a whole. Although the text is designed primarily for advanced college undergraduates, it has reference and collateral reading value for graduate students, and at the senior high school level will provide good resource material for able students and their teachers.

Hyman, Libbie Henrietta. *The Invertebrates* (6 vols.). I. *Protozoa through Ctenophora*, 1940. x+726 pp. II. *Platyhelminthes and Rhynchocoela*, 1951. viii+550 pp. III. *Acanthocephala, Aschelminthes, and Entoprocta*, 1951. viii+572 pp. IV. *Echinodermata*, 1955. viii+763 pp. V. *Smaller Coelomate Groups*, 1959. viii+783 pp. VI. *Mollusca I* (in preparation 1968). McGraw-Hill. illus. $16.50 ea. 40–5468. (C)
Originally conceived as an 8-volume set (of which 6 volumes have been published) to cover authoritatively and completely the taxonomy, morphology, physiology, embryology, and biology of all of the invertebrates. It is strictly a set of zoological treatises and does not deal with the diagnosis and treatment of human disease caused by various parasites and pathogens. An advanced set for all college libraries and for special technical collections. Some reference value to college undergraduates and teachers.

593 PROTOZOA

Curtis, Helena. *The Marvelous Animals: An Introduction to the Protozoa.* (Illus. by Shirley Baty.) Nat. Hist. Press, 1968. xvi+189 pp. $5.95. 67–15370. (JH–SH)
Introduces the protozoa by means of lucid and pertinent narrative, excellent light and electron micrographs, drawings, and diagrams. Chapters on movement, nutrition, reproduction, and behavior. Each of the major classes of protozoa is examined.

Hall, Richard P. *Protozoa, the Simpliest of all Animals.* (Holt Library of Science.) Holt, Rinehart and Winston, 1964. 123 pp. illus. $2.50. (JH–SH)
Information on morphology, ecology, and classification. Glossary and good illustrations.

Hegner, Robert W. *Big Fleas Have Little Fleas, or Who's Who Among the Protozoa.* Dover, 1968. viii+285 pp. illus. $2.00 (paper). 68–9783. (JH–SH)
Originally published by Williams and Wilkins (1938) it was based on a series of lectures delivered at Cornell University in 1937, and included material previously used at a Sigma Xi lecture at Syracuse University. The appendices include a summary of general facts about protozoa, a glossary, a bibliography, and an index. All high school and college zoology or biology students should read it; all libraries should shelve it.

Jahn, Theodore L., and Frances F. Jahn. *How to Know the Protozoa.* Little, Brown, 1949. 234 pp. illus. $3.75, paper $3.00. A51–3425. (JH)
A "pictured key" to the principal genera of the Protozoa, with information on general biology, collection, preparation for study, etc.

Kudo, Richard B. *Protozoology* (5th ed.). Thomas, 1966. xi+1174 pp. illus. $15.75. (C)

A well-known, comprehensive basic work on ecology, morphology, life history, and identification. Includes directions for collection, cultivation and observation. References and index.

594 MOLLUSCA

Abbott, R. Tucker. *American Seashells*. Van Nostrand, 1954. xiv+541 pp. illus. $16.50. 54–5780. (SH–C)
More than 1500 varieties of mollusks to be found in shallow waters from Labrador to Florida, and from Alaska to lower California and Central America are described. Identification data, life histories, and other interesting biological details are included. The book is of interest to amateur collectors and professional conchologists. Outstanding illustrations and bibliography.

Burch, John B. *How to Know the Eastern Land Snails*. Wm. C. Brown, 1962. 214 pp. illus. $3.50, paper $2.75. 61–18612. (JH–SH)
A "pictured key" guide to the land snails of the United States occurring east of the Rocky Mountain Divide. General information is included on distribution, habits, and economic importance.

Hoyt, Murray. *Jewels from the Ocean Deep: The Complete Guide to Shell Collecting*. Putnam's, 1967. 288 pp. illus. $5.95. 67–20439. (JH–SH–C)
A thorough guide on shells and shell collection intended for hobbyists, although there is much of interest and value for conchologists and malcologists. Written in clear, simple, readable language.

Johnstone, Kathleen Y. *Sea Treasure: A Guide to Shell Collecting*. Houghton Mifflin, 1957. xii+242 pp. illus. $5.00. 56–8270. (JH)
Devoted to encouraging the beginning conchologist and to giving helpful suggestions about collecting, classifying and storing his shells. Stories of many unusual shells and the creatures which make them are told. Although of little use in identifying specimens, this should answer most questions about collecting in general.

Lane, Frank W. *Kingdom of the Octopus: The Life History of the Cephalopoda*. Sheridan, 1960. xx+300 pp. illus. $7.50. 60–13164. (SH)
A popular natural history writer has surveyed the scientific literature to produce an accurate, fact-filled, and well-illustrated book on the octopus and his relatives—the cuttlefish and the squid. Appendices include anatomical diagrams, taxonomic list, glossary and bibliography.

Morris, Percy A. *A Field Guide to Shells of the Pacific Coast and Hawaii*. Houghton Mifflin, 1952. xx+220 pp. illus. $4.95. 52–8276. (JH–SH)
A counterpart of the previous book, covering the marine clams and snails that may be collected from San Diego to the Arctic Ocean, together with the common forms from Hawaii. Some of the rarer, minute and deep-water shells have been included.

Morris, Percy A. *A Field Guide to the Shells of Our Atlantic and Gulf Coasts* (rev. ed.). Houghton Mifflin, 1951. xix+236 pp. illus. $4.95. 51–12030. (JH–SH)
Nontechnical descriptions of the marine shells that may be collected along the East Coast from Maine to Florida. The number of individual species present

runs into the thousands; therefore, the author has selected the larger and more common forms with some additional selection of minute, rare and deep-water varieties.

Rogers, Julia Ellen. *The Shell Book* (rev. ed.). Branford, 1951. xxi+503 pp. illus. $9.50. 51–14048. (JH–SH)
This is a verbatim reprint of the 1903 original published by Doubleday. Because of changes in taxonomic nomenclature made in the interim it is somewhat antiquated. On the other hand the descriptive notes and the photographs are still good. Should be attractive to those who consider Abbott's book too expensive.

595.12 FLATWORMS

Smith, J. D. *The Physiology of Trematodes.* (University Reviews in Biology.) Freeman, 1966. xvi+256 pp. illus. $4.00 (paper). 66–23196. (C)
A useful source book for many students of parasitology. Well organized and written in a clear, direct style. Medical aspects are not directly emphasized, but most species considered are of medical or economic importance. All important topics are surveyed. The reference list is ample and well selected.

595.13 ROUNDWORMS

Lee, D. L. *The Physiology of Nematodes.* (University Reviews in Biology.) Freeman, 1965. x+154 pp. illus. $2.50 (paper). 65–7579. (SH–C)
"It is the aim of this book to present current knowledge of the physiology of nematodes (free-living as well as plant and animal parasite species) in a form which will be useful to undergraduates and to the University teachers."—Preface. This useful book for collateral reading and study makes up for the deficiencies in other literature mentioned by the author. A selected bibliography of 161 titles.

595.16 EARTHWORMS

Darwin, Charles. *Darwin on Humus and the Earthworm: The Formation of Vegetable Mould through the Action of Worms, with Observations on their Habits.* (Introd. by Sir Albert Howard.) Faber and Faber (dist. by Humanities), 1967. 153 pp. illus. $3.50. (C)
First published in 1881, culminating 40 years of research. Discusses anatomy, physiology, natural history and intelligence of the earthworm. Summarizes his observations on amount of humus moved through the digestive tract. Provides evidence of Darwin's ingenuity in devising experiment, and meticulousness of his experimental records.

595.2 ARTHROPODA

Carthy, J. D. *The Behavior of Arthropods.* (University Reviews in Biology.) Freeman, 1965. vii+148 pp. illus. $2.50 (paper). 65–7567. (SH–C)
This brilliant summary of current knowledge of arthropod behavior also dis-

cusses the physiological, anatomical, genetic, and evolutionary bases of behavior. Extensive bibliography. An excellent source book for secondary school courses in advanced biology and for collateral reading in college biology, zoology, entomology and physiology courses.

Cloudsley-Thompson, J. L. *Spiders, Scorpions, Centipedes and Mites: The Ecology and Natural History of Woodlice, 'Myriapods' and Arachnids.* Pergamon, 1958. xiv+228 pp. illus. $9.00. 57–14499. (SH–C)
The contents are mentioned briefly in the subtitle. Beginning students in entomology will derive much pleasure in reading this informal text which covers classification, distribution, feeding habits, life history, and other biological features.

Schmitt, Waldo L. *Crustaceans.* (Ann Arbor Science Library.) U. of Michigan, 1964. 160 pp. illus. $5.00, paper $1.95. 64–17438. (SH)
The author was curator of Marine Invertebrates at the U. S. National Museum for 33 years and travelled throughout the world to collect specimens and data on the subject of this book. This book explains reproduction, distribution, life history, and economic relationships.

595.4 ARACHNIDA

Comstock, John Henry. *The Spider Book.* (Rev. ed. by W. J. Gertsch.) Cornell U., 1948. 729 pp. illus. $12.50. 49–4798. (C)
A well-known book on the life-history and taxonomy of spiders that is a valuable reference for public and college libraries.

Gertsch, Willis J. *American Spiders.* Van Nostrand, 1949. xiii+285 pp. illus. $9.75. 49–11600. (SH–C)
Descriptions of most of the common species of spiders in North America, together with spectacular colored illustrations and interesting evolutionary and natural history notes, make up this interesting natural history book. It is intended mainly for recreational and collateral reading. Despite its age, it is a worthwhile acquisition for all libraries.

Kaston, B. J., and Elizabeth Kaston. *How to Know the Spiders.* Wm. C. Brown, 1953. vi+220 pp. illus. $3.50, paper $2.75. A54–1776. (JH–SH)
General natural history notes and pictured keys that aid in identifying the more common spiders; also suggestions for collecting and studying them.

595.7 INSECTS

Barker, Will. *Familiar Insects of America.* Harper & Row, 1960. xviii+236 pp. illus. $5.95; LB $5.11. 60–8525. (JH)
Natural history, evolution, reproduction, and other biological information on commonly-encountered insects makes up this interesting book for the beginning student and nonspecialist reader.

Bates, Marston. *The Natural History of Mosquitoes.* Harper & Row, 1965. xii +378 pp. illus. $2.45 (paper). 49–9326. (SH–C)
There is no more authoritative or complete basic study of the life cycle, habits, distribution, and relationships of mosquitoes to certain human diseases. A reprint of the 1949 original.

Borror, Donald J., and Dwight M. DeLong. *An Introduction to the Study of Insects* (rev. ed.). Holt, Rinehart and Winston, 1964. xii+819 pp. illus. $17.50. 54–5398. (SH)

Some entomologists think that this text replaces Comstock. It covers the same general territory, and may be more readable and attractive to the beginning student. But many will want Comstock on hand, too. Introductory chapters give general information on morphology, reproduction and development, and metamorphosis. Then follow separate sections devoted to each order. A section is devoted to Arthropoda other than insects, another section provides instruction on collecting and preserving insects and suggests insect study projects, thus making this a valuable reference for high schools with many college-bound students.

Chu, Hung-Fu. *How to Know the Immature Insects.* Wm. C. Brown, 1949. 234 pp. illus. $3.50, paper $2.75. A50–2933. (SH)

A pictured key guide for identifying the orders and families of many of the immature insects, with suggestions for collecting, rearing and studying them.

Comstock, John Henry. *An Introduction to Entomology* (9th ed.). Cornell, 1940. xix+1064 pp. illus. $12.50. 40–27640. (C)

Comstock's *Entomology,* along with Gray's *Manual of Botany* and the handbooks on amphibians and reptiles of Wright and Wright, are among the monumental works in systematics and natural history that probably will survive as "classics" in the same manner as the works of Cuvier, Buffon, et al., which represent the biology of an earlier century. The detailed anatomical descriptions and life history notes of Comstock are reliable, even though the taxonomic nomenclature may be somewhat out of date. Valuable for college libraries.

Evans, Howard Ensign. *Life on a Little-Known Planet.* (Illus. by Arnold Clapman.) Dutton, 1968. 318 pp. $7.95. 86–25771. (SH–C–P)

Dr. Evans initiates the reader into the secret universe of life that surrounds us; a universe seen and heard by many, but mysterious and little understood by most of us—the universe of insects and their relatives. Up-to-date discussions of insect life range into ecological, political, moral, and philosophical concepts. The reader is led to an appreciation of living nature and the realization that man must do everything possible to preserve an ecologically balanced world between man and man, and between man and nature.

Fabre, J. Henri. *The Insect World of J. Henri Fabre.* (Edwin Way Teale, ed.) Dodd, Mead, 1964. xvi+333 pp. $4.00; Fawcett, paper 60¢. 49–5858. (JH–SH)

J. Henri Fabre (1823–1915) had a tremendous interest in insects and his works describing his observations and experiences are classic examples of literature. The editor has selected and assembled here an intriguing collection of Fabre's writings. This book is good collateral reading not only for students interested in science, but also for students of English literature and composition.

Fox, Richard M., and Jean W. Fox. *Introduction to Comparative Entomology.* Reinhold, 1964. xiv+450 pp. illus. $10.00. 63–21380. (SH–C)

"This is a book about insects, myriapods and arachnoids written for zoologists. It is intended primarily to serve the needs of those who teach or study entomology as an academic subject and it assumes that the student using it has previously completed at least one full course in general zoology or general biology." —Preface. The text is encyclopedic in scope; it has a carefully selected bibliography. For all high school, college, and public libraries.

Frisch, Karl von. *Ten Little Housemates.* Pergamon, 1964. 146 pp. illus. $3.25; paper $2.45. 60–8059. (JH)
The author is a keen observer of insect behavior, best known for his work on bees. This book is about ten common insects that people hate, hunt, kill, or try to ignore: the house-fly, the gnat, the flea, the bedbug, the louse, the clothes moth, the cockroach, the silverfish, the spider and the tick. Good reading for all laymen.

Hutchins, Ross E. *Insects.* (Illus. by Stanley Wyatt.) Prentice-Hall, 1966. xii+324 pp. $6.95. 66–13038. (JH–SH)
Begins with a general summary of evolution and morphology, followed by an essay on instinct, intelligence and behavior. There are fascinating, succinct pieces on insect sounds and migrations. Also describes typical representatives of ecological groups: aerialists, those in the realm of water, hunters, farmers, builders, paper and tent makers, nectar gatherers, carpenters, miners, gall makers and pollinators. The black-and-white photographs are excellent. A necessary supplement for taxonomically and morphologically oriented textbooks. Also good for general recreational reading. Indexed.

Hutchins Ross E. *The World of Dragonflies and Damselflies.* Dodd, Mead, 1969. 127 pp. illus. LB $3.46. 69–15910. (EA–JH)
This life history and ecological study of dragonflies and damselflies is sufficiently limited in scope to facilitate handling in depth. It contains directions for those who wish to collect and mount these insects for study.

Jaques, H. E. *How to Know the Insects* (2nd ed.). Wm. C. Brown, 1947. 205 pp. illus. $3.25, paper $2.50. 36–15364. (JH–SH)
An illustrated key to the more common families of insects, with suggestions for collecting, mounting and studying them.

Kalmus, H. *101 Simple Experiments with Insects.* Doubleday, 1960. 194 pp. illus. $3.95. 60–51211. (SH)
Suggested experiments for the younger biology student and the amateur naturalist that will enable him to explore some aspects of the natural history and physiology of insects. Includes a list of biological supply houses.

Klots, Alexander B., and Elsie B. Klots. *Living Insects of the World.* Doubleday, 1959. 304 pp. illus. $14.95. 59–9100. (SH)
Striking color and monochrome photographs and an informative and nontechnical text provide a good introduction to entomology for everybody. Aside from descriptive accounts of representative species of the principal orders, a summary of insect morphology and biology, a bibliography and an index of insects enhance the educational value.

Lanham, Url. *The Insects.* Columbia, 1964. viii+292 pp. illus. $6.95. 64–14235. (SH–C)
The natural history of insects for the student and general reader who has some biological background. The organization of material makes an inter-

esting narrative for straight reading and still the index makes it as useful for reference as conventional textbook material. The general sequence deals with the place of insects in nature, form and function, ecology, and finally a "parade of representative forms." Bibliography.

Lutz, Frank E. *Field Book of Insects* (4th ed.). Putnam's, 1948. 510 pp. illus. $4.95. 18–5513. (JH–SH)

A well-known and popular field book which describes all of the principal families, many genera, and most of the common species of insects in the United States and Canada. Condensed information on natural history of aerial, aquatic, and terrestrial insects.

Rolston, L. H., and C. E. McCoy. *Introduction to Applied Entomology.* Ronald, 1966. v+208 pp. illus. $5.00. 66–20088. (SH–C)

Presents man's attempt to understand and control his worst enemy—the insects. A world-wide coverage is given to concepts, principles, life history, and ecology of a number of insect types. The scope of this coverage is wide and revealing.

Ross, Herbert H. *A Textbook of Entomology* (3rd ed.). Wiley, 1965. xi-+519 pp. illus. $9.95. 56–9827. (SH–C)

A good, conventional, well-recommended textbook of entomology which is more recent and less formidable than Comstock, yet contains information on systematics and identification lacking in Fox and Fox.

Simon, Hilda. *Insect Masquerades.* Viking, 1968. 95 pp. illus. $4.95; LB $4.31. 68–16070. (JH–SH)

A layman's exposition of some of the most interesting and unusual phenomena of nature. The four-color paintings that illustrate examples of insect "traps and trickery," "camouflage," "warning signals," and "nightmare insects," are beautiful and accurate.

Swain, Ralph B. *The Insect Guide: Orders and Major Families of North American Insects.* Doubleday, 1948. xlvi+261 pp. illus. $4.95. 48–7228. (JH–SH)

A descriptive guide to the families of North American insects occurring north of Mexico using pictures and nontechnical language. Includes suggestions for the amateur on collecting, preserving, and studying insects. Bibliography.

Teale, Edwin Way. *Grassroot Jungles: A Book of Insects* (rev. ed.). Dodd, Mead, 1944. xi+240 pp. illus. $6.50. 44–5481. (JH)

Some of the common insects to be found in lawn, garden, and field are described and studied with many personal experiences and observations by the author. The author's hobby is insect photography and his 150 illustrations cover a wide range of insect life. The enthusiastic, authoritative style and excellent illustrations are a stimulating combination.

Teale, Edwin Way. *The Strange Lives of Familiar Insects.* Dodd, Mead, 1962. xiii+208 pp. illus. $4.50. 62–18165. (JH)

Presents life-history sketches of 14 familiar insects: mayfly, cricket, chinch bug, housefly, dragonfly, praying mantis, lacewing fly, cicada-killer wasp, ladybird beetle, termite, ant, aphid, monarch butterfly, and paper-making wasp.

Wigglesworth, Vincent B. *The Life of Insects.* World, 1964. xii+359 pp. illus. $12.50. 63–14795. (SH–C)

Emphasizes natural history, behavior, physiology, and other details. Speciation, migration and relationship to man are considered. Descriptive catalog
by orders and excellent bibliography.

Wigglesworth, Vincent B. *The Principles of Insect Physiology* (6th ed.). Dutton, 1965. viii+741 pp. illus. $17.50. (C)
This sixth edition by the founder of modern studies in this field is authoritative, well written, and tastefully illustrated. Unfortunately new topics in
insect physiology developed during the past 10 years are not universally included, although topics covered in previous editions have been brought up to
date. Essential reading for advanced students in insect comparative physiology; may be useful for reference by students in beginning courses.

595.76 COLEOPTERA

Jaques, H. E. *How to Know the Beetles*. Wm. C. Brown, 1951. 372 pp. illus.
$4.75, paper $3.75. 52–4037. (SH)
A taxonomic key, accompanied by drawings, that will facilitate identification
of many of the common beetles. The user should be familiar with the rudiments of insect morphology.

Reitter, Ewald. *Beetles*. Putnam's, 1961. 205 pp. illus. $20.00. 61–12742. (SH)
Currently out of print, but so outstanding it is listed in the hope that the
publisher will reprint or that librarians may be able to find copies in book
stores. A truly beautiful oversize book that is a combination of photographer's art and a general descriptive text on the morphology, evolution,
classification, ecology, and geographical distribution. The fact that only six
of the species illustrated are North American is unimportant.

595.77 DIPTERA

Carpenter, Stanley J., and Walter J. LaCasse. *Mosquitoes of North America
(North of Mexico)*. U. of California, 1955. 359 pp. 127 plus. 288 text figs.
$10.00. 55–7555. (C)
Contains techniques for collecting and preparing for study; keys to genera
and species; descriptions of females, males and larvae; data on distribution,
bionomics and known medical importance. Plates of adult mosquitoes follow
the text. Bibliography of 770 items and index. Authoritative reference for
students and professionals.

Dethier, Vincent G. *To Know a Fly*. Holden-Day, 1963. viii+119 pp. $4.50,
paper $2.50. 62–21838. (SH)
An enjoyable, popular and humorous account of how and why a biologist
goes about his work, in understanding the physiology and behavior of the
common fly, and how the author began and conducts his research into the
physiology and psychology of flies. He concludes with general observations
on "what it's like" to be a scientist.

Oldroyd, Harold. *The Natural History of Flies*. Norton, 1965. xiv+324 pp.
illus. $8.50, paper $2.25. 65–7926. (JH–SH–C)
Enhanced by 32 excellent full-page enlarged photographs and 40 line drawings, a fascinating non-taxonomic account of the life histories of representa-

tive hexapods. The work has three major divisions: (1) a general account of flies, their eggs, larvae and pupae; (2) a survey of flies in 16 groups in an apparent ascending order to evolution; and (3) flies and man, swarms of flies, and on the past, present, and future of flies. An extensive and valuable bibliography.

595.78 LEPIDOPTERA

Ford, E. B. *Butterflies* (3rd ed.). Collins, 1957. xiv+368 pp. illus. $7.00. 47–3794. (SH)
A well-known descriptive natural history of butterflies by an internationally known authority who has considered butterflies not only as a collector's hobby, but has studied them and writes about them from the standpoint of evolution, geographic distribution, and natural history. Bibliography included.

Holland, W. J. *The Butterfly Book* (rev. ed.). Doubleday, 1931. xii+424 pp. illus. $15.00. 31–19409. (SH)
Originally published in 1898, and revised with the addition of new plates in 1931, this monumental work is subtitled, "A Popular and Scientific Manual, Describing and Depicting all the Butterflies of the United States and Canada." Undoubtedly many taxonomic changes in the interim have made it partially out of date, but the descriptive data are still good and the colored illustrations have not been surpassed.

Klots, Alexander B. *A Field Guide to the Butterflies of North America, East of the Great Plains.* Houghton Mifflin, 1951. xvi+349 pp. illus. $4.95. 51–10190. (SH)
A natural history and field guide to the butterflies that may be found east of the Great Plains from Greenland to Mexico. It deals also with life zones, ecology, principles of taxonomy, and geographic variation.

Klots, Alexander B. *The World of Butterflies and Moths.* McGraw-Hill, 1958. 207 pp. illus. $15.00. 58–3245. (SH)
Magnificent colored and monochrome illustrations with a well-written explanatory text which describes the biology and natural history of representative species of the second largest group of insects numbering perhaps 100,000 species.

Tazima, Yataro. *The Genetics of the Silkworm.* Prentice-Hall, 1964. xii+253 pp. illus. $10.95. 64–3417. (C)
The genetics of the silkworm is fundamental to the silkworm industry. This book summarizes the research reports that have been published in Japanese. This comprehensive treatment of the subject will interest all students of genetics.

595.79 HYMENOPTERA

Dines, Arthur M. *Honeybees from Close up.* (Photos. by Stephen Dalton.) Crowell, 1968. 114 pp. $6.95. 67–15407. (SH–C)
A well-known entomologist and beekeeper his combined talents with a

superb photographer to produce one of the most interesting life histories of the honeybee ever published. Covers the basics of beekeeping, also.

Evans, Howard Ensign. *Wasp Farm.* Nat. Hist. Press, 1963. viii+178 pp. illus. $3.95. 63–19968. (JH)

An informal narrative account of the author's observations on wasps, their private lives, their social organization, their ecology and other details of their biology. Good recreational reading for anybody. Unusual photographs.

Frisch, Karl von. *The Dance Language and Orientation of Bees.* (Tr. by Leigh E. Chadwick.) Harvard, 1967. xiv+566 pp. illus. $15.00. 67–17321. (C)

A monograph of von Frisch's 50 years of research on bees, their communication, methods of orientation, and sensory facilities. Presents a thorough understanding of the lives and communication methods of social insects.

Goetsch, Wilhelm. *The Ants.* U. of Michigan, 1957. 173 pp. illus. $4.50, paper $1.95. 57–7743. (SH)

The omnipresent ant was admired by Solomon for her industry and by Herodotus for her good mining. She keeps cows, weaves, raises slaves, and is served by them. There are many species of these colonial insects and the author discusses representatives from various parts of the world. This is an outstanding book of ant lore.

Hoyt, Murray. *The World of Bees.* Coward McCann, 1965. 254 pp. illus. $5.95. 65–13277. (JH–SH)

Suitable as a text for a course in agriculture and a trustworthy addition to a horticulture library or to the science collection of a school or public library. The average student will profit by a quick reading for he will learn interesting research information on bee culture. The well-written "Communication" chapter reveals that bees sense the location, direction, and distance of food supply sources. The experienced beekeeper will find a new approach to artificial insemination, multiple matings, and hybrid crossings of queens. The brief section on bee venom is factual and enlightening.

Lindauer, Martin. *Communication Among Social Bees.* Harvard U., 1961. ix+143 pp. illus. $4.75. 61–5579. (JH)

The author describes in detail the experiments by means of which he deciphered the "language" or technique of communication among bees. Tells of the progress that has been made in this field of research since the pioneer work of Karl von Frisch some 15 years prior to the appearance of this book. Bibliography.

Maeterlinck, Maurice. *The Life of the Bee.* (Tr. by Alfred Sutro.) Dodd, Mead, 1936. 278 pp. $3.00. 36–13847. (JH–SH)

Although some may consider this work out of date in the light of later research on social insects, it has long been considered a classic natural history essay and as such should be assigned reading for students.

Newman, L. Hugh. *Ants From Close-Up.* (Illus. by Stephen Dalton.) Crowell, 1967. xiv+112 pp. $6.95. 67–12407. (JH)

The life of the ant is told from the viewpoint of the scholar, with the aid of many rare photographs. Anatomy, physiology, and behavior are weaved into a most interesting account of the world's most social animal species.

Sudd, John H. *An Introduction to the Behavior of Ants.* St. Martin's, 1967. 200 pp. illus. $8.25, paper $3.95. 66–29852. (C)
Based on a study of many thousands of items in periodicals, this integrated account of the background and behavior of ants presupposes a basic knowledge of morphology and physiology. It discusses such topics as effectors and senses, navigation, nests, nutrition and food storage, parasites and predators mating, care of larvae and pupae, organization of work, learning, and social behavior. Selected bibliography and index.

596 VERTEBRATES

Barker, Will. *Familiar Animals of America.* Harper & Row, 1956. xv+300 pp. illus. $5.95; LB $5.11. 56–8770. (JH)
Comprehensive, nontechnical descriptions of over 50 species and families of North American animals. Tells the physical characteristics, diet, habits, and range of each animal; accompanied by black-and-white drawings.

Barrington, Ernest J. W. *The Biology of Hemichordata and Protochordata.* (University Reviews in Biology.) Freeman, 1965. vi+176 pp. illus. $2.50 (paper). 65–7586. (SH–C)
The lower chordates, including the marine sea squirts and amphioxus—the evolutionary precursors of the vertebrates—are considered. Deals with the behavior and physiological activities which cannot ordinarily be observed since the animals are seldom seen alive. List of 114 references and an index. Valuable to any reader who has had a basic course in biology.

Blair, W. Frank, A. P. Blair, P. Brodtkorb, F. R. Cagle, and G. R. Moore. *Vertebrates of the United States* (2nd ed.) McGraw-Hill, 1957. ix+819 pp. illus. $14.50. 56–9622. (C)
Pratt's manual became unavailable more than 10 years before the publication of this work, and this is its successor. Each major section has a separate author, yet they have worked together to provide coordination and a measure of uniformity. This is primarily a comprehensive handbook for identification of the fishes, amphibians, reptiles, birds, and mammals of the United States. The introduction provides a summary of chordate characteristics, vertebrate history, principles of classification and major features of vertebrate distribution.

Romer, Alfred S. *The Vertebrate Story* (4th ed.). U. of Chicago, 1959. vii+437 pp. illus. $7.50. 58–11957. (SH–C)
A general account of vertebrates, their evolution, comparative anatomy and embryology, including man as a representative vertebrate. Indispensable cultural background study for all students and particularly nonscience college majors.

Torrey, Theodore. *Morphogenesis of the Vertebrates* (2nd ed.). Wiley, 1968. x+600 pp. illus. $10.95. 62–8791. (C)
An integrated consideration of comparative vertebrate anatomy and physiology. This is a textbook for class use, and Romer's somewhat similar book is intended for collateral reading. College, large high school, and public libraries may wish to own both.

Young, John Zachary. *The Life of Vertebrates* (2nd ed.). Oxford, 1962. xv+ 820 pp. illus. $11.50. 62–2012. (C)

This readable yet substantial work has been widely acclaimed because it so successfully synthesizes anatomy, physiology, ecology, and evolution into an integrated account of animals with backbones. The revision was complete and incorporated new data throughout. For those with a background of an elementary course in biology or zoology.

597.42 GANOID FISHES

McCormick, Harold W., Tom Allen, and William E. Young. *Shadows in the Sea: The Sharks, Skates and Rays.* Chilton, 1963. xii+415 pp. illus. $10.00. 62–16260. (SH)

A readable account of the ganoid fishes, with personal observations, species descriptions, life history and distribution information, and reports on shark attacks. Also data on fishing for sharks, economic value of sharks and shark products, with an appendix on cookery and a bibliography.

597.5 TELEOST FISHES

Brown, Margaret E. (ed.). *The Physiology of Fishes.* Vol. I—*Metabolism.* Vol. II—*Behavior.* Academic, 1957. xiii+447; xi+526 pp. illus. (I) $17.00; (II) $18.00. 56–6602. (C)

Outstanding authorities in Canada, England, Holland, Scotland, and the United States have written the many chapters of these important volumes which will be useful to students of ichthyology, comparative morphology and physiology, endocrinology, biochemistry, animal psychology, and genetics. References at the end of each chapter. Background in biology, chemistry and physics needed. Recommended for college libraries, large public libraries, and special collections.

Eddy, Samuel. *How to Know the Freshwater Fishes.* Wm. C. Brown, 1957. 253 pp. illus. $3.75, paper $3.00. 58–3084. (JH–SH)

Pictured keys for identifying all of the freshwater fishes of the United States as well as some marine species which often enter fresh water. Useful, but not attractive or inspiring.

Grant, Leonard J. (ed.). *Wondrous World of Fishes.* National Geographic, 1965. 367 pp. illus. $8.75. 65–11482. (JH–SH)

An impressive array of talent has collaborated in the preparation of an illustrated overview of fishes and fisheries that begins with a general discussion of the morphology of fishes and a description of their environment. Then follow sections on recreational fishing, commercial fishing at sea, lake and stream fishes, and tropical fishes. There is a guide to fish cookery, an index, and a list of public aquaria. An excellent book for browsing or introducing a student to the scientific study of fishes and fisheries.

Hasler, Arthur D. *Underwater Guideposts: Homing of Salmon.* U. of Wisconsin, 1966. 155 pp. illus. $6.00. 66–11800. (C)

Very neatly summarizes, with supporting evidence, the present state of knowledge, observed and hypothetical, on the homing behavior of salmon.

The introduction briefly describes the life cycle of migrating salmonids and comments upon the survival value of homing phenomena. Part I deals with various hypotheses formulated to explain homing in salmon. Part II is concerned with the oceanic phase of salmon migration. The book is written in a clear and interesting style. Should be of interest to the layman, as well as the serious student. Bibliography and index.

Herald, Earl S. *Living Fishes of the World.* Doubleday, 1961. 304 pp. illus. $12.50. 61–6384. (JH–SH)
This systematically arranged work ranges from the primitive jawless species to highly specialized bony fishes. Includes details of structure, habits, range, life history and ecology. Glossary, bibliography, and index.

Hubbs, Carl L., and Karl F. Lagler. *Fishes of the Great Lakes Region.* U. of Michigan, 1964. 272 pp. illus. $7.95. 64–17435. (SH–C)
A complete guide to the identification of the fishes found in the Great Lakes basin. The basin includes the rivers, lakes, and streams of the entire Midwestern United States, and also southern Ontario and Quebec, New England, New York, New Jersey, Pennsylvania, Maryland, Delaware and Virginia. Includes keys to species and subspecies. Bibliography.

Innes, William T. *Exotic Aquarium Fishes* (19th ed.). Aquarium (dist. by Dutton), 1964. 541 pp. illus. $10.95. 35–8711. (JH–SH)
Since 1935 this book has been the most complete, accurate and widely-used handbook of professional and amateur aquarists. The 19th edition, edited by George S. Myers and with new technical annotations by Helen Simkatis, has updated the information.

Innes, William T. *Goldfish Varieties and Water Gardens.* Dutton, 1960. 385 pp. illus. $6.95. 50–8674. (JH–SH)
A handbook that tells everything an aquarist needs or wants to know about goldfishes, the varieties available, their care, and culture. Includes a section on aquatic plants, and on the construction and maintenance of outdoor pools.

Lagler, Karl F., John E. Bardach, and Robert R. Miller. *Ichthyology: The Study of Fishes.* Wiley, 1962. xiii+545 pp. illus. $12.95. 62–17463. (C)
An up-to-date and authoritative textbook on ichthyology that presupposes a foundation in zoology, chemistry and physics. It covers anatomy, physiology, reproduction, classification, ecology, genetics, and other basic biological considerations.

La Monte, Francesca. *North American Game Fishes.* Doubleday, 1945. xiv+202 pp. illus. $6.95. Agr. 46–2. (C)
A nontechnical guide that will facilitate identification of the freshwater and salt-water game fishes in North American waters. Of major interest to anglers and amateur naturalists.

Marshall, N. B. *The Life of Fishes.* (World Natural History Series.) World, 1966. 402 pp. illus. $12.50. 66–11276. (SH–C)
Scholarly and readable. Discusses morphology, physiology, ecology and natural history, reproduction and early life of marine and freshwater fishes. The third part is a most interesting consideration of the total aquatic environment which shows how fishes are adapted to the total complex of their particular living space. References are listed for each chapter—all excellent

primary sources. Drawings and photographs, some in magnificent color. The index is well made. No better general introduction to the entire subject of fishery biology.

Netboy, Anthony. *The Atlantic Salmon, a Vanishing Species.* Houghton Mifflin, 1968. 457 pp. illus. maps. $6.95. 68–23214. (SH–C)

Travel to all areas of North America and Europe to which *Salmo salar* originally was native, study of available published and unpublished material, and interviews with knowledgeable persons provided the basis for one of the most interesting and thorough studies of a single species of fishes ever produced. It explains the zoogeography, ecology, life history, economics, and the causes for its depletion or extermination of much of its former range. It tells of conservation programs now in effect to maintain it in still habitable waters. Good reading for laymen and students. Bibliography.

Norman, John R. *A History of Fishes* (2nd ed.). (Rev. by P. H. Greenwood.) (Illus. by W. P. C. Tenison.) Hill and Wang, 1963. xxxi+398 pp. $6.95. 63–11894. (SH–C)

Since publication of the first edition in 1931, this work has been an authoritative reference and assigned collateral reading book for those interested in any phase of fishes, fisheries, comparative anatomy, paleontology, evolution, comparative physiology, or zoogeography. It is organized by functions, systems, and disciplines so that the comparative aspect is always kept in mind by the reader. Concluding sections deal with economic matters.

Ommanney, Frances Downes. *A Draught of Fishes.* Crowell, 1966. 254 pp. illus. $6.95. 66–11055. (SH–C)

A comprehensive account of biological oceanography, fishing grounds, fishes, fishery resources, fishing methods, fish culture, and fishery products. An excellent background study and introduction to marine science. An appendix gives the scientific names of the fishes mentioned.

Perlmuter, Alfred. *Guide to Marine Fishes.* New York U., 1961. 431 pp. illus. $6.50. 60–14491. (SH)

A guide to the identification of marine fishes from Cape Cod to Cape Hatteras employing the unusual device of silhouettes accompanying the keys and descriptive notes. Recommended only for libraries within the general region to which it applies.

Schrenkeisen, Raymond M. *Field Book of Freshwater Fishes of North America North of Mexico.* Putnam's, 1963 [c. 1938]. 312 pp. $4.95. 38–27751. (JH–SH)

A useful handbook and well-known field guide for the amateur naturalist.

597.6 AMPHIBIANS

Bishop, Sherman C. *Handbook of Salamanders: The Salamanders of the United States, of Canada, and of Lower California.* Comstock, 1962. xiv+555 pp. illus. $12.50. 62–12199. (C)

A reprint of the 1943 original. Part I deals with general life history, reproduction and development, collection, preservation, and explains the use of keys and maps. Part 2 consists of keys and descriptive accounts that facilitate identification of species and subspecies. Bibliography. Despite its age it is still a fundamental reference for all professional and amateur herpetologists.

Cochran, Doris M. *Living Amphibians of the World.* Doubleday, 1961. 199 pp. illus. $14.95. 61–9491. (JH–SH)
The curator of Reptiles and Amphibians of the U. S. National Museum prepared this beautiful volume according to an inconspicuous taxonomic framework. It includes lively accounts of both the commonly encountered and the more unusual members of each amphibian family. A book for everyone.

Conant, Roger. *A Field Guide to Reptiles and Amphibians of the United States and Canada East of the 100th Meridian.* Houghton Mifflin, 1958. xv+366 pp. illus. $4.95. 58–6416. (SH)
Illustrated by photographs of living specimens, this is a complete and authoritative guide to the species of turtles, crocodiles, alligators, lizzards, snakes, salamanders, newts, frogs and toads in the area indicated in the title. Introductory notes on habitat, collection, transportation and care, as well as the appended glossary and bibliography, make it valuable to the amateur collector and natural history student.

Morris, Percy A. *Boy's Book of Frogs, Toads, and Salamanders.* Ronald, 1957. v+240 pp. illus. $4.50. 57–7484. (JH)
Amphibians are studied in considerable detail, and the physical appearances of several hundred species are described. Differences between amphibians and other animals are explained, as are distinctions between frogs and toads. Of special interest are the instructions for the capture, collection, and preservation of these animals.

Oliver, James A. *The Natural History of North America Amphibians and Reptiles.* Van Nostrand, 1955. xi+359 pp. illus. $7.95. 55–10534. (SH)
Personal field observations by the author supplemented by information obtained from the literature are embodied in this comprehensive book that includes life history, folk-lore, distribution, taxonomy, and other information on frogs, toads, salamanders, snakes, lizards and turtles. Good for general reading as well as for reference.

Porter, George. *The World of the Frog and the Toad.* (Living World Books.) Lippincott, 1967. 153 pp. illus. $5.95. 68–10419. (JH)
Simply but interestingly written, this book about the common frogs and toads of North America gives a brief description and résumé of the life cycle of each species presented in the familiar context of the four seasons. The pictures are a bit disappointing; few are action shots but in the main they are a series of static close-ups.

Smyth, H. Rucker. *Amphibians and Their Ways.* Macmillan, 1962. xv+292 pp. illus. $6.95. 61–10779. (JH)
An elementary introduction to the biology and natural history of amphibians for beginning naturalists. It includes information on North American frogs, toads, and salamanders as well as on interesting foreign species.

Stebbins, Robert C. *A Field Guide to Western Reptiles and Amphibians.* Houghton Mifflin, 1966. xiv+279 pp. illus. $4.95. 66–16381. (JH–SH–C)
Describes 207 species of western amphibians and reptiles through use of

species description, range maps, juvenile stages, and ecological orientation. A handy reference for quick field identifications and for brief life history notes.

Twitty, Victor Chandler. *Of Scientists and Salamanders.* Freeman, 1966. ix+ 178 pp. illus. $5.50. 66–24954. (C)
An interesting and scientifically valuable autobiographical account of his life long research with salamanders, including morphological, physiological, and taxonomic studies. An excellent bibliography is appended.

Wright, Albert Hazen, and Anna Allen Wright. *Handbook of Frogs and Toads of the United States and Canada* (3rd ed.). Cornell, 1949. xii+640 pp. illus. $12.50. 49–1510. (C)
A classical reference work on the classification, distribution, natural history, ecology, reproduction, and other biological information on amphibians of the United States and Canada, followed by keys and descriptions of families, genera, and species.

598.1   REPTILES

Barker, Will. *Familiar Reptiles and Amphibians of America.* Harper & Row, 1964. xi+220 pp. illus. $5.95. 62–14599. (JH)
Illustrated natural history notes on some of the more common snakes, lizards, turtles, tortoises, crocodiles, alligators, frogs, toads, salamanders and newts. Of interest to the amateur naturalist, younger student, or general reader.

Bellairs, Angus, and Richard Carrington. *The World of Reptiles.* American Elsevier, 1966. 153 pp. illus. $4.75. 66–17674. (JH–SH)
Written with scholarly precision and clarity, this is a good introduction to the origin, diversity, basic structure, physiology, natural history, and distribution of reptiles. Technical terms are explained.

Carr, Archie. *Handbook of Turtles: The Turtles of the United States, Canada, and Baja California.* Cornell, 1952. xvi+542 pp. illus. $12.50. 52–9126. (C)
Deals with all aspects of the biology, natural history, phylogeny, and related topics concerning 79 species and subspecies of turtles that inhabit the specified area. This book updates Pope's work published in 1939.

Carr, Archie. *So Excellent a Fishe: A Natural History of Sea Turtles.* Nat. Hist. Press, 1967. x+248 pp. illus. $5.95. 67–15371. (JH–SH–C)
A stimulating introduction to the field of vertebrate natural history, ecology, and behavior that leads into a thorough and interesting discussion of sea turtles. The author points out the many yet unchallenged research opportunities with sea turtles.

Ditmars, Raymond L. *The Reptiles of North America* (rev. ed.). Doubleday, 1936. xvi+476 pp. illus. $9.95. 36–27465. (SH)
The work is subtitled "A Review of the Crocodilians, Lizards, Snakes, Turtles and Tortoises Inhabiting the United States and Northern Mexico." It has been a favorite of amateur herpetologists and laymen since its original publication, and although professionals consider the taxonomy out of date, the basic natural history information is reliable and very readable.

Keeling, Clinton H. *Meet the Reptiles.* (Illus. by John Faulds.) Watts, 1965. 158 pp. illus. $3.95. 64–19476. (JH–SH)
An excellent literary narrative about reptiles in general, and some members of the Class in particular, written by the keeper of his own zoological gardens in Derbyshire, England. This is not a book to find out about or identify a particular reptile. Excellent supplemental reading. Indexed.

Minton, Sherman A., and Madge Rutherford Minton. *Venomous Reptiles.* Scribner's, 1969. xii+274 pp. illus. $7.95. 69–17042. (SH–C)
This is a desirable book on a fascinating subject. Fact is clearly distinguished from myth. The authors achieve their stated purpose of attempting to "present the venomous reptiles both as they are and as they have seemed to be." There is a good level of scholarship throughout the text and the appended tables contain information on reptile venoms, yields, and toxicity.

Morris, Percy A. *Boy's Book of Snakes: How to Recognize and Understand Them.* Ronald, 1948. viii+185 pp. illus. $4.50. 48–9569. (JH)
A handbook of the snakes of North America. Many aspects of snake life are treated in an introductory section; most of the book is devoted to identification of nearly 80 snakes. Poisonous and nonpoisonous snakes are grouped separately, and precautions which should be taken with each are given.

Morris, Percy A. *Boy's Book of Turtles and Lizards.* Ronald, 1959. vi+229 pp. illus. $4.50. 59–12121. (JH)
A guide to approximately 100 different kinds of turtles and lizards. Each is discussed with respect to physical appearance, habitat, diet, and peculiar traits. Introductions to the two divisions give complete descriptions of the probable evolution of each, as well as characteristics common to each family.

Oliver, James A. *Snakes in Fact and Fiction.* Macmillan, 1958. xv+199 pp. illus. $4.95; Doubleday, paper $1.25. 58–8154. (JH)
The author has tracked down fanciful and superstitious legends of giant, poisonous, swift, and crafty snakes. He also tells of how spurious fictional legends have been supplanted by actual accounts of professional explorers and biologists. Good recreational reading.

Pope, Clifford H. *The Giant Snakes: The Natural History of the Boa Constrictor, the Anaconda, and the Largest Pythons.* Knopf, 1961. xviii+290+vi pp. illus. $6.95. 61–16736. (JH)
An illustrated account for the nonspecialist reader of these spectacular reptiles, liberally seasoned with the author's own experiences and observations during his many years as field collector and investigator. A truly fascinating and instructive reading experience.

Pope, Clifford H. *The Reptile World: A Natural History of the Snakes, Lizards, Turtles, and Crocodilians.* Knopf, 1955. xxv+325+xiii pp. illus. $7.95. 54–12979. (JH–SH)
At the time it was written this book represented the most complete and up-to-date account of the earth's reptilian population. Even though herpetologists may feel that some details are out of date, these are insignificant. This comprehensive survey, for the lay reader, summarizes what is known about the biology, habits, reproduction, geographical distribution and ecology of reptiles. Bibliography.

Pope, Clifford H. *Turtles of the United States and Canada.* Knopf, 1939. xviii+337+v pp. illus. $5.95. 39-20221. (JH-SH)
A book primarily about the natural history and habits of turtles, including keys and diagnostic descriptions based on the author's own field experience, knowledge, and examination of available technical publications and popular articles. This book is more informal and more easily read than Carr's handbook; therefore, it is valuable for less mature readers even if the taxonomy is somewhat out of date.

Schmidt, Karl P., and D. Dwight Davis. *Field Book of Snakes of the United States and Canada.* Putnam's, 1941. xiii+365 pp. illus. $4.50. 41-25180. (JH-SH)
A thorough handbook of snakes of considerable value to the amateur herpetologist. Especially helpful are the sections on the collection, preservation, and study of snake specimens. A systematic account of the snakes of the United States and Canada, including information on size, distinctive markings, behavior and distribution, comprises most of the book.

Schmidt, Karl P., and Robert F. Inger. *Living Reptiles of the World.* Doubleday, 1957. 287 pp. illus. $14.95. 57-9783. (JH-SH)
A general descriptive review for the layman and beginning student in biology of the reptiles: turtles, tuatara, alligators, crocodiles, lizards, and snakes, with many colored photographs of living specimens.

Wright, A. Gilbert. *In the Steps of the Great American Herpetologist: Karl Patterson Schmidt.* (Illus. by Matthew Kalmenoff.) Evans (dist. by Lippincott), 1967. 127 pp. $3.95. 67-18531. (JH-SH)
A simple and fascinating account of one of America's foremost herpetologists, from his formative years to his invaluable professional contributions. This is also a brief guide to the education of aspiring herpetologists that concludes with information on collection and maintenance of amphibians.

Wright, Albert Hazen, and Anna Allen Wright. *Handbook of Snakes of the United States and Canada* (2 vols.). Cornell, 1957. xviii+1105 pp. illus. $25.00. 57-1635. (C)
The Wrights' comprehensive work on snakes will continue to be useful indefinitely in the zoological literature. It is the most complete reference for taxonomy, morphology, distribution, etc. While mainly of interest to professional herpetologists, it is needed for reference by students of comparative vertebrate morphology, paleontology, evolution, and other disciplines.

598.2 BIRDS

Adams, Alexander B. *John James Audubon: A Biography.* Putnam's, 1966. 510 pp. illus. $7.95. 66-15573. (SH-C)
Although not a masterpiece of biographical writing, this full-length, and in many ways definitive, portrait of John James Audubon is colorful, and holds the reader's interest. Bits of historical background are injected wherever needed to round out the narrative. As a life of Audubon, it probably is as good as any yet written and better than most. Index, bibliography, and chapter notes.

Alexander, W. B. *Birds of the Ocean* (rev. ed.). Putnam's, 1963. xiv+306 pp. illus. $4.95. 54-8706. (JH)

This useful guide to observation and identification has a comprehensive world scope and includes sections on each major oceanic family—gulls, pelicans, penguins, petrels, albatrosses, and many others. The distribution, life cycle, and overall characteristics of each family are discussed first, followed by brief descriptions of the appearance, dimensions, range, and important distinguishing features of the individual species.

Allen, Arthur A. *The Book of Bird Life* (2nd ed.). Van Nostrand, 1961. xxii+ 396 pp. illus. $9.75. 61–3177. (JH)

This excellent companion to a good field guide explains the principles of ornithology as well as methods of observation and study. The history and classification of the birds are first explained, followed by an examination of their distribution and migration, typical habitats, diets, special adaptations, and many other physical, behavioral, and economic aspects. The second section tells how to find or attract, photograph, trap, and band specimens for live study, including suggestions on keeping them as pets.

Amadon, Dean. *Birds Around the World: A Geographical Look at Evolution and Birds.* Nat. Hist. Press, 1966. x+175 pp. illus. $3.95. 64–13824. (JH–SH)

Much interesting information is presented here on ecology, natural selection and processes of evolution by a consideration of the distribution of the various kinds of bird life. Brief, clearly written and informative. A useful reference and background reading for students and adult bird watchers who desire to know how bird life may have evolved.

Audubon, John James. *The Birds of America.* Macmillan, 1953. xxvi+435 pls. $12.50. 37–28771. (JH)

This magnificent and complete series of 435 full-page color reproductions from Audubon's famous *Birds of America* is a must for every library. Each plate is accompanied by common as well as scientific names, and the range, habitat, and other identifying features and characteristic have been added by the editor. Many species were drawn twice from different angles, and the writeups for the additional pictures frequently emphasize the bird's beneficial attributes and the need for conservation.

Austin, Oliver L., Jr. *Song Birds of the World.* (Color illus. by Arthur Singer; edited by Herbert S. Zim.) Golden Press, 1967. 318 pp. $2.95 (paper).

Austin, Oliver L., Jr. *Water and Marsh Birds of the World.* (Color illus. by Arthur Singer; edited by Herbert S. Zim.) Golden Press, 1967. 223 pp. $2.45 (paper). (JH–SH–C)

Attractive pocket books that contain realistic color illustrations, descriptions of major groups, and brief narratives. Bird watchers and amateur ornithiologists may use these handy guides for primary field identifications.

Berger, Andrew J. *Bird Study.* Wiley, 1961. xi+389 pp. illus. $9.50. 61– 11513. (SH)

The physical, behavioral, and methodological aspects of ornithology are covered in detail in this text, emphasizing principles and general avian characteristics rather than specific facts or species. The anatomy, growth, and function of the living bird are fully examined as well as its migration, courtship, song, and other behavior, which are explained biologically rather than anthropomorphically. Many useful suggestions and general techniques for location, observation, and identification are also included, and the final chapter is devoted to ecologic and conservational problems.

Brodtkorb, Reidar. *Flying Free*. Rand McNally, 1964. 141 pp. illus. $2.95.
65–20332. (JH–SH)
Blind prejudice and ignorance on the part of his countrymen have led Mr.
Brodtkorb, a naturalist and journalist, not only to a series of rather spectacular adventures in an effort to arouse public opinion to the needs of conservation, but also to write a book about eagles to appeal to both adults and
children. Close-up photographs of golden and sea eagles add to the book's
value.

Darling, Lois, and Louis Darling. *Bird*. Houghton Mifflin, 1962. xx+261 pp.
illus. $5.75. 61–7608. (JH)
This attractively written and illustrated study of bird evolution, behavior,
and physiology covers many of the same subjects as Berger but is usually
nontechnical and more readable. The descriptions of social behavior, migration, and other activities avoid the extremes of humanization and rigid de-
terminism, showing that birds are frequently guided by instinct but nevertheless can and do learn. Detailed but clear diagrams and sketches illustrate the
discussions of body systems, organs, and processes.

Davison, Verne E. *Attracting Birds: From the Prairies to the Atlantic*. Crowell,
1967. 252 pp. illus. $6.95. 66–25431. (JH–SH–C)
A thorough and well-organized regional book for those interested in attracting and studying birds in North America. Contain much information
on the feeding habits of birds. A valuable guide.

Day, Albert M. *North American Waterfowl* (2nd ed.). Stackpole, 1959. xx+
363 pp. illus. $5.75. 59–3953. (SH)
The former director of the U. S. Fish and Wildlife Service describes the
efforts since the turn of the century to conserve and manage ducks, geese,
and other aquatic game birds. The methods used to study their migration
and nesting habits, the refuges and sanctuaries already established, flyway
patterns, life cycles, and other key survival factors are discussed. The important legislation which has been passed since the famous Lacey Act of
1900 is also described, with many shocking accounts of biased and ignorant
opposition, greedy and highly illegal practices, and the never-ending job of
law enforcement.

Forbush, Edward Howe, and John Bichard May. *A Natural History of American Birds of Eastern and Central North America*. Houghton Mifflin, 1939.
xxvi+554 pp. 97 pls. $12.50. 39–28977. (JH–SH)
Over 500 American birds found east of the ninety-fifth meridian are classified and described, and the majority are beautifully illustrated in 96 color
plates. The common and scientific names, identification, nesting habits, and
range of each is given, followed by a discussion of the bird's migration, diet
and other interesting characteristics. Personal experiences and observations
are frequently presented, and the calls, songs, or alarm cries of many species
are included.

Gilliard, E. Thomas. *Living Birds of the World*. Doubleday, 1958. 400 pp.
illus. $14.95. 58–10729. (JH)
This comprehensive and extremely attractive book combines a series of 400
excellent photographs (including 217 in full color) with an authoritative and
factual text. Over 1500 species are described in the discussion of the be-

havior and characteristics of all known families and subfamilies. A brief survey of fossil birds is also included, and the text is organized from lowest to highest on the scale of avian evolution.

Hanson, Harold C. *The Giant Canada Goose.* Southern Illinois U., 1965. xxiii+ 225 pp. illus. $9.75. 64–19798. (SH–C)
Practically everything known about this large goose is covered in chapters on behavior, food, enemies, migratory habits and other aspects. It is the best reference available on Canada Geese.

Hanzak, J. *The Pictorial Encyclopedia of Birds.* (Ed. by Bruce Campbell; Ned R. Boyajian, American Consultant.) Crown, 1967. 582 pp. illus. $10.00. 68–11449. (JH–SH–C)
A useful manual for bird watchers that contains over 1,000 photographs with accompanying annotations. An introductory essay provides general information on evolution, differentiation, morphology, and physiology, natural history, and geographic distribution.

Heinroth, Oskar, and Katharina Heinroth. *The Birds.* U. of Michigan, 1958. 181 pp. illus. $5.00, paper $1.95. 58–62521. (SH)
Nesting, mating, eating, and many other interesting bird habits are described, using mostly European examples. The authors examine many common questions and misconceptions about birds as they discuss how they fly, communicate, molt, migrate, and care for their young. The explanations are clear and nontechnical but contain a wealth of information. Morphology and physiology are also covered, including enlightening chapters on senses and intelligence.

Hochbaum, H. Albert. *Travels and Traditions of Waterfowl.* U. of Minnesota, 1967. viii+301 pp. illus. $2.95 (paper). 55–11707.
An explanation of how birds learn flight patterns and manifest migratory instincts. Their existing routes and cycles, travel habits, and special features and capabilities used in migration are fully discussed as a background for the general reader.

Jameson, William. *The Wandering Albatross.* Morrow, 1959. 128 pp. illus. $3.50. 59–11351. (JH)
With a wingspread of almost twelve feet, these strange and beautiful creatures, which may stay at sea for several years, are the largest living aquatic birds. The author tells of his encounters with them while on naval missions in the bleak and stormy south Atlantic and describes their life cycle, breeding habits, and flight, explaining in detail the gliding and soaring techniques which enable these comparatively weak birds to ride the winds for hours without flapping their wings. A chapter is also devoted to examining Coleridge's famous poem and his generally unknown invention of the fable concerning sailors and albatrosses. A well-researched and authoritative study.

Kaufman, John. *Wings, Sun, and Stars: The Story of Bird Migration.* Morrow, 1969. 160 pp. illus. $4.95. 69–11849. (JH)
Long a student of bird migration, Kaufman presents a scientifically accurate, well-organized summary of available knowledge. Migration is defined and related to the seasons, and various migration patterns are defined. Seasonal changes in weather and climate ("the outer clock") and coordinated instinct ("the inner clock") are explained. Encouragement and suggestions are of-

fered to bird clubs and individual bird watchers. The brief bibliography contains some good selections, and the index in which illustrations are noted with an asterisk is in adequate detail.

Kieran, John. *An Introduction to Birds* (3rd ed.). Doubleday, 1965. 77 pp. illus. $4.50. 65–13703. (JH–SH)
Since the appearance of the first edition in 1946, Mr. Kieran's guide for amateur ornithologists, enhanced by original paintings by Don Eckelberry, has been widely used. The revision involves changes in the text and a new introduction. The sequence of the 93 species described is not evolutionary and the descriptions are not those of a professional ornithologist. The birds are presented in order of comparative abundance.

Lanyon, Wesley E. *Biology of Birds.* Nat. Hist. Press, 1964. xii+175 pp. illus. $4.95, paper $1.25. 63–16626. (JH)
Following an introductory examination of the probable evolution and development of birds, the author discusses their special adaptations and structures for flying, important internal organs and systems, reproduction, growth, and means of survival. The amazing phenomena of bird navigation and migration is also covered in the light of recent experiments and discoveries. The appendix includes an index of common and scientific names used in the text and a bibliography by chapters.

Linduska, Joseph P. (ed.). *Waterfowl Tomorrow.* (Introd. by Stewart L. Udall.) U. S. G. P. O., 1964. xii+770 pp. illus. $4.00. 64–60084. (SH–C)
This anthology, prepared in the U. S. Bureau of Sport Fisheries and Wildlife, has special appeal to waterfowl specialists, sportsmen, laymen, and students, for it describes the waterfowl resources, their ecology and migrations, relations between waterfowl and competitors for their habitat (agriculture, industrialization, urbanization, etc.), conservation and management programs, and a forecast of the future of waterfowl. Excellent illustrations, biographical notes on contributors, and index.

McCoy, J. J. *The Hunt for the Whooping Cranes: A Natural History Detective Story.* Lothrop, 1966. xii+223 pp. illus. $4.95. 66–13212. (JH–SH)
Captivating tale of the search throughout the North American continent for the nesting grounds of the whooping crane. Gives meaning and a sense of urgency to the hunt by discussing how near to extinction the species was in 1937, when efforts were first made to protect the whooper in its wintering grounds. Glossary, references, and index.

McCoy, J. J. *Swans.* (Illus. by Giulio Maestro.) Lothrop, 1967. 160 pp. $3.95. 67–22595. (JH)
This excellent study includes general ornithological and zoogeographical information on swans in general, followed by specific ecological and life history accounts of the eight species. Conservation concepts and results of recent investigations are woven into the text.

McMillan, Ian. *Man and the California Condor: The Embattled History and Uncertain Future of North America's Largest Free-Living Bird.* Dutton, 1968. 191 pp. illus. $5.95. 67–11374. (JH–SH–C)
A highly personal account of the status of one of America's vanishing species, the California Condor. Of interest to ornithologists, conservationists, natur-

alists, and students, this book reveals much of the natural history of this raptorial bird and its struggle for survival.

McNulty, Faith. *The Whooping Crane: The Bird That Defies Extinction.* (Introd. by Stewart L. Udall.) Dutton, 1966. 190 pp. illus. $4.95. 66–11544. (SH–C)

McNulty examines the personal and political conflicts that have arisen out of the struggle to preserve the whooping crane. A fascinating story based on excellent reporting.

Mannix, Dan. *The Last Eagle.* (Illus. by Russell Peterson.) McGraw-Hill, 1966. viii+149 pp. $4.95. 65–28508. (JH–SH)

The principal character is a male whose life experiences are related from his hatching as an eaglet, then as a juvenile migrating to Florida, then through a period of four years as a wandering lone juvenile. Finally he returns with a mate and builds his own nest. A wide range of adventures is included. A worthwhile book, enhanced by good drawings.

Marshall, Alexander J. (ed.). *Biology and Comparative Physiology of Birds* (2 vols.). Academic, Vol. 1: 1960; Vol. 2: 1961. xii+518; x+468 pp. illus. $16.00 ea. 60–9073. (C)

A unique, authoritative, and comprehensive account of the classification, geographical distribution, embryology, osterology, morphology, physiology, endocrinology, mating and reproduction, migration, ecology, and other biological aspects of ornithology. Each chapter is written by a recognized authority. A most valuable reference set for teachers, college students, professional ornithologists, and others with sufficient biological background. References at the end of each chapter.

Peterson, Roger Tory. *A Field Guide to the Birds: Eastern Land and Water Birds* (rev. ed.) Houghton Mifflin, 1947. xxiv+290 pp. illus. $4.95. 47–5163. (JH–SH)

The author was awarded the Brewster Medal by the American Ornithologists' Union for this edition of his famous "Guide," covering over 700 species, subspecies, and other forms found east of the Rocky Mountains. It emphasizes the distinguishing characteristics of birds when seen at a distance and employs pattern drawings, field marks, and comparisons between species. The range of each is also defined, and the call or song is described whenever possible.

Peterson, Roger Tory. *A Field Guide to Western Birds* (rev. ed.). Houghton Mifflin, 1961. xxvi+366 pp. illus. $4.95; $2.95, (paper). 60–12250. (JH–SH)

This comprehensive and well-illustrated companion to the preceding volume covers 747 native and "marginal" species found west of the 100th meridian in the United States, Canada, Alaska, and Hawaii. The same system of identification is employed, but there are many more illustrations (658 in full color) and the descriptions include the bird's habitat and nests. Both guides are indispensable.

Pough, Richard H. *Audubon Land Bird Guide* (rev. ed.). Doubleday, 1949. xlii+312 pp. illus. $4.95. 50–5248. (JH–SH)

In a 48-page section of attractive color plates, 275 small land species living east of the Canadian Rockies and the 100th meridian in the United States are described and fully illustrated. The size, distinguishing marks, habits, voice, nest, and range of each is given, making this book more complete in natural history than Peterson's *Field Guide* but not as strong on means of recognition and identification. The foreword discusses various aspects of ornithology such as conservation, bird behavior, and seasonal movements, and a combined color and size key to small land birds is included.

Pough, Richard H. *Audubon Water-Bird Guide.* Doubleday, 1951. xxviii+352 pp. illus. $5.95. 51–10952. (JH–SH)
The same range, style, and form of description as the preceding, but including slightly more detailed size specifications of male and female and covering water, game, and large land birds—258 species in all. Over 600 illustrations, 485 in full color, amply supplement the text which contains information on identification as well as natural history.

Pough, Richard H. *Audubon Western Bird Guide: Land, Water and Game Birds.* Doubleday, 1957. xxxvi+316 pp. illus. $5.50. 56–7948. (JH–SH)
This handy supplement to Pough's other two guides treats in similar style over 200 species, both large and small, found in the western United States, Canada, and Alaska, including a full coverage of waterfowl and pelagic birds of the North Pacific and Bering Sea area. In addition it supplies the range and size of 411 marginal birds covered more fully in one of the author's eastern guides, along with page references to the descriptions and illustrations in those volumes.

Robbins, Chandler S., Bertel Bruun, and Herbert S. Zim. *Birds of North America: A Guide to Field Identification.* Golden Press, 1966. 340 pp. illus. $5.95, paper $2.95. 66–16454. (JH–SH)
Includes identification, geographical distribution, and other data on 700 species of birds distributed from the Mexican border to the Arctic Ocean. Approximately 2000 illustrations, mostly in color, are included in the text. Notations on bird songs are an added feature. References. Indexed.

Rourke, Constance. *Audubon.* (Keith Jennison Books.) Watts, 1966. 342 pp. illus. $6.95; to libraries $4.95. (SH–C)
The Keith Jennison Books are carefully selected examples of good literature published in 18-point type on 8½ by 11-inch pages in order to make them available to persons with limited vision and for those who find larger type books a relaxing reading experience. Equally useful for older students in high school with reading deficiencies with an appetite for good adult literature. This charming biography (originally published by Harcourt, Brace 1936) is an excellent portrait of one of America's most outstanding and best-known early naturalists, and a world famous artist-ornithologist.

Sanger, Marjory Bartlett. *World of the Great Heron: Saga of the Florida Keys.* (Illus. by John Henry Dick.) Devin Adair, 1967. x+144 pp. $10.00. 67–18236. (JH–SH–C)
A vivid description of Florida Key wildlife and its interrelationships with physical factors of climate, such as storms and hurricanes. Extensive coverage is given to the Great White Heron, along with information on other forms of wildlife.

Schorger, A. W. *The Wild Turkey: Its History and Domestication.* U. of Oklahoma, 1966. xiv+625 pp. illus. $10.00. 66–13426. (C)
Discusses the wild turkey from the historical, taxonomic, anatomical, and physiological points of view. Breeding, hunting, management and the attacks of predators is thoroughly examined. Well illustrated, this is an excellent reference for one with a background in the subject.

Simon, Hilda. *Feathers: Plain and Fancy.* Viking, 1969. 126 pp. illus. $4.53. 69–18259. (SH)
The five chapters of the book are titled, "What is a feather," "Growth and structure," "Form and function," "Colors and patterns," and "Unusual feathers and plumes." Treatment of the material throughout the book is accurate, although highly simplified. The greatest merit of the book is in the delightful illustrations, all of which were drawn by the author.

Stefferud, Alfred, and Arnold L. Nelson (eds.). *Birds in Our Lives.* (Foreword by Stewart L. Udall; Bob Hines, artist.) U. S. G. P. O., 1967. xiii+561 pp. illus. 29 cm. $9.00. (SH–C)
In 54 chapters by an even larger number of informed specialists, and with many good photographs and drawings, this ambitious work presents "a wide perspective of birds as they affect and are affected by people, other birds, and other forms of life and activities." Naturalists, ornithologists, sportsmen, conservationists, poultry raisers, and many others will use it as an authoritative background source and reference work. Indexed.

Terres, John K. *Flashing Wings: The Drama of Bird Flight.* (Illus. by Robert Hines.) Doubleday, 1968. xiv+177 pp. $4.95. 67–19083. (SH–C)
A simple, comprehensive, nontechnical consideration of nearly all aspects of bird flights, that includes personal anecdotes, information of underwater swimming, data on fossil birds, falconry, mechanisms such as gliding, hovering, flapping, soaring; speed and height information; and air maneuvering.

Van Tyne, Josselyn, and Andrew J. Berger. *Fundamentals of Ornithology.* Wiley, 1959. xi+624 pp. illus. $11.50. 59–9355. (C)
This advanced text was planned as a background for the late Dr. Van Tyne's graduate course and it requires a knowledge of zoology. It can serve as a useful reference at lower levels, covering evolution, social and individual behavior, special adaptations, distribution, and various other topics. The classification of the birds is fully explained, followed by a 167-page survey of all known families, giving general physical characteristics, range, natural history, and numerous references. Extensive bibliographies and glossary are included.

Wallace, George J. *An Introduction to Ornithology* (2nd ed.). Macmillan, 1963. xiii+491 pp. illus. $8.95. 63–11812. (SH–C)
Covers with less detail and theory much of the same material as Van Tyne and Berger, but contains more physiology and includes topics such as conservation and management, bird study in the past and present, and ornithological methods in the field and laboratory. A useful bibliography of foreign and regional technical publications and popular bird books is included.

Welty, Joel Carl. *The Life of Birds.* Saunders, 1963. xiii+ 546 pp. illus. $12.95. 62–11639. (SH)

A comprehensive and well-written study of avian biology, behavior, evolution, and natural history which is straightforward and suitable for beginning students and general readers alike and yet contains an impressive amount of information, from fascinating peculiarities and strange habits to fundamental principles of ornithology. Includes many recent experiments and discoveries such as the use of radar in studying bird migration, and graphs of interesting facts and statistics.

Wetmore, Alexander, et al. *Song and Garden Birds of North America.* National Geographic, 1964. 400 pp. illus. $11.95. 64–23367. (SH–C)

Wetmore, Alexander, et al. *Water, Prey, and Game Birds of North America.* National Geographic, 1965. 464 pp. illus. $11.95. 65–25605. (SH–C)

Dr. Wetmore assisted by some 20 others has prepared these outstanding, beautifully illustrated books. Each chapter begins with a general or introductory account which is followed by discussions of individual species. Accounts are nontechnical and often include personal experiences of the chapter author. Illustrations are reproductions of paintings or color photos. Each book contains a set of small hi-fi records of actual vocalizations of typical birds.

Williams, C. S. *Honker: A Discussion of the Habits and Needs of the Largest of Our Canada Geese.* Van Nostrand, 1967. ix+179 pp. illus. $7.50. 66–27530. (JH–SH)

This profusely illustrated volume deals with the breeding and other habits of the honker—the heavyweight of the Canada Geese. Clearly written, this is sprinkled throughout with occasional flashes of wry humor and has much to offer the general reader.

599 MAMMALS

Burt, William H., and R. P. Grossenheider. *A Field Guide to the Mammals* (rev. ed.). Houghton Mifflin, 1964. xxv+284 pp. illus. $4.95. 63–8125. (JH–SH)

Provides information on 378 species found in North America and surrounding waters. Contains up-to-date maps showing present distribution, beautiful and detailed color plates of each animal, and many other aids to quick and accurate identification through tracks, skully, teeth, nests, and distinguishing physical characteristics. Each description also gives the animal's habits, habitat, economic importance, and similar easily-confused species.

Burt, William H. *Mammals of the Great Lakes Region.* U. of Michigan, 1957. xv+246 pp. illus. $6.95. 57–5141. (JH–SH)

Despite the regional approach, this authoritative guide covers a majority of the common species found in North America and also contains useful tips on collecting and preparing specimens. Included are a complete worldwide classification of mammal orders and families, and a detailed key to the identification by sight or inspection of all species found in the Great Lakes region. The descriptions of each mammal follow the form of the author's *Field Guide* and are accompanied by regional and continental distribution maps. A glossary and tables of dentition, dimensions, and life history data are appended.

Burton, Maurice. *Systematic Dictionary of Mammals of the World.* (Illus. by David Pratt.) Crowell, 1962. 307 pp. $7.50. 62–10447. (SH)
Arranged according to the traditional sequences of classes, orders, suborders, and families with brief descriptions concerning characteristics, natural history, range, etc.

Cahalane, Victor H. (ed.). *The Imperial Collection of Audubon Animals.* (Original text, *The Quadrupeds of North America,* by John James Audubon and John Bachman; Foreword by Fairfield Osborn; Illus. by John James Audubon and John Woodhouse Audubon.) Hammond, 1967. xvi+307 pp. 150 co. pls. 32 cm. $25.00. 67–20565. (JH–SH–C)
Cahalane, a well-known mammalogist, has compiled and edited this exquisite revival of Audubon's work. Quite extensive coverage is given to many species. Recommended for all libraries and for private collections of those who enjoy and can afford fine books.

Davis, David E., and Frank B. Golley. *Principles in Mammalogy.* Reinhold, 1963. xiii+335 pp. illus. $10.00. 63–13545. (C)
Beginning with a survey of the distribution and natural history of the major and minor mammal orders, the discussion then turns to the processes of adaptation and evolution, explaining how and why they occur and examining in some detail the specific changes which have taken place during the history and development of modern mammals. Analytic aspects such as the determination of range, population dynamics, and food and energy cycles are also presented. Behavior and morphology of most mammals are described. Individual species are introduced primarily as examples; the major emphasis is on general principles.

Ewer, R. F. *Ethology of Mammals.* Plenum, 1968. xiv+418 pp. illus. $26.00. 68–21946. (SH–C)
Discusses the applicability of many ethological concepts to placental mammals, marsupials, and monotremes. The wealth of information about other previously unfamiliar forms represents a significant addition to behavioral literature.

Fitter, Richard. *Vanishing Wild Animals.* Coward-McCann, 1968. 48 pp. illus. LB $2.97. 68–23858. (JH–SH)
This book presents, in an interesting fashion, the story of the efforts of the IUCN and the World Wildlife Fund to save the endangered wild animals of the world. Thirty two colored plates of paintings by John Leigh-Pemberton are included. The three appendices are excellent. The first is an indexed list of extinct, endangered, and saved animals; the second is a list of national and international non-governmental conservation organizations; and the third, a bibliography.

Hamilton, William John, Jr. *American Mammals: Their Lives, Habits, and Economic Relations.* McGraw-Hill, 1939. xii+434 pp. illus. $9.95. 40–27054. (SH)
Although the species described in this book are often found only in the Western Hemisphere, many of the principles of mammalogy discussed have a wider application than the title suggests. The evolution of mammals, their

classification, anatomy, diets, habitats, and many other physical and behavioral aspects are covered. Chapters on destructive and predatory game, and on economic mammals in North America are also included.

Jordan, E. L. *Animal Atlas of the World.* Hammond, Inc., 1969. 224 pp. illus. 31.5 cm. $14.95. 75–83276. (JH–SH)

A good first reference for students and laymen interested in the general appearance, characteristics and geographic distribution of representative species of mammals. Each species account has a colored illustration, and outline map showing general distribution, and brief descriptive notes on morphological characteristics, habitat, size, and habits.

Morris, Desmond. *The Mammals: A Guide to Living Species.* Harper & Row, 1965. 448 pp. illus. $12.95. (JH–SH)

Approximately 300 representatives of the more than 4,000 living mammals are represented, each occupying a page with photography and brief text. Information on distribution and bibliography. For general reading and quick initial reference.

National Geographic Society. *Wild Animals of North America.* National Geographic, 1960. 400 pp. illus. $7.75. 60–15019. (JH–SH)

Discusses briefly the characteristics and evolution of mammals in general, then examines the common species in each of the major orders represented in North America. Their appearance, habits, distribution, and frequent personal accounts of observation are included. The beautiful illustrations alone—over 400 of them; more than half in full color—warrant the purchase of this outstanding book. The comprehensive text by 14 leading naturalists is enthusiastically presented.

Palmer, Ralph S. *The Mammal Guide: Mammals of North America North of Mexico.* Doubleday, 1954. 384 pp. illus. pls. maps. $5.95. (SH)

Based on published sources as well as the author's own field observations notes, sketches, and photographs. Provides common and scientific names, identification data, natural history notes, and information on status. Bibliography.

Peterson, Randolph L. *The Mammals of Eastern Canada.* Oxford U., 1966. xxxii+465 pp. illus. $15.95. 67–70014. (SH–C)

A compilation of 122 species of land and marine mammals of eastern Canada. Short descriptions of the orders and families are given. A valuable contribution to the literature on mammalogy.

Rue, Leonard Lee, III. *Pictorial Guide to the Mammals of North America.* Crowell, 1967. xiv+299 pp. illus. $7.95. 67–12408. (JH–SH–C)

Provides description, range maps, tracks, scientific names, average measurements, weights, habits and habitats, diet, breeding, enemies, and life. In addition a narrative elaboration of the data follows for the 65 species covered.

Sanderson, Ivan T. *Living Mammals of the World.* Doubleday, 1955. 303 pp. illus. $12.50. 55–10515. (JH)

A comprehensive, colorful account of the range, appearance, and identifying characteristics of practically all known species of living mammals. Descriptions of each major order and family portray not only their common and unifying features, but also the interrelationships between these groups and their probable evolution. Some may find the text too advanced in format or style, but anyone can benfit from the excellent photographs, which stimulate further reading.

Van Gelder, Richard G. *Biology of Mammals.* Scribner's, 1969. ix+197 pp. $5.95. 68–27782. (SH)
A concise, encyclopedic account of the biology of mammals. Facts about structures, functions and behavior are presented in sufficient detail to give an idea of the range and diversity of mammals. Most suitable for teachers who need a quick reference.

Walker, Ernest P., et al. *Mammals of the World* (2nd ed.). Johns Hopkins, 1968. Vols. 1 and 2; xlviii+1500 pp. illus. $30.00. Vol. 3 (Index); 769 pp. $15.00. 64–23218. (JH–SH–C)
Common and scientific names, habitat, range, relative abundance, size, color, and natural history details are included for each species. There is no other comparable comprehensive resource. This set belongs in all secondary school, college and public libraries as a reference work. Its rich, classified bibliography of some 40,000 entries will aid those who wish to read in greater detail.

Young, John Z. *The Life of Mammals.* Oxford, 1957. xv+820 pp. illus. $16.80. 57–12715. (C)
The most comprehensive work on comparative mammalian anatomy, physiology, and general biology currently in print. It is a major study and reference work intended primarily for students specializing in biology or medicine and hence should be in every college library. The emphasis is on the mechanisms by which life processes and reactions are regulated and controlled. One portion of the book is devoted to embryology.

599.2 MARSUPIALS

Breeden, Stanley, and Kay Breeden. *The Life of the Kangaroo.* Taplinger, 1967. 80 pp. illus. $5.50. 66–28744. (JH–SH)
The life cycle of the great gray kangaroo of Bribie Island, Australia, is well written and complemented with beautiful illustrations. Presents information on courtship, breeding, gestation, mother-infant relationships, growth and development, feeding habits, predator prey relationships, and social organization, in an interesting and readable manner, ideal for the secondary school student.

599.31 EDENTATA

Tirler, Hermann. *A Sloth in the Family.* (Tr. by Maurice Michael.) Walker, 1966. 47 pp. illus. $3.50. 66–28325. (JH–SH)
The author's efforts to tell the fascinating story of a much maligned animal are praiseworthy. He tells of a pet sloth, its capture, habits, adjustment to a new habitat, acquisition of a mate, and the birth of an offspring. The 32 colored illustrations show the animal in its natural habitat and in captivity. Although not intended as a scientific account, the observations recorded are worthwhile.

599.322 LAGOMORPHA

Rue, Leonard Lee, III. *Cottontail.* Crowell, 1966. xii+112 pp. illus. $5.95. 65–
27293. (JH–SH)
Drawing on the 57 sources listed in his bibliography and on his own extensive
observation of rabbits, Rue has produced a comprehensive, readable, and instructive book, illustrated with his own photographs, which range in quality
from competent to outstanding.

599.323 RODENTIA

Wilsson, Lars. *My Beaver Colony.* (Tr. by Joan Bulman.) Doubleday, 1968. xii+
154 pp. illus. $4.95. 68–22619. (JH–SH)
The author, a European biologist-naturalist, spent three years living close to
a colony of beavers in order to study their behavior under natural and controlled conditions. Since beavers live in lodges and underwater, observation
is difficult, but Wilsson presevered, and the results are delightful reading.

599.4 CHIROPTERA

Barbour, Roger W. and Wayne H. Davis. *Bats of America.* U. of Kentucky,
1970 [c.1969]. 286 pp. illus. $17.70. 73–80086. (C)
A single reference that facilitates identification of bats from the U.S.A. and
summarizes information on their life history, morphology, economic importance, and need for conservation if any. Includes keys for identification, a
guide to the study of bats, a glossary of scientific names and distribution
maps.

599.5 CETACEANS

Anderson, Harald T. (Editor). *The Biology of Marine Mammals.* Academic,
1969. xii+511 pp. illus. $21.50. 69–10320. (SH–C)
A distinguished group of active investigators in the biology of marine mammals have produced this book in which seven sections deal with physiological
activities of marine mammals to bring out their marked differences from land
mammals. Four additional sections deal with unique activities of marine
mammals. An excellent reference work.

Bailey, John. *The Wonderful Dolphins.* Hawthorne, 1965. 96 pp. illus. $2.95.
65–23631. (JH–SH)
A brief but useful summary of what is known about dolphines and of research in progress. The author is careful to distinguish between fact and
speculation, even in the tantalizing area of dolphin communication and
mimicry. The 47 illustrations, mostly photographs, are entertaining and
informative.

Barbour, John A. *In the Wake of the Whale.* (Surveyor Books.) Crowell-
Collier, 1969. 102 pp. illus. LB $3.95. 69–11397. (SH–JH)
This well written account of the whales as a biological resource of great
commercial value, and the over-exploitation of the most economically valuable
species to the point of near extinction, is a needed addition to young people's

literature. One important chapter is devoted to the biological cycle of the sea beginning with the diatoms. The gradual development of the whaling industry, its technical improvement and mechanization, and the eventual commercial exhaustion of the most valuable species, despite international agreements and stringent conservation regulations, are discussed.

Kellogg, Winthrop N. *Porpoises and Sonar.* U. of Chicago, 1961. xiv+177 pp. illus. $5.00, paper (Phoenix PSS518) $1.50. 61–11294. (SH)
A pioneer in the field of dolphin research explains some of his experiments with these remarkable creatures and their natural sonar, which in many respects is better than that of the Navy. An introductory discussion deals with porpoise taxonomy, intelligence, and other popular features, including several accounts of their playfulness and friendly disposition. The author then describes the procedures, results, and importance of his analysis of their signals and experimentation with their amazing powers to detect, avoid, and differentiate submarine objects by the sole use of their echo-ranging apparatus.

Matthews, Leonard Harrison (ed.). *The Whale.* Simon and Schuster, 1968. 287 pp. illus. 26 x 27.5 cm. $20.00. 68–12175. (C)
The most complete and definitive general review of the whale. This deals with mythology and folklore, biological characteristics and taxonomy, life history, ancient and modern whaling methods, scientific investigations, and whale products and their uses. The illustrations range from old wood cuts to remarkable color places, all of excellent quality. Ideal for background and reference.

Reidman, Sarah R., and Elton T. Gustafson. *Home is the Sea: For Whales.* Rand McNally, 1966. 264 pp. illus. $4.50. AC 66–10015. (JH–SH)
Readable and informative layman's account of life-history, morphology, fact, lore, and miscellany concerning cetaceans. This layman's book about whales, dolphins and porpoises is good recreational reading and an excellent first source for interested secondary school students and college undergraduates. The photographs and other illustrations have been selected carefully. A commendable feature is the tabulation of common and scientific names, size, distribution and primary distinguishing features of all aspects of cetaceans.

Scheffer, Victor B. *The Year of the Whale.* (Decorations by Leonard Everett Fisher.) Scribner's, 1969. 213 pp. $6.95. 68–57084. (SH–C)
This book is both a scientific account and a literary masterpiece that deserves two readings for full appreciation. The biography of the first year of life of a whale calf is enriched with information on the biology of sperm whales as well as a compedium of related history, fact, and lore. The reference notes in the back, listed by page and line indicate sources. There is an annotated bibliography of seven classical works on whaling, and an index.

Slijper, E. J. *Whales.* (Tr. by A. J. Pomerans.) Basic, 1962. 475 pp. illus. $12.50. 62–13870. (C)
The natural history and physiology of the whale family, which includes the largest and some of the smartest animals that have ever lived, are presented in a comprehensive study covering the anatomy, behavior, evolution, and major life processes of its important members. The author describes in detail their breeding and reproduction, habits, means of locomotion, migration, and distribution, including also several chapters on the history, methods, and

future of the whaling industry. A classification of the Cetacea and an extensive bibliography are appended.

599.725 EQUIDAE

Simpson, George Gaylord. *Horses, The Story of the Horse Family in the Modern World and Through Sixty Million Years of History.* Oxford, 1951. xxiv+ 247 pp. 32 pls. illus. $8.50. 51–12082. (C)
The horse is frequently used as an example in explaining the evolutionary process, but its development is never more than sketched in other books. This book presents the entire sixty-million year story, explaining in detail the structural changes which insured the survival and success of the family, and describing in full each of the horse's ancestors and relatives. On this solid foundation the final chapters discuss the patterns and causes of evolution in general. A unit on the modern horse is also included, covering its anatomy, domestication, and related wild species.

599.735 RUMINANTS

Colby, C. B. *Wild Deer.* Duel, Sloan & Pearce, 1966. 126 pp. illus. $3.95. 66–14958. (JH–SH)
Contains many facts concerning 19 of the world's most important deer species and subspecies. Each description is accompanied by a clear photograph Accurate information on distribution, habitat, reproduction and behavior are given plus important points concerning the conservation of some species, and their relations to man and to predators. A straightforward style: it treats animals realistically without sentimentality, yet makes pleasant reading.

Hoover, Helen. *The Gift of the Deer.* (Illus. by Adrian Hoover.) Knopf, 1966. viii+210 pp. $4.95. 66–25612. (JH–SH–C)
The story of a scientist-author and her artist husband who chose to live in the back country of northern Minnesota. The author records in beautiful prose the first hesitant acceptance of food and friendship by a starving buck and his actions through the remainder of that first harsh winter and for the next four years. Details, many hitherto unknown observations of feeding activities, maternal and paternal behavior, play and fear, communication, growth and development.

Spinage, C. A. *The Book of the Giraffe.* Houghton Mifflin, 1968. 191 pp. illus. $6.95. (SH)
This well-written and thoroughly documented work covers the paleontological record, the iconography, history, and distribution, and, to some extent, the unique anatomy and physiology of the giraffe.

Van Wormer, Joe. *The World of the American Elk.* (Living World Books.) Lippincott, 1969. 160 pp. illus. $5.95; LB $5.82. 77–86080. (SH–C)
The story of the elk's life is related in terms of the author's personal experiences, and based on references to other literature. Emphasis is placed on the need for management to maintain a suitable population in areas to which they have become adapted.

599.74 CARNIVORA

Adamson, Joy. *The Story of Elsa*. Pantheon, 1966. 318 pp. illus. $6.75. 66–3244. (SH)

A condensation of three previous books: *Born Free, Living Free,* and *Forever Free*. Elsa, an orphan lioness, was raised as a pet by the author and her husband, a Kenya game warden. She was taught later to hunt and adapt herself to her native environment to which she returned and survived. When Elsa died her cubs were resettled in the Senengeti National Park. A well-written story about man and beast, embodying an unusual study of a truly wild animal.

Dominis, John (Photographer) and Maitland Edey (Author). *The Cats of Africa*. Time, Inc., 1968. 192 pp. 48 col. pls. 27.5 x 27.5 cm. $12.95. 68–57805. (JH–SH–C)

Text provides an excellent introduction to the cats, their evolution and morphology, which is accompanied by two evolutionary trees, one for the carnivores and one for the specializations of cats. Descriptive text is followed by the color photographs with descriptive legends. The accounts of the lives of the cats includes detail of their hunting, courting, and child-rearing behaviors. Edey makes observations on the need for preserving the big cats. Includes an index and a bibliography.

Haas, Emmy. *Pride's Progress: The Story of a Family of Lions*. (Introd. by Joseph A. Davis, Jr.) Harper & Row, 1967. xii+116 pp. illus. $5.95. 67–15969. (JH–SH–C)

From over 3,000 of her own photographs of a pride of African lions at New York's Bronx Zoo, Miss Haas has selected the 168 that illustrate this story. Photographed over a period of three years (1963–66), Charlie and Princess raised three litters of cubs and the photographs and the accompanying explanatory narrative tell us much of their family life and behavior. Although written for older children and young adults, the text is sufficiently sophisticated for discriminating adult readers; the photographs will appeal to people of all ages. Thorough bibliography.

Haynes, Bessie Doak, and Edgar Haynes (eds.). *The Grizzly Bear: Portraits From Life*. U. of Oklahoma, 1966. xxi+386 pp. illus. $5.00. 66–10290. (SH–C)

This anthology begins with Henry Kelsey's journal note about "a great sort of a Bear," and continues with reports and tributes from better-known admirers of the grizzly, including Meriwether Lewis, Washington Irving, Bret Harte, Theodore Roosevelt, John Muir, William T. Hornaday, J. Frank Dobie, and many others.

Morris, Ramona, and Desmond Morris. *Men and Pandas*. McGraw-Hill, 1966. vii+223 pp. illus. $7.95. 66–28078. (SH)

Delightfully written and superbly illustrated, the first half deals with the history, habits, hunting, capture and zoo display; while the second half somewhat overplays the sexual behavior and anatomy of captive females.

Rue, Leonard Lee, III. *The World of the Red Fox*. Lippincott, 1969. 204 pp. illus. $5.95; LB $5.82. 69–16165. (SH–C)

An excellent account of the life of the red fox, this fine book is superbly illustrated with one of the finest collections of fox photographs ever assembled. Written for the layman, this volume contains a great amount of information on the red fox as well as personal accounts which give the reader a feeling of familiarity with the author. Valuable not only to school libraries, but to sportsmen, outdoor enthusiasts and biologists as well.

Verts, B. J. *The Biology of the Stripped Skunk.* U. of Illinois, 1967. 218 pp. illus. $7.95. 67–21857. (C)
This volume includes almost everything known about skunks. Contains extensive descriptions of methods and detailed references. An excellent reference on the subject.

599.745 PINNEPEDIA

Perry, Richard. *The World of the Walrus.* Taplinger, 1968. [c. 1967.] 162 pp. illus. $5.95. 68–11021. (SH–C)
A life history, ecological, and zoogeographical study of walruses gleaned from published and unpublished sources, representing material obtained by American, Canadian, Danish and Russian observers, explorers, and zoologists.

599.8 PRIMATES

Clark, W. E. Le Gros. *The Antecedents of Man: An Introduction to the Evolution of the Primates.* Quadrangle, 1960. vii+374 pp. illus. $6.00; Harper & Row (Torchbooks), paper $2.75. 60–8710. (C)
Following an introductory chapter on the evidence supporting the theory and process of evolution, the author classifies and surveys the living and extinct primate suborders and families. He examines in detail their teeth, skulls, limbs, brains, and other key features to uncover evidence of evolutionary relationships. The final chapter reviews the basic trends in the evolution of the order. Although occasionally technical, this book can be read with little background in physical anthropology or anatomy. Emphasizes the apes rather than prehistoric man.

Gray, Robert. *The Great Apes: The Natural Life of Chimpanzees, Gorillas, Orangutans and Gibbons.* Norton, 1969. 144 pp. illus. LB $4.14. (JH–SH–C)
An accurate, factual and sympathetic account of the day-to-day life of chimpanzees, gorillas, orangutans and gibbons. Interactions of family members is examined by presentations of the four genera, with a different members as the central figure. Emphasis is on the danger of extinction, and a plea is made for their right to survive.

Kummer, Hans. *Social Organization of Hamadryas Baboons: A Field Study.* U. of Chicago, 1968. vi+189 pp. illus. $8.95. 67–25082. (C)
In contrast to arboral savanna baboons, the Hamadryas species are organized "one-male units." Each unit consists of a dominant male, the females on which he rides herd, and their young. This work is the report of a year's field study in Ethiopia describing the social interactions of these units and their maintenance.

Reynolds, Vernon. *The Apes.* Dutton, 1967. 296 pp. illus. $10.00. 66–21302. (C)

A survey of the literature, from the evolutionary history of the apes to their precarious status in the modern world. Coverage that even a primatologist would find comprehensive. Shows how great are the challenges and opportunities for enterprising and responsible investigation.

Schaller, George B. *The Year of the Gorilla.* U. of Chicago, 1964. 260 pp. illus. $7.50, paper (Phoenix) $1.95; Simon & Schuster, paper (Ballantine) 75¢. 64–13946. (SH)

One of the leaders of the 1959–60 African Primate Expedition describes his extensive observations of the mountain gorilla, as well as the land, its people, and its abundant yet threatened fauna. The daily life and natural habits of these completely free-living apes was closely studied by the author and his wife, and they include many rare close-range photographs in their discussion of the almost intimate relationship they managed to establish with the animals. Although it reads like the exciting and dangerous adventure that it was, it is a detailed and factual study as well as an urgent plea for much-needed conservation.

Schrier, Allan M., Harry F. Harlow, and Fred Stollnitz (eds.). *Behavior of Nonhuman Primates* (2 vols.). Academic, 1965. Vol. I: xv+318 pp. $9.00; Vol. 2: xv+341 pp. $9.50. illus. 65–18435. (SH–C)

A considerable body of literature reporting on research with nonhuman primates has been developed during recent years. This survey, in two volumes, is devoted almost entirely to laboratory studies presented in 15 chapters written by 20 different authors. Brings together research results scattered widely in scientific journals. Suited as references for upper classmen, graduate students, and professionals.

Van Lawick-Goodall, Jane. *My Friends, the Wild Chimpanzees.* (Foreword by Leonard Carmichael; Photos by Hugo Van Lawick.) National Geographic Society, 1967. 204 pp. $4.25. 67–12051. (JH–SH–C)

A fascinating and superb report of Miss Goodall's observation of wild chimpanzees in the forests of Tanzania. Through her close contact with the chimpanzee, new insights into their behavior have been realized. Photographs and drawings are exceptional, the text has wide appeal.

600 APPLIED SCIENCE

Bronowski, Jacob, Gerald Barry, James Fisher, and Julian Huxley (eds.). *The Doubleday Pictorial Library of Technology: Man Remakes His World.* Doubleday, 1964. 367 pp. illus. $12.95. 63–13851. (JH)

A pictorial and systematic presentation of the development of modern technology. Readers will be fascinated by the succinct discussion of the monumental developments in technology, techniques of measurement, energy at work, resources from the earth and the sea, chemical technology, mineralogy, ceramics, food, textiles, building, transportation, communications and military technology. Glossary.

Clarke, Arthur C. *Profiles of the Future: An Inquiry into the Limits of the Possible.* Harper & Row, 1963. xv+234 pp. $3.95. 62–14563. (SH)

A very intriguing book and a good exercise in imagination. Although the reader must accept some "far-out" ideas—such as neutralizing gravity—in order for some of the proposals to work, most of the ideas are based on more widely accepted premises. For the most part this is not a detailed study of what is going to happen in the near future, but a more general and necessarily more sketchy survey of what may happen in the more distant future.

Derry, T. K. and Trevor I. Williams. *A Short History of Technology from the Earliest Times to A. D. 1900.* Oxford, 1961. xviii+782 pp. illus. $8.50. 61–3478. (SH)
The development of technology is examined with attention to the political and economic influences. In order to tell the story more naturally the authors did not adhere to a strictly chronological approach. However, they included historical surveys to avoid confusing the order of events. Interesting reading and surprisingly inclusive.

Hart, Ivor B. *The World of Leonardo da Vinci: Man of Science, Engineer and Dreamer of Flight.* Viking, 1962. xvi+374 pp. illus. $7.95. 62–17389. (SH)
The first half of the book is devoted to an attempt to draw a picture of Leonardo, the man, and of the world he lived in. Considerable space is given to a cultural and political examination of 15th century Italy and its science and technology. The second half deals with Leonard's scientific work and his ideas for achieving it. The author is an authority on da Vinci.

Leonardo da Vinci (ed. by Edward MacCurdy). *The Notebooks of Leonardo da Vinci.* Braziller, 1955. 1247 pp. illus. $7.50. 55–1485. (SH–C)
Leonardo da Vinci's original notebooks were a collection of his thoughts put down in the order in which they came. Here, various parts of his notebooks have been arranged under about 50 headings to give them continuity. One may see his wide interest range from the varied subjects of his writings. It should be noted that some of the scientific material may be incorrect by modern standards since Leonardo's original writing has not been changed. A good reference.

Meyer, Jerome S. *World Book of Great Inventions.* World Pub., 1956. 270 pp. illus. $5.95. 55–5290. (JH)
Traces the story of man's inventions from the days of primitive man to our modern age. Many of the most useful inventions of today are explained in concise discussions accompanied by clear diagrams. Also gives some background material on the factors affecting inventors of the different ages.

Mueller, Robert E. *Inventivity: How Man Creates in Art and Science.* Day, 1963. 193 pp. illus. LB $3.39. 63–7959. (JH)
Mueller trys to explain the elusive quality of inventiveness by writing from his experience and citing many examples in art, science and technology. He explains how ideas originate and are slowly developed into useful and needed inventions.

Singer, Charles, et al. (ed.). *A History of Technology* (5 vols.). Oxford, Vol. 1, 1954; Vol. 2, 1956; Vol. 3, 1957; Vol. 4, 1958; Vol. 5, 1958. 4011 pp. illus. $30.25 ea. A55–8645. (SH–C)
A detailed, five-volume study of technology from very early times to the beginning of this century. It is arranged chronologically, each volume dis-

cussing a different period. Topics are dealt with in greater detail and more subjects are included than in the shorter history by Derry and Williams. An excellent reference set for large school, college and public libraries.

Usher, Abbott Payson. *A History of Mechanical Inventions.* (rev. ed.) Harvard, 1959. xi+450 pp. illus. $10.00; paper (Beacon) $2.25. 52–10758. (SH)
Covers about the same subject matter as Meyer's book but has a more scholarly approach. Usher first investigates the historical, economic and cultural influences on technology, then he discusses the development of technology from ancient times to the present day. Usher puts more emphasis on machine tools, such as the lathe, that have made our modern age possible, while Meyers deals primarily with the "consumer products."

Weisberger, Bernard A., and Editors of American Heritage. *Captains of Industry.* American Heritage, 1966 (dist. by Harper & Row). 153 pp. illus. $4.95. LB $4.79. 66–17232. (JH–SH)
A helpful impression of the growth of American industry may be gained through a consideration of railroads (Vanderbilt), agriculture (McCormick, Armour, and Duke), oil (Rockefeller), finance (Morgan), steel (Carnegie), mining (Hill and Guggenheim), and automobiles (Ford). This readable work is appropriate for both school and public libraries, and nonspecialist adults will enjoy the attractive illustrated review. Brief list of additional readings and an index are included.

610 MEDICAL SCIENCES

Bleich, Alan R. *Your Career in Medicine.* Messner, 1964. 191 pp. illus. $3.95; LB $3.64. 64–20161. (JH–SH)
Discusses early preparation, medical school requirements, curriculum, and intern and residency programs. Special areas of practice discussed include: pediatrics, geriatrics, psychiatry, atomic medicine, and general practice. A list of approved medical schools, brief bibliography, and index are appended.

Calder, Ritchie. *The Wonderful World of Medicine.* Doubleday, 1969. 96 pp. illus. $3.95. 68–14989. (JH–SH)
This new edition has been updated and changed to a smaller format, making it much easier for the reader to handle, and removing some of the "kiddie-book" appearance. Unfamiliar terms in the text are clearly marked and appear in a glossary, includes cross references to others words.

Carlisle, Norman, and Jon Carlisle. *Marvels of Medical Engineering.* Advances in Science Series.) Sterling, 1966. 144 pp. illus. $3.95; LB $3.99. 66–25196. (SH)
Relates with the aid of vivid photographs, the record advances in the field of medical engineering, describing the background and the workings of: the heart-lung machine, high pressure oxygen chamber, laser beam, x-rays, use of isotopes in medicine, electronics, ultrasonics, thermography, computers, new advances in resuscitation, space medicine, as well as artificial kidney and man-made heart. Written in a relatively simple manner.

Dodge, Bertha S. *Hands that Help: Careers for Medical Workers.* (Illus. by Jane Clark Brown.) Little, Brown, 1967. xiv+247 pp. illus. $4.75. 67–19794. (JH–SH)

Describes the growing need for people to fill various technical roles in the allied health professions and the recent medical developments. Can be understood by anyone with a limited background.

Engeman, Jack. *Doctor: His Training and Practice.* Lothrop, 1964. 123 pp. illus. $4.95. 64–24913. (JH–SH)

A photographic account of the orientation, year-by-year progress through medical school, graduation, internship, residency, private practice, and specialization. Lists the name and addresses of medical schools of the United States, Canada, Puerto Rico, and the Philippines. A worthwhile supplement to other career-guidance books on the medical profession.

Hyde, Margaret O. *Medicine in Action.* McGraw-Hill, 1964. 160 pp. illus. $3.50; LB $3.28. 64–24602. (JH–SH)

The very readable text gives a clear overview of the medical profession including the work of doctors, nurses, technicians, pharmacologists, therapists, aids, psychiatrists, social workers, etc. A tabular appendix lists for each activity the length and nature of instruction or training, the place of employment, and the source of additional information obtainable by correspondence. Indispensable in school and public libraries.

Kitay, William. *The Challenge of Medicine.* Holt, Rinehart and Winston, 1963. 182 pp. $3.50; LB $3.27. 63–18617. (JH)

The history, opportunities, and professional requirements of the field of medicine. Comprehensive and up-to-date information is presented in a readable, well-organized format that will appeal to all levels of young adults contemplating a career in medicine.

Lopate Carol. *Women in Medicine.* Johns Hopkins, 1968. xvii+204 pp. $5.95. 68–19526. (SH–C)

This book should appeal to the woman who is thinking about a career in medicine as well as those responsible for medical school admissions who want to know feminine fears in regard to entering a field dominated by men. History of women in medicine, career guidance, women student's reactions to medical education, problems of specialization, dual careers of medicine and marriage, and the current prejudices of American medicine towards women are covered. Recommended for senior high and college counselors.

Nourse, Alan E. *So You Want to Be a Doctor* (rev. ed.). Harper & Row, 1963. xvii+173 pp. $3.95; LB $3.79; paper 60¢. 63–20316. (JH–SH)

Realistic advice to the high school student considering a medical education and career. Premedical requirements, application for admission to medical school, financial assistance, internships, and other important topics are covered. There is appropriate emphasis on the heavy demands and rewards of medical careers.

Paul, Grace. *Your Future in Medical Technology.* Rosen, 1962. 156 pp. $2.95; LB $2.79. 62–11575. (JH–SH)

Describes the wide variety of jobs in medical technology and the training, interest, duties, and rewards that are involved in them. Some of the specialties included are blood counts and blood banking, bacteriology, mycology, parasitology, enzymology, gas analysis, toxicology, biochemistry, and related biological sciences.

Riedman, Sarah R., and Elton T. Gustafson. *Portraits of Nobel Laureates in Medicine and Physiology.* Abelard-Schuman, 1963. 343 pp. illus. $4.95; LB $4.02. 63–18776. (SH)
Begins with a short description of Alfred Nobel and how the Nobel prizes are awarded, then tells about winners and their discoveries in various areas of medicine and physiology, such as infection, genetics, the circulatory system, and nutrition.

610.3 MEDICAL DICTIONARIES

Dorland, William A. N. *Dorland's Illustrated Medical Dictionary* (24th ed.) Saunders, 1965. xvii+1598 pp. illus. $13.00 (deluxe ed. $17.00). 0–6383. (C)
A complete dictionary of medical terms with two useful sections on "Modern drugs and dosage," and "Fundamentals of medical etymology." An alternative to Stedman's work. Larger libraries should own both.

Hoerr, Normand L., and Arthur Osol (eds.). *Blackiston's New Gould Medical Dictionary* (2nd ed.). McGraw-Hill, 1956. xxvi+1463 pp. illus. $13.50. 55–7269. (C)
A comprehensive dictionary of terms used in medical and allied sciences, including medical physics and chemistry, dentistry, pharmacy, nursing, veterinary medicine, zoology and botany, and medicolegal terms. With tables, charts, lists, and 252 illustrations, about half in color. An alternative to Dorland or Stedman, but less up-to-date.

Stedman, Thomas L., et al. *Stedman's Medical Dictionary* (21st ed.). Williams & Wilkins, 1966. xlvii+1680 pp. illus. $14.00. 61–11310. (C)
An authoritative medical dictionary that includes anatomical, bacteriological, chemical, dental, pharmacological, veterinary, and other special terms; a discussion of medical etymology; the most recent *official* anatomical terms and pharmacological preparations; and biographical sketches of figures in the history of medicine.

610.73 NURSING

Dietz, Lena Dixon. *History and Modern Nursing.* Davis, 1963. 365 pp. illus. $6.95. 61–11691. (SH–C)
A history of nursing which, rightly, is closely interwoven with the practice of medicine to give the interested reader and prospective nurse an overview of the profession in all of its facets. Aside from its informational value for career guidance it may serve as a very elementary introduction for nursing students.

Fream, William C. *Applied Human Biology for Nurses.* Williams & Wilkins, 1964. xi+408 pp. illus. $7.75. 64–7464. (SH–C)
A quick simple reference for student and graduate nurses and an excellent review for Nursing Board Examinations. Useful as collateral reading for beginning biology courses. A general survey of the characteristics of living things with an outline of their evolution and basic qualifications. Discusses in general terms the elements that comprise the human body, and then

proceeds to discuss them by systems. Clinical, pathological conditions are illustrated at appropriate times with simplified explanations relating to anatomy and physiology. Most of the medical situations encountered by nurses are covered.

Kay, Eleanor. *Nurses and What They Do.* Watts, 1968. 123 pp. LB $3.95. 68–10718. (JH–SH)

A young lady interested in learning about the nursing profession can find out everything they do, and through well-chosen examples, how it is done. A comprehensive coverage of all nursing fields is given. Ideal reference for school and public libraries and for school counselors.

Nourse, Allen E., and Eleanore Halliday. *So You Want to be a Nurse.* Harper & Row, 1961. 186 pp. $3.95; LB $3.79. 61–6189. (JH–SH)

A helpful, unclouded, undramatic look into just what nurses do and how they are prepared to do it. The emphasis is more on the practical training than on textbook instruction.

Ross, Janet S., and Kathleen J. W. Wilson. *Foundations of Anatomy and Physiology* (2nd ed.). Williams & Wilkins, 1966. vii+468 pp. illus. $10.50. 67–751. (C)

Written by nurses of long teaching experience for nursing students and instructors. Cells and tissues are first discussed, followed by chapters on specific organ systems. The text is nicely illustrated.

Stryker, Ruth Perin. *Back to Nursing.* Saunders, 1966. 312 pp. illus. $5.75. 66–18504. (SH–C)

Intended primarily as a textbook for a refresher course for registered and practical nurses, this is a handy manual that will also be useful for browsing by students who want to know more about the nursing profession.

610.9 MEDICAL SCIENCES—HISTORY

Harvey, William. *The Circulation of the Blood and Other Writings.* (Everyman's Library.) Dutton, 1963. xvii+236 pp. $2.45. 63–5486. (SH)

Harvey's publication of *Movement of the Heart and Blood in Animals* (1628) is rated the single most significant work in the history of medicine—it is the veritable foundation of modern physiology. This with *The Anatomy of Thomas Parr* (Harvey's report of the autopsy of a 152-year-old farmer) and *William Harvey's Last Will and Testament* make very interesting reading for students and laymen.

Inglis, Brian. *A History of Medicine.* World, 1965. xv+196 pp. illus. $12.50. 65–24165. (SH–C)

Traces the evolution of the treatment of disease and deformity up to modern times. Included is a demonstration of the progress in knowledge of anatomy and physiology, the use of drugs, and standards of cleanliness. Many of the outstanding examples of unorthodox medicine and quackery are described. Bibliography. There are 150 black-and-white plates and 9 in color.

Lee, Russell V., Sarel Eimerl, and The Editors of *Life. The Physician.* (Life Science Library.) Time, Inc., 1967. 200 pp. illus. $4.95. 67–20331. (JH–SH–C)

In these pictorial essays a physician and science writer review major events in medical history, discuss the development of modern medical education, reveal the complexities of medical practice and research, and thus provide broad insights useful for career guidance as well as informative reading. An appendix has notes on 31 physicians of the last 200 years whose lifesaving discoveries are incorporated in modern medicine. References.

Lutzker, Edythe. *Women Gain a Place in Medicine*. (The History of Science). McGraw-Hill, 1969. 160 pp. illus. $5.95. 69–17185. (JH–SH)
Presents an interesting account of the almost unbelievable trials and misfortunes encountered by five pioneer women in opening the medical profession to their sex.

Rapport, Samuel, and Helen Wright (eds.). *Great Adventures in Medicine* (rev. ed.). Dial, 1961. xxi+874 pp. $7.50. 61–16845. (SH–C)
A history of medicine from ancient epidemics, Biblical dietary and sanitary laws, and the Hippocratic Oath, to such recent developments as surgical transplants, cancer research, tranquilizing drugs, and the uses and hazards of atomic radiation.

Silverberg, Robert. *The Dawn of Medicine*. (Illus. by Frank Aloise.) Putnam's, 1966. 191 pp. LB $3.49. 67–14797. (JH–SH)
Presents a clear and most readable treatment of the mixture of science, magic, religion, and serendipitous discovery which characterized the dawn of medicine. Not only are patterns of medical history interwoven, it also focuses on the nature of science itself.

Thorwald, Jurgen. *Science and Secrets of Early Medicine*. Harcourt, Brace, 1963. 331 pp. illus. $12.00. 63–15319. (SH–C)
A well-written, profusely illustrated history of medicine in Egypt, Mesopotamia, India, China, Mexico, and Peru. Each section is prefaced with an account of cultural and historical background against which to understand medical theory and practice. Some of the medical knowledge of these ancient civilizations is remarkably contemporaneous and all of it is interesting reading.

610.92 MEDICAL SCIENCES—BIOGRAPHY

Binger, Carl. *Revolutionary Doctor: Benjamin Rush, 1746–1813*. Norton, 1966. 326 pp. $8.95. 66–15315. (SH–C)
Succeeds admirably in revealing the complex character of a man who was, among other things, a signer of the Declaration of Independence, a founder of Dickinson College, a practicing physician in Philadelphia, a teacher of chemistry on the medical faculty of the College of Philadelphia, and a pioneer in the field of psychiatry. A good biography of a distinguished American. Excellent collateral reading at the college level, of interest to high school students.

Carbonnier, Jeanne. *A Barber-Surgeon: A Life of Ambroise Pare, Founder of Modern Surgery*. (Illus. by Joseph Cellini.) Random, 1965. 186 pp. $3.75; LB $3.89. 65–20653. (JH–SH)
His struggles toward success, his failures, the jealousies of others, the attempts on his life, brief anecdotes of his contacts with the king, queen, and leading royalty, his development of the technique of vascular ligatures to

replace the cruel red hot iron, are clearly told. It is a small book, replete with inspiration for young students. Bibliography, no index.

Clapesattle, Helen. *The Doctors Mayo* (2nd ed.). U. of Minnesota, 1954. xiii+ 426 pp. illus. $8.50; Pocket Books, paper 50¢. 54–11771. (SH)
A widely-read and very good biography of the three American doctors whose names are synonomous with medical progress in the United States. Describes the growth of the small country town of Rochester, Minnesota, to a leading medical center of the world with the establishment of the Mayo Clinic.

Dunlop, Richard. *Doctors of the American Frontier*. Doubleday, 1965. x+228 pp. illus. $4.95. 65–13979. (JH–SH–C)
A dramatic account of the practice of medicine in the American Frontier during the nineteenth century. It tells of Ephraim McDowell who performed the first ovariectomy in medical history in backwoods Kentucky; about William Beaumont who utilized the opportunity provided by a gaping abdominal shotgun wound suffered by Alexi St. Martin to study the physiology of digestion; we learn how Daniel Drake battled diseases of the "interior valley of North America" (typhus, malaria, catarrh, diphtheria.) Other chapters relate work of the military doctors during the Indian Wars, of the medics of the mining camps, of the doctors that ministered to the cowboys, and many others.

Ghalioungui, Paul. *Magic and Medical Science in Ancient Egypt*. Barnes & Noble, 1965. 189 pp. illus. $5.00. 65–29851. (SH–C)
A sympathetic interpretation of the combined influences of magic and medicine in both ancient and modern times. Chapters are devoted to the different specialties of medicine and surgery, to hygiene, *materia medica,* and embalming. The illustrations are interesting and effective. Well organized and easy to read.

Heiser, Victor. *An American Doctor's Odyssey*. Norton, 1936. viii+544 pp. $9.50. 36–27369. (SH)
Dr. Heiser's autobiography (beginning with the Johnstown Flood of 1889 which left him an orphan, and ending with his 1934 letter to the Rockefeller Foundation resigning responsibilities in its International Health Division) records an "extraordinary happy and satisfactory life" in the public health branch of the medical profession. His travels to more than 45 countries, the assignments received, conditions encountered, and results obtained are brilliantly described in a book still important for perspective.

Ingle, Dwight J. *A Dozen Doctors*. U. of Chicago, 1963. vii+287 pp. $5.50. 63–20908. (SH)
Twelve autobiographical sketches by eminent men in biology and medicine, among them Sir Henry Dale, George Hevesy, and Russell M. Wilder. The narratives show some of the intellectual, temperamental, and environmental processes a scientist undergoes in his pursuits.

Maurose, Andre. *The Life of Sir Alexander Fleming: Discoverer of Penicillin*. Dutton, 1959. 293 pp. illus. $5.95. 59–5817. (SH)
In writing this book, the renowned French biographer took a course in bacteriology and persuaded a scientist friend to perform all of Fleming's experiments for him. This scientific background plus the careful collection of

facts and great insight that mark the work, make it an interesting narrative and a good chronicle of the power-plays, human frailities, and elements of chance that often surround important scientific discoveries.

611 ANATOMY

Berger, Andrew J. *Elementary Human Anatomy.* Wiley, 1964. xi+538 pp. illus. $8.95. 63–20627. (SH–C)
The naivete among many college graduates concerning human anatomy even though they have been exposed to some biology courses is at variance with the author's conviction that one of the broad objectives of a liberal arts education should be the acquisition of an elementary knowledge of the structure and function of the human body. This introductory anatomy book is the result. The first part deals with principles, and with chapters devoted to each of the nine systems. The second part has a chapter for each major region: head and neck, thorax, abdomen, pelvis, upper limb, and lower limb. References.

Dienhart, Charlotte M. *Basic Human Anatomy and Physiology.* (Illus. by Steven P. Gigliotti.) Saunders, 1967. x+247 pp. illus. $4.25. 67–12806. (SH–C)
A handy reference written to satisfy the basic needs of the large number of paramedical personnel and prospective teachers at the elementary and secondary level. The various systems are discussed individually with excellent illustrations to amplify and clarify the material.

Frohse, Franz, Max Brödel, and Leon Schlossberg. *Atlas of Human Anatomy* (6th ed.). (College Outline Series.) Barnes & Noble, 1961. 180 pp. illus. $4.50, paper $2.95. 61–10165. (SH–C)
The famous Frohse-Brödel colored anatomical wall charts are reproduced in miniature and completely labeled. Includes charts of the endocrine system by Schlossberg, and an excellent descriptive text by various authors. A standard and valuable guidebook for students, nurses, lawyers, insurance companies, and nonspecialist adults.

Gray, Henry. *Anatomy of the Human Body* (28th ed.). (Charles Mayo Goss, ed.) Lea & Febiger, 1966. 1458 pp. illus. $22.50. 59–12082. (C)
Since the publication of the first edition in 1858, "Gray's Anatomy," has been a fundamental text and reference in all English-speaking countries. Major periodic revisions have kept it current and have improved the presentations and illustrations. It begins with a discussion of embryology and topography, then continues with osteology, joints and ligaments, muscles and fasciae, then takes up the major systems one by one. References at the end of each chapter with a reference index, separate from the subject index, in the back.

Jacob, Stanley W., and Clarice Ashworth Francone. *Structure and Function in Man.* Saunders, 1965. xii+538 pp. illus. $7.00. 65–14505. (SH–C)
An ideal textbook for courses in human anatomy and physiology for superior high school students, college undergraduates, all paramedical specialties. Each organ system is described, using *Nomina Anatomica* terminology, in a clear concise style. Embryology, histology, anatomy, physiology, and common disorders of each functional system are extensively portrayed in 600 elegant new

illustrations. Outlined summaries, study questions, and suggested readings. Glossary definitions are concise and meaningful Comprehensive index.

Ramon y Cajal, Santiago. *Recollections of My Life*. (Tr. by E. Horne Craigie and Juan Cano.) M.I.T. Press, 1966. xi+638 pp. illus. $10.00. 66–3339. (SH–C)

The autobiography of the Spanish neuroanatomist, Santiago Ramon y Cajal, first published in English in 1937, has now been reprinted. There is nothing in this book that would be difficult for a high school student. However, because the scientific account is so schematic, its appeal will be greatest to those who already understand the issues and who use the microscope to untangle the connective patterns of the brain.

Samachson, Joseph. *The Armor Within Us: The Story of Bone*. Rand McNally, 1966. 192 pp. illus. $3.95. 66–10943. (JH–SH)

A remarkably complete, accurate, and entertaining introductory monograph on bone. Skeletal morphology is touched upon in a variety of contexts, but the often overlooked dynamic properties of this remarkable tissue are given greatest emphasis: bone growth and development, metabolism, disease, bone chemistry, research on bony tissues, bone transplantation. Well illustrated, good index. A delightful, authoritative, and comprehensive story of bone.

Woodburne, Russell T. *Essentials of Human Anatomy* (3rd ed.). Oxford U., 1965. xii+673 pp. illus. $15.00. 65–10004. (C)

An integrated systematic presentation of the subject by regions, and in this sense it is both regional and systematic. Within each region the order of presentation is from superficial to deep, corresponding to usual dissection procedures. Neuroanatomy, embryology, and special senses are covered where appropriate to the region under consideration. Enthusiastically recommended for medical students, physicians and for advanced students in physical education, physical therapy, and nursing.

612 PHYSIOLOGY

Asimov, Isaac. *The Human Body: Its Structure and Operation*. Houghton Mifflin, 1963. ix+340 pp. illus. $6.95; NAL (1964), paper 75¢. 62–14188. (JH)

A clear, forceful, and altogether excellent account of human physiology for students and for the interested laymen. Eleven chapters cover all the major organs and systems (except the brain which is dealt with in a companion volume by the same author, *The Human Brain*) with pertinent, well-placed illustrations.

Asimov, Isaac. *The Human Brain: Its Capacities and Functions*. Houghton Mifflin, 1963. xvii+357 pp. illus. $5.95; paper 95¢. 63–14549. (SH)

A lucid explanation of the brain and related organs and structures, including explanation of the brain and related organs and structures, including hormones of the pancreas, thyroid, adrenal cortex, and gonads; the nervous system; the specific parts of the brain; and the sensory system. The last chapter is a consideration of "mind" from the standpoint of biology.

Buddenbrock, Wolfgang von. *The Senses*. U. of Michigan, 1958. 167 pp. illus. $4.00, paper (Ann Arbor) $1.95. 58–5907. (SH)

The first third of this interesting physiological account of the senses discusses the general properties of sense organs, the idea of stimulus and response, and the place of sensory activity in living things. The remainder deals with specific senses, which in addition to the traditional five include color, heat, gravity and such internal senses as posture and the judgment of distance.

Burns, Neal M., Randall M. Chambers, and Edwin Hendler (ed.). *Unusual Environments and Human Behavior: Physiological and Psychological Problems of Man in Space.* Free Press, 1963. x+438 pp. $10.00. 62–15337. (C)
Those who have read *Space Biology* by Hanrahan and Bushnell may be interested in more detailed and technical insights which are offered here. This is an anthology of twelve individually written pieces, five to provide background orientation in psychophysiology, and seven that deal with specific problem areas. Each chapter concludes with a reference list. Readers should have completed basic work in physics and biology.

Carlson, Anton J., Victor Johnson, and H. Mead Calvert. *The Machinery of the Body* (5th ed.). U. of Chicago, 1961. xix+752 pp. illus. $6.50. 61–14536. (SH)
A basic textbook of human physiology covering all the major systems and organs of the body in 15 well-organized chapters. The first two deal with the science of physiology and protoplasm, and successive parts cover the circulatory, respiratory, digestive, excretory, and nervous systems, as well as muscles, sense mechanisms, disease defenses, body chemistry, and reproduction.

D'Amour, Fred E. *Basic Physiology.* U. of Chicago, 1961. xxii+642 pp. illus. $7.95. 61–5603. (SH–C)
This is a textbook written by an outstanding physiologist who has succeeded in retaining all of the fascination of his lectures. It is not a predigested presentation—students are asked to think and to solve problems. An excellent text for high school students in advanced biology courses or for those working on individual programs. Suitable for a cultural college course or for a standard elementary course for science-oriented undergraduates.

Easton, Dexter M. *Mechanisms of Body Functions.* Prentice-Hall, 1963. 371 pp. illus. $8.95 (text ed. $7.50). 63–7377. (SH–C)
The student who has had little or no training in the biological and physical sciences at the college level can use this book successfully. Since physiology is founded on physics, chemistry and anatomy, the necessary elementary concepts are discussed, as necessary, to clarify the principles of physiology. The discussion is arranged by bodily systems, with a final chapter on heredity and genetics. A good reference for larger high school and all college libraries.

Hanrahan, James S., and David Bushnell. *Space Biology: The Human Factors in Space Flight.* Basic, 1960. vi+263 pp. illus. $6.75; Wiley (Science Editions 270-S, 1961) paper $1.95. 60–12021. (SH–C)
Discusses the space vehicle and its necessary characteristics, G forces and weightlessness, radiation hazards, and the impact of astronautics. An epilogue adds data derived from the flights of Gagarin, Titov, Shepard, and Grissom which did not materially alter conclusions of previous research.

Horrobin, David F. *The Communication Systems of the Body.* Basic, 1964. viii+214 pp. illus. $4.95. 64–24590. (SH–C)
Presents a fresh concept of the interdependence of the nervous and endocrine systems in communicating information and action essential to survival. The approach to both anatomy and physiology is teleological; coupled with many simplfying analogies. Many chapters are high school level, but those on nerve cells, synapses, and reproduction (chapters 3, 12, 18) are too complex. The information is accurate, the writing lucid and interesting, and the index and line drawings invariably helpful. Excellent collateral reading for early college.

Luce, Gay Gaer, and Julius Segal. *Sleep.* Coward-McCann, 1966. 335 pp. $6.95. 66–13124. (SH–C)
A semi-popular account of modern sleep research and its practical implications for physical and mental health. Each chapter is accompanied by an extensive bibliography in which certain items are designated as especially appropriate for the general reader. Chapters on the effects of sleep deprivation, abnormalities of sleep, the effects of drugs, the nature of dreams, and learning during sleep. Interesting and lucid. Recommended as supplementary reading and as an excellent source of references to more technical literature.

McCulloch, Gordon. *Man and His Body: The Story of Physiology.* Nat. Hist. Press, 1967. 156 pp. illus. $4.95. AC 67–10040. (JH–SH)
Introduces physiology at its lowest level—cells; and then proceeds to logically develop the how and why of cell function, through each physiological system. The text is supplemented with beautiful illustrations and clear, fully-explained diagrams. Contains a tremendous amount of factual information.

McNaught, Ann B., and Robin Callander. *Illustrated Physiology.* Williams & Wilkins, 1964. viii+287 pp. illus. $6.75. 63–4417. (SH–C)
An illustrated introduction to physiology that makes extensive use of medical drawings and comparison diagrams. Virtually every page is organized around a picture or chart. Most useful as a supplement to standard introductory texts on human physiology. Beneficial to teachers in preparing demonstrations.

Morrison, T. F., et al. *Human Physiology.* Holt, Rinehart and Winston, 1966. xi+497 pp. illus. $6.60. (SH)
Designed to prepare students for college physiology, this text has a simplified format but the content is sophisticated. Following four introductory chapters, a systematic treatment of the body is given. Self-test questions and a vocabulary are included for student aid.

Nourse, Alan E., and The Editors of LIFE. *The Body.* (Life Science Library.) Time, Inc., 1964. 200 pp. illus. $6.60. 64–20219. (JH)
The two fundamental approaches to a study of the human body are comprehended in this attractive introduction: how the body is constructed and how it functions. Eight picture essays tell the story, and the final one: "The Making of a Doctor," is valuable career background for students and guidance personnel. The bodily measurements and bibliography in the appendices are interesting and useful.

Seeman, Bernard. *Your Sight: Folklore, Fact and Common Sense.* Little, Brown, 1968. xi+242 pp. illus. $5.95. 68–14740. (SH)

This is probably the best available resource on the eye and vision for the lay reader. The contents include discussions of the sense of sight, the eye as an organ, the many problems of a clinical and functional nature, eye hygiene and safety measures, visual aids, professional and organizational services, and resources for the blind.

Selzer, Arthur. *The Heart: Its Function in Health and Disease.* U. of California. 1966. xii+301 pp. illus. $5.95. 65–25023. (SH–C)
Directed to those laymen with a sincere interest in furthering their knowledge of the clinical aspects of cardiovascular disease. The text dwells at length on the approach of the physician to cardiac disease, emphasizing diagnosis and treatment of the more common cardiac conditions. Definitions are given of even the most basic physiological and medical terms and enough of the structure of the normal heart is presented to assure comprehension. This incites student interest and in view of its simplicity and generally accurate presentation would, by itself, be an ideal collateral high school text for students interested in further training in the biological sciences.

Sproul, Edith E. *The Science Book of the Human Body.* Watts, 1963. 232 pp. illus. $4.95; paper, Pocket Books (GC 174) 50¢. 55–5410. (JH)
A clear, perceptive explanation of the elements of human anatomy and physiology by a doctor and teacher of pathology. Easily understood by the junior high student, it is nevertheless a thorough, dignified presentation of this universally interesting subject, and probably especially valuable to the teen-ager who is discovering biology and becoming interested in how his own body works.

Tanner, James M., Gordon R. Taylor, and The Editors of *LIFE. Growth.* (Life Science Library.) Time, Inc., 1965. 200 pp. illus. $3.95. 65–26322. (JH–SH–C)
Eight picture essays discuss aspects of embryology, nutrition, metabolism, endocrinology and genetics involved in human growth and interspersed are some of the anthropological, sociological and psychological aspects. Historical notes and findings of recent research are carefully explained. Colored photographs show various stages in the development of the human fetus. Ideal collateral reading for high school students and many parents will want this book at home for reading and discussion with their growing children. Suggested additional readings and a good analytical index round out an indispensable addition to all secondary school, public, and college libraries.

Taylor, Norman B., and Margaret G. McPhedran. *Basic Physiology and Anatomy.* Putnam's, 1965. xvii+648 pp. illus. $6.95. 65–16348. (C)
A basic and elementary textbook of human anatomy and physiology suitable for undergraduates in a liberal arts curriculum, for nurses, medical technicians, therapists, or others with the background of college biology and introductory chemistry courses. Basic biological principles are reviewed in the introduction. There is a glossary, and the abundant illustrations are well labelled.

Tievsky, George. *Ionizing Radiation: An Old Hazard in a New Era.* Thomas, 1962. xix+154 pp. illus. $8.00. 61–17027. (C)
A clinical radiologist, in consultation with geneticists, physicists, radiation biologists, and others, has written a factual appraisal of problems presented by

increasing amounts of environmental radiation. An authoritative work for those who wish readable general information.

Toronto, Alan F. *Structure and Function of the Heart.* (Science Resource Series.) Heath, 1964. vii+112 pp. illus. $1.32 (paper). 64–7432. (SH–C)
Presented without recourse to mathematics and physics, it is a descriptive account of the embryology, anatomy and physiology of the heart. A glossary is provided for beginners. Presents the principal ideas about the structure and function of the heart in a clear, interesting fashion. Discussions are largely limited to the human heart; comparative studies are not mentioned. A good reference for high school students and beginning nonbiology majors in college.

Van Bergeijk, Willem A., John R. Pierce, and Edward E. Davis, Jr. *Waves and the Ear.* (Science Study Series.) Doubleday, 1960. 235 pp. illus. $1.45 (paper). 60–5948. (SH)
Three research scientists at Bell Telephone Laboratories describe the physics of sound, the complex physiology of hearing, the relation of sound to the nervous system, and the problems of communication and speech in this concise, interestingly written book.

Vroman, Leo. *Blood.* Nat. Hist. Press, 1967. xii+178 pp. illus. $4.95. 67–12892. (JH–SH)
An entertaining, highly personalized account reflecting the biophysical aspects of the blood. Recommended for collateral reading, the aim is to introduce and stimulate interest in the subject area.

Woodburn, John H. *Know Your Skin.* (Illus. by Lee Ames.) Putnam's, 1967. 160 pp. LB $3.49. 67–24183. (JH–SH)
Deals with the basic functions of the human integument: anatomy, texture, and color; cytology, texture, and color of hair; blushing and blanching; common skin infections; skin poisons and bites; and relation of tactile stimulation to emotional and affective states.

612.015 MEDICAL BIOCHEMISTRY

Asimov, Isaac. *The Chemicals of Life.* Abelard-Schuman, 1954. xi+159 pp. illus. $3.50; NAL, paper 75¢. 54–10220. (JH–SH)
A clear explanation of the elements of biochemistry by the well-known science writer who is also a biochemist. Deals with proteins, enzymes, vitamins, hormones, and how all these work to maintain the complex chemical reactions of the body.

Asimov, Isaac. *Life and Energy.* Doubleday, 1962. 380 pp. illus. $5.95. 61–12491. (SH)
A biochemist presents a background of the basic principles of motion, energy, heat, electricity—all dealt with in physics. On that foundation he explains the physiology and endocrinology of processes in the human body. An historical and descriptive account intended for the layman.

Asimov, Isaac. *The Living River.* Abelard-Schuman, 1959. 232 pp. $3.95. 59–11650. (SH)
An interesting and fairly detailed study of the biochemistry and biophysics of the human circulatory system. Describes how blood can temperature-condi-

tion the body, keep each cell supplied with food and oxygen, remove waste, carry chemicals from their point of production to where they are used, and protect the body against germs and toxins.

Brown, J. H. U., and S. B. Barker. *Basic Endocrinology for Students of Biology and Medicine* (2nd ed.). Davis, 1966. x+219 pp. illus. $4.50 (paper). 66–22430. (C)
A text for students of physiology and biochemistry and for first-year medical science courses consisting of 20 or 30 lectures. Covered are the individual endocrine glands, their anatomy, embryonic origin, histology, function, secretory products, activities of the hormones, and in some cases, the relationships of the endocrine glands to disease status. Considers briefly general activities of hormones, secretion, transport and excretion, hormone assays, and experimental procedures. Each chapter has several selected references.

612.3 NUTRITION

Altschul, Aaron M. *Proteins: Their Chemistry and Politics.* Basic, 1965. xii+337 pp. illus. $7.50. 64–15931. (SH–C)
Clearly understandable to the intelligent high school student or the average college freshman. Background facts and definitions are lucidly stated. The irrational resistance to needed changes and the wanton waste of millions of tons of food protein in a protein-hungry world are placed before the reader with facts and figures to support each statement. Includes discussion of possibilities to multiply available food protein. An unemotional informative appraisal of the forces of hunger and need.

Bogert, L. Jean. *Nutrition and Physical Fitness* (8th ed.). Saunders, 1966. xi+613 pp. illus. $7.50. 60–7460. (C)
A basic text that covers all aspects of nutrition with emphasis on practical considerations of meal planning, special diets, and the processes involved in the book's use of food. Also deals with the specific elements of nutrition: fats, carbohydrates, proteins, vitamins, and minerals.

Chaney, Margaret S. *Nutrition* (7th ed.). Houghton Mifflin, 1966, xvi+534 pp. illus. $7.50. 60–16139. (C)
An introduction to food and how the human body uses it. Successive chapters deal with the metabolism of fats, proteins, the "energy balance," vitamins and minerals, and nutrition during special periods of life. Assumes some knowledge of organic chemistry and physiology.

Fleck, Henrietta, and Elizabeth Munves. *Introduction to Nutrition.* Macmillan, 1962. viii+656 pp. illus. $7.50. 62–7189. (SH)
Covers all aspects of basic nutrition, including two introductory chapters on the cultural meaning and historical development of food and nutrition. Each element of nutrition is discussed—fats, carbohydrates, proteins, vitamins, minerals—and the principles are applied to special groups and diets.

Guthrie, Helen Andrews. *Introductory Nutrition.* Mosby, 1967. viii+464 pp. illus. $7.85. 67–10507. (SH–C)
A good reference for high schools and general libraries on the science of proper body nourishment. Intended and recommended as a beginning college text emphasizing basic principles of nutrition and applied nutrition.

McHenry, E. W. *Basic Nutrition* (rev. ed.). Lippincott, 1963. xvii+409 pp. $6.25 (text ed. $5.75). 62–11874. (SH–C)
A standard textbook in nutrition with 14 chapters covering methods of investigation, energy requirements, fats, carbohydrates, proteins and amino acids, nutrient elements, vitamins, the nutritive value of foods, special diets, the evaluation of nutritional conditions, and the causes and prevention of malnutrition.

Schifferes, Justus J. *What's Your Caloric Number? A Prudent Guide to Losing Weight.* Macmillan, 1966. 184 pp. illus. $4.95. 66–20824. (SH–C)
The problem of maintaining "ideal" weight in adults is considered in this excellent, nontechnical, well-documented book. Reducing methods presented are based on currently accepted principles of nutrition. Topics include a detailed procedure for determining caloric needs for individuals, diet plans, psychological aspects of eating and weight reduction, values of exercise, and nutrition facts and fallacies.

Wilson, Eva D., Katherine H. Fisher, and Mary E. Fuqua. *Principles of Nutrition* (2nd ed.). Wiley, 1965. xii+483 pp. illus. $7.95. 59–11816. (C)
Covers all the nutritional elements, including an extensive treatment of vitamins and a chapter on antibiotics as nutritional factors; digestion, metabolism, energy needs and energy values of food; and the application of nutritional principles to the selection of diets, both normal and for special groups and conditions.

612.6 HUMAN REPRODUCTION AND EMBRYOLOGY

Arey, Leslie Brainerd. *Development Anatomy: A Textbook and Laboratory Manual of Embryology* (7th ed.). Saunders, 1965. 695 pp. illus. $10.00. 65–12317. (SH–C)
Since publication of the first edition in 1924, it has gradually assumed a place of high honor and widespread usage among embryological textbooks and college laboratory manuals. Embodied in the new edition are substantial revisions of the text made since 1954. Some new illustrations have been added and some of the older ones have been redrawn. Part I, "General development," and Part II, "Special development," are accounts of human embryology which are of great interest to biologically-oriented readers. Part III, "A laboratory manual of embryology," has been reset in a larger type face, corresponding to that in the two earlier divisions, and is devoted to directions for the study of whole amounts and serial sections of chick and pig embryos of various stages.

Demarest, Robert J., and John J. Sciarra, M.D. *Conception, Birth, and Contraception; a Visual Presentation.* (Introd. by Mary S. Calderone, M.D.) Blakiston, 1969. 129 pp. illus. 28.5cm. $8.95. 69–13667. (C)
This is an excellent direct factual presentation of the subjects, and will be a very useful basic reference for adults and teachers. It is highly recommended for teachers' reference collections, college libraries, and science and technology collections of public libraries. The 61 large-sized illustrations are the major feature of the book and the accompanying straightforward facts explaining the illustrations deal with the essentials of the anatomy and physi-

ology of conception, embryonic and fetal development, birth, and contraception.

Knepp, Thomas H. *Human Reproduction: Health and Hygiene*. (Preface by Richard V. Lee, M.D.) Southern Illinois U., 1967. ix+102 pp. $3.50. 67–11701. (JH–SH)
Presents factual material about the human reproductive system, human embryology, birth, veneral disease, and other pertinent information accurately, concisely, and in an interesting fashion.

Lader, Lawrence, and Milton Meltzer. *Margaret Sanger: Pioneer of Birth Control*. Crowell, 1969. 174 pp. illus. $4.50. 72–81955. (JH–SH)
Margaret Sanger realized the need for birth control in the early 1900's. She fought the laws and established birth-control clinics, having travelled abroad to locate effective birth-control methods. The authors tell the story of her struggle and eventual success in a interesting fast-moving narrative.

613.8 ADDICTIONS AND HEALTH

Diehl, Harold S. *Tobbaco and Your Health: The Smoking Controversy*. (McGraw-Hill Series in Health Education.) McGraw-Hill, 1969. xvi+271 pp. illus. $4.95. 69–13216. (SH–C)
The evidence that smoking is an important health hazard has been accumulating for many years. This book summarizes this evidence for the lay reader. It is neither a diatribe, nor a presentation of pro and con designed to allow the reader to draw his own conclusion. The author believes that cigarette smoking is a serious health hazard. A series of appendices provides references for further reading, a study of the death rate among matched pairs of smokers and non-smokers, advice on how to give up cigarette smoking as well as other information.

Hyde, Margaret O. *Mind Drugs*. McGraw-Hill, 1968. 150 pp. $4.50. 68–9553. (SH–C)
In collaboration with psychiatrists, psychologists and pharmacologists, the author summarizes up-to-date information on the widespread use of mind-altering drugs. The psychological dependency or physical addictive characteristics of these drugs are described, but by and large, the drugs of abuse are neither condemned nor praised. There is no need to do so for the stories speak articulately for themselves. Contains a glossary of technical terms. The book will be of interest to teachers, counselors, parents, and to young people themselves.

Lingeman, Richard R. *Drugs from A to Z: A Dictionary*. McGraw-Hill, 1969. 277 pp. $6.95. 68–30559. (C)
An alphabetical listing of the drugs of abuse (hallucinogens, opiates, barbiturates and other central nervous system depressants, and general stimulants employed illegally for their pleasurable and euphoric properties). Roughly 1,100 slang or otherwise esoteric names, initials or expressions dealing with the drug addict, hippie, chemist, pharmacologist, and pharmacist are defined for the layman. It is more than a dictionary because it often explains in depths. No syllabification nor pronunciation guides are provided. It is replete with important information which will be useful to teachers, psychologists,

physicians and others who desire further insight into the underground, alien world of the escapist drug abuser.

Ochsner, Alton. *Smoking and Your Life*. Messner, 1964. 144 pp. $3.00. 64–23119. (JH–SH)

In the third edition of a book, first published in 1954 under the title *Smoking and Cancer*, an eminent thoracic surgeon reviews all of the available information concerning the effects of smoking on health. It includes data and answer's questions raised in the U. S. Surgeon General's report. The excellent bibliography contains many references to primary sources in periodical literature.

Winn, Mitchell (Editor). *Drug Abuse: Escape to Nowhere*. Smith Kline & French Labs. (Dist. by NEA, Wash., D. C.), 104 pp. illus. $1.00 (paper). 66–6404. (JH–SH)

This is a book for students and their teachers which describes educational approaches to the prevention of drug abuse. Since techniques for preventing drug abuse on the elementary, secondary, and college levels. There are chapters on problems of identifying drugs and those who use them, the effects of those drugs, and methods of therapy for drug abusers. References, a list of recommended films, technical definitions and a glossary of slang terms are included.

614. PUBLIC HEALTH

Hobson, William (Editor). *The Theory and Practice of Public Health* (2nd ed.). Oxford, 1965. xi+401 pp. illus. $16.00. 65–8047. (C)

This survey of public health practices was written by 36 British specialists and is an introduction for the student contemplating or just beginning study for a career in the public health professions. The treatment is sufficiently general to acquaint the layman with the nature and scope of public health problems and practice. Curiously, however, the book contains no reference to birth control or family planning, and hence the concern of public health professions with population control is not reflected.

Lasagna, Louis. *Life, Death and the Doctor*. Knopf, 1968. xi+322+xii pp. $4.50. 68–19221. (SH–C)

A popular view of important medical issues that pleads for the medical professions to assume a greater role in the public issues created by new medical advances. He criticizes medical education and medical practice as unsatisfactory for the community-wide medicine as practiced and demanded by modern society. The ethical and social issues of artificial organs, organ transplants, abortion, birth control, artificial insemination and genetic control are discussed. The book is suitable for students seeking encouraging views of modern medicine and society in rapid transition.

Tunley, Roul. *The American Health Scandal*. Harper & Row, 1966. 282 pp. $4.95; Dell (paper) 75¢. 66–10641. (SH–C)

A book of the highest importance that takes the wraps off American health and medical care and, in a most dispassionate manner, explains why we don't have the best health and medical care in the world, what we can learn from medical programs in other countries, how the American family can obtain the best in medical care today, and what needs to be done to insure that medical care becomes a right for every citizen as in the case with compulsory free education. A wealth of objective fact.

614.7 SANITATION AND ENVIRONMENTAL COMFORT

Barry, Gerald, J. Bronowski, James Fisher, and Julian Huxley. *Health and Economics: Man's Fight Against Sickness and Want.* (Illus. by Hans Erni.) Doubleday, 1965. 363 pp. $12.95. 66–10037. (JH–SH–C)
Explains by text and pictures the fundamentals of medical science from which public health services draw their rationale, and the principles of economic science. Equally divided into sections on health considered as a public health activity and on economics as a science. Each chapter is divided into a series of topics concisely discussed and pictorially illustrated within the limits of two facing pages. A 30-page appendix attempts to summarize how health contributes to the economic well-being of society and how medical science and economic science have evolved to their present stages of development. Remarkable both for clarity of expression and the quality and relevance of the illustrations.

Carr, Donald E. *Death of the Sweet Waters.* Norton, 1966. 257 pp. illus. $5.95. 66–12796. (SH–C)
The first five chapters review the history of water supplies, water utilization, and pollution in relation to public health, pollution control, and water management in Asia, Africa and Europe. The remaining two-thirds of the book are devoted to the water supplies and resources of the United States, the pollution problems and their management and mismanagement, and appraisals of various proposals for the control of pollution and the restoration or provision of adequate supplies of potable water. Written with colorful phraseology, conviction and candor.

Herber, Lewis. *Crisis in Our Cities: Death, Disease and the Urban Plague.* Prentice-Hall, 1965. xii+239 pp. illus. $5.95. 65–12920. (SH–C)
An analysis of the public-health problems in contemporary cities: air pollution, water pollution, disease, and physical and emotional stresses. Water pollution chapters review inadequate treatment of human wastes, acid mine wastes, stable detergents, and the slow contamination of ground water, providing an excellent review of the scope of the problem, the strong evidence linking pollution to disease, and the urgency of adopting measures to control what is cast into air or water. A third group of chapters considers increases in cardiovascular disease, neuroses and other ailments that many be traced to the tensions of city life.

Lewis, Howard R. *With Every Breath You Take.* Crown, 1965. xvii+322 pp. illus. $5.00. 64–23821. (JH–SH–C)
Alerts the lay public to one of the hazards of our environment in a manner both entertaining and factually accurate. Covers the nature, sources, effects and control of air pollution and concludes with suggestions for action programs by the individual and the community. Illustrations are well-chosen. For further reading, and advanced college and professional purposes, the reader should seek the same references Mr. Lewis used and cited.

Perry, John. *Our Polluted World: Can Man Survive?* Watts, 1967. 213 pp.
$4.95. 67–10988. (JH–SH)
The subject is air and water pollution. The effect of man on his environment and the prospects for his survival are examined. Presents a naturalist's
view of the ecological consequences of an expanding technological society.

Stewart, George R. *No So Rich As You Think.* (Illus. by Robert Osborn.)
Houghton-Mifflin, 1968. vi+248 pp. $5.00. 67–25450. (JH–SH–C)
Portrays the waste disposal problem from its inoffensive and natural beginnings to its present monstrous magnitude. This book is cogent now and
basic for the future.

Still, Henry. *The Dirty Animal.* Hawthorne, 1967. 298 pp. $5.95. 67–14861.
(JH–SH–C)
An exhaustive example of reporting which documents the thousands of facts
in each of the major problem areas of pollution: air, water, and land. The
author provides more than criticism; an abatement program for both the
public and private sectors of the economy is suggested. Should stimulate
and concern.

615 THERAPEUTICS AND PHARMACOLOGY

Andrews, Sir Christopher. *The Common Cold.* Norton, 1965. 187 pp. illus. $5.50.
65–11004. (SH–C)
Basic information on virology is blended well with critical reviews of clinical studies and the whole seasoned just enough with personal anecdotes from
the author's career to make the book enjoyable but not undignified. The
three main topics covered are research on common cold viruses, natural
history of colds, and the control of colds. A brief annotated bibliography
lists research papers as well as books. Authoritive debunking of myths about
the common cold is here, but there is much more: the excitement, discourage-
ment, and unpredictability of scientific research.

Baldry, P. E. *The Battle Against Bacteria.* Cambridge, 1965. ix+102 pp. illus.
$4.50. paper $1.95. 65–15311. (SH–C)
A history for the general reader of the development of antibacterial drugs.
Tells the intrinsically interesting stories of Pasteur, Koch, Ehrlich, Lister,
Jenner, Fleming, Florey, and Waksman as highlights. In the midst of these
biographies is a welcome reminder that the work of great men is always
built on that of countless others who are seldom named.

Cooley, Donald G. *The Science Book of Modern Medicines.* Watts, 1963. 228
pp. $4.95. 63–21756. (JH–SH)
As recently as 1940, almost 90 percent of the prescriptions that are filled
today could not have been filled because some essential ingredient did not
then exist. This is a report of the diligent research that uncovered the new
ingredients, and of the tremendous advances that have resulted from their
use in modern drugs and medicines.

Deno, Richard A., Thomas D. Rowe, and Donald C. Brodie. *The Profession of
Pharmacy: An Introductory Textbook.* Lippincott, 1966. xii+256 pp. illus.
$6.50. 58–59992. (C)
Describes the historical development, ethical standards, organizations, litera-

ture, and current problems of pharmacy. Chapters cover such topics as pharmaceutical education, retail pharmacy, research, manufacturing pharmacy, promotion and distribution of drugs, and pharmaceutical legislation. Valuable for anyone contemplating a career in pharmacy.

Fiennes, Richard. *Man, Nature and Disease.* NAL (Signet T2653) 1964. xiv+ 268 pp. illus. paper 75¢. 65–1240. (SH–C)
The history and nature of infectious diseases are discussed together with descriptions of the principal kinds of infectious agents: bacteria, fungi, viruses and animal parasites. Heredity, environmental factors, social conditions, and population dynamics contributing to infectious disease are emphasized. Records known facts accurately and concisely.

Kreig, Margaret. *Black Market Medicine.* Prentice-Hall, 1967. x+304 pp. illus. $5.95. 67–18920. (SH–C)
A worthwhile narrative based on FDA documentation and other authentic sources that reveals legal and illegal methods by which organized crime produces and markets low-potency, contaminated, outdated, and intentionally mislabeled drugs. Even though costs are high for materials and resarch, the list price of ethical preparations may be exhorbitant. This has been publicized and the consumer seeks a lower priced product. Thus a climate exists in which the illegal drug counterfeiter finds a lucrative market for his substandard wares. A good book for every citizen.

Kreig, Margaret B. *Green Medicine.* Rand McNally, 1964. 462 pp. illus. $6.95. 64–14403. (SH)
The author has interviewed hundreds of scientists in the United States and abroad, corresponded with many others, searched the literature and accompanied a scientific expedition to the Amazon to assemble the background for her book. Here is a narrative account of the search for medicinal plants throughout the world, and of the research activities involved in preparation, testing, and critical evaluation. A readable book that is an outstanding example of good popular science writing.

Kremers, Edward, and George Urdang. *History of Pharmacy* (3rd ed.; rev. by Glenn Sonnedecker). Lippincott, 1963. xii+464 pp. illus. $9.50. 63–20827. (C)
The authors have presented a sociohistorical view of pharmacy as a profession. This work, since the first edition in 1940, has remained as the best work of its kind in English and shows the truly international character of pharmacy. It was developed within the total political, cultural, and technological setting of each historical period. The first part traces the early and Medieval European history; the second tells of the rise of pharmacy as a profession in representative countries of Europe; the third is devoted to pharmacy in the United States; and part four tells of the discoveries and other contributions to knowledge, humanity, science and industry.

Lewis, J. J. *An Introduction to Pharmacology* (3rd ed.). Williams & Wilkins, 1964. xvi+1048 pp. illus. $11.50. 65–7372. (SH–C)
Although essentially written for medical students, it is not a text on therapeutics. Pharmacology, pharmacodynamics, toxicology, insecticides, and chemotherapeutic agents are of such general interest that the book will be useful for beginning college and even high school students interested in

modern concepts of drug action. The text is condensed, well written, and easy to read. It emphasizes the chemical nature of drugs, structure activity relationships, and the mechanisms by which these compounds act.

Lyght, Charles E. (ed.). *The Merck Manual of Diagnosis and Therapy* (10th ed.). Merck, 1961. xvi+1907 pp. $7.50. 61–31760. (C)
The general purpose is to provide physicians and others with organized facts so as to facilitate accurate diagnosis and promote effective treatment. Etiologic, physiologic, pathologic and other background material is included to facilitate definitive diagnosis. Part I is the diagnostic and therapeutic information organized in 21 homogeneous sections. Part II presents clinical, nursing, laboratory, pediatric, and other procedures. A standard reference for all college and public libraries.

Modell, Walter, Alfred Lansing, and the Editors of *LIFE*. *Drugs*. (Life Science Library.) Time, Inc., 1967. 200 pp. illus. $3.95. LB $4.95. 67–25859. (JH–SH–C)
Traces the origins and history of modern drugs, their uses and misuses, discusses some categories of drugs in detail. Provides historical orientation and discusses both popular and professional interest in drugs and their use.

Poole, Lynn and Gray Poole. *Electronics in Medicine*. McGraw-Hill, 1964. 160 pp. illus. $3.75. 64–19215. (C)
Written in a brief and popular style, which carries the reader through existing and speculative applications of medical electronics. The use of cardiac pacemakers, of television to intensify the image on the X-ray fluorescopic screen, the seemingly limitless possibilities in the bio-medics of computers, the significance of ultrasound in medical diagnosis, the control of epilepsy by electronics and many other subjects are considered in a concise manner. The book provides physicians who seek a quick, succinct survey of current developments with a general nonspecialist source of knowledge. Should be interesting to the lay public.

Stecher, Paul G., et al. (eds.). *The Merck Index of Chemicals and Drugs: An Encyclopedia for Chemists, Pharmacists, Physicians, and Members of Allied Professions*. (8th ed.). Merck, 1968. 1713 pp. $15.00. 68–12252. (C)
An indispensable reference work for all college and larger public libraries that contains about 10,000 descriptions of chemical entities and a cross index of approximately 42,000 entries. Supplements include data on organic "name" reactions, radioactive isotopes, poisons, calories in foods, weights and measures, refractive index of liquids, etc.

Stevenson, R. Scott (ed.). *The Universal Home Doctor*. Prentice-Hall, 1965. ix+612 pp. illus. $8.50. 65–14935. (JH–SH)
Has a wide appeal and should interest those who wish to understand normal and usual problems of the body, as well as those who are concerned about problems of disease relating either to themselves or others. It deals, not only with common, down-to-earth problems, but also with the commoner diseases which afflict society, such as appendicitis and cancer. Defines medical terms which might baffle a layman encountering them for the first time. Contains good general diagrams of the human anatomy and descriptions on how the various parts function. An appendix covers the main types of accidents and first aid.

Prescott, Frederick. *The Control of Pain*. Crowell, 1964. xiv+146 pp. illus. $4.50. 65–12493. (SH–C)
Written for the layman, a simple and effective historical account of the development of modern methods for relieving pain. The discussion on the mechanism of pain and on the use of anaesthetics to control pain at all levels in the practice of medicine will give comfort and assurance to lay persons. A good chapter on drug addiction is included. Will afford valuable background to nonprofessionals working with physicians or nurses, or to anyone interested in learning more about the subject.

Waksman, Selman A. *The Conquest of Tuberculosis*. Univ. of Calif. Press, 1964. xiv+241 pp. illus. $5.00. 64–21065. (SH–C)
A story, much of it in the first person, of the struggle against a disease. There is probably no person alive today who is more familiar with the facts concerning the conquering of "the great white plague" than the author. The facts are well presented, the story line is well developed and the material is essentially free of errors. Contains a limited index.

Zinsser, Hans. *Rats, Lice and History*. Little, Brown, 1935. xii+301 pp. $6.50; (Bantam) paper 75¢. 36–17603. (JH)
A classic in popular medicine, this absorbing book is a biography of typhus, a virulent disease that has plagued man for 1500 years. The first two-thirds of the work set the stage by considering epidemics, parasitism in man and other animals, the bacteria, the problem of the rat and the louse, the influence of epidemi-disease on military and political history; then typhus is dealt with directly, from its early appearance to what modern medicine has accomplished in its control.

616 MEDICINE

Burnet, Sir Macfarlane. *Natural History of Infectious Disease* (3rd ed.). Cambridge, 1962. xi+377 pp. $6.50. 62–51191. (SH)
Considers both the nature of infectious disease as part of the general picture of how life developed, and the relationship of infectious disease to man. Most of the chapters of this edition have been extensively rewritten with new material from the fields of antibiotics and synthetic insecticides, immunology, chemotherapy, bacterial genetics and others. Good but fairly difficult reading for the layman.

Dubos, René, Maya Pines, and the Editors of *LIFE*. *Health and Disease* (Life Science Library). Time, Inc., 1965. 200 pp. illus. LB $4.95. 65–18287. (JH–SH–C)
Well-illustrated historical account of human health, the cure, prevention of disease and epidemics, and the decrease of mortality. It also is an up-to-date consideration of some of American's major health problems. The social aspects of medicine including pollution of air and water, sanitation, nutrition and food supply, and human psychology are all considered. Can be used to good advantage as assigned reading in all "health" or "personal and family living" courses. Good background or "refresher" reading for adults. Well-selected bibliography, analytical index.

Faust, Ernest Carroll, Paul C. Beaver and Rodney C. Jung. *Animal Agents and Vectors of Human Disease*. (3rd ed.). Lea & Febiger, 1968. ix+461 pp. 186 text figs. 10 pls. $11.50. 68–18866. (C)
The first chapter discusses the phenomena of parasitism and provides a classification of animal parasites and vectors, followed by general discussions of pathogenesis, diagnosis, epidemiology, prevention and control. Then follow sections devoted to each of the phyla represented among parasites and vectors. A good textbook and general reference for premedical and biological students.

Hirsh, Joseph and Herman Zaiman. *Vectors and Victims: Being a Collection of Essays about Flies without Zippers and other Nuisances of Man*. (Preface by Chauncey D. Leake.) Charles C. Thomas, 1965. x+70 pp. $4.25. 65–12380. (SH–C)
Nine essays on enteric diseases, parasites and their reservoirs, and the disease-harboring capacity of man's sometimes close associates: the dog, the pig, and the goat. Facetious language and amusing satire are employed in preparing surprisingly entertaining and scientifically accurate discourses. Perhaps it is too sophisticated for students below senior high level, yet teachers of younger students can draw upon it for factual and illustrative material to enliven their presentations in biology and health courses.

Lamb, Lawrence E. *Your Heart and How to Live With It*. Viking, 1969. 247 pp. illus. $5.95. 69–15662. (SH–C)
Presents the basic physiology of the cardiovascular system in terms easily understood by the layman, and explains the meaning of medical terminology for many common diseases and maladjustments of the systems. Should be of interest to students who desire information on the heart and blood vessels as well as those who are suffering from some type of heart disease or malfunction.

Luce, Gay Gaer, and Julius Segal. *Insomnia: The Guide for Troubled Sleepers*. Doubleday, 1969. viii+370 pp. $6.95. 69–10937. (SH–C)
Recent research on sleep and dreaming has provided a major scientific breakthrough. This is a survey of current scientific work and a theraputic handbook, it does an excellent job of summarizing and synthesizing sleep research, making practical applications of the research, providing sound therapeutic guidelines, sounding words of caution, and reinterpreting many popular myths about sleep patterns.

McGrady, Pat. *The Savage Cell*. Basic Books, 1964. xvi+432 pp. $8.50. 64–22643. (SH–C)
A good account of current controversies and recent advances in the study of cancer. Gives vivid accounts of contemporary ideas concerned with epidemiologic and hereditary aspects of cancer. Explains in simple terms the possible roles in the etiology of cancer of food, drugs, air pollution, occupation, tobacco, radiation, emotions, and hormones. In addition the reader is introduced to the possible roles of chromosomes, genes and viruses as etiologic factors in cancer. Finally, accounts are given of current techniques employed in cancer detection and treatment. A good addition to high school and public libraries.

Ross, Walter. *The Climate is Hope: How They Triumphed Over Cancer*. Prentice-Hall, 1965. 184 pp. $3.95. 65–25256. (SH–C)
The purpose of the book is threefold. First, to demonstrate by the personal stories of former patients and the explanations of nationally-known physicians that cancer is a curable disease. Secondly, to educate the reader to the importance of early detection and prompt treatment. Thirdly, to warn against nostrums and treatments advertised as "cancer cures." Educational value for students and the public in general. It may even inspire the student to look for additional reading beyond the field of popular medical reporting.

Rouché, Berton. *Curiosities of Medicine: An Assembly of Medical Diversions, 1552–1962*. Little, Brown, 1963. x+338 pp. $5.95; paper (Berkeley) 60¢. 63–13979. (JH-SH)
A collection of twenty-one essays on a variety of medical subjects, by such contributors as René and Jean Dubois, Robert Lindner, Ashley Montagu, Samuel Hopkins Adams, and others less well-known but equally adept. Some of the titles are: "The beginning, name, nature, and signs of the sweating sickness," "Glue-sniffing in children," "Stomach cancer in Iceland," "The case of the perilous prune pit," and "On yawning."

Roueché, Berton. *Eleven Blue Men*. Little, Brown, 1954. vii+215 pp. $4.50; (Berkeley) 75¢. 54–5121. (JH)
Twelve superbly written narratives of medical detection that originally appeared in *The New Yorker*. Describes diseases and epidemics from trichinosis, gout, and smallpox to an outbreak of tetanus among heroin addicts, a case of leprosy in New York, and a smog in a Pennsylvania industrial town that struck down half the inhabitants. Worthwhile reading for anyone interested in public health or medical research.

Roueché, Berton. *The Incurable Wound*. Little, Brown, 1957. 177 pp. $4.50; Berkeley (paper) 50¢. 58–5653. (JH)
Six more well-written accounts of medical detection that first appeared in *The New Yorker*, and which cover the discovery of rabid bats in the United States, a strange story of amnesia, the extraordinary history of aspirin, the workings of the Poison Control Center in the New York City Health Department, and some of the bizarre side effects of treatment with the little understood hormones, cortisone and ACTH.

Sutton, Maurice. *Cancer Explained*. Hart, 1967. 91 pp. illus. $3.95. 67–23614. (SH–C)
The current status of cancer treatment is described in layman's language. This can be useful in alerting the young reader to the danger signals of cancer and as a warning against the use of cigarettes.

Sutton, P. M. *The Nature of Cancer*. Crowell, 1966. 159 pp. illus. $3.50. 65–21208. (SH–C)
Starts with a brief discussion of cells and growth and follows with two short chapters on disease, and on tumors and cancers. The rest of the book is concerned with descriptions of various cancers in man; cancers in experimental animals induced by chemicals, hormones, and viruses; heredity and cancer;

prevention, diagnosis and treatment of cancer; and several theories on the genesis of cancer. Service to enlighten the uninformed and presents the subject of cancer more authoritatively than usual in a popular science book.

Williams, Greer. *The Plague Killers.* Scribner's, 1969. xii+345 pp. $6.95. 68–27796. (JH–SH–C)

In preparing this straightforward story of the conquest of hookworm, malaria and yellow fever, the author had access to unpublished archives of the Rockefeller Foundation. He has conducted many interviews with persons involved in the research that developed the methods for combating these widespread diseases. There have been many other popular accounts of the conquest of yellow fever but none in so factual and well-documented depth as we find here. With this solid book in their collections, school and public librarians probably will decide that they can discard a few inadequate juveniles dealing with malaria and yellow fever.

617 SURGERY

Boylan, Brian Richard. *The New Heart.* Chilton, 1969. ix+221 pp. $4.95. 70–80444. (SH–C)

Covers contemporary medical and surgical treatments of heart disease and is particularly timely during this period of widely publicized heart transplants. Ancient beliefs are reviewed, medical procedures and surgical techniques used in the treatment of heart disease are discussed, and some practical therapeutic and preventive measures that are now being introduced or considered are presented.

Longmore, Donald. *Spare-Part Surgery; The surgical practice of the future.* (Edit. and illus. by M. Ross-Macdonald.) Doubleday, 1968. 192 pp. $5.95; paper $2.45. (SH–C)

A clear exposition of mechanical and transplanted limbs, artificial kidneys, heart-lung preparations, breathing machines, and organ transplants. Describes a plan for a World Tissue Service to aid in compatible transplantation of human organs.

Malgaigne, Joseph F. *Surgery and Ambroise Pare.* (Tr. and ed. by Wallace B. Hamby.) U. of Oklahoma, 1965. xxxi+435 pp. $10.00. 65–11231. (SH–C)

The original, published in 1840 (as well as this excerpted one-volume translation), is divided into three major parts: (1) "History of Western Surgery from the Sixth to the Sixteenth Century;" (2) "Surgery During the First Half of the Sixteenth Century;" (3) "Ambroise Pare." The section on Pare is much more than a biography—it is also an account of and excerpts from his writings supplemented by citations and excerpts from the writings of his contemporaries. Recommended for larger secondary school collections, and for all public, college, and medical libraries.

Nourse, Alan E. *So You Want to be a Surgeon.* Harper & Row, 1966. xiv+171 pp. $4.95; LB $4.43. 66–13920. (JH–SH)

Similar to the author's *So You Want to be a Doctor*, this addition to the publisher's series of career books briefly recounts the history of surgery, surveys the work of surgical interns and surgical residents, mentions some of the specialties, and summarizes the necessary education and training.

Richardson, Robert G. *Surgery: Old and New Frontiers.* (Rev. ed. of *The Surgeon's Tale.*) Scribner's, 1968. x+310 pp. illus. $7.95. 68–27790. (SH)
This book is of interest to the general reader and should be particularly useful to the student contemplating a medical career. It is a history of surgery, from the discovery of anaesthesia 120 years ago to the developments in transplants surgery. The story of the pioneers in surgery, their successes, their failures, the discarded and revised techniques, and revived procedures, is told against the background of the modern surgeon, the leader of a team of specialists who now command a battery of projections of future developments.

Riedman, Sarah R. *Masters of the Scalpel: The Story of Surgery.* Rand McNally, 1962. 320 pp. illus. $4.50; paper $2.25. 62–8049. (JH)
The book opens with the dramatic story of a case of open-heart surgery, and follows with a historical account of the development of surgery from a papyrus account dating back to the seventeenth century B.C. These are semibiographical vignettes of the most prominent figures in the evolution of surgical techniques. Sixty photographs are included as well as an index. There is no bibliography.

Rosenberg, Nancy, and Reuven K. Snyderman. *New Parts for People; The Story of Medical Transplants.* Norton, 1969. 126 pp. illus. $4.25. 68–22723. (JH–SH)
This timely and clearly written book is an excellent presentation of the history, development, and present state of medical transplants. It gives credit to early scientists, and describes rejection and the immune reaction. The last chapter covers legal and moral aspects which have developed concerning these scientific advances.

617.6 DENTISTRY

Boucher, Carl O. *Current Clinical Dental Terminology.* Mosby, 1963. xxv+501 pp. $11.00. 63–7357. (C)
A glossary of terms in 21 areas of dental practice that will be useful as a supplement to standard medical dictionaries. At the end is a list of words according to dental specialty that should be valuable to practitioners and technicians in professional or subprofessional specialties.

Bremmer, Maurice David K. *The Story of Dentistry.* (3rd ed.). Dental Items of Interest, 1954. xv+462 pp. illus. $7.50. 55–17010. (SH)
A history and description of dentistry that includes anthropological facts and theories about dentition; the historical development of dentistry in Egypt, Greece, Rome, and other civilizations. The story goes on to medieval and renaissance dentistry in Europe; the roots of modern dentistry and its growth in the United States since colonial times; some major technical advances such as fillings, dentures, and braces; and some specific economic and commercial problems related to these techniques.

Greenberg, Saul N., and Joan R. Greenberg. *So You Want To Be A Dentist.* Harper, 1963. xv+168 pp. $3.95; LB $3.79. 63–8130. (JH–SH)
A good guide to dentistry for the young person considering it as a career. Gives prerequisites, curriculum requirements, information on state board

examinations, a description of specialty areas, a sample aptitude text for dental schools, a list of United States dental schools with tuition and fees, and some of the problems of setting up a practice. Although some of the earlier information is out of date, the book is worthwhile.

Permar, Dorothy. *A Manual of Oral Embryology and Microscopic Anatomy* (4th ed.). Lea & Febiger, 1968. 150 pp. illus. $6.50. 63–12349.
An advanced text that covers the embryology, histology, and anatomy of the oral cavity in detail. Deals with the development and structure of tooth enamel, dentin, tooth plup, cementum, oral mucous membrane and salivary glands, tooth eruption, and the shedding of primary teeth. Suitable for students with fundamental training in the biological sciences, and especially good for "science projects."

Scott, James H., and Norman B. B. Symons. *Introduction to Dental Anatomy* (4th ed.). Williams and Wilkins, 1964. xi+388 pp. illus. $9.50. 61–1834. (C)
A basic text in dental anatomy that will be useful to students with some background in comparative anatomy and a good foundation in the biological sciences. Five main sections cover the form and arrangement of the teeth; the development of face, teeth, and jaws; the histology of the dental and parodontal tissues; the functional anatomy of the oral cavity; and comparative dental anatomy.

Woodforde, John. *The Strange Story of False Teeth*. (Foreword by James Laver. Universe, 1970. xii+137 pp. illus. $4.95. 73–97597. (JH–SH)
A fascinating historical and technological history of the use and fabrication of artificial teeth. Initially developed only to improve personal appearance the crude items were worn with considerable discomfort and pain, and were not useful for eating. Modern dentures are often more satisfactory than the individual's own teeth. It is a story that will interest almost everyone.

617.7 OPHTHALMOLOGY

Gregg, James R. *Your Future in Optometry* (Careers in Depth). Rosen, 1963. 160 pp. $2.95; LB $2.79. (SH)
Portrays a day in the work life of a practitioner. Covers also required skills and training, and lists colleges where training is offered.

Snyder, Charles. *Our Ophthalmic Heritage*. Little, Brown, 1967. xii+170 pp. illus. $12.00. 67–16750. (SH–C)
This is an anthology of 37 readable and well-documented, essentially biographical, accounts of the lives of persons intimately related to the history of ophthalmology and, to some extent, physiological optics. It is valuable collateral reading for students of medical science, ophthalmology, optics, and optometry, and serves as a reliable reference to anyone interested in the history of science, especially of the visual sciences.

617.96 ANESTHESIOLOGY

Davison, M. H. Armstrong. *The Evolution of Anesthesia*. Williams & Wilkins, 1965. 236 pp. illus. $8.75. 65–29902. (C)

The first chapter is a chronology of general and medical world history. Subsequent chapters deal with the various anesthetic agents and the methods of administering them. The use of medical terms will require the reader to refer to a medical dictionary occasionally. Mention of primary sources is included in the text rather than in a separate bibliography. Important is the expression of the author's philosophy that the development of anesthesia was preceded by the development of humanitarianism.

620 ENGINEERING

Amstead, B. H., and Wilburn McNutt. *Engineering as a Career Today.* Dodd, Mead, 1967. ix+207 pp. illus. $3.75. AC 67–10325. (JH–SH)
Opens with a general discussion of engineering and engineers, telling what an engineer is and what he is not. The opening chapters contain an account of some of the various technical and nontechnical functions of engineers, and discussion of the training required. Five chapters describe various engineering specialties. Advises the high school student on the choice of career and on the procedures involved in choosing and applying for admission to a college. A list of accredited colleges of engineering showing the degree programs and estimated costs.

Armytage, W. H. G. *A Social History of Engineering.* M.I.T. Press, 1966. 378 pp. illus. $10.00. 66–17829. (SH–C)
Surveys the history of technological development, and the manner in which social life has been affected thereby. The scope is world-wide with a slight emphasis on British achievement. Recommended as a very good factual account of technological development but it is less successful when the author discusses the impact of such development on society. Good specific and general bibliographies.

Bishop, R. E. D. *Vibration.* Cambridge Univ. Press, 1965. vii+120 pp. illus. $5.50; paper $1.95. 64–21527. (JH–SH)
A talented explanation of the subject of mechanical vibrations, describing the nature of vibrations, what they mean to engineers, and how they may be analyzed. A considerable number of demonstrations and illustrations add interest, clarity, and vitality to the explanations. Its engineer's emphasis on useful application should be especially appealing to many young readers.

Calder, Ritchie. *The Evolution of the Machine.* (The Smithsonian Library.) American Heritage (dist. by Van Nostrand), 1968. 160 pp. illus. $4.95. 68–17249. (JH–SH)
Covers the evolution of the machine through an emphasis of the separate factors leading to, or preventing, different lines of development. Includes chronology of important steps in the evolution of the machine, a collection of some of the more interesting but less successful ideas patented in the United States, and a collection of brief biographies.

Clauser, H. R. (Editor-in-chief). *The Encyclopedia of Engineering Materials and Processes.* Reinhold, 1963. xi+787 pp. illus. $27.50. 63–13448. (C)
A useful reference to the metals, woods, plastics and rubbers used in industry. Includes many charts with important data about these materials and some explanatory diagrams. Alphabetical arrangement and cross-references facilitate research.

De Camp, L. Sprague. *The Ancient Engineers.* Doubleday, 1963. 408 pp. illus. $4.95. 62–15901. (SH)

The development of engineering is traced to Galileo's time. In addition to accounts of the Egyptians, Greeks, and Romans (often found in books on this subject), the author has included interesting material on the Chinese, Indians and Byzantines. There are some excellent photographs.

Glegg, Gordon L. *The Design of Design.* Cambridge, 1969. 93 pp. illus. $4.95. 69–12432. (SH–C)

The author describes three elements of an engineer's job, the inventive, the artistic, and the rational. The burden of his text is an analysis of these three elements, with a final discussion of "safety margins." Although the book is oriented to engineers, much of its value to all readers is in the parallels one readily discovers in other forms of creative and innovative human enterprise.

Jones, Franklin D., and Paul B. Schubert (eds.). *Engineering Encyclopedia* (3rd ed.). Industrial, 1963. 1431 pp. illus. $15.00. 63–10415. (SH–C)

A more general engineering reference than Clauser's and probably more suitable for the high school or beginning engineering student. Includes a greater number of articles which usually are briefer and less technical than Clauser's.

Kemper, John Dustin. *The Engineer and His Profession.* Holt, Rinehart and Winston, 1967. viii+248 pp. illus. $4.95. 67–11812. (SH–C)

An excellent description of the engineering profession for those interested in an engineering career, for engineering students and recent graduates.

Kirby, Richard Shelton, et al. *Engineering in History.* McGraw-Hill, 1956. vii+530 pp. illus. $10.00; paper $5.95. 55–8287. (SH)

Many of the world's great engineering projects from ancient times to modern days are described. The historical background for these projects shows that engineers encounter and solve financial and cultural problems in addition to the technical ones. Interesting reading; well illustrated with photographs and diagrams.

Manchester, Harland. *Trail Blazers of Technology: The Story of Nine Inventors.* Scribner, 1962. 215 pp. illus. LB $3.31. 62–17731. (JH)

The stories of nine great inventors of the 19th and 20th centuries are told in simple language. Descriptions of their personal problems make the accounts interesting and human. Suitable for junior high outside reading and references.

Pollack, Philip. *Careers and Opportunities in Engineering.* (Rev. by John D. Alden.) Dutton, 1967. 224 pp. illus. $4.95. 67–11381. (JH–SH)

An authoritative work that offers an historical sketch of engineering, and explains the various professional and technical opportunities in the engineering sciences. Describes the aptitude and education needed, and the kinds of work each specialist performs. The appendices list engineering societies, universities and colleges that offer curricula in engineering, technical institutes, and references.

Smiles, Samuel. *Selections from Lives of the Engineers; with an Account of Their Principal Works.* (Ed. by Thomas Parke Hughes.). M.I.T. Press, 1966. ix+447 pp. illus. $10.00. 66–19360. (SH–C)

Samuel Smiles describes the professional lives of a group of British engineers,

most of whom lived during the Eighteenth Century. The three included in
the edition (Brindly, Rennie and Telford) contributed much to the emergence of Britain as a mercantile power. Rennie was unique in having university training, Brindley having apprenticed as a millwright and Telford as
stone mason. The nature of these accomplishments is related, and the local
environment before and after construction, and the financial and political
maneuvers required to initiate them, are mentioned. Notes by the editor
elaborate portions of the text.

Smith, Ralph J. *Engineering as a Career.* (2nd ed.). McGraw-Hill, 1962. x+
394 pp. illus. $5.95; paper $4.50.; instructor's manual $1.25. 62–18802.
(SH–C)
Keyed for the college student who has not finalized his major field of specialization, this career guide is fairly sophisticated and comprehensive. In
addition to guidance information there is a section on college training and
one of the engineering sciences with previews into the problems to be encountered.

Stirling, Nora. *Wonders of Engineering.* (Illus. by Emil Weiss.) Doubleday,
1966. 127 pp. $3.25. AC 66–10378. (JH)
Twelve well-chosen examples are used to describe the civil engineer's art
from conception through design to completion. The works are as diverse as
Stonehenge, the Great Eastern Steamship and the Palomar Telescope. The
chapters on the Aqua Marcia Aqueduct and the Simplon Tunnel are highlights. Explains well the problems encountered by the engineers, and how
they were solved in light of the state of the art. The drawings are good and
often complement the text, clearly showing some intricate design. The vocabulary is not difficult, though not held down.

The Way Things Work: An Illustrated Encyclopedia of Technology. (Tr. from
the German.) Simon & Schuster, 1967. 590 pp. illus. $8.95. 67–27972. (JH–
SH–C)
Designed to inform the reader as to the physical, mechanical, chemical, electrical, electronic, or other principles and processes by which devices, implements, machines, etc., operate. Clear, concise diagrams aid in explaining the
operations. A useful reference book for libraries and personal ownership.

Whinnery, John R. (ed.). *The World of Engineering.* McGraw-Hill, 1965. vii+
304 pp. illus. $5.95. 64–25375. (SH–C)
A well-executed series of essays written by acknowledged experts for the
inspiration and guidance of prospective engineers. A rather sophisticated
vocabulary is required to read the entire book, and even some of the mathematics may be beyond the range of some high school students. There is so
much merit in the presentation that these may be minor difficulties. A good
choice for high school libraries.

621.3 ELECTRICAL ENGINEERING

Buban, Peter, and Marshall L. Schmitt. *Understanding Electricity and Electronics.* McGraw-Hill, 1962. x+446 pp. illus. $8.75; text ed. $6.60. 60–
13761. (SH)
The student with no previous electronic training will find this book valuable.

Begins with information on basic procedures such as soldering and the use of tools, and explains the operation of many common electrical appliances. Good diagrams and experiments assure comprehension.

Carter, Robert C. *Introduction to Electrical Circuit Analysis.* Holt, Rinehart and Winston, 1966. xii+500 pp. illus. $10.50. 66–13291. (SH–C)
For a two-year technical-institute curriculum, presents basic circuit theory for engineering technicians, with emphasis on physical interpretations. There are numerous solved examples and problem sets (odd-numbered problems have answers). The student is expected to have had college-level algebra, basic electricity in physics, and (possibly concurrently) plane trigonometry. Calculus is needed for the solution of some problems. The glossary of terms in informative and precise. There is an appendix on slide rule techniques.

Eaton, J. R. *Beginning Electricity.* Macmillan, 1952. viii+365 pp. illus. $6.50. 52–10507. (SH)
More advanced than Buban and Schmitt, but easily understandable to the high school student. Introduces the basic concepts of electricity and magnetism. Describes in detail modern motors, generators, transformers and power supply systems.

Graham, Kennard C. *Interior Electric Wiring: Residential* (6th ed.). Amer. Tech. Soc., 1962. 311 pp. illus. $4.75. 61–17508. (SH)
The electrician will find this book helpful both in telling him the correct way to do a job and the easiest way to do it. The regulations which the electrician must follow are explained. Includes helpful diagrams and many tricks to get the job done quickly.

Hubert, Charles I. *Operational Electricity.* Wiley, 1961. x+530 pp. illus. $10.50. 61–11517. (C)
A good introduction for the beginning student in electrical engineering. First the author explains AC and DC current and electricity in general, then he presents the basic formulas and concepts necessary in current design. Finally, the reader is shown how to use this knowledge in setting up useful apparatus.

Jackson, Herbert W. *Introduction to Electric Circuit* (2nd ed.). Prentice-Hall, 1965. xiv+479 pp. illus. $9.95. 59–14511. (C)
Designed to teach electrical theory to the student in the two-year technical school but it can be understood by anyone with a knowledge of algebra. Explains completely the important concepts of electric circuits. Good for a person preparing for a career in electronics because it emphasizes theory rather than practical applications.

Kogan, Philip, and Joan Pick. *The Silent Energy: Foundations of Electrical Technology.* (Foundations of Science Library; The Physical Sciences.) Sampson Low, Marston and Co. (dist. by Ginn), 1966. 127 pp. illus. $5.25. 66–17977. (JH–SH)
Essentially a treatment of the technological aspects of electricity, it provides a source of good material for non-science and science students. The illustrations are plentiful and for the most part excellent. In addition to more standard topics like the telephone and magnets, the book discusses three-phase supply, servomechanisms, automatic telephone exchanges, thermo-electric effects, tunnel diodes, and other applications of science.

Lewis, Floyd A. *The Incandescent Light.* Shorewood, 1961. 128 pp. illus. $2.95. 61–14122. (JH)
An account of the discovery and development of the Edison incandescent lamp, showing his patience and determination to produce a lighting system using electricity. Includes some biographical material on Edison's life during this period.

Oldfield, R. L. *The Practical Dictionary of Electricity and Electronics.* Amer. Tech. Soc., 1959. viii+216 pp. illus. $5.95. 58–59540. (SH)
Clear, concise definitions and good diagrams are provided for many terms in basic electronics, radio, television and radar. Appendices include electrical formulas, and explanation of electrical symbols, and other useful tables. A good reference for students who find *Van Nostrand's Encyclopedia of Physics and Electronics* too advanced.

Ress, Etta Schneider. Signals to Satellites . . In Today's World. Creative Educational Society, 1965. 176 pp. illus. $5.95. 64–8100. (JH–SH)
A well-written book on various forms of communication aimed at the junior high level, interesting enough for adults to enjoy and appreciate. Begins in prehistoric days of the cave men and ends with some of the latest satellites for verbal, written and pictorial communications. Electricity is introduced historically and all the developments of telephone, telegraph, cables, radio, television, motion pictures and newspapers are discussed. The illustrations are good.

Sharlin, Harold I. *The Making of the Electrical Age: From the Telegraph to Automation.* Abelard-Schuman, 1963. 248 pp. illus. $5.95. 63–16297. (JH)
The history of electricity is discussed with special attention to the telegraph, electric motor, central power station and dynamo. Concentrates on the very important basic discoveries rather than the more spectacular later achievements such as radar and television.

Smith, Ralph J. *Circuits, Devices, and Systems: A First Course in Electrical Engineering.* Wiley, 1966. xiii+776 pp. illus. $11.95. 66–17612. (SH–C)
This excellent text is more than a "first course in electrical engineering." The work can serve as an introduction to "engineering science," or to what might be termed the rationale of "philosophy" of engineering. The organization is fairly standard, although unusually comprehensive. A calculus background, but only a sketchy familiarity with college physics, is required. A complete grasp of the methodology taught in this text will go a long way to assure the rare quality of problem-solving competence in tomorrow's engineers.

Steinberg, William B., and Walter B. Ford. *Electricity and Electronics Basic* (2nd ed.). Amer. Tech. Soc., 1961. 262 pp. illus. $4.50. 60–5308. (JH)
A good introduction to electronics for the young student. Explains many of the common electrical devices found in the home and gives instructions for making simple electrical gadgets.

Timbie, William H. *Elements of Electricity* (4th ed.). Wiley, 1953. vii+631 pp. illus. $8.95. 536444. (C)
A very comprehensive text for a beginning course in electrical engineering also useful as a reference. Important concepts and terms are emphasized

with bold face type and a summary of each chapter is provided. Includes problems and appendix.

621.329 LASERS AND MASERS

Brotherton, Manfred. *Masers and Lasers: How They Work, What They Do.* McGraw-Hill, 1964. xvi+207 pp. illus. $8.50. 63–23249. (SH–C)
The author had firsthand experience in the development of these extraordinary and very useful electronic devices. His exposition of the physical principles and details of operation is outstanding. Intended for the layman.

Brown, Ronald. *Lasers: Tools of Modern Technology.* (Doubleday Science Series.) Doubleday, 1968. 192 pp. illus. $5.95; $2.45 (paper). (SH–C)
This popular book for the layman discusses the principles, types, uses, and science of lasers. As nonmathematical and nontechnical an exposition for the layman as the nature of the subject allows.

Carroll, John M. *The Story of the Laser.* E. P. Dutton, 1964. 181 pp. illus. $3.95. 64–11086. (JH–SH)
A satisfactory, elementary, nonmathematical explanation of lasers, and a description of their present and potential usefulness. For the student who wants to "build one" there are general directions, but he will need an outstanding electronics expert as a mentor, and some of the components are expensive.

Klein, H. Arthur. *Masers and Lasers.* Lippincott, 1963. 184 pp. illus. $3.95. 63–18676. (JH)
Popular interest in lasers and masers has created a need for laymen's literature. The theory, history, types and uses of lasers and masers are explained. Some interesting experiments are described showing ways in which these devices may be utilized in the future. Photographs and diagrams are included.

Marshall, Samuel L. (ed.). *Laser Technology and Applications.* McGraw-Hill. 1968. x+294 pp. illus. $14.00. 67–23479. (C)
The design and construction of lasers, their chronological development, an overview of the physics of laser theory, and the principles of laser action are presented in the collected papers of a number of professional contributors, many of whom are pioneers in the field.

Nehrich, Richard B., Jr., Glenn I. Voran, and Norman F. Dessel. *Atomic Light: Lasers—What They Are and How They Work.* Sterling, 1967. 104 pp. illus. $3.95. 67–27761. (SH)
Clearly presents the theory and applications of the laser beam. Virtually every conceivable use to which laser beams can be put is outlined, and in some cases, discussed quite thoroughly.

Patrusky, Ben. *The Laser: Light That Never Was Before.* Dodd, Mead, 1966. 128 pp. illus. $3.50; LB $3.23. 66–16292. (JH–SH)
A very well-written and illustrated book on the development and engineering accomplishments required for the development of the laser. Important additions to the literature are: (a) a classification of lasers by method of emission, and (b) a classification of the laser in terms of its potential benefit to mankind.

Scientific American, Readings from. *Lasers and Light.* (Introd. by Arthur L. Schawlow.). Freeman, 1969. vi+376 pp. illus. $10.00; $5.95 (paper). 77–80079. SBN 7167–0985–6. (SH–C)
Historical, aesthetic, and psychological discussions are included as well as ones on the physics, chemistry, and biology of light. Particularly recommended as collateral reading for bright students of optics and electronics who want to know more.

Stehling, Kurt R. *Lasers and Their Applications.* World, 1966. xi+201 pp. illus. $6.00. 66–18464. (SH–C)
Although the author has concentrated on applications—past, present, and future—of the laser and/or maser, he has given excellent coverage to the physical problems encountered by various research groups at various frequencies and wave lengths. Altogether, a very well-written book.

Thorp, J. S. *Masers and Lasers: Physics and Design.* St. Martin's, 1967. 311 pp. illus. $8.50. 67–24827. (C)
Written for undergraduate and graduate study, this reports on the rapid expansion in maser and laser research and development. The technical and electronic aspects of preparing and characterizing crystals is emphasized.

621.359 FUEL CELLS

Halacy, D. S., Jr. *Fuel Cells: Power for Tomorrow.* World, 1966. 160 pp. illus. $4.50; LB $4.28. 66–13907. (JH–SH)
The organization and content of this book are excellent and the history of fuel cells is concise but adequate. Explanations of basic principles are correct and clear. The most significant section to many potential readers will be the instructions for making a demonstration fuel cell—complete with safety rules and a list of materials and supplies.

Hart, A. B., and G. J. Womack. *Fuel Cells: Theory and Application.* (Modern Electrical Studies.) Chapman and Hall (dist. by Barnes & Noble), 1967. xii+372 pp. illus. $11.50. (C)
Fuel cells are described as those devices in which an electrical potential is produced by the chemical use of an externally supplied fuel. A chemical engineering viewpoint of fuel cell development is given.

621.38 ELECTRONIC ENGINEERING

Berens, Stephen, and Jack Berens. *Understanding and Troubleshooting Solid State Electronic Equipment.* Chilton, 1969. 176 pp. illus. $7.95. 69–13147. (SH–C)
This is primarily for the use of technicians, but is useful to anyone who likes to do his own troubleshooting of solid state circuitry. An elementary introduction to electron theory of matter aids in understanding semiconductor circuits, and basic diagrams simplify more complex circuits.

Carroll, John M. *Careers and Opportunities in Electronics.* Dutton, 1963. 191 pp. illus. $3.95. 63–9849. (JH)
Valuable to the student who wishes to know more about electronics as a career. Considers the work of the electronics engineer, scientist and tech-

nician in the fields of communications, broadcasting, military and industrial electronics. Includes facts about the field in general and a list of colleges offering electrical engineering.

Carroll, John M. *Secrets of Electronic Espionage.* Dutton, 1966. 224 pp. illus. $3.95. 66–11556. (JH–SH)

Mr. Carroll, in an intriguing book, has shown the development of electronic spies from the first crystal radio set used in World War I in a continuing story of each system, its countermeasure, its countercountermeasure, and so forth, up through the "spy in the sky" satellites of today. The reader will find his imagination stimulated to predict the next step in this continuing behind-the-scenes war against obsolescence. Indexed.

Elgerd, Olle I. *Control Systems Theory.* (McGraw-Hill Electrical and Electronic Engineering Series.) McGraw-Hill, 1967. 562 pp. illus. $12.50. 66–20716. (C)

Covers the field of control systems theory in amazing breadth and reasonable depth without becoming boring or leaping far ahead of the reader. Numerous examples and consistent emphasis on the practical significance of theoretical results. Clear, appropriate illustrations of high quality. Assumes a strong mathematics background. A college freshman or sophomore might have considerable difficulty with this book, but more advanced students should gain much from it. Indexed.

Evans, Walter H. *Introduction to Electronics.* Prentice-Hall, 1962. vii+518 pp. illus. $13.50. 62–7842. (C)

The student with a background in elementary calculus and electrical circuits will find this book useful. Studies the circuits found in radios, televisions, computers and radar equipment, and explains many compounds including vacuum, gaseous, and tunnel diodes, transistors, and photoelectric cells.

Johnson, J. Richard. *How to Build Electronic Equipment.* Rider, 1962. vi+290 pp. illus. $3.95. 62–10429. (JH)

Provides much information useful to anyone who builds or repairs electronic equipment. Includes basic topics such as necessary tools and soldering, as well as chassis layout, coil winding, wiring and testing. Many photographs and drawings aid in the presentation.

Kiver, Milton S. *Transistors* (3rd ed.). McGraw-Hill, 1962. vii+528 pp. illus. $9.00; $6.50 text ed. 62–15145. (SH)

A complete study of transistors beginning with theory and continuing with circuit design, applications and service. Includes experiments with transistors to illustrate properties of their circuits. Useful as a reference or for general information.

Latham, Donald C. *Transistors & Integrated Circuits.* Lippincott, 1966. ix+128 pp. illus. $4.50. 65–21654. (SH–C)

Written so high school students can develop a good basic understanding of the manufacture and use of transistors and their more advanced relatives, integrated circuits. It provides an up-to-date education in these new techniques and includes a few basic circuits. A "must" for every electronic engineer who wants to catch up with modern technology.

Mandl, Matthew. *Fundamentals of Electronics* (2nd ed.). Prentice-Hall, 1965. xiv+674 pp. illus. $14.60; $10.95 (text ed.). 65–17802. (SH)

A well-written and complete coverage of the fundamentals of tubes, transistors, circuit theory, test equipment, solid state theory, basic physics, etc. There is very little mathematics and no derivations of equations. The wide coverage limits the discussion of each topic to some degree but it is adequate as an introduction to the subject matter. It is well illustrated, with excellent review questions for each chapter, practical problems to help the student understand the material, and answers at the end of the book.

Mandl, Matthew. *Industrial Control Electronics.* Prentice-Hall, 1961. x+344 pp. illus. $11.50. 61–11816. (C)
An advanced work useful to the engineering student. Explains various signaling and measuring devices used in control circuits, as well as the circuits themselves. Includes a chapter on computers in industry. Illustrated with photographs, drawings and circuit diagrams.

Mann, Martin. *Revolution in Electricity.* Viking, 1962. 171 pp. illus. $5.00. 62–8095. (JH)
This book presents a good description of the theories involved in solid state physics, clearing the way for student comprehension of the world of semiconductors. The products and achievements that future electronic devices may make possible are also discussed.

Overhage, Carl F. J. *The Age of Electronics.* McGraw-Hill, 1962. ix+218 pp. illus. $8.95. 62–17372. (SH)
A collection of nine articles by different authors concerning certain developments in electronics such as radar, transistors, masers, and space relays. The major part of the book is on the advanced high school level but some is college material. Interesting collateral and background reading.

Page, Robert Morris. *The Origin of Radar: An Epic of Modern Technology* (Science Study Series). Doubleday, 1962. 196 pp. illus. paper $1.25. 62–12922. (SH)
Relates the development of radar from the first crude idea to its use in World War II and carefully explains the improvements that were necessary to construct the plan position radar and other advancements. Contains photographs of early radar apparatus.

Pierce, John R. *Electrons and Waves: An Introduction to the Science of Electronics and Communication* (Science Study Series). Doubleday, Anchor S38, 1965. 226 pp. illus. paper $1.25. 64–25265. (JH–SH–C)
Beginning with a resumé of electronics in the world, the author traces the principles and applications of electronic technology. The reader needs basic algebra, and preferably trigonometry, for complete comprehension.

Pollack, Harvey. *Experimental Electronics for Young People.* Rider, 1962. viii+128 pp. illus. $3.45. 62–13394. (SH)
Provides the student with a series of interesting and worthwhile experiments in electronics which start with basic principles and gradually become more difficult. Includes photographs and diagrams.

Romanowitz, H. Alex. *Fundamentals of Semiconductor and Tube Electronics.* Wiley, 1962. xii+620 pp. illus. $8.95. Lab. Manual $2.95. 62–8787. (C)
A good college-level text covering the use of vacuum tubes and transistors in various types of circuits. Begins with basic circuit theory so that a student with very little previous electronic experience can easily follow the book. Includes questions, problems and an extensive appendix.

621.384 RADIO

Collins, A. Frederick. *The Radio Amateur's Handbook* (12 ed.). Crowell, 1970. vi+374 pp. illus. $5.95. 69–18669. (SH)
A valuable aid to the beginning or experienced radio amateur. Covers fundamentals of construction techniques, elementary electronic theory, and FCC regulations, plus information on more complicated apparatus. The concluding chapters deal with mobile transmitters, transistors, tunnel diodes and solid-state devices. Glossary and index included.

Dunlap, Orrin E., Jr. *Communications in Space: From Wireless to Satellite Relay* (rev. ed.). Harper, 1964. 175 pp. illus. $5.95. 62–9909. (JH)
The development of electronic communication is traced from the time Maxwell developed his fundamental formulas of electromagnetic waves to today's space communications. Gives the history of radio, television, teletype and radio-photo, comparing the different communication systems.

Marcus, Abraham and William Marcus. *Elements of Radio* (5th ed.). Prentice-Hall, 1965. 672 pp. illus. $8.04. 65–10441. (JH–SH)
Designed for a one-year course, primarily for training technicians, rather than for students of physics and electronics. It begins with an introduction to electricity and electronics, devoid of mathematics, considers next the crystal set, then goes onto vacuum tube circuits, transistors, etc. Each section concludes with a summary, a glossary, and problems.

Morgan, Alfred. *The Boy's Fourth Book of Radio and Electronics*. Scribner's, 1969. 227 pp. illus. $4.50; LB $4.05. 70–85274. (JH–SH)
Introduces radio and electronics amateurs and enthusiasts to the technology of solid state physics, semiconductors and transistors. Explains the background and practical applications and offers instructions for constructing solar-powered radio receivers, a galvonometer, a transistor motor, and other devices. These can be constructed by amateurs and sources of materials are listed. One chapter tells how to become an amateur radio operator. A glossary is included.

621.388 TELEVISION

Buchsbaum, Walter H. *Fundamentals of Television*. Hayden, 1964. x+291 pp. illus. $8.95; paper, Rider (0391) $4.95. 64–7788. (SH–C)
Although intended as a text for courses in TV servicing it emphasizes basic fundamentals. It does describe some servicing techniques, but for the practioner additional information on servicing is necessary, particularly for color receivers. The principles of electronics, electricity, and magetism, and a foundation of high school mathematics should be prerequisites for students using this textbook. Teacher's guide and laboratory manual available.

Fink, Donald G., and David M. Lutyens. *The Physics of Television* (Science Study Series). Doubleday, 1960. 160 pp. illus. paper $1.25. 60–5925. (JH)
Presents a clear and basic explanation of the operation of television comprehensible to a person with no electronic training. Stresses how a television works (including color TV and light theory)—rather than how to repair a receiver.

Kerman, Stephen D. *Color Television and How It Works* (rev. ed.). Sterling, 1965. 68 pp. illus. $2.95; LB $2.99. 62–12592. (JH)
Describes briefly how television began and then explains how black-and-white and color television pictures are produced on a home receiver. Many illustrations, almost one per page. To make this material understandable to the young reader much has been omitted. Therefore, this book might best be used in conjunction with an early physics course where the instructor would be available to answer questions.

Kiver, Milton S. *Television Simplified* (6th ed.). Van Nostrand, 1962. vii+637 pp. illus. $9.95; $7.96 text ed. 62–4853. (SH)
Begins with descriptions of television transmission and reception, with attention given to TV cameras, antennas and amplifiers. Then it studies the various receivers in detail and concludes with UHF converters and color television. Valuable to the student planning to be an electronics engineer or technician.

Margolis, Art. *Make Your Own TV Repairs.* Arco, 1968. 104 pp. illus. $3.50. 67–26561. (SH–C)
A thorough consideration of the TV system. Chapters specialize on specific systems sections starting with the antenna, going on to the high voltage section, the various controls and their associated circuitry, and the sound section. Coverage is given to picture tube replacement, color TV adjustment and repair. Knowledge of electronics is required.

Martin, Albert V. J. *Technical Television.* Prentice-Hall, 1962. xv+557 pp. illus. $12.50. 62–53095. (C)
Discusses the operation of television in detail; circuits and components, some of the differences in receivers, and various proposed schemes for color television. Intermediate in difficulty between books written for technicians and professional engineers. Well illustrated.

Rowland, John. *The Television Man: The Story of John L. Baird.* Roy Publishers, 1966. 143 pp. $3.50. AC 67–10138. (JH–SH)
The reader is sitting beside the inventor from the first reproduction of a shadowy image to color television, having excellent resolution, suffering with his frustrations and applauding at his successes. A fascinating story of the life and heartbreaks of an inventive genius, who continued to work both in good times and in bad to make his visions into reality.

621.389 SOUND RECORDING AND REPRODUCING SYSTEMS

Wescott, Charles A. and Richard F. Dubbe. *Audels Practical Guide to Tape Recorders.* Audel, 1967. 277 pp. illus. $4.95. (SH–C)
Examines the history, theory, test equipment and test procedures of magnetic recording, the tape transport system, drive motors, etc. Addressed to service technicians, hobbyists, and professional recordists.

621.4 HEAT AND PRIME MOVERS

Hall, Newman A., and Warren E. Ibele. *Engineering Thermodynamics.* Prentice-Hall, 1960. x+643 pp. $16.85; $13.00 text ed. 60–12244. (C)

A comprehensive textbook on thermodynamics which could be understood by a student doing independent study. A knowledge of calculus is required. Contains problems and explanatory diagrams.

621.43 INTERNAL COMBUSTION ENGINES

Boumphrey, Geoffrey. *Engines and How They Work.* Watts, 1967. 255 pp. illus. $6.95. 60-10809. (JH)
Supplies simple but accurate descriptions of the operation of virtually all types of modern engines. Bearings and gear mechanisms are explained and several important historical engines are discussed. Illustrated with helpful diagrams and drawings, many in color.

Cummins, Clessie L. *My Days with the Diesel.* Chilton, 1967. x+190 pp. illus. $5.95. 67-11846. (SH–C)
While previously little known, the contributions of Clessie L. Cummins in the design and manufacture of high speed automotive diesels are a vital part of American technology. Cummins has written his autobiography in a direct, simple, narrative fashion. Easy to read, it holds one's attention even when dealing with technical or financial matters which, handled less skillfully, might easily bore. An excellent selection of photographs, no index.

Sandfort, John F. *Heat Engines* (Science Study Series). Doubleday, 1962. xxiv+292 pp. illus. paper $1.45. 62-14688. (SH)
A PSSC collateral reading book that traces the development of heat engines from James Watts' steam engine to today's gas turbine. In addition, the basic concepts of thermodynamics are presented.

Wilson, Charles Morrow. *Diesel: His Engine Changed the World.* (Illus. by Denny McMains.) Van Nostrand, 1966. v+181 pp. $4.75; LB $4.53. 66–16910. (JH–SH)
The very readable biography of Rudolph Diesel will teach young readers the virtues of thrift, perseverence, hard work, and love for one's fellows. It is entertaining throughout and the few sketches in the text add to the reader's enjoyment. The technical problems involved in the design and production of engines are discussed at a level that young readers can understand and appreciate.

621.48 NUCLEAR ENGINEERING (See also 539 MOLECULAR, ATOMIC and NUCLEAR PHYSICS)

Groueff, Stephane. *Manhattan Project: The Untold Story of the Making of the Atomic Bomb.* Little, Brown, 1967. xii+372 pp. illus. $6.95. 67-11231. (SH–C)
A very well told and fascinating account of the wartime atomic bomb program. The scientific and technical details are described only to the extent necessary to make the story intelligible. The primary interest of the author is in the men who carried out the project, the incredible technical difficulties they faced, and the decisions they made. The people involved come alive in skillful portrayals. Bibliography.

Hogerton, John F. *The Atomic Energy Deskbook*. Reinhold, 1963. xiii+673 pp. illus. $11.00. 63–13445. (SH–C)
A useful encyclopedia of terms used in atomic energy. Alphabetical listing of many subjects encountered in both peacetime and military applications make it a good reference work. Articles are not difficult to read and important terms are in bold face type.

Mann, Martin. *Peacetime Uses of Atomic Energy*. (rev. ed.). Viking, 1961. 191 pp. illus. $5.00; $1.45 paper. 61–7038. (JH)
Mann designed his book to show some of the interesting atomic programs rather than to discuss the problems they create, as in Calder's book. Explains fascinating projects which may be possible in the future as well as some which are already in use. Not indexed, unfortunately.

Stokley, James. *The New World of the Atom*. Washburn, 1957. xvi+288 pp. illus. $5.50. 57–6603. (SH)
A history of the beginning of the atomic age giving a good account of the development of military and peaceful uses of atomic energy. Includes the important ways nuclear energy was utilized up to 1957.

621.49 SOLAR ENERGY ENGINEERING

Halacy, D. S., Jr. *The Coming Age of Solar Energy*. Harper, 1963. 241 pp. illus. $4.95. 63–15951. (JH)
A good survey of the many methods of utilizing solar energy. The reader will be amazed by the variety of schemes proposed to capture the energy of the sun. Although there is some repetition, the author knows his subject well.

Hoke, John. *Solar Energy*. (Foreword by Hubert Humphrey.) (First Books Series.) Watts, 1968. 83 pp. illus. LB $2.65. 68–10336. (JH–SH)
Reviews the history of solar energy, its uses in space exploration and its actual and potential use on the ground. Solar cells, motors, electrical generation, a solar furnace, and a solar heating plant are among the applications described.

621.59 CRYOGENIC ENGINEERING

Allen, Richard J. *Cryogenics*. Lippincott, 1964. 160 pp. illus. $3.95. 64–19041. (JH–SH)
An introduction to the study and practical applications of low temperature physics. Explains processes for liquifying gases, and explains cryogenic techniques in electronics, nuclear physics, surgery, rocketry, preservation of foods, etc. Well-selected photographs and drawings complement the text.

Bell, J. H., Jr. *Cryogenic Engineering*. Prentice-Hall, 1963. xii+411 pp. illus. $15.25. 63–17252. (C)
An advanced work on the field of cryogenic engineering (low temperature processes). Includes material on the theories of cryogenics and on the use of liquid gases in industry. There are many charts and diagrams.

McClintock, Michael. *Cryogenics*. Reinhold, 1964. x+270 pp. illus. $10.75. 64–16625. (C)

Intended for the student or engineer not aquainted with the subject, hence less advanced thán Bell's text. It is organized under the major topics of refrigeration, insulation, thermometry, liquid helium, mechanical properties of solids, magnetic phenomena, thermal and transport properties, superconductivity and cryogenics applications. The appendix contains a temperature conversion graph. An ideal book for general study and reference.

621.9 MACHINE TOOLS

Rolt, Lionel T. C. *A Short History of Machine Tools.* M.I.T. Press, 1965. 256 pp. illus. $7.50. 65–12439. (SH–C)
This well-documented and nontechnical evolutionary account of the development of machine tools will interest almost every layman, engineering student, and professional person. This is good reading for mechanical trades students and apprentices also. The account progresses from simple rotating spindles and potters' wheels to present-day lathes, boring mills, grinders, milling machines, broaching machines, and highly automatic single and multiple machine tools. The reader needs an elementary background in the principles of mechanics.

623.4 MILITARY ORDNANCE

Blow, Michael, and The Editors of AMERICAN HERITAGE. *The History of the Atomic Bomb.* (Consultant: William W. Watson.) American Heritage (dist. by Harper & Row), 1968. 156 pp. illus. $5.49. 68–23895. (JH–SH)
An excellent historical account of the steps in physical and chemical research, including the discovery of transuranium elements, that led to the first sustained nuclear reaction. Then follows the technological history of the manufacture of fissionable materials, the construction and testing of the atomic bomb, its first military use, and the subsequent development leading to peaceful uses of nuclear energy.

Buehr, Walter. *Firearms.* Crowell, 1967. 186 pp. illus. $4.95. 67–18525. (JH–SH)
Covers the period of technological history from the invention of gunpowder to the development of atomic age weaponry. It describes both the conventional and the unusual and bizarre. Should be of interest to all young readers.

Cleator, P. E. *Weapons of War.* Crowell, 1968. 224 pp. illus. $5.95. 68–20091. (SH–C)
The first part is a history of weapons and weaponry. The last part, beginning with the invention and use of gunpowder and shot and culminating with the perfection and use of rocketry, and of the use of fission and fusion devices, is more absorbing. Excellent list of British and American references. Indexed.

Peterson, Arnold L. (ed.) *Encyclopedia of Firearms.* Dutton, 1964. 367 pp. illus. $13.50. (SH–C)
Articles have been contributed by 47 specialists on firearms, which present condensed data from previous publications and original information dealing primarily with small arms, with some data on light ordnance such as machine

guns and swivel guns. History and development, as well as descriptive details are included, and most articles contain additional references. Text figures and photographs are well-chosen.

623.51 BALLISTICS

Lowry, E. D. *Interior, Ballistics: How a Gun Converts Chemical Energy into Projectile Motion.* (Chemistry in Action Series.) Doubleday (Anchor AMC5), 1968. xi+174 pp. illus. $1.45 (paper). (SH–C)
The three major problems of ballistics are analyzed: size and geometry of propellant granules, relation of projectile weight and recoil, and maximum velocity of a projectile. The section on ballistics of rockets will be of much current interest. Principles of chemistry and thermodynamics of previous centuries will be more understandable to readers with this means of presentation. Some knowledge of chemistry is desirable.

624 CIVIL ENGINEERING

Boardman, Fon W., Jr. *Tunnels.* Walck, 1960. ix+144 pp. illus. $3.75. 60–9372. (SH)
First there is a discussion of caves and the history of tunnels followed by material on canal, railroad and automobile tunnels. Concludes with examples of tunnels designed for special purposes such as water supply and defense.

Coon, Martha Sutherland. *Oahe Dam, Master of the Missouri.* Harvey House, 1969. 124 pp. illus. $4.50; LB $4.39. 69–10752. (JH)
An outstanding account of the site exploration, survey, planning, and construction of the multiple purpose Oahe Dam near Pierre, South Dakota. Presents the complete story of the project in a logical sequence written in an uncluttered and nontechnical style. It is a resource book for social studies, career guidance, and general information.

Cullen, Allan H. *Rivers in Harness: The Story of Dams.* Chilton, 1962. 175 pp. illus. $3.95. 62–18207. (JH)
Since ancient times, man has dreamed of building dams to control rivers and irrigate farmlands. Later, he discovered he could use the falling water to supply power. This book traces the development of dams from the simple structures of ancient times to the complex systems of today.

Gies, Joseph. *Bridges and Men.* (Illus. by Jane Orth Ware.) Grosset & Dunlap, 1966. xv+343 pp. illus. paper $2.95. (JH–SH–C)
The reader is brought face to face with bridges and men devoted to bridge construction from earliest prehistoric attempts to span chasms, streams, gullies, valleys and harbors to the latest achievement of the Verrazano-Narrow Bridge at the entrance to New York Harbor. One shares the successes and failures of bridges and men alike. Beautifully illustrated; contains bibliography and apppendix of useful facts on bridges.

Jacobs, David, and Anthony E. Neville. *Bridges, Canals & Tunnels.* (The Smithsonian Library.) American Heritage (dist. by Van Nostrand), 1968. 160 pp. illus. $4.95. 68–54195. (JH–SH)
A history of civil engineering in the United States as related to population

increases, innovations and expansion of transportation is told through the construction of canals and waterways, and the great tunnels and bridges. Basic construction methods and engineering principles are described. Excellent illustrations, many of historic value.

Hammond, Rolt. *Civil Engineering Today.* Oxford, 1960. x+229 pp. illus. $4.80. 62–6441. (JH)
Contains extensive material on foundation, bridge, tunnel, canal and highway engineering. Describes in detail many great engineering projects, including how serious difficulties were overcome. Presents the prospective civil engineer with a good review of what his future work might entail.

Overman, Michael. *Roads, Bridges, and Tunnels; Modern Approaches to Road Engineering.* Doubleday, 1968. 191 pp. illus. $5.95. 68–18088. (JH–SH)
History and techniques of civil engineering are presented in a manner that demonstrates how basic mathematical procedures, as well as scientific principles, and business methods are involved in the design and construction of highways, tunnels and bridges. There is also a history of the profession and excellent explanatory diagrams and colored illustrations.

Sandström, Gösta. *Tunnels.* Holt, 1963. 427 pp. illus. $6.95. 63–21876. (SH)
Pictures some tunnels man has built and how they serve him. Covers approximately the same material as Boardman, but is more up-to-date and gives much greater detail on tunneling methods and on difficulties that have been encountered. Includes the story of the Mt. Blanc Tunnel.

Smith, H. Shirley. *The World's Great Bridges.* Harper, 1952. x+180 pp. illus. $3.95. 54–89993. (JH)
An historical account of bridge construction from the time of the Romans to the present day. Includes 43 photographs of important bridges.

625   RAILROAD AND ROAD ENGINEERING

Hellman, Hal. *Transportation in the World of the Future.* Evans (dist. by Lippincott), 1968. 187 pp. illus. $4.95. 68–30803. (JH–SH)
The author describes a vast array of revolutionary transportation equipment and control systems, many already in the design phase, which are soundly based on current engineering principles. High speed rail travel, computer controlled automobiles and highways, and rapid transit pneumatic tunnels and subways are described and illustrated in excellent detail. A thorough bibliography is included, along with a good index.

Ross, Frank, Jr. *Transportation for Tomorrow.* Lothrop, Lee & Shepard, 1968. 160 pp. illus. $4.50. 68–14067. (JH)
Describes newly developed and potential methods for improved transportation of individuals and commercial products. Advanced automotive designs, computer-controlled highways, high-speed railways, tube trains, surface-effect vessels, rocket air transports, and VTOL aircraft, are among the devices described. Engineering and operating details are described for lay readers.

Stanton, Robert Brewster. *Down the Colorado.* Univ. of Okla. Press, 1965. xxv+237 pp. illus. $5.00. 65–10109. (JH–SH–C)
A previously unpublished account of a railroad survey through the canyons of the Colorado River made by civil engineer Robert Stanton during the

years 1889 and 1890. Stanton's survey ran from Colorado to California and he reported that the route was feasible from an engineering point of view. Good collateral reading.

627 HYDRAULIC ENGINEERING

Burchell, S. C., and The Editors of HORIZON Magazine. *Building the Suez Canal.* Harper & Row, 1966. 153 pp. illus. $4.95; LB $4.79. 66–21551. (JH–SH)
Emphasized de Lesseps' efforts to achieve his dream of 40 years, a channel from the Mediterranean to the Red Sea. There are 133 illustrations, 31 in full color. The text is accurate and readable. Young people will find the book as attractive as others in *Horizon* Caravel series. The bibliography lists several recent books which dig deeper into the technical, political and economic implications of the Suez Canal.

Carlson, Carl Walter, and Bernice Wells Carlson. *Water Fit to Use.* John Day, 1966. 127 pp. illus. LB $3.86. 66–15093. (JH–SH)
The conservation, use, and management of water resources so that they may serve the needs for domestic water supplies, power production, and industry are complex and necessary tasks that should concern every citizen. The authors have done a good job. The treatment is clear and concise, the illustrations abundant and explanatory. Solid reading for all high school students and elementary and secondary teachers.

McNickle, L. S., Jr. *Simplified Hydraulics.* McGraw-Hill, 1966. xi+196 pp. illus. $10.00. 65–25048. (SH–C)
Written at the level of the troubleshooter or beginning student, and in practical step-by-step fashion, discusses and fully illustrates the "how" and "why" of efficient hydraulic system operation, from the reservoir through working components. Highly valuable to technicians and mechanics in all fields using hydraulic equipment, to engineers desiring a quick review of modern hydraulic systems, and to all others seeking basic knowledge of present-day industrial hydraulics. Each chapter contains a good set of study-aid questions, answered in the back of the book. Good index and a bibliography.

Popkin, Roy. *Desalination: Water for the World's Future.* (Foreword by Stewart L. Udall.) Praeger, 1968. xv+235 pp. $6.50. 68–16091. (SH–C)
A facts and figure summary of the present and potential role of desalination in the solution of the world's water problem. The author has compiled a chronological, technological, geographic, and socio-economic history of desalination. Includes a description of desalination methods and research.

Wright, Jim. *The Coming Water Famine.* Coward-McCann, 1966. 255 pp. $5.00. 66–13127. (SH–C)
Succinctly describes our escalating misuse of water, an indispensable resource, and the measures that should be and are being used to alleviate this misuse. Directs attention to some of the abuses that are out of hand. It is masterfully written by one who knows the facts and should be required reading for all public officials, teachers, members of service clubs, chambers of commerce, conservationists, industrialists and students. Indexed, no bibliography.

Zimmerman, Josef D. *Irrigation.* Wiley, 1966. xvi+516 pp. illus. $17.00. 65–27648. (SH–C)
Covers every imaginable aspect of irrigation engineering from the building of dams to the laying of plastic pipe. More than a highly informative handbook on engineering, for he also considers politics, economics, sociology, and every aspect of agriculture from agronomy to zoology. Examines a wonderfully complex variety of problems from the use and misuse of sewage to the size of standpipes. All of the information is detailed, precise, and clear. Ample, appropriate, and properly located diagrams and illustrations. He presents information useful for working in every sort of society from the most highly industrialized to the most primitive. References to the literature are included in footnotes.

628 SANITARY AND MUNICIPAL ENGINEERING

Camp, Thomas R. *Water and Its Impurities.* Reinhold, 1963. vi+355 pp. illus. $18.00. 63–21623. (C)
A survey of water supply and waste disposal, discussing the quality of water necessary for drinking, swimming, industry and wildlife. Shows in detail purification and sewage-treatment procedures. Also includes extensive material on the corrosiveness of water and the decomposition of organic matter by bacteria.

Carhart, Arthur H. *Water—or Your Life.* (2nd ed.). Lippincott, 1959. 322 pp. $4.95. 59–10129. (JH)
The twin problems of water shortage and pollution are still largely detached and unreal to many Americans, whose "curiosity about water resources ends at the bathroom faucet." In this factual and alarming study the author explains the unbelievable mass of water uses and abuses which combine to create this very real situation. Also discussed are some possible solutions, the progress made so far, and the past and future of water shortages. A few highly enlightening descriptions of the precarious sanitation maintained in public water supply systems, as well as the miraculous conversion of virtual sewage into drinking water help to stress the immediacy of the problem.

Hardenberg, W. A., and Edward R. Rodie. *Water Supply and Waste Disposal.* International, 1961. xiii+503 pp. illus. $8.50. 60–14571. (C)
Water purification, water supply, sewage treatment and sewage disposal are covered from an engineering point of view. Photographs and drawings illustrate equipment used in such systems. Examples of design problems are given with solutions and many useful charts are included.

Hirshliefer, Jack, et al. *Water Supply: Economics, Technology, and Policy.* U. of Chicago, 1960. xii+378 pp. illus. $7.50. 60–14355. (C)
An advanced discussion of the nation's water supply, dealing extensively with economic and legal aspects. The future water supply is investigated and possible new methods of purification, distribution, and reuse are examined as to economic feasibility.

Nordell, Eskel. *Water Treatment for Industrial and Other Uses* (2nd ed.). Reinhold, 1961. xi+598 pp. illus. $14.00. 61–11891. (C)
The topic of water supply and treatment is explored in detail. Covers al-

lowable limits for impurities in water for both private and industrial use as well as ways of removing harmful agents. Contains many charts, both in the text and in the appendices. Primarily intended for operating engineers in charge of domestic water supplies and for industrial engineers, but much of the text is useful as reference material to the educated layman.

629.1 AERONAUTICS

Ahnstrom, D. N. *The Complete Book of Helicopters* (rev. ed.). World, 1968. 175 pp. illus. $5.95. 54–8173. (JH–SH)
A complete coverage of the history of the helicopter, with chapters on their early history, their diverse uses, the theory of helicopter flights, becoming a helicopter pilot, and future developments in helicopter technology. Well illustrated, this work also contains a bibliography and list of helicopter flight schools.

Allen, John E. *Aerodynamics: A Space Age Survey.* Harper & Row, 1966. 128 pp. illus. $2.95; paper $1.25. 62–14561. (SH–C)
A potpourri of phenomena, from large-scale natural phenomena to industrial and other non-aeronautical situations. Aeronautical aspects usually associated with aerodynamics such as the airplane and missile are also covered. Excellent and extensive illustrations. Introductory calculus course.

Becker, Beril. *Dreams and Realities of the Conquest of the Skies.* Atheneum, 1967. 276 pp. illus. $5.38. 67–18991. (JH–SH–C)
Clearly points out the influence of the times and moods of the people on the flying machine and how it affected civilization. Depicts man's attempt to achieve a steerable balloon and to sustain stable flight in heavier-than-air craft. Concentration is on the nineteenth and twentieth centuries, with a summary of the last few decades' advancements.

Bernardo, James V. *Aviation in the Modern World.* Dutton, 1960. 352 pp. illus. $5.95. 59–5824. (SH)
A presentation of the influence of airplanes, rockets and missiles on the world both in wartime and peacetime. Explains the principles of flight, air regulations and air navigation. Includes an appendix of sources of aviation information, many photographs, and career guidance material.

Blacker, Robert D. *Basic Aeronautical Science and Principles of Flight.* American Technical Soc., 1958. xi+242 pp. illus. $5.95. 58–10423. (JH)
Begins with a short history of aviation and a survey of its effect on mankind. Then explains the theory of flight and the construction of aircraft and aircraft engines. Concludes with a description of a typical flight in a light airplane.

Caras, Roger A. *Wings of Gold.* Lippincott, 1965. 224 pp. illus. $4.95. 65–15250. (JH–SH)
The history of the development of naval aviation is examined with emphasis on the men behind the machines. As with most histories of aviation, the evolution of ideas is treated first, followed by a struggling infancy, and finally a survey of the hardware in the arsenal. As a reference of naval aviation, this is adequate.

Davis, Clive E. *The Book of Air Force Airplanes and Helicopters.* Dodd, Mead, 1967. 112 pp. illus. $3.75. AC 67–10618. (SH–C)

The primary function and usefulness of this book is for nontechnical reference and superficial familarity with the aircraft of the United States Air Force. General descriptive information is presented as appropriate to each aircraft, without any attempt at critical comparisons or analyses.

Fay, John S. *The Helicopter and How it Flies* (2nd ed.). Pitman, 1967. x+ 105 pp. illus. $3.95. (SH)

This book explains the operation of helicopters, while Ahnstrom's dealt with their history and uses. The various mechanisms necessary for helicopters to fly properly and smoothly are described using clearly-drawn diagrams. Gives the student a lucid explanation of the forces involved in helicopter flight.

Glines, Carroll V., and Wendell F. Mosley. *The DC-3* (rev. ed.). Lippincott, 1966. 203 pp. illus. $5.50. 66–11160. (SH–C)

An informal history of a sentimental old airplane—The Gooney Bird. Many who grew up with the airlines will relish this, many who worked with this sound and sturdy aircraft during their military service will reminisce on reading familiar descriptions and anecdotes.

Green, William. *MacDonald Aircraft Handbook.* (Illus. by Dennis I. Punnett.) Doubleday, 1966. 608 pp. $5.95. 66–21006. (JH–SH–C)

An extremely compact, accurate recognition guide. Many variations of aircraft are not included.

Hart, Clive. *Kites: An Historical Survey.* (Forward by Charles H. Gibbs-Smith.) Praeger, 1967. 196 pp. illus. $12.50. 67–20415. (JH–SH–C)

A comprehensive coverage of the history of kites. Interesting, beautifully printed and illustrated, this is a treasure that secondary school, college, and public libraries should acquire and which many individuals will wish to own.

Loening, Grover. *Takeoff into Greatness: How American Aviation Grew So Big So Fast.* (Forward by Donald D. Douglas.) (Industries of America Series.) Putnam's, 1968. 256 pp. illus. $4.95; LB $3.94. 68–15064. (JH–SH)

An interesting, useful, and well-written history of American aviation by one of America's outstanding and pioneering aircraft designers. Emphasis is on the period from Kitty Hawk to the late 1920's. The many photographs, some rare, are of good quality.

Mason, Herbert Molloy, Jr. *Bold Men, Far Horizons: The Story of Great Pioneering Flights.* Lippincott, 1966. 197 pp. illus. $4.95. 66–18447. (JH–SH–C)

Focuses on a small group of men who were pioneers in long distance aviation. Described vividly are: (1) the first nonstop flight across the Atlantic; (2) the first nonstop flight across the United States; (3) the first circumnavigation of the globe; (4) the traverse of Africa from Cairo to Capetown; (5) the first east-to-west crossing of the Atlantic; (6) the first flight across the Pacific from California to Australia; (7) the recordbreaking world flight by Post and Gatty; and (8) the nonstop flight from Moscow to California over the North Pole.

Rolt, L. T. C. *The Aeronautics: A History of Ballooning 1783–1903.* Walker, 1966. 267 pp. illus. $5.95. 66–23941. (JH–SH–C)

Begins with a detailed, scholarly, but occasionally lighthearted account of the birth of lighter-than-air flight in late 18th Century France, proceeds to accounts of the first daring aeronautics and their ascensions, and goes on to describe some of "The great showmen" of ballooning, the history of "Balloons at war," the impact of ballooning upon the science of the earth's atmosphere, "The navigable balloon," and "Balloons at play."

Shapiro, Ascher H. *Shape and Flow: The Fluid Dynamics of Drag.* (Science Study Series.) Doubleday, 1961. xv+186 pp. illus. $1.25 (paper). 61–12581. (JH)
Presents some of the basic concepts of fluid dynamics and how they are used to reduce drag. Several experiments are pictured which show exactly what would not be expected. Then the author discusses the Reynolds' number and the laws of fluid dynamics and the unexpected experimental results are explained in the last chapter. Photographs of the experiments are valuable.

Stever, H. Guyford, James J. Haggerty, and The Editors of *Life. Flight.* Time, Inc., 1965. 200 pp. illus. $3.95. 65–24362. (SH–C)
The eight chapters of this beautifully illustrated volume concern the dreams of flight, the essentials of flight, aerodynamics, propulsion, structure, navigation, aeronautical research, and future plans and programs. Many relatively complex topics are clearly explained, often through the use of photographic essays.

Von Kármán, Theodore, and Lee Edson. *The Wind and Beyond: Theodore von Kármán—Pioneer in Aviation and Pathfinder in Space.* Little, Brown, 1967. 376 pp. illu. $10.00. 67–11227. (SH–C)
This excellent book about a great man is filled with nuggets of von Kármán's personal philosophy of science, engineering, education, the conduct of research, etc., that are invaluable. Rather than proceeding in a strictly chronological manner, several chapters flash back as various facets of his life or work are covered. This has the advantage of covering a particular facet in a logical and continuous fashion.

629.133 AMATEUR ROCKETRY

Stine, G. Harry. *Handbook of Model Rocketry—NAR Official Handbook* (2nd ed.). Follett, 1967. 303 pp. illus. $6.95, paper $4.95. 65–18968. (JH–SH–C)
The official handbook of the National Association of Rocketry, this is an indispensable acquisition for all public and secondary school libraries. It traces the history of rocketry, describes the construction of model rockets, their engines, basic principles of aerodynamics, launching and recovery, model rocketranges, clubs, and contests. The appendices include Sample NAR Section Bylaws, a list of model manufacturers, bibliography, glossary, and index.

629.1334 UNIDENTIFIED FLYING OBJECTS

Condon, Edward U. (Project Director). *Final Report of the Scientific Study of Unidentified Flying Objects; Conducted by the University of Colorado under contract to the United States Air Force.* (Daniel S. Gillmor, Editor; Introduction by Walter W. Sullivan.) Dutton, 1960. xxiv+967 pp. illus. $12.95; paper, Bantam (YZ 474) $1.95. (SH–C)

This work has concluded again, as have previous studies, that there are logical explanations for practically all of the many hundreds of reported sightings and that absolutely none of them are "ships from outer space." This compendium is so complete and detailed that it supersedes all previous books and analyses as a reference work.

Vallee, Jacques. *Passport to Magonia: From Folklore to Flying Saucers.* Regnery, 1969. 384 pp. illus. $6.95. 76–8851. (JH–SH)

Jacques Vallee has performed a unique service in the creation of this compendium on celestial apparitions. He treats "flying saucers" from the point of view of folklore. A vast volume of anecdotal material has arisen from the sightings and imaginations of people around the world. These make fascinating reading and one is struck by the similarity between ancient sightings and what is currently reported as UFO's. Are we to believe that these were truly seen and reported as having been seen? Vallee provides an answer which clears the picture of chaff and highlights the role of folklore. Most of what is discussed is folklore. Today, it would be most difficult to sell the idea of elves and fairies. We consider ourselves a sophisticated society who insist on verification for what is reported. There is an appendix of the most celebrated global sightings.

629.2 MOTOR VEHICLES

Black, Stephen. *Man and Motor Cars.* Norton, 1966. 373 pp. illus. 67–15819. (C)

Ergonomy, the biological study of design, is the focal point of the work. Safety is the foremost consideration of automobile design, which is evaluated from the point of view of physics, anatomy, physiology, and psychology. Useful in design and engineering courses and as collateral reading for some psychology courses at the college level.

Crouse, William H. *Automotive Mechanics* (5th ed.). McGraw-Hill, 1965. 616 pp. illus. $8.85. (SH–C)

The essentials of automotive mechanics are explained in a clear, layman oriented fashion with good diagrams and photographs. The operation, maintenance, and service of all automotive systems are explained.

Donovan, Frank. *Wheels for a Nation.* Crowell, 1965. 352 pp. illus. $6.95. 65–23376. (JH–SH)

Touches on most aspects of the automobile, beginning with the origins of the automobile in the 1890's and the men who made the first motor cars, and progressing to current and future research programs. Some aspects of the history are sketchy; however, generally the treatment is informative and valuable.

Fanning, Leonard. *Men, Money and Automobiles: The Story of an Industry.* World, 1969. 184 pp. illus. $4.50. 68–14698. (JH–SH)

This is a newspaperman's story of the automobile industry. It is an accurate, thumbnail account of the leading individuals whose enterprises have survived. There are perhaps barely enough of the mechanical details that fascinate the teen-agers to sugar-coat the biography and economics.

Georgano, G. N. (Editor). *The Complete Encyclopedia of Motor Cars, 1885–1968.* Dutton, 1968. 640 pp. illus. $19.95. 68–22674. (SH–C)

Lists every car ever built in any part of the world from 1885–1968, over 4,100 cars in all. This is valuable as a reliable reference: there are almost 2,000 black-and-white photographs, plus 48 pages of beautiful color photos. Entries are crisp, factual, and complete. A three part index and several appendices supplement this marvelous work.

Kearney, Paul W. *Highway Homicide.* Crowell, 1966. x+181 pp. illus. $5.95. 66–18823. (JH–SH–C)
Hits hard at the myth that the driver is the cause of all accidents. Well documented evidence indicates more than half of highway casualties are the result of improper vehicle design, and points out the problem of the drinking driver. Comparable to Ralph Nader's *Unsafe at Any Speed* (Grossman, 1965).

Lent, Henry B. *The Look of Cars.* Dutton, 1966. 158 pp. illus. $4.95. 66–11384. (JH–SH–C)
One of the finest books on automobile styling produced for the general reader. Many well-chosen illustrations clearly define "good design." Well written, factual, and accurate material that describes the styling process.

Nader, Ralph. *Unsafe at any Speed; The Designed-in Dangers of the American Automobile.* Grossman, 1965. 365 pp. illus. $5.95 (Simon and Schuster, paper $1.00). (JH–SH)
The revelations in this book concerning the hazards and defects automobiles in relation to human safety, have been responsible for improved engineering design, elimination of hazards, and in the adoption of life conserving features that are now legally required in American automobiles.

Rae, John B. *The American Automobile: A Brief History.* U. of Chicago, 1965. xiv+265 pp. illus. $5.95; paper $1.95. 65–24981. (SH–C)
Not only a history of the birth and development of the "horseless carriage" but also a well-written, careful demonstration of how the developments in the automotive industry have produced still-accelerating changes in sociology, economics, urbanization, industrialization, mass transportation, and all other aspects of human life and enterprise. Many will view the excellent photographs with nostalgia.

Smith, Leroi. *How to Fix Up Old Cars.* Dodd, Mead, 1968. x+210 pp. illus. $4.50. 68–54450. (SH)
The author clearly outlines the pros and cons of used car purchasing, repairing, and rebuilding. He offers excellent advice in the selection of car models worth rebuilding, and describes those which are likely to retain resale value. Bargain hunting in wrecking yards is well covered. A center spread of photographs is supported by clear, informative diagrams throughout. An index and two appendices are included.

629.4 ASTRONAUTICS

Adams, Carsbie C., and Wernher Von Braun. *Careers in Astronautics and Rocketry: Training and Opportunities in the Space and Missile Fields.* McGraw-Hill, 1962. xv+252 pp. illus. $6.95. 61–18306. (JH–SH–C)
Following a brief history of astronomy and rocketry, the various fields that comprise astronautics are outlined. The various training programs and career availability areas are also described.

Associated Press, Writters and Editors of the. *Footprints on the Moon.* (Manuscript by John Barbour.) Macmillan, 1969. 216 pp. illus. 31 cm. $7.95. 71–93584. (JH–SH)

The Associated Press has planned, prepared and published a most interesting and useful documentary account of all of the major events of manned space flight, including programs of the U.S.A. and the U.S.S.R. The color photographs supplement the dramatic narrative of both the success and failure.

Bernardo, James V. *Aviation and Space in the Modern World.* Dutton, 1968. 382 pp. illus. $7.95. 68–17290. (SH)

Covers almost everything imaginable in the aerospace field; history of aviation, maps, airports, aircraft, rockets, satellites, research, aerodynamics, meteorology, etc. Extensive bibliographies, career reference sources, autovisual aids lists and publishers addresses make this a valuable reference.

Butler, S. T. and H. Messel (eds.). *Apollo and the Universe; Selected Lectures on the U. S. Manned Space Flight Program and Selected Fields of Modern Physics and Cosmology.* Pergamon, 1968. 422 pp. illus. $7.00: $5.50 (flexicover). 67–30286.

Six lectures delivered to a group of outstanding high school students as a part of the 1967 Summer Science School of the University of Sydney (Australia). These lectures were all written by eminent men in their field, especially for the fourth-year high school student. The book, however is not for light reading by an average high school science student, but it will be very appealing to the outstanding high school science students, college students, science-conscious laymen, and professionals.

Clarke, Arthur C. (ed.). *The Coming of the Space Age.* Meredith, 1967. xiv+302 pp. $6.95. 67–11025. (SH–C)

A combination of history, science fiction, scientific speculation, journalism, and theology. Beginning with the success of Sputnik and the failure of Vanguard, Clarke lets the workers in the field speak for themselves. Technical reports are mingled with tongue-in-cheek satire, irony is interwoven with spiritual innocence, science fiction is mingled with hard-headed technology. Although the technical information is not intellectually demanding, the philosophical purposes of the book and the linguistic styles require a fairly mature and well-informed reader.

Clarke, Arthur C. *The Promise of Space.* Harper & Row, 1968. xxi+325 pp. illus. $8.95. 68–17042. (SH–C)

Professional writing skill, careful checking of facts, a history of personal contact with the space program, and enthusiasm for space travel are combined in this marvelous product of a decade of space reporting. Anyone interested in the history, the science and the practical harvests of space exporation should read this book. Condensed bibliography of primary sources included.

Corliss, William R. *Propulsion Systems for Space Flight.* McGraw-Hill, 1960. ix+300 pp. illus. $12.50. 59–14442. (C)

An evaluation of propulsion systems which may be used in space flight, that first studies the performance of the present chemical rockets and their possible improvements and then investigates the future of nuclear, electrical and other high energy drives. The reader should have a background of elementary calculus and physics.

Cortright, Edgar M. (Comp.). *Exploring Space with a Camera.* NASA (dist. by Supt. of Docs., U.S.G.P.O.), 1968. 214 pp. illus. 29.5 cm. $4.25. 68–60027.

During the first ten years of space exploration many thousands of photographs were obtained, from which those reproduced in the book in black-and-white and in color were selected. The appendix is a series of colored illustrations and brief descriptions of spacecraft. The detailed list of contents and index make it a valuable reference work which is needed by all secondary school, college, and public libraries.

Craig, Richard A. *The Edge of Space: Exploring the Upper Atmosphere.* (Science Study Series, S55.) Doubleday, 1968. 159 pp. illus. $3.95, paper $1.25. (SH)

Explains how the upper atmosphere is explored by balloons, satellites, rockets, and ground-based measurements, how meteorologists have learned about sunspots, atmospheric tides, and the aurora borealis; and how communications and space flight are affected by the upper atmosphere.

David, Heather M. *Wernher von Braun.* Putnam's, 1967. 255 pp. LB $3.64. AC 67–10141. (JH–SH)

An excellent biography that extends beyond the life of this famous space pioneer and is, therefore, ideal for anyone intrigued with space flight. Traces von Braun from his youth in Germany to his preeminence in the United States space program.

Glasstone, Samuel. *The Book of Mars.* NASA (dist. by Supt. of Docs., U.S.G.P.O.), 1968. 315 pp. illus. $5.25. 65–62244. (SH–C)

Although somewhat over-shadowed by the fly-by mission of Mariner 6 and 7, this volume will be of great value in providing the necessary in-depth background data and history of Mars. Topics presented range from the historical background of Mars, the planet in relation to the solar system, the various physical properties of Mars and her atmosphere, the possibility of Martian life and finally the exploration of Mars. Physics and chemistry are required.

Glasstone, Samuel. *Sourcebook on the Space Sciences.* Van Nostrand, 1965. xviii+937 pp. illus. $7.95. 65–7824. (SH–C)

Enables the user to dig below the surface of popular and superficial literature on rockets, satellites and space probes. The systematic arrangement of major topics, prefaced by a discussion of the purposes of space studies and the historical background, makes the book well suited for self-study. The text discusses, in order, orbits and trajectories; propulsion and power; guidance and information systems; practical applications; the sun and the solar system; the earth and its environment; the plants; the universe; and finally, man in space. An indispensable work for all public, secondary school, college, and many special libraries.

Hunter, Maxwell W., II. *Thrust into Space* (Library of Science.) Holt, Rinehart and Winston, 1966. 224 pp. illus. $2.95; paper $1.96. 65–23276. (SH–C)

Almost every aspect of space propulsion from the early Goddard rockets to solar sailing is covered. Neatly organized, beginning with the fundamentals of rocket motion and working through the velocity spectrum up to the velocity of light in well-defined increments. The treatment is up to date and physical explanations (with a minimum of equations) are very understandable. History sprinkled throughout the book makes it interesting and enjoyable. Of special value are the many, well-organized data charts on propellants, planets, the solar system, comets, etc. A glossary of space and space-related terms and a selected bibliography. An excellent collateral reference for a space flight course or as an introduction to someone interested in space propulsion and space missions.

Hymoff, Edward. *Guidance and Control of Spacecraft.* (Library of Science.) Holt, Rinehart and Winston, 1966. 176 pp. illus. $2.95; paper $1.96. 65–23277. (SH–C)
Begins with a historical survey of the early applications of guidance and control, then explains the fundamental details of present-day systems. Topics considered are navigation, propulsion, the gyroscope, inertial guidance, and ballistics and trajectories. The diagrams and pictures are excellent and numerous.

Kash, Don E. *The Politics of Space Cooperation.* Purdue U., 1967. 137 pp. $4.95. 67–64072. (SH–C)
Treats international scientific cooperation as nongovernmental and intergovernmental scientific activities which have political implications. Space cooperation policy is analyzed from a conservative approach and an intuitive approach, followed by goals of cooperation, patterns of cooperation, and efforts in the United Nations.

Ley, Willy. *Rockets, Missiles, and Men in Space.* Viking, 1968. xvii+557 pp. illus. $10.95. 67–20676. (SH–C)
An expansion and updating of earlier volumes, this is a description of the early German rocket development by the Verin für Raumschiffahrt. The mathematics of rocket performance are simple and reduced to valid physical concepts.

The McGraw-Hill Encyclopedia of Space. McGraw-Hill, 1968. 831 pp. illus. $27.50. 68–26673. (SH–C)
An objective. world-wide view of different nations' space efforts is presented by over 150 of the world's foremost authorities on space. The text is fairly advanced and is supplemented with over 1,200 photographs, drawings, and diagrams. A valuable reference work.

Marks, Robert (ed.). *The New Dictionary and Handbook of Space.* Praeger, 1969. 531 pp. illus. $10.00. (Bantam, paper $1.95). 73–94221. (SH–C)
With definitions of 50,000 terms, about one-third of them having special meanings in space engineering, this book will be a useful reference for students interested in space exploration. A majority of the terms come from college-level courses in elementary astronomy, chemistry, and physics.

Meitner, John G. (ed.). *Astronautics for Science Teachers.* Wiley, 1965. vii+381 pp. illus. $8.95. 65–16419. (SH–C)
The 10 chapters were written by scientists, engineers, or teachers with scientific backgrounds. It ranges briefly through the history of astronautics to date, and contains a projection of astronautical developments into the early 1970's. Appropriate and informative for teachers, amateur astronomers, students in senior high school, and those taking elementary descriptive and astronomy college courses.

NASA. *Earth Photographs from Gemini III, IV, and V.* NASA (dist. by Supt. of Docs., U.S.G.P.O.), 1967. 266 pp. illus. $7.00. 66–62098. (SH–C)
This beautiful collection of color photographs represents the best of three Gemini flights. Each photograph is accompanied by an explanatory caption that describes the location of the photograph, and the significant meteorological and geological information. A comprehensive appendix lists all 550 photos taken during the flights.

Naugle, John E. *Unmanned Space Flight.* (Holt Library of Science.) Holt, Rinehart and Winston, 1965. 175 pp. illus. $2.95, paper $1.96. 65–23279. (JH–SH–C)
This authoritative book provides depth of exposition and explanatory detail. After a brief introduction to the solar system and solar observations, there is a discussion of the exploration of interplanetary space and one on exploring the earth's environment. Next come chapters on geodesy, exploration of the moon and planets, and on space astronomy. Concludes with a glossary, a bibliography, and an index.

Newlon, Clarke (Comp.). *The Aerospace Age Dictionary.* Watts, 1965. 282 pp. $5.95. 65–11382. (JH–SH)
More than 3000 entries related to missiles and rockets, meteorology, aerodynamics, control and guidance, propulsion, NASA and military commands, and other related items. Appendices include conversion factors, units of measurement, and capsule biographies of astronauts and leading personalities in aerospace programs and industries.

Ordway, Frederick I., III, James P. Gardner, and Mitchell R. Sharpe, Jr. *Basic Astronautics: An Introduction to Space Science, Engineering, and Medicine.* Prentice-Hall, 1962. xiii+587 pp. illus. $15.50. 62–16319. (C)
The first of three parts decribes outer space and some of the questions we hope to solve by invading it. The second part investigates rockets and space structures, while the third covers space medicine. Includes many photographs, charts, and diagrams.

Posin, Daniel Q. *Exploring and Understanding Rockets and Satellites.* Benefic, 1967. 96 pp. illus. $1.95. 66–26021. (JH)
Discusses the history and present-day developments of rockets and satellites. Technical terms are clearly defined; questions, thought problems, and suggested experiments conclude each chapter.

Sanger, Eugene. *Space Flight: Countdown for the Future.* (Tr. and ed. by Karl Frucht.) McGraw-Hill, 1965. vii+301 pp. $6.95. 64–8622. (JH–SH–C)
Sparkles with zeal and enthusiasm, and is convenient for supplementary reading. Useful as background for space enthusiasts who will, hopefully, go on to other literature. The primary value of the book is its stimulus to the imagination. The technical description covers a broad spectrum from systems currently in use to entirely hypothetical and extremely speculative possibilities. The emphasis is largely on propulsion systems.

Sharpe, Mitchell R. *Living In Space: The Astronaut and His Environment.* (Doubleday Science Series.). Doubleday, 1969. 192 pp. illus. $5.95; paper $2.45. 69–12122. (SH–C)
This is a thorough report on the technical and biological aspects of life sup-

port in space that comes on strong for continuation of the manned space program.

Shelton, William. *American Space Exploration: The First Decade.* Little, Brown, 1967. xii+367 pp. illus. $5.95. 67–21179. (SH)
Absorbing narrative of U. S. Space program, 1957–67. Good illustrations, lengthly glossary of terms, complete log of American space flights during 10-year period.

Von Braun, Wernher. *Space Frontier.* Holt, Rinehart and Winston, 1967. viii+ 216 pp. illus. $4.95. 67–12579. (JH–SH–C)
A collection of the author's monthly columns which have appeared in *Popular Science.* This compilation constitutes a sweeping portrayal of the considerations and disciplines involved in the exploration of space.

629.45 SPACE TRAVEL

Clarke, Arthur C., and the Editors of *LIFE. Man and Space.* Time, Inc., 1964. 200 pp. illus. $3.95. 64–25368. (SH–C)
Oriented toward the layman and the family library, this covers early studies and myths of space travel, rocketry, pioneers of the space program, satellite launchings, challenges of lunar and planetary flights, and the problems of space travel. The text is augmented with numerous high quality photographs and illustrations.

DuBridge, Lee A. *Introduction to Space.* Columbia, 1960. xii+93 pp. illus. $3.50, paper $1.45. 60–8186. (SH)
In four lectures entitled "Attaining a space orbit," "The space laboratory," "The solar system" and "The universe," the author effectively introduces some of the basic concepts of space and space travel. This small book contains much information clearly explained, plus interesting photographs.

Dunlap, Orrin E. *Communications in Space* (rev. ed.). Harper & Row, 1964. 206 pp. illus. $5.95. 64–18103. (JH–SH)
A chronological and historical treatment of the research, development, and production of modes of communication which led to the modern science of space communication. The revised edition includes new chapters on lunar communication, satellite systems, communications satellites, space communication laboratories, maser and laser communications, and new research advances.

Faget, Max. *Manned Space Flight.* (Holt Library of Science.) Holt, Rinehart and Winston, 1965. 176 pp. illus. $2.95 (text ed., paper $1.96). 65–23272. (JH–SH–C)
A no-nonsense account for lay readers of the essentials of astronomy and aeronautics involved in space flight, and lucid explanations of the design, construction and operation of manned spacecraft. The well-labelled and accurate photographs and explanatory diagrams are superior to those that have appeared to date in any other layman's account.

Freedman, Russell. *2,000 Years of Space Travel.* Holiday, 1963. 256 pp. illus. $3.95. 63–25034. (JH)
For more than 2,000 years man has wondered about the heavenly bodies—

what they are and if he can visit them. Here, the growth of our knowledge about the universe is traced from the Babylonians to modern times as well as the speculations about space travel that have accompanied it. Many of the great science fiction space travel stories are summarized.

Gatland, Kenneth. *The Pocket Encyclopedia of Spaceflight in Color: Manned Spacecraft.* (Illus. by John W. Wood and Tony Mitchell.) Macmillan, 1967. 256 pp. $3.50. 67–22617. (SH–C)
This small pocket encyclopedia is readable and entertaining in its format and scope. Beautifully illustrated with diagrams and photographs of recent space vehicles and space-related information, this is concise yet has the feeling of the narrative.

Hoyt, Mary Finch. *American Women of the Space Age.* Atheneum, 1966. 88 pp. illus. LB $3.41. 64–19556. (JH–SH)
Based on interviews, this contains brief vignettes of some 30 women who have made significant contributions in basic research and developmental activities that have contributed to the U. S. space program. Many of the important tasks in basic medical and physiological research, nutrition, mathematics, design of space suits and fittings for space craft, astronomical research, and other fields of endeavor, have been and are being performed by women scientists and engineers.

Lauber, Patricia. *Big Dreams and Small Rockets: A Short History of Space Travel.* Crowell, 1965. 71 pp. illus. $3.75. 65–13137. (JH)
Although dealing with the early history of space travel and its origins in literature, this brief history of rockets is valuable for the young reader. With the rapid pace of current space research, many contemporary developments are not included; however, as a history of the background to the space age, this is sufficient.

Lewis, Richard S. *Appointment on the Moon: The Inside Story of America's Space Venture.* Viking, 1968. xiv+434 pp. illus. $10.00. 68–22871. (SH–C)
The development of the American space program from its early beginnings in World War II Germany up to the first Saturn 5 launch is traced for the general reader. Intermingled with the man to the moon theme are discussions of unmanned probes. The text is clear and interesting and is garnished with excellent photographs.

Odishaw, Hugh (ed.). *The Challenges of Space.* U. of Chicago, 1962. xviii+379 pp. illus. $6.95. 62–19627. (C)
A collection of articles by more than 20 authors discussing aspects of the space program varying from flying telescopes and ion drive to military space research. Each author knows his topic well so the student often gets more facts than a general book by one author can give. The articles are generally well written and nontechnical.

Shelton, William. *American Space Exploration: The First Decade.* Little, Brown, 1967. xii+367 pp. illus. $5.95. (SH–C)
Traces the first 10 years of space exploration, beginning with a recount of early missiles and rocket tests at Cape Kennedy and then proceeds through the many programs tested: Vanguard, Explorer, Pioneer, Discoverer, Tiros, Ranger, Mariner, Surveyor, and others. The manned space program is covered in detail.

Von Braun, Wernher, and Frederick I. Ordway, III. *History of Rocketry and Space Travel*. (Introd. by Frederick C. Durant, III; Illus. by Harry H-K Lange.) Crowell, 1967. xi+244 pp. 28 cms. $14.95. 66–22417. (JH–SH–C)
A complete and authoritative background and historical work on astronautics. Begins with a review of man's efforts to understand the universe some 5,000 years ago. Then follows "A thousand years of rocketry" that introduces the third and fourth chapters on "Pioneers of space travel" and "The legacy of the pioneers," which include chronological summaries of early tests in the United States and elsewhere. Succeeding chapters deal with the use of rockets in World War II, postwar rocketry, and the developments leading up to and culminating in space exploration programs. A handsome book with profuse illustrations both color and black and white. Probably the best book of this nature yet produced, will long endure as a fundamental reference work.

Walters, Helen B. *Hermann Oberth: Father of Space Travel*. Macmillan, 1962. xxix+169 pp. illus. $4.00. 62–21204. (JH)
The story of Oberth's struggle for rockets to be used in space travel is told in this biography. His early ideas and experiments are presented, followed by his rocket work during World War II and his more recent work with new ideas and improved designs. Diagrams of some of his space inventions are included.

630 AGRICULTURE

Borgstrom, Georg. *The Hungry Planet: The Modern World at the Edge of Famine*. Macmillan, 1965. xx+487 pp. illus. $7.95. 65–20200. (SH–C)
A thorough study is presented of population expansion and unrealized food-production potentials. An assessment of population in relation to tilled land, and a discussion of the "calorie swindle" versus meaningful protein requirements, the nutrition problems of various areas of the world. Evaluates the resources of the sea, synthetic nutrients, water needs, air pollution, increasing urbanization, the dilemmas of modern technology, and concludes with a penetrating chapter, "Food or Moon Rockets?" References, good index.

Krebs, Alfred H. *Agriculture in Our Lives*. (2nd ed.). Interstate, 1964. xi+696 pp. illus. $5.50. 64–12602. (SH)
Introduces the basic concepts of agriculture to young people, describing careers, livestock and their care, farm crops and how they are improved, diseases of plants and animals, soil weeds and insects, marketing, and how agriculture is changing through the application of modern scientific techniques. The suggested activities at the end of each chapter may be a good source of science project ideas for younger students.

Pearson, Lorentz C. *Principles of Agronomy*. Reinhold, 1967. xiv+434 pp. illus. $9.50. 66–28454. (C)
A valuable contribution to the growing number of publications on the productions of food, feed, and fiber. Presents principles of crop production and recent advances in agriculture.

Sanders, H. C. (ed.). *The Cooperative Extension Service*. Prentice-Hall, 1966. xii+436 pp. illus. $11.35; $8.50. (text ed.). 66–11247. (SH–C)

Intended primarily for field extension service workers and specialists. Covers every educational aspect of extension work ranging from planning, on to individual contacts, meetings, presentations on radio, television, and in the press, and even to tests and program evaluations. Useful to train those entering the extension service, and those going into 4–H Club work. Attention is concentrated on how to teach or reach the public. Reliable, accurate, and practical.

Soth, Lauren. *An Embarrassment of Plenty: Agriculture in Affluent America.* Crowell, 1965. 209 pp. illus. $5.95. 65–26437. (JH–SH)
The evolution of the various economic, political, and social conditions in American agriculture are explained in a historical narrative. Land policy, westward expansion, food and nutrition, science and technology, and education at all levels are explored briefly and accurately. The economic dilemma of the farmers and the involved role of government are explained simply and convincingly. The evolution of price support programs and the development of government scientific activities are especially well handled. Ends with a discussion of alternative future programs for the United States, and a suggestion of the possible changes in developing countries. The best introductory book on this subject.

U. S. Department of Agriculture. *After a Hundred Years: The Yearbook of Agriculture, 1962.* U. S. G. P. O., 1962. xiii+688 pp. illus. $3.00. Agr. 62–298. (SH)
A history of the Agriculture Department and the story of agriculture in the United States over the past 100 years, with numerous photographs illustrating progress in farming. Main topics include plants, conservation, forests, animals, insects, technologies, markets, and economics. The text is simple in style and will be useful to all students, especially for beginning ideas for projects and experiments relating to agriculture.

U. S. Department of Agriculture. *Century of Service: The First 100 Years of the United States Department of Agriculture.* (Centennial Committee.) U. S. G. P. O., 1963. xv+560 pp. $2.75. 63–175. (SH)
Tells how agriculture in the United States has progressed from an economy of scarcity to an economy of abundance in the space of a hundred years, and why less than 9% of our labor force is engaged in agriculture today, compared with 20% to 40% in much of Western Europe, 45% in the Soviet Union, and over 70% in much of the world. Presented in historical form, this information will be useful to those interested in American history and government, as well as students of agriculture.

U. S. Department of Agriculture. *A Place to Live: The Yearbook of Agriculture, 1963.* U. S. G. P. O., 1963. xxiii+584 pp. illus. $3.00. Agr. 63–468. (SH)
Written "to inform all Americans about the effects of urbanization and industrialization on rural America and the need for plans and action so that people will have a proper place to live." Describes many natural and social changes that have occurred as a result of growing cities, and gives suggestions for action to guide these changes. Though general in scope, some of the articles are fairly technical and require knowledge of the fundamentals of agriculture and economics.

631 FARMING

McColly, H. F., and J. W. Martin. *Introduction to Agricultural Engineering.*
McGraw-Hill, 1955. x+553 pp. illus. $11.50. 54–11763. (SH)
This elementary text is also intended as a handy guide and reference for
teachers of vocational agriculture, farm machinery dealers, agricultural representatives for industrial organizations, and farmers. Many important fundamentals of physics, mathematics, and mechanics are explained simply and well-
illustrated, and among the many other topics covered are the history and use
of farm machinery, planning and building a home and other structures, soil
and water management, and the use of electricity. Chapters include review
questions and problems as well as suggested references.

Neubauer, Loren W., and Harry B. Walker. *Farm Building Design.* Prentice-
Hall, 1960. xii+611 pp. illus. $14.65 (text ed. $11.00). 61–8229. (C)
A detailed study of the design, structure, and materials used in the planning
and construction of homes, barns, bins, silos, and many other common farm
buildings and structures. Includes estimation of cost and other economic
considerations. Although much of the material is somewhat specialized and
of use only to those with architectural or engineering backgrounds, there
is also considerable information of use to the farmer and interested student.

U.S. Department of Agriculture. *Power to Produce: The Yearbook of Agriculture, 1960.* U. S. G. P. O., 1960. xii+480 pp. illus. $2.25. Agr. 60–362.
(SH)
Summarizes and explains the use of power in agriculture, particularly that
derived from the internal combustion engine and electric motors. Describes
the workings, agricultural aircraft, barn inventions, and an assortment of
harvesters, cutters, diggers, vibrators, and cultivators. Profuse photographs
and an even, clearly written text make it valuable for reference or straight
reading, even by city-dwellers.

631.4 SOIL AND SOIL CONSERVATION

Alexander, Martin. *Introduction to Soil Microbiology.* Wiley, 1961. x+471 pp.
illus. $9.95. 61–11512. (C)
Examines the biological and biochemical elements and dynamics of the soil,
covering the characteristics, ecology, function, and relative abundance of soil
bacteria, fungi, protozoa, algae, etc., as well as analyzing carbon, nitrogen,
and mineral transformations. The basic perspectives of each of the parent
disciplines—microbiology soil science and biochemistry—are stressed throughout, and the pertinent classical and recent developments in these fields are
frequently mentioned. References to primary sources are listed after every
chapter.

Allen, Shirley Walter and J. W. Leonard. *Conserving Natural Resources.* (3rd
ed.). McGraw, 1966. ix+370 pp. illus. $9.95. 58–13851. (SH)
Presents American natural resources and ways they can be protected, dealing
not only with soil, forest, land, water, animal and mineral resources, but
also with the social element in their control, and the idea of human powers
as resources.

Bennett, Hugh Hammond. *Elements of Soil Conservation.* (2nd ed.). McGraw, 1955. x+358 pp. illus. $5.00. 53–12044.

This broad and well-rounded survey deals with the national organizations and methods of soil conservation as well as its vital importance, causes, effects, and present extent in the United States. The control of stream, sheet, and gully erosion is explained, and sections are also devoted to the prudent use and management of water, flood control, and the related aspects of wildlife conservation. Each chapter contains study questions. Clear, well-chosen photographs and diagrams illustrate types of damage as well as the means used to remedy them.

Buckman, Harry O., and Nyle C. Brady. *The Nature and Properties of Soils.* (7th ed.). Macmillan, 1969. xviii+653 pp. illus. $10.95. 69–10539. (C)

This excellent book should have wide acceptance as a college text and should be of interest to many readers who want to know something about soils. Coverage is far-reaching, most of the topics one might want to consider are presented: inorganic silicate matter, organic deposits, minor elements, liming materials, fertilizers, and green manure.

Clawson, Marion, R. Burnell Held, and Charles H. Stoddard. *Land for the Future.* Johns Hopkins, 1960. xix+570 pp. illus. $8.50. 60–9917. (C)

Examines the present, past, and future uses of land in the United States, and covers six major categories—urban, recreational, agricultural, forestry, grazing, and miscellaneous. Their probable growth and competition within the next 40 years is predicted on the basis of certain reasonable and carefully-stated assumptions, and the overall conflict of growing land needs and fixed national area is also discussed. Although the writing is generally non-technical, considerable supporting data is included in the text and appendices.

Dasmann, Raymond F. *Environmental Conservation.* Wiley, 1959. x+307 pp. illus. $6.50. 59–6761. (SH)

A well-written conservation text that covers all the specific natural resources, but stresses the need for an environmental, ecological approach. Material is treated in relation to history, present governmental and private practices, and suggested plans and necessities for the future. Convenient references at the end of each chapter, and frequent charts and illustrations.

Held, R. Burnell and Marion Clawson. *Soil Conservation in Perspective.* Johns Hopkins, 1965. xiii+344 pp. illus. $7.50. 65–22946. (SH–C).

Deals with such broad topics as history, philosophy, politics, population, and scientific accomplishments as applied to proper care of the land and water resources. Discusses the overall problems of food supply. The development of our conservation program is traced. The contributions of various governmental agencies are discussed. Graphs and tables are used effectively.

Kellogg, Charles E. *The Soils That Support Us.* Macmillan, 1941. x+370 pp. illus. $6.95. 41–16797. (JH)

A simplified but very informative introduction to the variety, characteristics, and economic importance of soils, which can be used as a reference in elementary earth science or agricultural classes. The author first explains how soils are formed, their basic composition, and their gradual change due to biological factors, water, and man. The classification and distribution of various soil types is then discussed showing why each is best for certain crops, and the remaining chapters deal with the use, conservation, and improvement of the land.

Millar, C. E., L. M. Turk, and H. D. Foth. *Fundamentals of Soil Science.* (4th ed.). Wiley, 1965. x+491 pp. illus. $9.95. 65–21451. (SH–C)
Starts with the fundamental concept of a soil, its stages of development, requirements for plant growth, including physical, chemical, mineralogical and biological properties affecting plant growth. Follows with a modern classification of soils, fertilizer treatments and practices including farm manuring, soil erosion, soils of arid regions and irrigation of such regions. Most effective as a text in a college course in soil science. A useful high school reference.

Osborn, Fairfield. *Our Plundered Planet.* Little, 1948. xiv+217 pp. $4.50; paper, $1.95. 48–6167. (JH)
The ever-increasing threat to civilization and life itself due to the abuse and mismanagement of vital soil and water resources is described in a strongly-worded and vivid account. Part I discusses man's place in nature and the universe, emphasizing the need for a more modest and ecological perspective on all life and our own existence. In Part II the author describes the waste and exploitation of the land wherever man has farmed it in recent centuries and points out the continuing process today, with its almost inevitable consequences. Although dated in some aspects, this book has great historical value. Some of its predictions have proved to be too conservative.

Sears, Paul B. *Deserts on the March.* (3rd ed.). U. of Oklahoma, 1959. xiii+ 178 pp. $3.75. (SH)
Following a survey of the waste and decline of the land in many areas of Europe and Asia the author examines the history of our own country and points out with occasional biting sarcasm the ignorance, greed, and mismanagement which has already deprived us of soil, forests, water, grasslands, and wildlife. The complicated ecologic relationships involved are stressed throughout, showing how even "natural" wasting forces and pests take over only after man has tipped the balance of nature, and emphasizing the need for a broader understanding of nature as well as careful, reasoned action.

Stallings, J. H. *Soil: Use and Improvement.* Prentice-Hall, 1957. viii+403 pp. illus. $7.56. 57–5050. (JH–SH)
Covers completely but nontechnically the history, process, and control of erosion. The first section deals briefly with some of the fatal effects of soil erosion on ancient civilizations and early America. The second section analyzes the action of raindrops, flowing water and wind, and shows how they can break up and transport the soil when unchecked. Methods of surveying, testing, and classifying land capability are also presented, including chapters on the importance of plant cover and other ecologic principles. The final section surveys the farming techniques and special control measures used for protection and improvement.

Udall, Stewart L. *The Quiet Crisis.* (Foreword by John F. Kennedy). Holt, 1963. xiii+208 pp. illus. $5.00. 63–21463. (JH–SH)
A history of the utilization and competition for the resources of the United

States (forests, minerals, topsoil, fisheries, and wildlife) leading to their diminution, waste, and destruction. The contributions of men like Thoreau, Marsh, Powell, Muir, Olmstead, Pinchot, the Roosevelts and others to the conservation movement and action programs are discussed. Important background reading for all American students.

Waksman, Selman A. *Soil Microbiology*. Wiley, 1952. vi+356 pp. illus. $8.00. 52–9965. (C)
Introduces the study of microorganisms in soil, including such topics as decomposition of plant and animal residues, humus, the properties and transformations of nitrogen and carbon dioxide, control of disease-producing organisms in soil, fertility, and other developments in the field. An introductory chapter describes the history of work with soil and some important contributors to its study.

Worthen, Edmund L., and Samuel R. Aldrich. *Farm Soils: Their Fertilization and Management*. (5th ed.). Wiley, 1956. vii+439 pp. illus. $5.50. 56–5066. (SH–C)
Although it deals more with the fundamental principles of soil management and fertilization than with specific recommendations, this handy guidebook includes information of the soil regions of 22 eastern States and contains suggestions for planting on certain soils. The effect of various plant nutrients and fertilizer ingredients is covered, explaining how they work as well as how to apply them and evaluate commercial brands and recommendations. Crop rotation, erosion, and water control are also explained, and the use of special farm machinery and attachments is briefly discussed. Technical terminology is avoided throughout, and a glossary of the few terms not in frequent use is appended.

632.3 PLANT DISEASES

Carefoot, Garnet L., and Edgar R. Sprott. *Famine on the Wind: Man's Battle Against Plant Disease*. Rand McNally, 1967. 231 pp. $5.95. 67–21605. (SH–C)
Provides an historical background concerning the development and utilization of ten plant commodities, including a discussion of economic and sociological factors and implications, as well as information on fungi, bacteria, and viruses that cause damage to the crops.

Stakman, E. C., and J. George Harrar. *Principles of Plant Pathology*. Ronald, 1957. xi+581 pp. illus. $10.00. 57–9298. (C)
A comprehensive study of the classification, development, and treatment of plant diseases, explaining the conditions, causes, and effects of infection and the nature of the relationship between host and pathogen. Deals with the many problems caused by resistant species, the spread of infection, and incomplete scientific knowledge. Some important international plant disorders are discussed in detail, but the emphasis is on the occurrence and treatment of all diseases rather than on specific varieties. The work of many leading plant pathologists is referred to frequently, and a separate index of names is included.

632.58 WEEDS

Muenscher, Walter Conrad. *Weeds*. (2nd ed.). Macmillan, 1955. xvi+560 pp.
   illus. $10.00. 55–14427. (SH–C)
   The best known identification guide and reference work on the control of
   obnoxious weeds. Describes nearly 600 species, including common and
   botanical names, description of range and native habitat, methods of propagation, means of dissemination, type, poisonous qualities (if any), and the
   soil where each weed is most frequently found. Illustrations show the whole
   plant, roots, seeds, flowers, and fruit. Keys and tables aid identification.
   Bibliography.

632.7 PLANT PESTS—INSECTS

Anderson, Roger F. *Forest and Shade Tree Entomology*. Wiley, 1960. vii+428
   pp. illus. $9.95. 60–11714. (C)
   Designed to help forestry and entomology students understand forest and
   tree insect problems and the methods used to prevent damage to trees and
   wood products. The first fifth of the book explains the basic structure, physiology, classification, development, and ecology of insects in general; the remainder is devoted to individual treatment of the important forest pest
   species, including description, characteristics and methods of control.

Peairs, Leonard Marion and Ralph Howard Davidson. *Insect Pests of Farm,
   Garden, and Orchard*. (6th ed.). Wiley, 1966. viii+661 pp. illus. $17.50.
   56–6485. (C)
   The general morphology, physiology, and development of insects, as well as
   natural and artificial methods of controlling them are discussed, including
   tables and formulas for preparing and applying insecticides. The text then
   deals with specific orders, families, and species injurious to various types of
   domestic plants and animals, crops, stored products, and household materials.
   Their range and appearance, life cycle, and several recommended means of
   treatment and control are mentioned. Explains classification of insects and
   lists their common and scientific names.

Pfadt, Robert E. *Fundamentals of Applied Entomology*. Macmillan, 1962. x+
   668 pp. illus. $9.95. 61–10775. (SH)
   This complete text can be used by agriculture students with no background
   in entomology as well as by those whose knowledge of insects is more theoretical than practical. After a survey of the history of entomology, insect
   structure, function, growth and classification, there follow several chapters on
   chemical control and application, a discussion of the various hosts (such as
   livestock, crops, stored grains, and man), the insects which affect them, the
   nature of their injury, methods of control, and more details on representative
   species.

Rolston, L. H. and C. E. McCoy. *Introduction to Applied Entomology*. Ronald,
   1967. 208 pp. illus. $5.00 62–20088. (C)
   An excellent presentation of man's attempts to understand and control insect
   depredations. The initial presentation of broad concepts and principles is followed by a survey of important orders and their ecology and life history,

then there is a discussion of geographical and seasonal distribution. Economic management, environmental management, chemical control, application methods and control programs are discussed. References and index included.

632.95 PESTICIDES

Gunther, F. A., and L. R. Jeppson. *Modern Insecticides and World Food Production.* Wiley, 1960. xv+284 pp. illus. $10.50. 60–50965. (C)
The taxonomy and structure of insects and the methods, materials, and importance of controlling them are discussed in general, covering briefly chemical, physical, cultural, and biological techniques. The text then examines the effectiveness and application of various insecticides and acaricides, as well as the problems of resistance, residues, and other harmful effects. The remaining chapters discuss the composition and use of specific fumigants, sprays ,and compounds.

Whitten, Jamie L. *That We May Live.* Van Nostrand, 1966. vii+251 pp. $6.95. 66–31955. (SH–C)
Provides a balanced account of the place of pesticides in today's world. The central thesis is that if pesticides are used in moderation according to specified guidelines, there is no real hazard to man. Concludes that "the highest standard of living in the world is possible in this country because it takes so few people to produce food for the rest, thus leaving a great proportion of the population free to provide for us the other products we use and consume." Anyone who has read *Silent Spring* (Houghton Mifflin, 1962) should read this book also.

633 CROPS

Elting, Mary and Michael Folsom. *The Mysterious Grain.* (Illus. by Frank Cieciorka.) Evans (dist. by Lippincott), 1967. x+118 pp. $4.50. 67–10832. (JH–SH)
The story of corn is presented in a most interesting manner that keeps the reader in suspense and motviates him to learn more about the origin of this important grain.

Hughes, Harold DeMott, and Edwin R. Henson. *Crop Production: Principles and Practices.* (rev. ed.). Macmillan, 1957. xii+620 pp. illus. $8.95. 57–7270. (SH)
Historical, political, and economic aspects are included in this broad and well-organized study of the methods, problems, and determining factors in the successful production of field crops. The growth and structure of plants and the management of crops in general is examined first, followed by chapters on the planting, cultivation, harvesting, and improvement of corn, grains, tobacco and many other important crops. Review questions and problems are included.

Leonard, Warren H., and John H. Martin. *Cereal Crops.* Macmillan, 1963. vii+824 pp. illus. $10.00. 63–9596. (SH–C)
An up-to-date and factual reference intended for those with some scientific background. A section on general principles, including economics, botany,

harvesting machinery, processing, and marketing, introduces this comprehensive examination of the science production of cereal crops. The text then deals with corn, wheat, rye, barley, oats, rice, sorghum, and millets, covering their distribution and history of cultivation in the United States. Also discusses structure, growth, and varieties, and the entire process of commercial production and experimental improvement.

Peterson, Rudolph, F. *Wheat: Botany, Cultivation, and Utilization.* (World Crops Series.) Interscience (Wiley), 1965. xxiv+422 pp. illus. $16.00. 65–8175. (SH–C)

Wheat is clearly described in its broad world aspects to give the reader a general background of information about its botany, culture and utilization. Includes the morphological development, physiology, cytology and genetics, origin and history of the species, and the major established varieties. Culture includes distribution and conditions of production, diseases, breeding methods, seeding, growing, harvesting, and storing the grain. Utilization covers a world basis. A glossary and comprehensive bibliography.

U. S. Department of Agriculture. *Seeds: The Yearbook of Agriculture, 1961.* Supt. of Docs., U. S. G. P. O., 1961. xiv+591 pp. illus. $2.00 Agr 61–260. (SH)

A collection of articles on the production, life processes, certification, testing and marketing of seeds. There is a 48-page section of good photographs and many diagrams in the text, making the book a valuable one for reference and for reading by anyone interested in agriculture. Many valuable appendices and a long, explicit glossary.

Walden, Howard T., II. *Native Inheritance: The Story of Corn in America.* Harper & Row, 1966. xx+199 pp. illus. $6.95. 66–20769. (SH–JH–C)

Devoted largely to the economic and industrial importance of maize in the United States. Designed for popular consumption as well as professional reference. Underscores the utilization and cultural effects of this unique grass. Bibliography and index.

Wheeler, W. A. *Forage and Pasture Crops.* Van Nostrand, 1950. xi+752 pp. illus. $9.75. 50–9881. (C)

A guide to the culture, special characteristics, climate and soil adaptations, and forage and conservation uses of the grasses and legumes of the United States. Four sections deal with crops for forage and their management; special problems and uses of legumes; special problems of grasses; and a list of tables on seed and adaptations of grasses and legumes, and selected reference sources.

Wheeler, W. A., and D. D. Hill. *Grassland Seeds.* Van Nostrand, 1957. xx+734 pp. illus. $15.00. 57–8145. (C)

Comprehensive reference material about the grass and legume seeds used for forage, pasture, and soil conservation planting. This first half covers general processes such as seed formation, testing, disease and inoculation, harvesting, and marketing; part two deals with specific crops and their seed problems. A long section of appendices lists terms, tables, and statistics on forage crops and field seed, and standards and organizations of the industry.

Wilsie, Carroll P. *Crop Adaptation and Distribution.* Freeman, 1962. vii+448 pp. illus. $9.00. 61–17385. (C)

Plant distribution and change in a given area is analyzed and explained through adaptation and genetics, conditions of temperature, moisture, and other atmospheric factors including amount of light, types of soil, and biotic competition or interference. General climatology and the modified Koppen system of classification is then explained, after which representative crops of the tropics, subtropics, and intermediate climates are studied, showing their adaptation, growing requirements, and consequent world production centers. Chapters are also included covering the origin of cultivated plants and the use of ecological theory to maximize productivity.

Wilson, Harold K., and A. Chester Richer. *Producing Farm Crops*. Interstate, 1960. 336 pp. illus. $5.75. (text ed. $4.50). 60–5733. (JH–SH)
Although it covers some of the same material as Hughes and Henson, this book is somewhat simpler and is written for the farmer and farm boy in addition to its use as a text in vocational agriculture courses. The techniques of producing various common crops for cash, forage, pasture, etc., are explained, including where they are grown and why, and the means of breeding, protecting, and nourishing. Many basic principles of sound and economical farming are also presented.

Wolfe, T. K., and M. S. Kipps. *Production of Field Crops: A Textbook of Agronomy*. (5th ed.). McGraw, 1959. x+653 pp. illus. $10.50. 58–13894. (SH–C)
In addition to the regional distribution, methods, and problems of growing specific cash, forage, and other field crops, the text covers many fundamental principles of production, such as seeds and seeding practices, crop rotation, fertilization, the botany of common species, and the improvement of domestic breeds. Some crops (e.g. corn, wheat, tobacco) are discussed at length, while for others the description lists origin, history, appearance, varieties, adaptation, and uses in a "field book" style.

634 FRUIT CULTURE

Darrow, George M. *The Strawberry; History, Breeding and Physiology*. (Introd. by Henry A. Wallace.) Holt, Rinehart and Winston, 1966. xvi+447 pp. illus. $15.00. 66–12155. (SH–C)
Examines the history, morphology, physiology, genetics, breeding, and industry of strawberries. Useful for all horticulturists; every articultural college and experiment station should own a copy. Due to the reasonably nontechnical presentation, this beautiful work can be read by senior high school students and above.

Eck, Paul and Norman F. Childers (eds.). *Blueberry Culture*. Rutgers Univ., 1966. xx+378 pp. illus. $15.00. 66–18880. (C)
A collection of technical papers on all phases of blueberry culture. Blueberries are discussed extensively from their basic botany, through all stages of growing and processing, to their sale. Recommended for those interested in the subject, not for general library use.

Schneider, G. W., and C. C. Scarborough. *Fruit Growing*. Prentice-Hall, 1959. 307 pp. illus. $8.20. 60–8215. (JH)
A basic text for fruit growers and students of fruit production. Explains the

botany and physiology of fruit crops, flower bud formation and pollination, fruit growth and quality, and the special problems related to growing each of these crops: apples, pears, peaches, cherries and plums, grapes, bush fruits, and strawberries. Useful glossary and list of publications.

634.9 FORESTRY

Allen, Shirley Walter, and Grant William Sharpe. *An Introduction to American Forestry*. (3rd ed.). McGraw, 1961. viii+466 pp. illus. $9.75. 60–6956. (C)
A thorough basic text that deals with the social, scientific and engineering aspects of forestry in a useful, well-organized format. Covers the essentials of geography, biology and ecology as related to forests, as well as forest measurement, harvesting, protection from fire, insects, and disease, government policies and practices, and forestry education.

Bruére, Martha Bensley. *Your Forests*. Lippincott, 1957. x+163 pp. illus. $4.50. 57–6782. (JH)
An elementary introduction to forests and their study. Describes forests and foresters, what the forest produces, the management and care of forests, fires and how they are prevented and extinguished, harvesting, and other interesting topics. Emphasizes what man can do to enhance and protect his forest land.

Coombs, Charles I. *High Timber: The Story of American Forestry*. World Pub., 1960. 223 pp. illus. $4.95. 60–12898. (JH)
Describes modern forestry in a clear readable style with numerous photographs. Deals with trees and their structure, the development of modern forestry, lumbering and other commercial uses of forests, the relation of water and wildlife to trees in forests, the dangers of fire, insects and disease, and some career opportunities in the forestry of tomorrow.

Dowdell, Dorothy, and Joseph Dowdell. *Tree Farms: Harvest for the Future*. Bobbs-Merrill, 1965. 164 pp. illus. $4.00. 65–26498. (SH–C)
Offers information for the effective management of tree farms. Some of the subjects are fire control, diseases and insects, losses from animals, tree improvement through the application of genetics, the various silvicultural systems for harvesting the crop, watersheds, forest influences, and the many uses of the forest. A very valuable primer in forestry not only for forestry students and tree farmers, but also for vocational counselors, elementary school teachers, farmers, and the general reader.

Forbes, Reginald D., and Arthur B. Meyer (eds.). *Forestry Handbook*. Ronald 1955. ix+1170 pp. $15.00, 55–6815. (C)
A complete and authoritative reference guide to forestry and related fields, with numerous charts, tables, graphs, outlines, and other concise information sources. Covers forest measurement, geology and soils, forest management, silviculture, protection against fire, insects and disease, watershed management, logging and surveying, the chemistry and physics of wood, the economics and finance of forestry, and other practical topics.

Freeman, Orville L., and Michael Frome. *The National Forests of America*. Putnam's 1968. 194 pp. illus. 31x24 cm. $12.95. 68–22261. (SH)
Presents a well-illustrated history of the national forest, the national forest

movement and the U.S. Forest Service. Also included are glimpses of the forests and their resources, including wildlife. Various appendices are the primary reference assets of the book, the text and illustrations are good for reading for recreation and general enlightenment.

Goor, A. Y., and C. W. Barney. *Forest Tree Planting in Arid Zones*. Ronald, 1968. vi+409 pp. illus. $15.00. 68–31432. (C)
The authors describe techniques and species used in afforestation under differing conditions of aridity and soil, as influenced by the purpose of the plantation. Represents the combined efforts of foresters, ecologists, soil scientists, plant physiologists and specialists in other allied sciences. An appendix includes conversion factors, a summary of seed handling, a summary of nursery practices and site requirements of trees.

Hanaburgh, David H. *Your Future in Forestry*. Rosen, 1961. 159 pp. $2.95; LB $2.79. 61–10806. (JH)
The aim of the Careers in Depth Series is to help the student interested in a field to evaluate his own abilities and ambitions "in light of the requirements, the advantages, and the drawbacks of such a career." Information on employment opportunities and educational assistance, sources of further information, and a brief self-evaluation test are included, and the nature of forestry work in various organizations and fields is described.

Harrison, C. William. *Forests: Riches of the Earth*. Messner, 1969. 191 pp. illus. $3.95; LB $3.64. 69–13044. (JH)
Discusses the basic botany, geographical distribution, uses, and abuses of the world's forest resources. Draws heavily on historical anecdote. The illustrations are good and add interest to the narrative.

McCormick, Jack. *The Living Forest*. Harper, 1959. viii+127 pp. illus. $4.95, LB $4.43. 57–11791. (JH)
A brief introduction to the elements of the forest and forest ecology, with interesting information and many pictures on such topics as tree diseases, insects, life on the forest floor, galls, stumps, forest weather, and the relation of man to the forest. Answers many practical problems in forest management and scientific forestry.

Platt, Rutherford. *The Great American Forest*. (Prentice-Hall Series in Nature and Natural History.) Prentice-Hall, 1965, xii+271 pp. illus. $6.95. 65–25253. (SH–C)
A most fascinating story of our fabulous forests. Wilderness tracts, discovery and the first forests give a colorful description from the ice age up to the settlement of the continent by man. Paleontological evidence is given as to probable origin, and then on the subsequent migration of the deciduous and coniferous forests. The sections on wood, river of sap, and the leaf power are concerned with the many wonderful life processes of a tree. An interesting chapter on the tallest, oldest and strangest trees. A captivating story of the great American forests destined to be one of the great nature classics.

Richardson, S. D. *Forestry in Communist China*. John Hopkins, 1966. xvi+237 pp. illus. $6.95. 65–26176. (SH–C)
The first part is concerned with economic, agricultural and industrial problems in relation to the political and administrative mechanisms that affect

the practice of forestry. Information on wood factories, seasoning and preservation of wood, and the pulp and paper industry. A chapter on the organization of education and research institutes in forestry is included. Of special interest is the chapter on the impact of Michurinist biology on science, forestry and arboriculture. Of interest to teachers, professional foresters, arboriculturists and silviculturists.

Shirley, Hardy L. *Forestry and Its Career Opportunities.* (2nd ed.). (The American Forestry Series.) McGraw-Hill, 1964. ix+454 pp. illus. $9.95. 63–16468. (SH)
Virtually every aspect of the forestry industry and related fields is discussed in detail for benefit of prospective foresters. Written principally for students and guidance counselors. Indexed.

Stoddard, Charles H. *Essentials of Forestry Practice.* (2nd ed.). Ronald, 1968. ix+362 pp. illus. $7.00. 68–30894. (SH)
A broad study of forestry as a vocation, its methods, and the principles on which it is based. Silviculture and other fundamentals of economical logging, conservation, and management are fully explained. Chapters are included on the processing of wood products, the distribution and classification of forests, and the organization and administration of forested areas. The appendix includes a glossary and tables of important commercial timber species, giving their characteristics and uses.

U. S. Department of Agriculture. *Trees: The Yearbook of Agriculture.* U. S. G. P. O., 1949. xiv+944 pp. illus. $2.75. Agr. 55–2. (SH)
The management and use of trees, woods, and forests comprise the main theme of this authoritative and comprehensive work. It combines discussions of general aims and principles with first-hand accounts of specific projects, problems, and conditions in various parts of the country. Recreation, conservation, lumbering, and many other important aspects are examined. Controlling fire, insects, and disease are covered. An extensive guide to tree identification and many other aids are appended.

Wolff, Leslie. *Science and the Forester.* Criterion, 1961. 194 pp. illus. $3.95. 61–12798. (JH)
An inside view of scientific forestry, seen through the eyes of a young forester as he goes about his duties. Describes how a healthy forest is maintained through new plantings and felling of old diseased trees, insects and fungi which attack trees and how they are controlled, the problem of the destructive forest fire, fossil studies, and laboratory work on wood and forest products.

635   HORTICULTURE

Abraham, George. *The Green Thumb Book of Indoor Gardening.* Prentice-Hall, 1967. 304 pp. illus. $6.95. 67–19294. (SH–C)
Contains an immense amount of information for those who want to grow plants indoors. The characteristics of a large number of plants for growth indoors are given, along with cultural facts that a grower would want to know. Practical information on soils, watering, fertilizers is included, often with reference to particular plants which may be sensitive or have special requirements.

Behme, Robert Lee. *Bonsai, Saikei and Bonkei.* Morrow, 1969. 255 pp. illus. $9.95. 68–56412. (SH–C)
A complete and informative book on bonsai, the art of creating natural beauty in miniature by dwarfing trees. Saikei, or employing tiny living plants in tray landscapes, and bonkei, tray landscapes made from artificial materials, is also included. A brief history of the art, how to buy, wire, and train bonsai, their potting and repotting, and a number of growing techniques are outlined.

Blake, Claire. *Greenhouse Gardening for Fun.* Barrows (dist. by Morrow) 1967. 256 pp. illus. $6.95. 67–15151. (JH–SH–C)
Presents the basic fundamentals for growing plants in the private or home greenhouse. Well-written chapters on greenhouse types and styles, climate control, soils and fertilizers, growing equipment, and how to make green house gardening a remunerative hobby. Others describe benced, potted and hanging plants, bulbs, vegetables, herbs as well as how to increase plants. Lists the generic and specific names along with the common names of plants, includes an agenda for the hobbyist for each month of the year. Color plates, Index.

Christopher, Everett P. *Introductory Horticulture.* McGraw, 1958. viii+482 pp. illus. $9.50. 58–6677. (C)
After describing the field of horticulture and the fundamental principles of plant growth, this introduction gives detailed accounts of the raising of vegetables, commercial flowers, fruits, home landscaping, and arboriculture. There are also discussions of storage and marketing procedures, and horticultural shows. Each chapter has review questions and related readings.

Christopher, Everett P. *The Pruning Manual.* Macmillan, 1954. xx+320 pp. illus. $6.95. (SH)
This handy guide for use with or without previous knowledge of botany or horticulture combines the relevant principles of plant structure and growth with instructions on pruning a wide variety of trees, shrubs, and plants. Chapters on grafting, special effects, and various fundamental tools are indexed, and the treatment of damaged, diseased, and weakened trees is also explained.

Denisen, E. L. *Principles of Horticulture.* Macmillan, 1958. ix+509 pp. illus. $8.85. 58–6866. (SH)
Introduces the general principles of horticulture and how they may be applied to the home garden. Describes basic plant processes, seed growing, grafting and budding, pruning, soil and water management, pest control and other problems, then discusses the application of horticultural principles to lawns, vegetable gardens, ornamental plants, and fruit plantings. There is a special chapter on hobbies in horticulture and a glossary of clearly defined terms.

Dowdell, Dorothy, and Joseph Dowdell. *Careers in Horticultural Sciences.* Messner, 1969. 222 pp. illus. $3.95. 69–12109. SBN 671–30267–X, Trade; 671–32068–8 MCE. (JH–SH)
Covers thoroughly every aspect of the horticultural industry. Chapters start with short personal narratives, include a good general discussion of the field, and conclude with useful names and addresses. The appendix lists colleges and universities offering work in the several areas.

Edmond, J. B., T. L. Senn and F. S. Andrews. *Fundamentals of Horticulture.*
(3rd ed.). McGraw, 1964. xiv+476 pp. illus. $9.50. 62–21114. (SH)
This comprehensive text in horticulture is divided into three parts: a study of
the fundamental processes of plants; the application of these processes to
horticultural practice in soils, plant propagation, fertilizers, pruning, and
other areas; and a description and discussion of principal crops, including
fruits, vegetables, and commercial flowers. Emphasizes new developments
in biochemistry as related to hormones, vitamins and enzymes.

Graff, M. M. *Flowers in the Winter Garden.* Doubleday, 1966. xii+203 pp. illus.
$4.95. 66–17418. (SH–C)
A worthwhile addition to horticultural literature. Anyone interested in the
choice of plants for a protected, specially-prepared plot for winter bloom will
derive pleasure and profit from Mrs. Graff's pioneering study. The section
on "Summer dress for winter bulbs" will be of even more general interest. It
is refereshing to read plant descriptions so much in contrast to the "never-
mention-the-faults" writing which is universally and exclusively the stuffing
of catalogs.

Janik, Jules. *Horticultural Science.* Freeman, 1963. xi+472 pp. illus. $8.50.
63–7192. (C)
The first section of this introductory text covers the biology of horticulture,
the structure, classification, growth, and development of plants. The second
part deals with the technology of plant growing, while the last part describes
the horticultural industry and some of its problems. The book is more concerned with fundamental principles than practical applications, although the
student is encouraged to apply the material himself.

Knott, James E. *Vegetable Growing* (5th ed.). Lea & Febiger, 1955. 358 pp.
illus. $5.00. 55–7846. (JH)
A basic text in vegetable production, the first half being devoted to general
principles, such as classification, seedage, soils, fertilizing, and storage, and
the remainder to descriptions of individual crops, arranged according to
"warm season" and "cool season" crops and vegetable type.

Kraft, Ken, and Pat Kraft. *Fruits for the Home Garden.* (Illus. by Kathleen
Bourke.) Morrow, 1968. 28+287 pp. illus. $6.95. 68–23114. (SH)
A readable and useful handbook intended primarily for the home gardener
that will also appeal to readers in general because of the organization in
chapters devoted to each of the important fruits, nuts, and berries. There is
historical information as well as culture requirements and other useful details.

Mahlstede, John P., and Ernest S. Haber. *Plant Propagation.* Wiley, 1957.
x+413 pp. illus. $8.25. 57–5924. (C)
A complete introduction and guide to the principles of sexual and asexual
plant propagation and their practical application. Explains the methods of
natural and controlled seed production and development, as well as propagation through roots and bulbs, cuttings, and grafting. The use of special
tools, materials and structures, the specific procedures employed with common trees and plants, and the fundamental principles of plant structure and
growth are also covered. Tables of data on the propagation of various other
trees, shrubs, vines and herbaceous plants are appended.

Shoemaker, James S. *Vegetable Growing* (2nd ed.). Wiley, 1953. vii+515 pp. illus. $6.95. 52–13795. (SH)

After two general chapters on seed production and perennial crops, this basic reference work in horticulture describes the special characteristics of and procedures for growing all the more popular vegetables, including corn, root, legume, bulb and vine crops, and potatoes, mushrooms and herbs. The history and classification of each type is given, and a special chapter is devoted to insecticides and other forms of pest control.

MacGillivray, John H. *Vegetables Production.* McGraw, 1952. viii+397 pp. illus. $10.50. 53–465. (C)

The first half of this text in vegetable production deals with the general principles of horticulture, such as classification of vegetables, production areas, soils, seed characteristics, fertilizers, preservation, and insect and disease control. The remainder is devoted to specific types of vegetables, arranged according to environmental needs. There is special reference to western crops, although all the major crops are covered.

Wilson, Helen Van Pelt, and Leonie Bell. *The Fragrant Year: Scented Plants for Your Garden and Your House.* Barrows (dist. by Morrow), 1967. xiii+ 306 pp. illus. $10.00. 67–15150. (SH)

The authors' choice of plants for their gardens has been based primarily on scent as a "fourth dimension" of the garden. Their discussion includes shrubs, trees, annuals, and perennials; they offer seasonal selections of fragrant plants for the garden; there is a chapter on window gardens, and also one offering a "fragrant schedule" for the living room. The illustrations are excellent black-and-white drawings by Mrs. Bell. Selected bibliography on fragrant plants, good index. Has greater appeal to gardeners than to students, since the descriptive detail is intended for them and not for the botany student.

636 ANIMAL HUSBANDRY

Acker, Duane. *Animal Science and Industry.* Prentice-Hall, 1962. ix+502 pp. illus. $8.50. 63–7622 (C)

An efficiently organized introduction to animal science and livestock industry. Material is presented by topic rather than by species, with chapters covering nutrients, ruminant nutrition, animal reproduction, heritability, breeding programs, livestock markets, processing meat animals, wool and mohair, dairying, and other basic subjects. Appendices give production data for States, grading systems for species, and a valuable glossary.

Anderson, Arthur L., and James J. Kiser. *Introductory Animal Science.* (4th ed.). Macmillan, 1963. x+800 pp. illus. $9.95. 63–9586. (C)

A comprehensive standard reference-text that has sections on all the major farm animals, cattle, swine, sheep, horses, and mules. Describes marketing requirements and methods, processing, merchandising, feeding, management, and breeding, with an appendix on livestock record associations. (Previous editions titled *Introduction to Animal Husbandry.*)

Cole, H. H. (ed.). *Introduction to Livestock Production, Including Dairy and Poultry.* Freeman, 1962. xi+787 pp. illus. $8.75. 62–8851. (C)

Forty authors have contributed articles to this collection of readings that relate nutrition, genetics, physiology, and veterinary medicine to the management of livestock. Topics include livestock products, types and breeds of livestock, inheritance and genetic improvement, nutrition, classification, grading, and livestock diseases.

Davis, Richard F. *Modern Dairy Cattle Management*. Prentice-Hall, 1962. vi+264 pp. illus. text ed. $6.95. 62–9285. (C)

Describes modern dairy methods, underlying theories pertaining to dairying and milk production, livestock physiology, business aspects of milk production, and the progress of the dairy industry and its role in our economy. A good reference for the professional dairyman or for anyone considering a career in dairying or animal husbandry.

Frandson, R. D. *Anatomy and Physiology of Farm Animals*. Lea & Febiger, 1965. 501 pp. illus. $12.50. 65–19429. (SH–C)

Gross and microscopic anatomy and physiology of the horse, with appropriate additional material for the cow, sheep, pig, and dog. Illustrations are numerous, most of them from other standard books. An appendix table of the size, shape, and other characteristics of the organs of the horse, cow, sheep, pig, dog, and cat is an excellent feature. The physiological accounts are current.

Griffen, Jeff. *The Pony Book*. (Illus. by Jeanne Mellin Herrick.) Doubleday, 1966. 287 pp. $7.50. 65–23795. (JH–SH)

A detailed, well-written, and attractively illustrated book that describes the various breeds of ponies, with discussions of the miniature donkey and burro included. There are chapters dealing with races and shows; also purchasing, care, and training of the animals. A worthwhile book for any reader interested in the subject.

Hart, Susanne. *Life With Daktari: Two Vets in East Africa*. Athneum, 1969. 224 pp. illus. $7.95. 74–75746. (SH–C)

The author relates her experiences in treating and studying a variety of East African mammals such as lions, cheetahs, rhinoceros and antelopes. Throughout, a deep concern is expressed for wild animals; their well-being, conservation and value in research. Her husband plays a prominent part in her story and pioneered in the development of immobilization techniques with large animals.

Hammond, John. *Farm Animals: Their Growth, Breeding and Inheritance* (3rd ed.). Edward Arnold, 1960. viii+322 pp. illus. $7.50. 61–1835. (SH)

A scientific approach to better livestock production, through the application of basic principles of genetics, fertility, and growth, is outlined in this well-illustrated and factual text. Horses, cattle, sheep, pigs, and poultry are studied individually, describing their reproductive cycle and development, and explaining the practical applications in breeding, nutrition, milk and meat production, etc. The last half of the book is devoted to the general study of genetic applications in selective and developmental breeding.

Johnson, Norman H. *The Complete Puppy and Dog Book*. (In collaboration with Saul Galin.) Atheneum, 1965. xvii+494 pp. illus. $7.95. 65–23660. (JH–SH)

Describes the principal breeds, the growth of the puppy by stages from birth to maturity, nutrition, training and showing, ailments, first aid, mating, old age, services, and other information that will be of interest to dog owners and breeders. 122 black-and-white photographs of various breeds. Of most use in public libraries or as a personal acquisition—probably not needed in most school libraries.

Kays, John M. *Basic Animal Husbandry.* Prentice-Hall, 1958. xiii+430 pp. illus. $8.50. 58–5490. (C)

The breeding, feeding, management, marketing and judging of cattle, swine, sheep and horses is explained, emphasizing the great advances and modern standards which have made animal husbandry a true science. The slaughter, processing and commercial produce of many breeds is also covered, describing and clearly illustrating the commercial grades and varieties. Useful and interesting reading for farmer, layman and consumer.

Morrison, Frank B. *Feeds and Feeding, Abridged: The Essentials of the Feeding, Care, and Management of Farm Animals, Including Poultry* (9th ed.). Morrison, 1958. vi+696 pp. illus. $4.75. 58–1215. (C)

The ninth edition of this basic agricultural text (first published in 1917) has much additional material, especially that related to nutrition and the use of antibiotics, arsenic, and hormones. It first presents the fundamentals of nutrition and practical feeding information, next it describes and analyzes the important foodstuffs, and finally it discusses the feeding and management of various classes of livestock.

Naether, Carl A. *The Book of the Domestic Rabbit: Facts and Theories from Many Sources, Including the Author's Own Experience.* David McKay, 1967. x+128 pp. illus. $4.95. 67–20182. (JH–SH)

Considerable information on the characteristics of domestic rabbits is provided here, together with practical recommendations concerning their selection, feeding and management. Photographs accompany most of the descriptions of the some 30 different breeds and varieties. Best used as collateral reading for students and for hobbyists.

Willman, H. A. *A 4-H Handbook and Lesson Guide* (2nd ed.). Cornell, 1963. xiv+314 pp. illus. $5.95. 63–11491. (JH)

Describes the essentials of 4-H work for leaders, county-extension agents and teachers, including purposes, procedures and program planning; a list of fifteen subject-matter lessons on agricultural topics; and a guide to recreational activities. Four useful appendices give tables and general information lists, colleges and agricultural organizations, and plans for developing a dog obedience course.

Wing, James M. *Dairy Cattle Management: Principles and Applications.* Reinhold, 1963. xvii+394 pp. illus. $10.00. 62–21381. (C)

Covers the basic dairy topics of feeding, lactation, breeding management, etc., and includes material on body reactions, effects of weather, silage, and haymaking. Stresses scientific principles that can be applied to modern dairy farm management and gives a "Ready Reference Handbook" of practical information at the end.

636.088 LABORATORY ANIMALS

Mossesson, Gloria R., and Sheldon Scher. *Breeding Laboratory Animals.* Sterling, 1968. 128 pp. illus. 20.3 cm. $5.95. 68–8758. (SH–C)

This pocket size manual is complete, for it begins with a discussion of the usefulness of research animals in public health, clinical medicine, biomedical, and genetic research. Illustrated chapters are devoted to the principal laboratory mammals: the mouse, the rat, the golden hamster, the guinea pig, the rabbit, and the gerbil. A glossary, tables of weights and measures and equivalent measures, and an index conclude the book. A practical handbook for commercial breeders, amateurs, and teachers of secondary school and college biology courses.

636.5 POULTRY

Biester, H. E., and L. H. Schwarte (eds.). *Diseases of Poultry* (5th ed.). Iowa State U., 1965. xiv+1382 pp. illus. $18.00. 65–10568. (SH–C)
There is no more comprehensive text and reference book currently in print on the poultry anatomy, physiology, genetics, pathology and therapeutics. Valuable to students, poultrymen, veterinarians, pathologists, public health workers and others. All chapters have been revised and updated, many have been completely rewritten. Each contains additional references. In appropriate localities, high schools, colleges, public libraries, and special libraries will find it in demand.

637 DAIRY INDUSTRY

Hammer, Bernard W., and Frederick J. Babel. *Dairy Bacteriology* (4th ed.). Wiley, 1957. ix+614 pp. illus. $9.95. 57–10806. (C)
The first half of this text is devoted to problems of milk production and dairy product manufacturing, while the last ten chapters deal with the bacteriology of specific products and operations. Includes much new information from recent research, especially that relating to antibiotics, bacteriophages and enzymes.

Judkins, Henry F., and H. A. Keener. *Milk Production and Processing*. Wiley, 1960. viii+452 pp. illus. $8.50. 60–10317. (SH)
Explains the properties of milk and its testing, handling and processing. Chapters cover dairy farm operation, breeds, milk secretion, handling of herds and record-keeping, tests for milk and how they are applied and some special products such as ice cream and cheese. Lengthly appendices give information on many practical problems of the dairyman.

639.39 CULTURE OF REPTILES

Kauffeld, Carl. Snakes: *The Keeper and the Kept*. Doubleday, 1969. xii+248 pp. illus. $5.95. 68–22516. (JH–SH–C)
Information on collection, housing, feeding, treatment, and management of snakes and their diseases is related in this book by the Director and Curator of New York's Staten Island Zoo. Appendix includes: food of 35 groups of snakes, discussion of scientific and common names, and herpetological societies. For amateur and professional herpetologists.

639.9 WILDLIFE CONSERVATION (See also 333 CONSERVATION)

Fischer, James, Noel Simon, and Jack Vincent. *Wildlife in Danger.* (Foreword by Harold J. Coolidge and Peter Scott, Preface by Joseph Wood Krutch.) Viking, 1969. 368 pp. illus. $12.95. 69–12903. (SH–C)
The "Red Data Books," a looseleaf series published by the International Union for Conservation of Nature and Natural Resources, are the recognized world guides to endangered wildlife species. Their severly structured information has here been recast in encyclopedic style with the addition of much background information. Several prefaces and introductions explain why and how various species are threatened with extinction, why it matters, and what can be done about it. A standard reference work in its field.

Laycock, George. *America's Endangered Wildlife.* Norton, 1969. 226 pp. illus. $4.95. 68–22850. (SH–C)
Nearly all animals presently on the official list of rare and endangered species in the United States have their precarious positions explained in this book. It indicates what is presently being done to bring these animals into viable population and points out the most recent thinking on what must be done to prevent recurrence of these tragedies. Illustrated with excellent photographs. An appendix contains an annotated, systematically arranged list of rare and endangered American Wildlife, followed by a list of names and addresses of organizations concerned with their preservation.

McClung, Robert M. *Lost Wild America; The Story of Our Extinct and Vanishing Wildlife.* (Illus. by Bob Hines.) Morrow, 1969. 240 pp. $5.95. 69–13397. (JH–SH)
Vignettes of about 70 extinct or endangered animal species are narrated here with excellent chapters on current problems in environmental quality control. Good illustrations maintain an awareness of the animal under discussion. The book and its bibliography could serve many students as a stimulating start in a research project on wildlife management or concerning the influence of man on other species. There is a good index.

Murphy, Robert. *Wildlife Sanctuaries: Our National Wildlife Refuges—A Heritage Restored.* (Foreword by Stewart L. Udall.) Dutton, 1968. 288 pp. illus. 28x22.5 cm. $22.50. 67–20543. (SH–C)
A remarkably thorough and attractive story of the initial exploitation of America's wildlife and the extermination of certain species, which was followed by the biological surveys and inventories of the late 1800's and early 1900's. The result was initiation of large scale conservation programs. The illustrations are numerous and of excellent quality.

Stoutenburg, Adrien. *Animals at Bay: Rare and Rescued American Wildlife.* (Illus. by John Schoenherr.) Doubleday, 1968. 159 pp. illus. $3.50. 68–17781. (JH)
The story of man's efficient, greedy and sometimes totally irrational predation of wild animal life in North America is told in a simple, matter-of-fact manner. The emphasis is on the positive steps to be taken to rescue endangered species. The message is clear: the responsibility remains with man.

655 PRINTING AND PUBLISHING

University of Chicago Press. *A Manual of Style* (12th ed., revised). U. of
Chicago, 1969. ix+546 pp. $10.00. 6–40582. (SH–C–P)
Those engaged in professional writing, editors, printers, and publishers know
about this latest edition of the *University of Chicago Style Manual*, as it is
popularly known, for they have been awaiting it eagerly. We include it in this
book list because it is an indispensable guide and reference work that belongs
in all secondary school, public, college and university, and special libraries.
There are samples of styles of book type, a good bibliography, and many other
features that writers and editors need for constant reference.

660 INDUSTRIAL TECHNOLOGY

Alexander, Guy. *Silica and Me: The Career of an Industrial Chemist*. (Chemistry in Action Series.) Doubleday, 1967. 111 pp. illus. $3.95. 67–12889.
(SH–C)
An interesting account of a man's initial experiences as an industrial research
chemist. Designed to give the interested laymen a thorough introduction to
the many different facets of the chemical industry.

DeGarmo, E. Paul. *Materials and Processes in Manufacturing* (2nd ed.). Macmillan, 1962. viii+929 pp. illus. $11.95. 62–7279. (C)
The engineering student will find this work a good introduction to the
techniques of manufacturing. Covers materials, casting and forming, machin-
ing, welding and other processes and techniques. Many of the materials,
processes, and machines are evaluated and the importance of economy in
production is emphasized.

Feder, Raymond L. *Your Future in Chemical Engineering*. Rosen, 1961. 158
pp. $2.95; LB $2.79. 61–10807. (SH)
After a glimpse of the chemical industry and a discussion of the work of
chemical engineers, Feder looks into the qualifications and education necessary in the chemical engineering profession; also, he tells of the variety of
jobs available and how to apply for them. Appendices list professional societies, suggested reading and accredited schools of chemical engineering.

Hey, H. D. (ed.). *Kingzett's Chemical Encyclopaedia: A Digest of Chemistry &
Its Industrial Applications* (9th ed.). Van Nostrand, 1966. xi+1092 pp.
illus. $32.95. 67–2238. (SH–C)
This ninth edition reflects 15 years' progress since publication of the eighth
edition in 1952. It is not intended to serve the specialist in his own field
but will serve him, as well as many students, nontechnical individuals, industrialists and others seeking information on chemistry, process, products
and new materials. Belongs in all libraries: school, city, college and industrial.

Kent, James A. (ed.). *Riegel's Industrial Chemistry*. (rev. ed.). Reinhold, 1962.
xii+963 pp. illus. $21.50. 62–13257. (C)
Since the publication of the first edition in 1928 this textbook has been widely
used and is a principal reference in chemical technology. It deals first with
basic industrial problems of the chemical industry such as economics, water

supply, waste disposal, and fuels. Then follow chapters on products and groups of products, each written by individual specialists. References and index.

Pepler, Henry J. (ed.). *Microbial Technology*. Reinhold, 1967. x+454 pp. illus. $14.00. 67–26866. (SH–C)
Presents the current state of development of new microbiological ventures, updates the older processes, and includes operations and products by-passed in other technical surveys of industrial microbiology. Discusses developments, processes, and scientific innovations.

Roberson, Edwin C., and Roy Herbert. *Fuel, the Conquest of Man's Environment*. Harper & Row, 1963. 128 pp. illus. $2.50. 62–20918. (JH)
A popular general history and discussion of fuels and fuel technology which is interesting collateral reading for beginning students despite its British orientation.

Rusinoff, Samuel E. *Manufacturing Processes: Materials and Production* (3rd ed.). Amer. Tech. Soc., 1962. 753 pp. illus. $10.75. 61–18823. (SH–C)
Describes the machines and techniques used in the fabrication of metals and plastics. Although designed for the engineer, it is useful to college and high school students interested in manufacturing in general. Many machines and processes are illustrated.

Sax, Irving. *Dangerous Properties of Industrial Materials*. (3rd ed.). Reinhold, 1968. vii+1251 pp. $35.00. 68–54199. (SH–C)
This has become the standard handbook on the various hazards associated with the use of chemicals in industry. About three-quarers of the book consists of an alphabetic listing of substance, their hazard ratings, and recommended counter-measures. Cross-references are provided to synonyms for the compounds described. Should be available for convenient references in all schools and colleges where laboratory courses are taught, as well as in industrial establishments.

Shreve, R. Norris. *Chemical Process Industries* (3rd ed.). McGraw-Hill, 1967. 905 pp. illus. $18.50. 66–20721. (SH–C)
Discussions of specific industries or chemical processes occupy 38 of the 40 chapters. Starting with basics of water, fuels, coal chemicals and gases, the treatises develop such specialty industries as industrial carbon, ceramics, cements and related materials, glass, the several inorganic heavy chemical process industries, nuclear, explosive and propellent industry, photographic products, the coating industries, leather and related materials, agrichemical, fragrances and food additives, oils, fats, waxes, detergents, fermentation industry, wood, paper, plastics, films and fibers, rubber, petroleum, petrochemicals, dyes and intermediates, and pharmaceutical products. It is excellent reference for engineers, to the undergraduate student as a textbook or for collateral reading, and to the superior high school student for reference.

Taylor, F. Sherwood. *A History of Industrial Chemistry*. Abelard, 1957. xvi+467 pp. illus. $7.50. 57–8025. (SH)
A well-known science historian has prepared this popular account which begins with an introductory survey and then in the first part discusses the industries of metallurgy, cements and mortars, ceramics, glass, pigments, mineral

acids, combustibles, dyes, cleansing agents, pharmaceuticals, sugars and al-
cohols. The second part concerns the scientific chemical industries. Bibliography.

663.2 WINES AND WINE MAKING

Austin, Cederic. *The Science of Wine.* American Elsevier, 1968. 216 pp. $6.75.
68–26815. (C)
This is a textbook of wine making and of the manufacture of ethyl alcohol.
The reader should have a background of a basic general work in organic
chemistry and some knowledge of biochemistry. Sections cover the major
constituents of wine, the microbiological and chemical "disorders" of wine
and the chemistry of wine.

Waugh, Alec and the Editors of *TIME-LIFE* Books. *Wine and Spirits.* (Foods
of the World series.) Photos by Arie deZanger. Time, Inc., 1969. [c.1968].
208 pp. (Accompanying recipe book, 96 pp.) $6.95. 68–55300. (SH–C)
The discussions of wines and spiritous liquors exquisitely explain their history, geography, technology, and variety. There are interesting digressions
about food and apertifs, a guide to serving wines, a rating of the best
vintages of Bordeau and Burgundy, a glossary and index.

666 CERAMICS AND GLASS

Berlye, Milton K. *The Encyclopedia of Working with Glass.* Oceana Publications, 1968. xi+270 pp. illus. $12.50. 67–25903. (SH–C)
A comprehensive and informative reference work that covers, in the first
part, all of the detailed techniques and operations of working with glass. In
the second part are discussions of activities and products using glass. Industrial, architectural, and artistic uses are covered. Glossary, bibliography, and
index.

Diamond, Freda. *The Story of Glass.* Harcourt, 1953. ix+246 pp. illus. $4.50.
53–7864. (JH)
Glass was used by the ancients and can tell us much about their way of
life. Man has since developed countless uses for many forms of glass. The
author describes these events and show man's dependence on glass.

Hammesfahr, James E., and Clair L. Stong. *Creative Glass Blowing.* Freeman,
1968. 196 pp. illus. 9 col. pls. $8.00. 68–14225. (SH–C)
Glass blowing now seldom occurs in laboratories in the United States because work formerly done by indivdual craftsman is done by machines. However, there is occasional need in a laboratory to assemble a special apparatus
or experimental set-up, or to make repairs. Students in chemistry, physics,
or other disciplines should know about this handbook, which is historical
in its introduction but practical in its well-illustrated discussions of equipment, techniques and methods. For hobbyists there is a section on ornamental glassware.

Phillips, Charles J. *Glass: Its Industrial Applications.* Reinhold, 1960. iv+252
pp. illus. $8.50. 60–11580. (C)
The first six chapters cover the history of glass, its production and its phys-

ical and chemical properties. The following nine chapters describe its use in packaging, electronics, scientific apparatus, architecture, fiber-glass and other areas. A glossary of glass terms is included.

Shand, E. B. *Glass Engineering Handbook* (2nd ed.). McGraw-Hill, 1958. x+484 pp. illus. $15.00. 58–10006. (C)
The four sections: glass technology, glass manufacture, applications, and fibrous glass cover many concepts in glass important to industry. Contains many charts and tables. It is more technical and is designed more for the students, technicians and engineers than Phillips (above) which is primarily descriptive.

Mitchell, Lane. *Ceramics: Stone Age to Space Age* (Vistas of Science Series). McGraw-Hill, 1963. 128 pp. illus. $2.50; LB $2.63. (Scholastic, paper 50¢). 63–14661. (JH)
A very simple introduction to ceramics, their composition, fabrication, and uses. Contains suggestions for projects and experiments.

668.4 PLASTICS

Arnold, Lionel K. *Introduction to Plastics.* Iowa State Univ., 1968. 205 pp. illus. $6.95. 68–12020. (C)
Prepared for use in a one-quarter course in chemical engineering, this will also be useful for reference and valuable to those wishing a general view of the field. Chapters are devoted to the production of plastic materials and finished products as well as to applications.

Cook, J. Gordon. *The Miracle of Plastics* (Science for Everyman Series). Dial, 1964. 272 pp. illus. $4.95. 62–12317. (JH–SH)
The history, properties, methods of production, and applications of the more important plastics. Illustrations, both drawings and photographs, are numerous and well-chosen. No prior knowledge of chemistry is required since much detailed, technical information not fully explained in the text is contained in a 20-page reference section. Useful glossary and a trade name index of the best-known plastics. Useful for reference and collateral reading.

Griff, Allan L. *Plastics Extrusion Technology.* (2nd ed.). Reinhold, 1967. ix+ 278 pp. illus. $7.95. 62–17523. (C)
A discussion of extrusion methods of forming useful plastic objects. Although designed for the engineer, students interested in plastics or manufacturing techniques will find the material informative. Diagram of many machines and machine parts are included.

Kaufman, Morris. *Giant Molecules: The Technology of Plastics, Fibers, and Rubber.* (Doubleday Science Series). Doubleday, 1968. 192 pp. illus. $5.95; paper $2.45. (SH–C)
This well-written study of polymer chemistry begins with a brief but interesting history of the subject then examines what a polymer is, the chemistry involved, and the technology of producing useful articles from raw materials. The stereo chemistry of polymers is also examined. An eight page summary table itemizes many of the topics central to the discussion of polymers.

Lappin, Alvin R. *Plastics: Projects and Techniques.* McKnight & McKnight, 1965. 136 pp. illus. $4.00. 64–22304. (JH–SH)

Descriptions of 48 projects that can be made with plastics, each accompanied by a photograph, dimensional sketch, and list of materials. The second section is devoted to the equipment and techniques to be used in making the various articles, followed by a bibliography and index. The book is intended for the amateur hobbyist or for the student in industrial arts classes. It contains no information on the chemistry of plastics nor on commercial or industrial uses and techniques.

Melville, Harry. *Big Molecules.* Macmillan, 1958. 180 pp. illus. $4.95. 58–12856. (SH)
Tells how long chains of atoms are made and used in plastics and other products. Describes the chemical make-up of large molecules and the significance of the arrangement of atoms in them. Some background in chemistry is necessary.

Newman, Thelma R. *Plastics as an Art Form.* Chilton, 1964. xxi+338 pp. illus. $12.50. 68–18240. (SH–C)
Written primarily for artists and sculptors who are using plastics, this unusual and unique book is of value to scientists and science teachers because of its excellent descriptions of materials and techniques that will be useful in the preservation of specimens, the preparation of exhibits and models, etc. Step-by-step photographs outline procedures and the various plastics are described. Trade names, manufacturers, and sources of supply are listed. Glossary and index.

Oleesky, Samuel S., and J. Gilbert Mohr. *Handbook of Reinforced Plastics of the Society of the Plastics Industry, Inc.* Reinhold, 1964. viii+640 pp. illus. $25.00. 64–15205. (C)
Reinforced plastics represent the fastest growing and forward-looking segment of the plastics industry. This valuable handbook, although intended for engineers has much value as a special reference work for the inquisitive college student and layman with an appropriate background in mathematics and the physical sciences. It describes reinforced plastics, resins, catalysts, promoters, reinforcing materials, and fillers. Also covered in detail are molding methods, molds, tooling, finishing methods, design, test methods, safety and hygiene.

Simonds, Herbert R., and James M. Church. *A Concise Guide to Plastics* (2nd ed.). Reinhold, 1963. xii+392 pp. illus. $12.00. 63–13447. (C)
Discusses the properties, forms, manufacture, processing, applications and future of plastics. One chapter contains statements of 52 plastics manufacturers describing the history, products, production and profits of their companies. Concludes with an index of trade names which lists chemical components and manufacturers.

669 METALLURGY

Biringuccio, Vannoccio. *Pirotechnia.* (Tr., with introd. and notes by Cyril S. Smith and Martha T. Gnudi.) M.I.T. Press, 1966. xxvi+477 pp. illus. paper $3.45. 59–6712. (SH)
Biringuccio's book, originally published in 1540, was the first printed book to deal in breath and depth with all aspects of metallurgy. It was unusual

practice at that time for any craftsman to reveal all of the details of his trade or profession. Of great historical value because there was little progress in metallurgy from the time of Biringuccio until the end of the Eighteenth Century. An excellent resource for anyone interested in the history of science and technology.

Burt, Olive W. *The First Book of Copper.* Watts, 1968. 86 pp. illus. $2.65. 68–10469. (JH)
An outstanding introduction to the folklore, history, mining, ore dressing, smelting, fabrication, and uses of copper. Representative Fifteenth Century woodcuts and outstanding photography supplement the text. A valuable and interesting book for general reading and collateral reference.

Clark, Donald S., and Wilbur R. Varney. *Physical Metallurgy for Engineers.* (2nd ed.). Van Nostrand, 1962. xv+629 pp. illus. $9.50. 62–825. (C)
A textbook directed to the basic training of students in physical metallurgy. It concentrates on the selection, treatment, fabrication and use of metals and alloys whereas the book by Robert E. Reed-Hill (cited below) is concerned with laying a scientific and theoretical foundation. Libraries, therefore, need both. After presenting background material, the book covers all of the various aspects of metallurgy, casting, fabrication, soldering, brazing and welding. A special chapter is devoted to metallurgy in nuclear engineering. The appendices contain important reference tables and graphs.

Fisher, Douglas Alan. *Steel: From the Iron Age to the Space Age.* Harper & Row, 1967. viii+200 pp. illus. $4.95; LB $4.79. AC 66–10122. (JH–SH)
A rare contribution to young people's educational literature. Men, machines, economics, plus fortunate accidents serve to fashion a fascinating story about steel and its importance.

Frier, W. T. *Elementary Metallurgy* (2nd ed.). McGraw-Hill, 1952. x+258 pp. illus. $5.50. 51–12607. (SH)
A good introductory text on the subject of metallurgy understandable to the high school student. Deals primarily with the properties and methods of producing iron and iron alloys but there is also one chapter on nonferrous alloys and another on testing metals. Has an appendix of metal data tables.

Parr, J. Gordon. *Man, Metals and Modern Magic.* Iowa State U., 1958. 237 pp. illus. $4.95. 58–8631. (JH)
A narrative account of the ancient and modern history of metallurgy that should be read by students as a preface to Sullivan's *The Story of Metals* and Rogers' *The Nature of Metals.*

Reed-Hill, Robert E. *Physical Metallurgy Principles.* Van Nostrand, 1964. x+630 pp. illus. $11.95. 64–915. (C)
An introductory college textbook that explains basic metallurgical phenomena and the properties of metals suitable for the student with a basic foundation in physics, chemistry and mathematics. The approach is basically theoretical. Recommended for reference collections of college libraries and technology collections in public libraries.

Reinfeld, Fred. *Uranium and Other Miracle Metals* (2nd ed.). Sterling, 1959. 128 pp. illus. $3.95. 59–10862. (SH)
Since the advent of the atomic age, uranium has become the world's most

precious metal. Many persons want to know the answers to fundamental questions found in this book concerning the nature, appearance, natural occurence, mining, processing, industrial importance and uses of uranium and transuranium metals. The condensed account also reviews the duties and powers of the U. S. Atomic Energy Commission.

Rogers, Bruce A. *The Nature of Metals* (2nd ed.). Iowa State U., 1964. x+ 324 pp. illus. $5.50; Amer. Soc. for Metals $5.00; M.I.T. paper $2.95. 64–13375. (JH)
An elementary introduction to metallurgy which begins with a consideration of the characteristics of common and uncommon metals, the periodic table, and how to identify metals. Alloys and crystallography are explained, as well as atomic movements and their effects in solid metals, and other facts and processes inherent in the science and technology of metals.

Sullivan, John W. W. *The Story of Metals.* Iowa State U., 1951. x+290 pp. illus. $5.50. Amer. Soc. for Metals. $5.00 51–2333. (JH)
An historical account of the origin of metals in an explanation of the evolution of the earth, the formation of mineral deposits and the composition of the earth's crust. A chronological story follows of man without metals, of the bronze age, the early iron age, and of the development of mining and metallurgy. Finally there are discussions of modern iron and steel, copper, zinc, lead, tin, aluminum, magnesium, nickel, chromium, tungsten and titanium. The conclusion deals with the future of metals. Glossary.

Watson, Aldren A. *The Village Blacksmith.* Crowell, 1968. 125 pp. illus. $6.95. 68–21615. (JH–SH)
The history of American metallurgy begins with colonial ironworks and the village blacksmith who, in addition to shoeing horses, hand-crafted all of the tools, building hardware, household utensils, agricultural implements, wagons, carriages and sleds that were involved in the self-contained local economies. Aldren, who is his own illustrator, has provided a good record of the life and times of the village blacksmith which will be read by many with nostalgia, but should be read by young people as part of their general education.

676 PULP AND PAPER INDUSTRY

Britt, Kenneth W. (ed.). *Handbook of Pulp and Paper Technology.* Reinhold, 1964. vii+537 pp. illus. $25.00. 64–15592. (C)
A concise and readable presentation of all of the essential facts concerning the technology of pulp and paper. Although intended primarily as a reference manual for paper and pulp manufacturers, the presentation is intelligible to any reader who has had a basic chemistry course. The historical information, process diagrams, suggestions on forest management, and other topics make the volume valuable as a general reference in libraries.

677 TEXTILES

Potter, Maurice David, and Bernard P. Corbman. *Fiber to Fabric* (3rd ed.). McGraw-Hill, 1959. x+342 pp. illus. $5.48. 58–13887. (SH)
An introductory guide to textiles for the student and for the layman without technical training. In separate chapters the discussion deals with textile

fibers, testing fibers, spinning, weaving, finishing, dyeing, decoration, cotton, linen, wool, silk, cellulosic fibers, nylon and the other synthetic fibers, fiberglass, knitting, minor textiles, floor coverings, and care of fabrics. A good book for informative general reading and for reference.

678.2 RUBBER

Le Bras, Jean. *Introduction to Rubber.* (Tr. by J. H. Goundry.) Hart, 1969 [c.1968]. 112 pp. illus. $9.75.
An informative, interesting account of the development and industrial use of natural and synthetic rubber. Describes the discovery and industrial development rubber, the development of plantation rubber and limited exploitation of plants other than *Hevea,* the composition and chemical and physical properties of rubber, synthetic rubbers, rubber technology, and rubber industry.

681 PRECISION MECHANISMS AND RELATED MACHINES

Bruton, Eric. *Clocks and Watches 1400–1900.* Praeger, 1967. 209 pp. illus. $10.00 67–14706. (SH–C)
An interesting historical and descriptive account of clocks and watches, aided by many photographs and drawings. The characteristics and development of various movements and cases will interest the student of chronometry, and the descriptive detail will enable the collector to date and authenticate specific timepieces. The appendices are important reference sources: (1) "Places where ancient vertical clocks made in England may be seen." (2) "Where to see collections of clocks and watches" [in England]. (3) "Some eminent and better known makers of clocks and watches." (4) "A chronology of horological inventions."

Greve, John W., and Frank W. Wilson. *Handbook of Industrial Metrology.* Prentice-Hall, 1967. xii+492 pp. illus. $16.00. 67–12084. (SH–C)
Primarily a reference book on principles, techniques, and instrumentation design and application for physical measurements in the manufacturing industries. It is intended as a textbook to prepare young men to qualify as registered professional engineers. An excellent handbook for practicing members of the American Society of Tool and Manufacturing Engineers and for those who must understand the nature of the work done by these professional engineers.

Kirk, Franklyn W., and Nichols R. Rimboi. *Instrumentation.* (2nd ed.). Amer. Tech. Soc., 1966. 263 pp. illus. $6.95. 62–11498. (SH)
A careful study of many common types of measuring instruments used in homes and industrial plants. Clear, complete and concise explanations of their operation are accompanied by easily understandable diagrams. A very good introductory book because the authors stress the basic operating principles and construction of the instruments.

Michel, Henri. *Scientific Instruments in Art and History.* (Tr. by R. E. W. Maddison and Francis R. Maddison.) Viking, 1967. 208 pp. illus. 29 cm. $18.50.
A historical anthology of illustrations (mostly in color) of some of the finest

and artistically elaborate early scientific instruments in existence. Five major sections: (1) basic measuring, drawing and calculating instruments; (2) topographical and navigational instruments; (3) astronomical instruments; (4) instruments of gnomonics and chronometry; and (5) instruments of the physical sciences. A bibliography, a list of important collections, a subject index, and an index of makers, designers and collectors. An interesting, although specialized, book for libraries that can afford it. It will appeal to readers interested in the history of science and technology.

711   URBAN PLANNING

Bacon, Edmund N. *Design of Cities*. Viking, 1967. 296 pp. illus. $15.00. 67–23826. (C)
An excellent treatment of the development of the design of cities in a sequential manner. Emphasis is on design philosophy as it changes in relation to changes in the social structure.

Gallion, Arthur B., and Simon Eisner. *The Urban Pattern* (2nd ed.). Van Nostrand, 1963. x+435 pp. illus. $10.50. 63–24088. (C)
An examination of the processes by which cities are planned and built. Zoning, open space, off-street parking, urban renewal, new towns, planning as a government function, and measures for control of density and obsolescence are of the topics discussed for the challenged citizenry of an expanding America.

Gibberd, Frederick. *Town Design* (5th ed.). Praeger, 1967. 372 pp. illus. 28.5 cm. $22.50. 67–13491. (C)
Considers all aspects of planning for the urban sciences. The informative and imaginative, well-illustrated book is essential for all architectural libraries, college and large public libraries.

Halprin, Lawrence. *Freeways*. Reinhold, 1966. 160 pp. illus. $12.50. 66–22687. (JH–SH–C)
This interesting pictorial summary of urban transportation problems, culminating in the construction of modern freeways, reveals that highways and freeways are not only civil engineering projects, but must be planned, executed, and maintained in conjunction with economic, cultural and demographic complexes in the urban areas. The approach is informative and cultural, not engineering technology.

Johnson, James H. *Urban Geography: An Introductory Analysis*. Pergamon, 1967. xv+188 pp. illus. $5.50; paper $4.00. 67–21274. (SH–C)
The geographer has emerged in recent years as an important member of urban study groups confronting complex problems. His work, like that of the social scientist, the economist, the environmental and systems engineer, the architect and planner, has new relevance for students at every level. The author has developed a clear, concise, and effectively illustrated text, which will be of great usefulness to college students, especially those in elementary professional urban geography courses.

Johnson-Marshall, Percy. *Rebuilding Cities*. Aldine, 1966. vi+390 pp. illus. $15.00. 64–21388. (SH–C)
The introduction and opening chapter on the historical background of city

growth provides a good beginning, followed by an analysis of the component elements in urban planning and design. The chapter on "Visions and designs" demonstrates man's imaginative spirit in city design from the Renaissance to our own *avant garde*. The second half deals more specifically with planning tools and legislation and employs as examples, London, Coventry, and Rotterdam. The concluding chapter, on the city of the Twenty-first Century, is more universal, and is well worth reading as a separate piece. Portions of the book would be a useful reference material in senior high school. Several instructional areas in the universities may welcome it as a text resource.

Lewis, David (ed.). *The Pedestrian in the City.* Van Nostrand, 1966. 299 pp. illus. $18.50. 66–3836. (SH–C)
More than 50 artists, architects, designers, photographers, and writers have contributed to this fascinating pictorial work that demonstrates the urgent need in many parts of the world for solving complex and multi-faceted problems involving pedestrians and transportation. Contents deal with (1) problems of separating pedestrians and traffic as illustrated in five major cities, (2) discussions of theoretical aspects, and (3) ethnological aspects of pedestrian and transportation problems.

Lynch, Kevin. *The Image of the City.* Harvard, 1960. vii+194 pp. illus. $8.50. M.I.T. Press, paper $2.95. 60–7362. (SH)
This book about the appearance of cities and of the elements of good urban planning emphasizes the importance of site development and utilization in all architectural planning. The book is a product of the Joint Center for Urban Studies of M.I.T. and Harvard. It will interest students of engineering, architecture, sociology, economics, and ordinary people who read for general enlightenment.

Lynch, Kevin. *Site Planning.* M.I.T. Press, 1962. 248 pp. illus. $10.00. 62–13231. (C)
An introduction to the art of site planning, or the arrangement of groups of structures on the ground. The analysis of site and purpose, land use and circulation design, visual form, climate, site engineering, landscaping, housing, shopping centers, industrial estates, light, noise and even the air are some of the many aspects to be considered by the planner and worked into harmonious unity.

Mayer, Albert. *The Urgent Future: People—Housing—City—Region.* McGraw-Hill, 1967. xiii+184 pp. illus. 28 cm. $16.50. 66–26580. (C)
Deals with the importance of accepting a moral and emotional commitment to create a more worthy, multipurpose environment for all people. An incentive to discussion and more thoughtful consideration of planning problems.

Meyerson, Martin. *Face of the Metropolis.* Random, 1963. 249 pp. illus. $7.50; paper $2.95. 61–13841. (SH)
Analyzes 70 examples of urban design or development in the hope of stimulating a wider and more sophisticated public interest. This book was sponsored by Action, Inc., the National Council for Good Cities, whose purpose is to help individuals and groups make their cities better places in which to live and to earn a livelihood.

Mumford, Lewis. *The City in History.* Harcourt, 1961. xi+657 pp. illus. $11.50; $8.50 text ed. 61–7689. (C)

"This book opens with a city that was, symbolically, a world: It closes with a world that has become, in many practical aspects, a city. In following through this development I have attempted to deal with the forms and functions of the city, and with the purposes that have emerged from it; and I have demonstrated, I trust, that the city will have an even more significant part to play in the future than it has played in the past, if once the original disabilities that have accompanied it through history are sloughed off."— Preface.

720 ARCHITECTURE

Blake, Peter. *The Master Builders*. Knopf, 1960. xxix+400 pp. illus. $8.00. 60–10276. (SH)
Biographies of three masters in architecture, Le Corbusier, van der Rohe, and Wright. A unifying theme runs through the three sections and uses the works of these men as milestones in its development.

Coles, William A., and Henry H. Reed, Jr. (eds.). *Architecture in America: A Battle of Styles*. Appleton, 1961. xviii+412 pp. illus. paper $3.25. 61–5653. (C)
"This book presupposes no prior knowledge of architectural history or criticism on the part of the student. It is a self-contained introduction to architectural thought. For this reason it opens with an extensive anthology of major statements, ranging from antiquity to the present, on the nature of architecture, its aims, and the problems of architectural theory and practice . . ."—Preface. Among those represented by statements are Vitruvius, Palladio, Wren, Veblen, Sullivan, Wright, LeCorbusier, Gropius, Saarinen and Zevi.

McLaughlin, Robert W. *Architect: Creating Man's Environment*. Macmillan, 1962. xvi+201 pp. illus. $3.50. 61–15183. (SH)
The director of the School of Architecture at Princeton University describes the role of the architect, the practice of architecture, and the scientific and artistic aspects of the profession. The student oriented toward an architectural career will benefit from his discussion of professional schools, the establishment of a practice, the work of successful architects, and his hypothetical "Week in the life of an architect."

Rasmussen, Steen Eiler. *Experiencing Architecture*. M.I.T. Press, 1962. 245 pp. illus. $10.00; paper $2.95. 62–21637. (SH)
"It is not enough to see architecture; you must experience it. You must observe how it was designed for a special purpose and how it was attuned to the entire concept and rhythm of a specific era." Professor Rasmussen develops this theme for that sympathetic and knowledgeable group of amateurs which he believes is the basis of competent professionalism.

Rogers, W. G. *What's Up in Architecture*. Harcourt, Brace & World, 1965. 192 pp. illus. $3.95. 65–21702. (JH–SH)
Concerns the basic concepts and problems facing the modern architect. The author begins with a descriptive "walk" that makes the reader aware of the

past and then of the present scene. He attempts to show how new materials and methods of construction have resulted in our contemporary buildings. Then follows a fairly good account of the lives of five architects involved in the modern architecture movement: Henry Hobson Richardson, Louis H. Sullivan, Frank Lloyd Wright, Le Corbusier, and Ludwig Mies van der Rohe. The latter part of the book is concerned with new directions and urban planning. Bibliography.

Roth, Richard. *Your Future in Architecture*. Rosen, 1960. 159 pp. $2.95; LB $2.79. 60–11116. (JH)
Recommended for the younger interested student considering an architectural career this straightforward survey of the profession also has valuable appendices: a self-evaluation test; list of schools of architecture; addresses of state architectural registration boards; AIA chapter and state organizations; and a suggested reading list.

720.9 HISTORY OF ARCHITECTURE

De Cenival, Jean-Louis. *Living Architecture: Egyptian*. Grosset & Dunlap, 1964. 192 pp. illus. $7.95. (SH–C)
Relates the architecture of an ancient culture to contemporary 20th century thought. One section deals with the technical problems encountered and how the ancient Egyptian architects solved them. Accompanying drawings and diagrams are simple, direct and effective, and the photographs are magnificent.

Giedion, Sigfried. *Space, Time and Architecture* (4th ed.). Harvard, 1962. xiviii+778 pp. illus. $12.50. 62–14459. (SH)
A fundamental book of interest to everyone (since buildings are a part of human life) portraying the growth of architecture and its traditions, and describing the interrelationships with other human activities.

Jacobus, John M. *Twentieth-Century Architecture: The Middle Years 1940–65*. Praeger, 1966. 215 pp. illus. $18.50. 66–12526. (C)
An explanation of the trends of contemporary architecture in a historical perspective. Illustrated with excellent photographs and helpful sketches. Chapters are: (1) "A genealogy for contemporary architecture". (2) "War and postwar developments: 1940–50". (3) "Wright, Gropius, Mies van der Rohe, and Le Corbusier: their 'late' styles". (4) "The early 1950's; variations on familiar themes". (5) "Crisis and reorientation: theatres, churches, concert halls, and arenas in the middle and later 1950's". (6) "New patterns of order: the architecture of the 1960's." A textbook for college students or a reference work for others interested in architecture.

Leacroft, Helen, and Richard Leacroft. *The Buildings of Ancient Greece*. Scott, 1966. 40 pp. illus. $3.50; LB $2.63. 66–12056. (JH–SH)
The primary value of this short simple introduction to Greek architecture is in the accurate and excellent illustrations (many in color). These are clearly labelled, well executed (often showing plan and elevation in a single drawing), and graphically illustrate the complex and sometimes confusing archi-

tectural terminology. The standard differences between Doric, Ionic, and Corinthian styles are illustrated. Short, well-illustrated sections are devoted to various types and classes of buildings. There is a short, interesting section on town building. Good collateral material (illustrating types of buildings and towns in the Classical Greek Period) in either an "Introductory classics" or "History of western civilization" course.

Burchard, John, and Albert Bush-Brown. *The Architecture of America: A Social and Cultural History*. Little, 1961. x+595 pp. illus. $15.00. Abridged ed. paper $3.95. 61–5736. (C)

An exceptionally comprehensive and well-documented history of American architectural advancement from 1600 to 1960. The work was sponsored by the American Institute of Architects to celebrate its centennial in 1957.

Fletcher, Banister. *A History of Architecture on the Comparative Method* (17th ed.). Scribner, 1961. 1033 pp. illus. $17.95. 2–17274. (C)

Displays the characteristic features of architecture of each country by comparing the buildings of various periods and by considering the influences of geography, geology, climate, religion, history, and sociology which have contributed to the evolution of particular styles. Special drawings serve as keys of characteristic architectural features, dimensions, ornamentation, etc. Not only the architect, but also students of history and classical architecture, and/or interested layman will find this book a veritable treasure. For all college and public libraries.

Hamlin, Talbot. *Architecture through the Ages*. Putnam, 1953. li+684 pp. illus. $9.50. 53–11395. (SH)

The purpose of this one-volume work (which Lewis Mumford called "the most adequate summary of architectural history with which I am acquainted") is to show that buildings are not isolated objects existing arbitrarily in one or another of the recognizable set "styles," but rather as the results of the ways of living, governing, worshiping, and doing business at the time they were built.

Hamlin, Talbot. *Benjamin Henry Latrobe*. Oxford, 1955. xxxvi+633 pp. illus. $15.00. 55–8117. (C)

A definitive biography of America's first professional architect. He was in charge of design and construction of the U. S. Capitol from 1808 to 1817, worked on the White House, built the old Baltimore Cathedral and St. John's Church in Washington. He was originator of the Greek Revival style in the United States, and pioneered in the introduction of steamboats on the Ohio and Mississippi Rivers.

Joedicke, Jürgen. *A History of Modern Architecture*. Praeger, 1959. 243 pp. illus. $12.00; text ed. $9.00. 59–7459. (SH)

A beautifully illustrated survey of Twentieth Century architecture. Presents lucidly the history of modern architecture, the creativity displayed by the pioneers in the field, and an analysis of Western architecture today.

McCullum, Ian. *Architecture USA*. Reinhold, 1959. 216 pp. illus. $7.95. 59–16224. (SH–C)

A cross-section of the best of American architecture as selected by an

Associate of the Royal Institute of British Architects and presented through brief biographies of 32 architects with photographs showing their work.

Morrison, Hugh. *Early American Architecture: From the First Colonial Settlements to the National Period.* Oxford, 1952. xiv+619 pp. illus. $17.00. 52-7831. (SH)
An historically important and interesting account of architecture in the Colonies, beginning in St. Augustine in 1565 and ending in San Francisco in 1848. Woven into the narrative are details of Old World background and origins, structural and decorative details, and the methods of early builders and craftsmen. Bibliography.

Jacobs, Herbert. *Frank Lloyd Wright: America's Greatest Architect.* Harcourt, Brace & World, 1965. 223 pp. illus. $3.95. 65-25306. (JH–SH)
The author, on the basis of his personal acquaintance with Frank Lloyd Wright for a quarter of a century, has written an exceptionally readable account of the architect's life. It is a warm and human account covering his tribulations, triumphs, and glories. The first good account for younger readers. The photographs are well-selected; some have never been published previously.

Naden, Corrine J. *Frank Lloyd Wright: The Rebel Architect.* (Immortals of Engineering.) Watts, 1968. 147 pp. illus. LB $2.95. 68-10349. (JH–SH)
Attempts to weave the story of Wright the man—a stormy, rebellious, opinionated, arrogant, and tragic figure—with Wright the architect. His architectural creations and innovations dominate the story. Many of his best known buildings are illustrated and described in detail.

Rosen, Sidney. *Wizard of the Dome: R. Buckminster Fuller, Designer for the Future.* Little, Brown, 1969. x+169 pp. illus. $4.95. 74-77452. (SH–C)
A light and airy dome to cover an entire city, a dome two miles wide and a mile high, under which people could in fact live with no threats of air pollution, no individual air conditioners, no snow removal, no umbrellas, no colds —possibly an idyllic kind of living existing only in the mind of some far out dreamer. Not so. The principle of the "geodesic" dome is really a summary of Fuller's ideas of construction "to do the most with the least". This geometric process, now universally accepted and acclaimed, was not patented until 1954 when Fuller was 60; this after most of his earlier ideas and inventions including the Dymaxion house and car, and certain energy conservation and prefabrication systems were so far in advance of their time that they never did bring him the acclaim he deserved. Now his domes of various sizes and uses are built in many parts of the civilized world.

Wright, Olgivanna Lloyd. *Frank Lloyd Wright: His Life, His Work, His Words.* Horizon, 1966. 224 pp. illus. $7.50. 66-26703. (SH)
Frank Lloyd Wright's wife is the author of this biography that deals with his personal life and accomplishments, and includes quotations from his writings that tell some aspects of the background, the philosophical spirit of his design, and the practical facets of construction. The accompanying 115 photographs and drawings enrich the text. A summary of Wright's architectural innovations and a chronological listing of his buildings and projects are worthwhile appendices.

721 ARCHITECTURAL CONSTRUCTION

Condit, Carl W. *American Building Art: The Twentieth Century.* Oxford, 1961.
xviii+ 427 pp. illus. $15.00. 61–8369. (SH)
A discussion of many of the great architectural and engineering projects and
structures that reveal the ingenuity, combinations of skills, and teamwork
involved. Particularly important are the building demands resulting from
the railroads, the automobile, the airplane, the need for electric power, water
supplies, irrigation, etc. Provides good collateral reading for studies of
American technology.

Faber, Colin. *Candela: The Shell Builder.* Reinhold, 1963. 240 pp. illus. $17.50.
62–14294. (C)
Although Candela did not invent the concrete shell nor did he pioneer in the
use of the hyperbolic paraboloid, he has exploited and developed that shape
with unusual skill and versatility. This important book not only describes
practical applications, but also explains the basic mathematics and construction details.

Marks, Robert W. *The Dymaxion World of Buckminster Fuller.* S. Ill. Univ.,
1966. 232 pp. illus. $10.00. 60–5487. (C)
Fuller has doggedly attacked the major problems of shelter and transportation with an uncompromising but successfully demonstrated idea: to get from
any type of structure the maximum net performance per unit of energy input.
This is the "Dymaxion" idea—a principle that underlies all of his inventions, developments, and discoveries. Most of this book is devoted to illustrations that are especially stimulating.

Nervi, Pier Luigi. *Structures* (Tr. by Guiseppina and Mario Salvadore). Mc-
Graw, 1956. ix+118 pp. illus. $9.50. 56–10881. (SH)
Nervi began designing unique reinforced concrete structures over 40 years
ago, combining his excellent mathematical knowledge with sound intuition and
great creative skill. In this day of cooperative or group enterprise in the
construction of great buildings, Nervi represents a master craftsman who
conceived, designed, engineered, and supervised the construction of his own
creations.

Parker, Harry, Charles Merrick Gay, and John W. MacGuire. *Materials and
Methods of Architectural Construction* (3rd ed.). Wiley, 1958. xi+724 pp.
illus. $11.95. 58–8213. (C)
A standard manual of information concerning the common materials used
in construction, of methods of combining them in building construction, and
of modern design formulas and working unit stresses. It is particularly useful
for the beginning architecture student, or for reference by the layman who
may wish to become familiar with some phases of architectural practice.

Saarinen, Aline B. (ed.). *Eero Saarinen on His Work.* Yale, 1962. 108 pp.
illus. $17.50. 62–16241. (SH)
A selection of buildings dating from 1947 with statements by the architect.
This oversize superbly illustrated book will be treasured by those who admire
the work of Eero Saarinen who has conceived some of the most notable
architectural forms—including American Embassies in Oslo and London, air-
line terminals in New York and Washington, the Jefferson National Expan-

sion Memorial Arch in St. Louis, as well as beautiful and comfortable chairs for offices and homes.

Salvadori, Mario, and Robert Heller. *Structure in Architecture*. Prentice-Hall, 1963. 370 pp. illus. text ed. $12.50. 63–7992. (C)
"In this thoughtfully written book, Professor Salvadori endeavors to eliminate one of the most serious gaps between theory and practice in the field of structures. His aim is to build a bridge between the more or less conscious intuition about structure, which is common to all mankind, and the systematic knowledge of structure, which gives a fair representation of physical reality on the basis of mathematical postulates."—From the Foreword by Pier Luigi Nervi.

Wright, Frank Lloyd. *The Natural House*. Horizon, 1954. 223 pp. illus. $7.50. NAL (MQ 753) paper 95¢. 54–12278. (SH)
This book is convincing evidence that Wright did not build only for the rich. Herein he tells about and illustrates his "Usonian" house which are moderate-cost dwellings of infinite variety. Included are actual photographs, floor plans, and sketches.

Zuk, William. *Concepts of Structure*. Reinhold, 1963. 80 pp. illus. $5.95. 63–11425. (SH)
This is not a "how-to-do-it" book, yet its purpose is to explain the basic principles of structural design, and the functional analysis of structural elements and components. Particularly valuable to the student is the translation of geometric and physical principles into functional details of architectural design.

770 PHOTOGRAPHY

Baines, Harry. *The Science of Photography*. Wiley, 1967. 322 pp. illus. $8.25. 58–4658. (SH)
This history of photography describes the basic physical and chemical principles and processes involved, and presents a review of physical optics, and other fundamentals. This is a scientific discussion for professional and amateur photographers in as simple language as the nature of the subject permits.

Boucher, Paul E. *Fundamentals of Photography*. (4th ed.). Van Nostrand, 1963. xiv+535 pp. illus. $10.00; text ed. $7.50. 63–24071. (C)
A well-balanced and comprehensive guide recommended by physicists and outstanding professional experts. Its central theme is to provide basic physical and photo-chemical theory, and through laboratory exercises, practical experience in the use of photographic apparatus and darkroom procedures. Science majors, particularly those who intend to do research or teach, need the fundamental training this book provides. It covers all aspects of photography and the appendices include tables, formulary and glossary.

Ceram, C. W. (Pseud. of Kurt W. Marek.) *Archaeology of the Cinema*. Harcourt, Brace & World, 1965. 264 pp. illus. $6.50. 65–19106. (JH–SH–C)
This is a book about the prehistory and early history of motion pictures, starting in 1832 with the invention of the first devices giving the illusion of movement to 1897 when the film industry could be said to have been started.

The illustrations consist of a wide selection of old prints, photographs of equipment, scenes from films, and pictures of individuals associated with the development of motion pictures.

Feininger, Andreas. *The Complete Photographer.* Prentice-Hall, 1965. viii+ 344 pp. illus. $8.95. 65–12017. (C)

For the would-be professional. The author's approach to technical details is that of the artist rather than that of the scientist. There are 32 pages of plates, half in color. A praiseworthy feature is the numerous callouts in the margin referring the reader to related material on other pages—a sort of concordance. Discusses color photography as well as monochrome.

Glyck, Zvonko. *Photographic Vision.* (Lester Kaplan, Editor-in-chief, Amphoto.) Chilton and Amphoto, 1965. 165 pp. illus. 31 cms. 163 b. and w., 88 col. illus. $12.50. 65–23147. (SH–C)

By the presentation of one of the most varied, excellent, and unusual collections of photographs, the author seeks to impart to the photographer, whether he be amateur or professional, a sense of creativity and of artistry. The chapters have the quality of literary essays, yet each one in concert with the accompanying illustrations of experimental photographs, designs, and abstraction, is a learning adventure. It was designed to instill in every photographer a sense of artistry by indicating how he can arrange and control every element of his picture when he has learned to visualize, in advance, the result he desires.

Newhall, Beaumont. *Latent Image: The Discovery of Photography.* (Science Study Series.) Doubleday, 1967. xii+148 pp. illus. LB $4.95; paper $1.25. 67–122460. (JH–SH)

The fascinating story of the early beginnings of photography which discusses the work of the early pioneers of the photographic process. Designed as collateral reading for science students.

Rhode, Robert B., and Floyd H. McCall. *Introduction to Photography.* Macmillan, 1965. 278 pp. illus. $5.95. 65–15572. (SH–C)

A good background book for students and practitioners of pictorial photography who wish to go beyond the standard instructions that accompany photographic equipment at the time of its purchase. Discusses both the technical and the artistic aspects of photography. While the ideas are presented lucidly, only the older high school student or the college freshman—or an adult with equivalent educational background—are likely to have the patience to study a book so packed with worthwhile information.

Skoglund, Gosta C. *Colour in Your Camera.* Chilton, 1969. 164 pp. illus. $8.95. 69–16643. SBN 240–44743–3. (JH–SH–C)

The techniques of color slide photography are given in this book useful for the experienced amateur photographers and serious beginners. Attention is given to many different photographic situations from sceneries to closeup, and film characteristics under various lighting conditions are explained. Many illustrated examples are provided.

Sussman, Aaron. *The Amateur Photographer's Handbook* (7th ed.). Crowell, 1965. x+400 pp. illus. $5.95; text ed. $4.50. 61–18664. (JH)

A comprehensive guide for the amateur that explains techniques, ideas and procedures. Explains equipment, lighting, exposure, development of the negative, the making of prints, enlargements, slides, etc. Includes an excellent section on color photography.

Thomson, C. Leslie. *Colour Films: The Technique of Working with Colour Materials* (4th rev. ed.). Chilton, 1969. 278 pp. illus. $10.95. 69–1143. (SH)
A satisfactory compromise between the "how-to-do-it" book and introductory technical treatment of the subject of color photography. Completely up to date in providing practical information about the principles and practice of color photography, with an emphasis on laboratory practice. For the advanced amateur and above.

771 PHOTOGRAPHY: EQUIPMENT, SUPPLIES, CHEMISTRY

Dolan, Edward F., Jr. *The Camera.* Messner, 1965. 191 pp. illus. $3.95; LB $3.64. 65–21610. (JH–SH–C)
An accurate and vivid history of black-and-white photography and description of the modern still camera, with a chapter on motion-picture photography. The romance of photographic reportage is illustrated by examples from Fenton and the Crimean War (1855) to Glenn and his oribital satellite. Proves that popular technical writing can be both exciting and accurate. Bibliography and index.

John, D. H. O., and G. T. J. Field. *Photographic Chemistry.* Reinhold, 1963. ix+330 pp. illus. $11.50. 63–24072. (C)
A discussion of photographic chemistry for those with a practical knowledge of photography and a foundation of elementary physics and chemistry. The author holds that the diversified applications of photography in scientific research and technology emphasize a need for understanding the basic chemical processes inherent in photography. The discussion is limited to monochrome photography.

Lincoln, Marshall. *Electronics for Photographers.* Chilton, 1965. 159 pp. illus. $6.95. 66–23322. (JH–SH)
A very simple and straightforward exposition for the novice. Chapters include: "Resistance, voltage, and photoflash circuits," "B-C flash units," "Synchronizers," "Slave flash units," "Light meters and automatic cameras," "Timers and enlarging meters," and "Specialized equipment." Additional chapters furnish the reader with a very basic introduction to some electronic principles. Glossary, tables of data and formulas.

Neblette, Carroll B. *Photography: Its Materials and Processes* (6th ed.). Van Nostrand, 1962. viii+508 pp. illus. $16.50. (C)
An authoritative and comprehensive work on all phases of photography. It begins with a history of photography, then presents a review of physical optics. The construction and operation of cameras, shutters, synchronizers, and flash devices are clearly presented. Then follows a consideration of chemistry, negatives, papers, solutions, and techniques in accurate and informative detail. A comprehensive discussion of all phases of color photography, electrophotography and other special topics is included.

910 GEOGRAPHY

Cheney, T. A. *Land of the Hibernating Rivers: Life in the Arctic*. Harcourt,
Brace, 1968. 121 pp. illus. $2.95. 68–13814. (JH–SH)
Organic adaptations to the colder parts of the northern hemisphere are the
subject of this short, but interesting, book. The author, a geologist-geog-
rapher, explorer, and alumnus of the Byrd expedition of 1946, describes a
spectrum of bleak environments, beginning with the taiga of eastern Siberia
and ending with the polar pack, with intermittent stages in the tundra of
Alaska and the ice cap of Greenland. Recommended for informative and
entertaining reading.

Church, R. J. Harrington. *West Africa: A Study of the Environment and Man's
Use of It*. (Geographies for Advanced Study Series.) Wiley, 1968. xxix+
543 pp. illus. $8.95. 68–9201. (C)
Examines the physical characteristics of West Africa, its resources, and
presents a fully integrated analysis of each of West Africa's fifteen political
subdivisions. Although somewhat subjective, this book can be used by personnel responsible for developing West Africa and by graduate students in
advanced seminars.

Church, R. J. H., John I. Clark, and H. J. R. Henderson. *Africa and the Islands*.
Wiley, 1964. xiv+494 pp. illus. $9.75. (JH–SH)
This is a new geography of Africa. It is the joint effort of specialists in
certain areas and aspects, authors who have been in Africa on many occa-
sions and for long periods, who have taught in varied types of universities,
training colleges and schools in Africa and Europe. The broad picture of
Africa is given in Part I. (Africa as a Whole) ; Part II (Regional Studies)
fills in concisely the essential details; the Conclusion presents "Political,
social and economic trends." Reading lists are appended to the chapters.

Cohen, Saul B. (ed.). *Problems and Trends in American Geography*. Basic,
1967. xiv+298 pp. illus. $6.50. 67–28500. (C)
Lucid, concise, and provocative, these nineteen essays by eminent geographers
have been selected for their scope and depth from a series of overseas Voice of
America broadcasts. This should be required for undergraduate geography
majors and for graduate students in the social sciences.

Cotter, Charles H. *The Astronomical and Mathematical Foundations of Geography*. Elsevier, 1966. ix+244 pp. illus. $7.00. 66–22459. (SH–C)
For students of the history of science, as well as of geodesy, geology, and
physical geography, there are concise accounts of historical theories and
practices which are difficult to find elsewhere. Coverage is particularly good
on ideas about the size and shape of the earth, the geometry of measuring
it, navigational terms and techniques, and the basic elements of surveying.
Useful for reference and collateral reading.

Cressey, George B. *Asia's Lands and Peoples*. (3rd ed.). McGraw-Hill, 1963.
xix+ 663 pp. illus. ($14.50; text ed. $10.50). 62–22087. (SH–C)
This examination begins with the geology, the land forms, climate, soil,
mineral resources, and then gives attention to the people of Asia, their livelihood, and their political problems. A realm-treatment starts at the East
Asian, followed by the Southeastern, the South Asian, the Southwest Realm

and finally by the Soviet Realm. The late Professor Cressey states in his preface, "Readers are at least entitled to know where the author has been. These travels represent nine visits to Asia, a decade of residence, and nearly a half million miles of travel."

Cressey, George B. *Crossroads: Land and Life in Southwest Asia.* Lippincott, 1960. xiv+593 pp. illus. ($13.50.; text ed. $9.75). 60–11518. (SH–C)
Three principal ideas form the theme—the crossroads character of Southwest Asia, the role of water in its economy, and the way in which man is changing the landscape. The abundance of photographs, maps, tables, and especially the reading lists at the end of each chapter make this a valuable reference. However, the text (rightly described as "vigorous prose") lifts it out of the usual dry textbook category; it might be called a documentary film in words.

Dohrs, Fred E., and Lawrence M. Sommers (eds.). *Introduction to Geography: Selected Readings.* Crowell, 1967. 390 pp. illus. paper $3.95. 67–12699. (C)
Contains 30 articles, arranged in six sections that deal with evolution of geographic thought, the geographer's tools, and the techniques of geography. Additional readings (keyed to selected geography texts and correlated with the volume) are listed in tabular form to aid the instructor, providing considerable flexibility both in depth of subject and level of instruction. The volume's greatest appeal will probably be to the upper division undergraduate and the graduate student.

Durand, Loyal, Jr. *Economic Geography.* Crowell, 1961. xii+578 pp. illus. $7.75. 61–9094. (SH)
A text organized by products or groups of products with similar or interchangeable final uses. A single product is carried through from its raw material source to its final manufacture and use. For instance, corn is dealt with by a discussion of its origin, the growing of corn, environmental requirements and regions of production, world trade, the American Corn Belt, its natural and cultural landscape, the crop season, and the animal economy dependent upon the Corn Belt. Also contains chapters concerned directly with the manufacturing of the world.

Finch, Vernon C., et al. *Elements of Geography: Physical and Cultural.* (4th ed.). McGraw-Hill, 1957. x+693 pp. illus. $9.50. 56–12264. (C)
A mature text for college level, but probably the best in this field of general geography. Part I analyzes the nature, origin, and distribution of the important elements of the physical or natural earth; Part II describes the distribution of mankind; Part III deals with the cultural earth. Concludes with appendices which give additional climatic data, a treatment of map projections, a brief description of the systems of American land survey, and a selected list of United States topographic quadrangles.

Gresswell, R. Kay. *Physical Geography.* (Praeger Surveys in Geography.) Praeger, 1967. viii+504 pp. illus. $7.00. 67–21756. (C)
This is a well written and comprehensive survey of physical geography. Meteorology, climatology, weathering and rivers, ice and glaciation, coasts and the sea, volcanoes and vulcanism, soils and natural vegetation are topics covered. An excellent introduction undergraduate text for geography majors.

Hance, William A. *The Geography of Modern Africa.* Columbia, 1964. xiv+653
   pp. illus. $13.50. 64–14239. (SH–C)
   Presents the major features of the economy of Africa, analyzes the handicaps
   and attributes that affect economic development, and assesses some of the
   potentialities for growth. The five introductory chapters deal with Africa
   in general, the remaining examine individual countries, particular attention
   being paid to tropical Africa. Includes a bibliography, a subject index of
   geographical names, and many maps and tables. The author has studied
   Africa since 1941, taught courses about the continent, and made four field
   trips, including an extended one in 1962–63.

Hatherton, Trevor (ed.). *Antarctica.* Praeger, 1965. xvi+511 pp. illus. $18.50.
   65–20823. (SH–C)
   This excellent encyclopedia summary covers all phases of Antarctica, ranging
   from natural sciences through sociological implications of man living on that
   continent. Each of the 18 topics is covered by an expert. The organization
   and presentation are ideal. Scientific terminology is to the point, and no
   sacrifice in definitiveness has been made. The timely, excellent illustrations
   and well-balanced chapter bibliographies offer the reader guidance in pursuing
   each subject of interest to any depth he may desire. A must for every library;
   offers the school and college student, as well as the public, the data now commonly sought after because of the popular interest in the Antarctic.

Houston, J. M. *The Western Mediterranean World: An Introduction to Its
   Regional Landscapes.* (Praeger Advanced Geographies.) Praeger, 1967.
   xxxii+800 pp. illus. $13.50. 67–25037. (C)
   Deals with the general features of climate, geologic structure, landforms,
   vegetative cover, population, and settlement types of the western Mediterranean.
   Richly illustrated, this volume will appeal to geography majors.

James, Preston E. *Introduction to Latin America: The Geographic Background
   of Economic and Political Problems.* Odyssey, 1964. xvi+366 pp. illus. $5.50.
   64–19194. (SH–C)
   A new revised text organized in five parts as follows: An Overview; The
   Spanish Countries of Mainland Latin America; Brazil; The Antillean Countries and Colonies and The Guianas; General Conclusions and Appendices.
   There is a bibliography, an index, and guides to pronunciation, both Spanish and Portuguese. The industrial and the democratic revolutions are described, and the appearance against a background of varied lands and peoples
   of the new urban centers. (This is a condensed version of the author's
   Latin America, 1959.)

Monkhouse, F. J. *The Geography of Northwestern Europe.* Praeger, 1966. xv+
   528 pp. illus. $9.50. 66–15014. (SH–C)
   An interesting physio-economic geography of 11 mainland countries of Northwestern Europe located east of a line extending from the North Cape of
   Norway to the Bay of Biscay and as far east as the Russian satellites. East
   Germany is the only Communist-bloc country considered. Following an initial chapter on the historical background of each state, Monkhouse analyzes
   the geographic conditions of the individual countries. A final chapter deals
   with the groupings within which most of the states are now aligned. Further
   readings, a short bibliography, and an alphabetical index.

Osborne, R. H. *East-Central Europe: An Introductory Geography.* Praeger, 1967. 384 pp. illus. $7.50. 67–21456. (C)
A suitable text for a college level introductory course dealing with seven of the eight socialist states (East Germany excluded) located between the USSR and Western Europe; collateral reading for advanced and graduate college students, professors, scholars, and businessmen. Contains more than 50 maps, many statistical tables, an excellent bibliography, and a guide to the pronunciation of geographic names. The individual countries are analyzed in separate chapters.

Philbrick, Allen K. *This Human World.* Wiley, 1963. 500 pp. illus. $7.95. 63–9432. (SH–C)
"Geography provides an approach to knowledge of the world which can help each of us develop more meaningful relationships to it . . . . The plan of *This Human World* is the interweaving of several threads . . . . The first five chapters describe the world systematically in three main subdivisions— physical, cultural, and organization. Chapters 6–11 are arranged around the analysis of Europe. Chapters 12–14 deal with the Americas and their focus on the United States and Canada. Then the division of the old-world Eurasian land mass into the Communist Block and the regions of the Eurasian perimeter is treated in Chapters 15–19. Chapter 20 summarizes world regional organization." —Preface. Includes a large number of original world maps.

Powers, William E. *Physical Geography.* Appleton-Century-Crofts, 1966. ix+566 pp. illus. $8.50. 66–10964. (SH–C)
Designed as collateral reading for a year's course on the freshman level. Notable for its clear, concise, balanced presentation. The subject matter is discussed under five major headings: "Physiography: the development of landforms"; "Physiography: regional description of the United States and Southern Canada"; "The weather"; "Climates"; "Earth resources". Illustrated by photographs, diagrams, maps, and cross-sections. The end-papers show the physiographic provinces of the United States.

Symthe, James M., Charles G. Brown, and Eric H. Fors. *Elements of Geography.* (rev. ed.). St. Martin's 1966. xiii+466 pp. illus. $4.60. 66–7702. (SH)
An excellent text for high school survey courses or an introduction to the principles of geography. The descriptive charts, graphs, pictures and photographs are very good, and all inclusive. There is a vocabulary study for each chapter.

Stamp, Sir Dudley (ed.). *Dictionary of Geography.* Wiley, 1966. xvi+492 pp. illus. $10.00. (JH–SH–C)
A compendium of geographical information which is otherwise obtainable only from a variety of sources; a convenient reference tool for libraries and geographers. The entries include a selection of geographic terms, societies, journals, awards, and biographic sketches of eminent geographers. Brief accounts of countries and of the world's principal physical features and urban centers. List of selected geographical books published in Britain, with a few additions from Canada and the United States. A useful compendium for quick reference, but the serious scholar must still go, in many cases, to standard works in which more details are available.

Warntz, William. *Geographers and What They Do*. Watts, 1964. ix+149 pp. illus. $3.95. 64–11916. (JH)

This book is a valuable preliminary survey of the science of geography. It outlines the many careers open to those who had major training in geography and related disciplines, outlines the personal qualifications and academic preparation, and in an appendix lists government agencies that employ geographers.

Watson, J. Wreford. *North America: Its Countries and Regions*. (rev. ed.). Praeger, 1968 [c.1967]. 881 pp. illus. $12.00. 67–271758. (C)

This revision shows new economic and social trends within North America—improved agricultural methods, transportation, marketing, urbanization. No other single book on the regional, economic, and cultural geography is as complete, authoritative, and valuable.

White, C. Langdon, Edwin J. Foscue, and Tom L. McKnight. *Regional Geography of Anglo-America* (3rd ed.). Prentice-Hall, 1964. xvii+524 pp. illus. $9.95. 64–10071. (SH–C)

An outstanding introductory college text and good source book for secondary school students. The first three chapters are entitled "Anglo-America and its regions," "The city and industrial geography," "The American manufacturing region." The fifteen remaining chapters are devoted to analysis of sequential regions. A selected bibliography appears at the end of each chapter. Appendix A describes "The physical background of Anglo-America"; Appendix B, "The occupants of Anglo-America."

Zelinsky, Wilbur. *A Prologue to Population Geography*. Prentice-Hall, 1966. ix+150 pp. illus. paper $1.95. 66–10948. (C)

A worthy contribution toward the geographic integration and interpretation of world cultural phenomena of concern to social scientists in several disciplines. Three main parts: (1) "What does a population geographer study–" (2) "Distributions of the world's population," (3) "Toward a typology of population regions." Supporting the text are seven world maps and a diagram showing some techniques for mapping population. Geographers will value the book; other social scientists will appreciate the enhancement of their subject matter by geographic interpretation.

910.4 EXPLORATION

Caras, Roger A. *Antarctica: Land of Frozen Time*. Chilton, 1962. xi+209 pp. illus. $6.00. 62–16642. (SH)

A member of the Explorers Club, amateur naturalist, anthropologist, and semi-professional photographer sets forth the amazing technology of modern times that is making the wilderness of ice give ground and describes the scientific equipment used in Antarctic research and exploration. Eight appendices contain useful information for researchers.

Carrison, Daniel J. *Christopher Columbus: Navigator to the New World*. (Immortals of History Series.) Watts, 1967. xi+178 pp. illus. $2.95. LB $2.21. 67–10333. (JH–SH)

An extremely well-written work. Points out for young readers the significance of courage, perseverance, and effort. It points out, too, a number of life's perils. Good black-and-white pictures and maps accompany the book, as does a chronology of Columbus, a selected bibliography, and an index.

Chapman, Walker. *The Loneliest Continent: The Story of Antarctic Discovery.* N. Y. Graphic Society, 1964. 279 pp. illus. $4.95. 64–20272. (JH–SH–C) The story of Antarctic discovery and exploration is telescoped into a meaningful and coherent account despite the rapidity and brevity with which the accomplishments of some 100 Antarctic explorers are related. Its chief value lies in the overview of Antarctic exploration and mapping from references to the unknown southland by Greek philosophers 400 B. C. through Operation Deep Freeze in 1956. Pictures or sketches of several explorers, their ships and their equipment enhance the value and interest. A good bibliography leads the serious scholar to many supplementary books. Better for collateral reading than reference.

Fuchs, Vivian, and Edmund Hillary. *The Crossing of Antarctica.* Little, Brown, 1958. xv+328 pp. illus. $7.50. (SH) This is the story of the long exploratory journey across the last untraversed continent led by Fuchs, and of the work of the support party led by Hillary. It shows how dog-sleds, tractors, aircraft and daring and resourceful scientists, technicians, engineers, mechanics, and airplane pilots cooperated in this massive undertaking. Excellent illustrations.

Hakluyt, Richard. *Voyages* (8 vols.). (Everyman's Library, nos. 264, 265, 313, 314, 338, 339, 388 and 389). Dutton, 1962. $2.25. ea. 62–51174. (C) Richard Hakluyt was born in 1553, took holy orders, in 1593 became Archdeacon of Westminster, died in 1616, and was buried in Westminster Abbey. His famous chronicle, *Principal Navigation, Voyages and Discoveries of the English Nation,* is a basic work of superb scholarship. In it an ancient time comes alive once more for the appreciative student with a broad interest in geography and the history of exploration.

Hapgood, Charles H. *Maps of the Ancient Sea Kings.* Chilton, 1966. 315 pp. illus. $14.50. 65–24459. (SH–C) The first half of this book is fascinating and has great merit as a description of an intellectual exercise. The author and his students, worked for several years with a number of medieval navigational maps to convert them to a modern grid system. The results are a series of remarkably up-to-date outlines of the coasts of the world. Hapgood dates their original construction no later than 6,000 years ago. The last half of the book is concerned with the author's speculations regarding the culture of the people who made maps, with glaciation, and the earth's shifting crust. Recommended for the cartographic discussions, rather than for the latter part.

Hillary, Sir Edmund, and Desmond Doig. *High in the Thin Cold Air.* Doubleday, 1962. xi+254 pp. illus. $6.95. 62–15860. (SH) Included because of across-the-board appeal of "the Abominable Snowman" legend. Contains magnificent photographs of the top of the world, accounts of climbing (in this instance for scientific evaluations of sustained high altitude effects on human beings), and of brave men reacting to emergencies.

Keating, Bern. *The Grand Banks.* (Photos by Dan Guravich.) Rand McNally, 1968. 128 pp. col. pls. 29 x 24 cm. $9.95. 68–54710. (JH–SH) This is a series of related stories of the hydrographic and biological characteristics, of the pre-Columbian discoverers and their successors who established the fisheries and settled the land, of the fishermen and their families, of the fishing activities and methods on the banks, and a concluding essay on the dwindling abundance of the resources due to over-exploitation of the most desirable species.

Masselman, George. *The Atlantic: Sea of Darkness.* McGraw-Hill, 1969. 127 pp. illus. $4.95. 68–58509. (JH–SH)
Here, for the first time is the complete story of the exploration of the Atlantic which began with northern expeditions by the Phoenicians, included those of Father Brendan and the Irish monks, and ended with the Viking voyages to America 300 years before Columbus. The second phase was the South Atlantic expeditions of the Portuguese which culminated in rounding the Cape of Good Hope and entering the Indian Ocean by Diaz (1488). Columbus' three voyages are well-known and he was followed by Vespucci who explored the Coast of Venezuela and returned to Lisbon via Sierra Leone on the African Coast. There are references for those who wish more detail.

Powell, John Wesley. *Down the Colorado* (Diary of the First Trip Through the Grand Canyon, 1869). (Photographs and Epilogue by Eliot Porter, 1969; Foreword and Notes by Don D. Fowler.) Dutton, 1969. 168 pp. 36.5 cm. $30.00. 70–77353. (SH–C)
It is very appropriate that John Wesley Powell's diary of his first trip through the Grand Canyon of the Colorado River be published in a striking example of the very best in modern book manufacture to commemorate the centennial of that historic venture. Rereading Powell's diary a hundred years after it was written with the accompanying illustrations is a vivid armchair adventure. The book is a treasure for any individual or library who can afford it.

Siple, Paul. *90° South.* Putnam, 1959. 384 pp. illus. $5.75. 59–11029. (SH)
The absorbing story of scientists isolated in six months' darkness at the South Pole; includes a history of polar explorations as well as very human details about cramped living in such bitter weather.

Weems, John Edward. *Peary: The Explorer and the Man.* Houghton Mifflin, 1967. ix+362 pp. illus. $6.95. 67–10925. (JH–SH–C)
A biographic novel, but its narrative character is dominant. The detailed documentation, including many quotations from previously unavailable personal papers of Admiral Peary, renders this work useful as a secondary source of information, for those concerned with research on Arctic exploration. Written by a man with obviously intense admiration for Peary. Reader attention is maintained by a rapid succession of fascinating narratives that vividly portray significant events in Peary's youth as well as his achievements as an explorer.

912 ATLASES

C. S. Hammond & Company. *Ambassador World Atlas.* Hammond, 500 pp. illus. Text, maps, tables, index. 10½ x 14½ in. $14.95. Map 57–267. (JH–SH)
Provides detailed maps of every country, topographical and resource maps of the continents, human and physical geography, individual state and pro-

vincial maps. A stratospheric view of the earth, polar areas, maps, a section on outer space and the solar system, and an illustrated gazeteer of the world are other features.

C. S. Hammond & Company. *Contemporary World Atlas* (New Perspective Edition). Hammond, 1967. 256 pp. illus. Text, maps, indexes. 9 x 11 inc. $7.95. Map 67–11. (JH)
An inexpensive condensed atlas for younger students and for personal ownership. Indexes and other pertinent data are included on the same pages or the pages facing the maps. The text and maps are alphabetically arranged and a convenient gazeteer index begins the book, with information on area and population for each area listed. Includes political, typographical and economic maps.

National Geographic Society. *Atlas of the World.* Nat. Geog. Soc., 1965. (Revision in 1970). 304 pp. including 115 map pages. $18.75 (deluxe ed. with slipcase, $24.50). Map 62–24. (SH–C)
The outstanding reason for owning this atlas is not only general continental or regional maps, but also the large number of relatively unique maps, many of them insets found in no other atlas. There is an extremely informative double-page spread on Antarctica, a two-page north polar "Global view of the World" highlighting major relief features, a two-page map of Alaska, two pages of the most popular U. S. national parks, six pages of Pacific island inset maps including the Hawaiian Islands and two pages of West Indian island inset maps. Scattered through the atlas are many other relatively large-scale island inset maps of obscure or little-known areas.

Rand McNally & Company. *Goode's World Atlas,* (12th ed.). Edward B. Espenshade, Jr., Editor. Rand McNally, 1964. xii+288 pp. primarily maps and index. 9.75 x 11.36 in. $9.95, $7.50 (text ed.). Map 64–13. (JH–SH–C)
An excellent choice for personal and home use, especially because of size. This is one of the best and most used general reference atlases. Includes planet and earth-sun relations, world maps, resource maps, regional and special maps, world comparison table, principal countries, glossary, pronouncing index.

Rand McNally. *New Cosmopolitan World Atlas.* Rand McNally, 1968. xl+248 pp. and 128 pp. illus. glossaries and indexes. 14.5 x 11.25 in. $16.95, del. ed. $22.50. Map 68–3. (JH–SH)
An outstanding atlas, especially for American use and Western Hemisphere coverage. Opens with a special satellite and space section, continues with physical, political and comparative world maps, historical maps, selected world information, United States information, glossary of foreign geographical terms, map terminology explanation of the index and abbreviations, and index to political-physical maps.

Reader's Digest Great World Atlas. Funk and Wagnalls, 1965. 232 pp. maps, text, tables, illustrations, index. 11 x 16 in. $17.00. Map 65–10. (JH–SH)
An impressive atlas with great visual impact. In the home it should awaken or renew interest in the earth sciences. In the school library it probably will become the atlas most extensively used. There are four parts entitled "The universe and the earth," "The countries of the world." "The world about us," and "Indexes." Outstanding features are maps created from three-di-

mensional models, views of the earth as it would appear 25,000 miles in space, U. S. Navy photo-reliefs of the ocean floors, and color photographs of a relief model of the United States.

The World Book Atlas. Field, 1970. xx+392 pp. $30.45 (discount to libraries). 70–86815. (JH–SH)
Designed to supplement and complement *The World Book Encyclopedia.* Maps and charts are arranged in eleven groupings: Space; World; Europe; Asia; Africa; Australia; New Zealand and Pacific Islands; Polar Regions; Latin America; Canada; United States. Physical, political and historical features are subheaded under each region except space and polar. Concludes with a world travel guide, population tables, and index.

913.031 ARCHAEOLOGY—GENERAL

Bacon, Edward (ed.). *Vanished Civilizations of the Ancient World.* McGraw-Hill, 1963. 360 pp. illus. $28.50. 63–14869. (SH)
Each of the 14 sections in this profusely illustrated, oversized work deals with a great civilization of the past as revealed by modern archaeology. The diversity and depth of such topics as the Ainu of Neolithic Japan, the Mayas of Mexico, the megalith builders of Europe, and the Easter Islanders make this an outstanding survey of present archaeological knowledge. For large libraries with ample funds.

Baldwin, Gordon C. *The Riddle of the Past.* Norton, 1965. ix+149 pp. illus. $3.75; LB $3.48. 65–13341. (JH–SH)
A meticulously precise and accurate introduction. Covers the basic vocabulary of the archeologist, methods of locating prehistoric sites, critical field excavation techniques and techniques of laboratory analysis. In a realistic fashion the author does not neglect the essential routine work for the sake of the exciting moments of exceptional discoveries. Important collateral techniques in related disciplines are also considered. Concludes with practical suggestions as to the training needed to become a professional archeologist and also stresses the role of the amateur.

Bibby, Geoffrey. *The Testimony of the Spade.* Knopf, 1956. xviii+414 pp. illus. $7.95. 56–8916. (SH)
Traces "the movements and flowering culture of our early ancestors from the Alps north to Scandanavia, and from Russia to Ireland—from the cave dwellers of France to the "barbarian" tribes described by Caesar and Tacitus and the Norse sagas." Some of the great archaeological finds of northern Europe and the men who made them are discussed, with anecdotes and some philosophizing.

Braidwood, Robert J. *Prehistoric Men* (7th ed.). (Illus. by Susan Richert Allen and Philipp Herzog.) Morrow, 1967. 181 pp. $5.00. 67–29295. (SH–C)
Briefly introduces the discipline of archeology and its history. Emphasis is given to the gradual appearance of culture as reflected in more complicated stone tools, the rise of village farming communities, and the eventual establishment of civilized nations.

Ceram, C. W. (Pseud. of Kurt W. Marek). *Gods, Graves, and Scholars* (2nd ed.) (Tr. by E. B. Garside and Sophie Wilkins.) Knopf, 1967. xvi+441+xiv pp. illus. $7.95. 67–11119. (SH–C)
No other popular account of archeology has been more widely read than the original edition. Interesting accounts of the explorations of great archeologists who uncover the past history of mankind are gathered. In this revised edition, the author has expanded his original with the addition of new chapters to up-date the content and to expose new research techniques.

Ceram, C. W. (Pseud. of Kurt W. Marek). *The Secret of the Hittites, the Discovery of an Ancient Empire.* Knopf, 1956. xxi+281+x pp. illus. $5.95. 53–9457. (SH)
Twenty centuries before Christ, the Indo-European Hittites descended into Asia Minor and established a great nation that conquered Babylon and fought successful wars with Egypt. Virtually nothing was known about these people until this century. Only in the past few years have archaeologists begun to decipher and read their language. An intriguing factual account that reads like an adventure story.

Cottrell, Leonard. *Lost Cities.* Rinehart, 1957. 251 pp. illus. $4.50; Grosset, paper $2.25; LB $3.36. 57–9627. (SH)
The Assyrian cities of Nimrud and Nineveh, ancient Hittite sites in Turkey, Babylon, Ur, and other cities of Babylon are explored in all the suspense of their discovery. Also included are findings at Macchu Picchu, the last great capital of the Inca Empire, and Chichen Itza, sacrificial site of the Mayas.

Davidson, Marshall B. (ed.). *The Horizon Book of Lost Worlds.* (Narrative by Leonard Cottrell). Amer. Heritage (Doubleday), 1962. 431 pp. illus. $17.95. 62–19438. (SH)
A soundly-written and lavishly illustrated oversize book that describes nine of the more familiar ancient civilizations: the Egyptian, Mesopotamian, Indus Valley, Cretan, Mycenaean, Hittite, Etruscan, Khmer, and Mayan. Valuable for insights into ancient crafts and technology.

De Borhegyl, Suzanne. *Ships, Shoals, and Amphoras: The Story of Underwater Archaeology.* Holt, 1961. 176 pp. illus. LB $3.59. 61–9050. (JH)
With the invention of the Aqualung in 1943, the science of underwater archaeology was born. The author, a diver, takes readers on dives to Greek and Roman galleys, explorations for the mythical island of Atlantis, underwater excavations of a sacred well of the ancient Maya in Mexico. Suggestions on how to do your own underwater exploring.

Deetz, James. *Invitation to Archaeology.* Nat. Hist. Press. 1967. x+150 pp. illus. $4.50; paper $1.25. 67–10384. (SH–C)
A very up-to-date conceptualization of archeology as a subdiscipline of cultural anthropology. Focuses upon the salient theoretical domains of archeology. An excellent text for an introductory course in anthropology. Among the nine interesting and informative chapters are "Dating methods," "The analysis of form," and "Space and time." A final chapter, "Archaeology tomorrow," includes a typical description of a type of artifact, and a list of selected readings. Bibliography and index.

Hawkes, Jacquetta (ed.). *The World of the Past.* Knopf, 1963. Vol. 1, 601 pp.; Vol. 2, 709 pp. illus. $20.00 set. 63–12396. (C)

An anthology of writings on archaeology by scientific experts, as well as popular and literary commentators. The result is a work that is somewhat difficult in parts, but sufficiently diverse so that it offers absorbing reading for everyone. Some of the sections are: "About archaeology;" "The old Stone Age and the evolution of man;" "Mesopotamia and Palestine;" "The Egyptian World;" "Asia Minor;" "Greece and Italy;" "Britain and Europe."

Heizer, Robert F. (ed.). *A Guide to Archaeological Field Methods* (3rd ed.). Nat. Press, 1958. ix+162 pp. $7.95; paper $6.00. 58–26907. (C)

A valuable hand book that covers in a direct, explanatory style such topics as a real site survey, preparation for and methods of excavation, recording data, collecting artifacts, stratigraphy, photographic records, state and federal regulations concerning archaeological sites, and chronological methods. Includes an appendix on archaeology as a career.

Hole, Frank and Robert F. Heizer. *An Introduction of Prehistoric Archeology.* Holt, Rinehart and Winston, 1965. x+306 pp. illus. $8.50. 65–11842. (SH–C)

Concentrates on both field investigation and the interpretation of artifacts, rather than on the substantive reconstructions of prehistory. A brief history of archeology is followed by concise, evaluative descriptions of the techniques by which the archeologist goes about his work in field and laboratory and in the preparation of his reports. The contributions of other scientific disciplines to dating, environmental reconstruction and the technical analysis of artifacts are reviewed. Great value as a reference book and as a guide to archeological literature or perhaps as a supplemental text.

Piggott, Stuart (ed.). *The Dawn of Civilization: The First World Survey of Human Cultures in Early Times.* McGraw-Hill, 1961. 404 pp. illus. $28.50. 61–11703. (SH)

The beautiful pictures and rewarding, though fairly difficult, text of this oversized book present a comprehensive and exciting introduction to archaeology, Prominent scholars trace the history of man from a half-animal to his earliest civilizations in all parts of the world. For large libraries with ample funds.

Pyddoke, Edward (ed.). *The Scientist and Archaeology.* Roy, 1963. xiii+208 pp. illus. $6.95. 64–13618. (C)

A rather technical, well-written book which gives insight into the work and techniques of the archeologist. Chapters are by authorities (all British) and describes various techniques and specialties. Excellent elementary text or a career guidance reference for college students.

Robbins, Maurice, and Mary B. Irving. *The Amateur Archaeologist's Handbook.* Crowell, 1966. xiv+273 pp. illus. $6.95. 65–27148. (SH–C)

Several introductory archeology handbooks have been issued within the past several years—this is one of the best. The text and some of the charts presume some previous knowledge or interest. Illustrations of selected projectile point types and other artifacts are well-done. There are good examples of record forms and ideas for excavation controls and what to photograph. Contains over 110 illustrations.

Silverberg, Robert. *Frontiers in Archeology*. Chilton, 1966. x+182 pp. illus. $4.95. 66–15760. (SH–C)

Six archeological sites or cultures are chosen to illustrate the process of discovery, investigation and interpretation of the remains: Jericho and Ugarit in the Middle East, Anyang in China, Zimbabwe in east Africa, the Aztec civilization of Mexico, and the prehistoric remains of Easter Island. A well selected bibliography. A reliable, well-organized and readable book. Conveys some of the fascination of archeology without distorting the facts.

Silverberg, Robert. *Lost Cities and Vanished Civilizations*. Chilton, 1962. ix+177 pp. illus. $3.95; Bantam, paper 50¢. 62–8846. (JH)

Exciting tales of the adventures and everyday life of six ancient cities: Pompeii, Homer's Troy, Knossos of Crete, Babylon, Chichen Itza of the Mayas, and Angkor in the jungles of modern Cambodia.

Suggs, Robert C. *Modern Discoveries in Archaeology*. Crowell, 1962. 117 pp. illus. $3.50. 62–7745. (JH)

An exceptionally well-written account of recent advances in archaeology, including accurate and intriguing descriptions of the Carbon-14 dating method, the decipherment of the Linear-B script of ancient Crete, and the Dead Sea Scrolls. Also tells of some new Neanderthal finds in Iraq, very early bison hunters in North America, and Leakey's findings of the first human tool-makers in East Africa.

913.32 ARCHEOLOGY—EGYPT

Cottrell, Leonard. *The Lost Pharaohs*. Holt, 1961. 250 pp. illus. $5.00; Grosset, paper $2.25. 61–5299. (SH)

An archeological survey of Egypt, including the land itself, its history, some important "digs," and the work of the Egyptologists in piecing together evidence to uncover the story of this civilization that began over 3,000 years before Christ.

Cottrell, Leonard. *The Secrets of Tutankhamen's Tomb*. N.Y. Graphic Society, 1964. 139 pp. illus. $4.50; Dell, paper 50¢. 64–20445. (JH–SH–C)

A brief, factual, vivid account of the discovery and excavation of Tutankhamen's tomb, of the archeologists who accomplished it, and of the methods used. Also describes Tutankhamen's role in Egyptian history, but it is *not* a survey of Egyptian archeology or history. Cottrell writes clearly, with no technical jargon and with interesting details to support his general statements.

Emery, Walter B. *Lost Land Emerging*. Scribner's, 1967. xii+334 pp. illus. $7.95. 67–15489. (SH–C)

An account of the growth of knowledge about the Nubia region and the salvage operations to save many ancient monuments threatened by the Aswan High Dam. Serves as collateral reading for an introductory course in ancient history.

Greener, Leslie. *The Discovery of Egypt*. Viking, 1966. vii+216 pp. illus. $6.95. 67–10215. (SH–C)

The reader is escorted on a historical tour of the centuries-long discovery of

Egypt's fascinating history. Beginning in 1250 B.C. and extending to the present day there is a spotty but continuous record of explorers, historians, soldiers, and monks who visited Egypt and perceived, however dimly, a new vista in mankind's past. The book is a series of well-written accounts of episodes along the route of discovery.

Honour, Alan. *The Man Who Could Read Stones: Champollion and the Rosetta Stone.* (Illus. by Anthony Aviles.) Hawthorn, 1966. 190 pp. $3.25. 66–15250. (JH–SH)
Fictitious conversations and certain events are used to present the biography of Jean-François Champollion. Warmly written and well-researched, serves both as an introduction to Egyptology and as a collateral reference book for the younger reader. Annotations for the writings of Champollion, a translation of the text of the Rosetta Stone and other Egyptian texts, an excellent index and a fair bibliography.

Horizon Magazine, Editors of, and Jacquetta Hawkes. *Pharaohs of Egypt.* American Heritage (dist. by Harper & Row), 1965. 153 pp. illus. $4.95; LB $4.79. 65–27220. (JH–SH)
The reigning kings are treated in historical perspective—from the unification of Upper and Lower Egypt in 3100 B.C. to the closing days of the New Kingdom. The volume is well written, exceptionally well illustrated (black-and-white and color), has two good maps, an index, and a suggested reading list.

913.33 ARCHEOLOGY—PALESTINE

Anati, Emmanuel. *Palestine Before the Hebrews.* Knopf, 1963. xx+453 pp. illus. $8.95. 62–12898. (SH)
A summary of the cultural history of man in Palestine, a story that goes back 600,000 years and ends 3,200 years ago when the Hebrews invaded Canaan. A fairly detailed account, technical in places, but with many photographs, charts, and drawings that make it understandable and interesting to the layman. Herein one may discern the rudiments of modern technology.

Archaeological Institute of America (Comp.). *Archaeological Discoveries in the Holy Land.* Crowell, 1967. xv+224 pp. illus. $12.50. 67–23673. (JH–SH–C)
Details much of the extremely fruitful Palestinian archeological investigations initiated after World War II. The articles are grouped into three major areas: "Before the Bible," "Biblical cities and Temples," and "Post-biblical Palestine."

Deuel, Leo (ed.). *The Treasures of Time.* World Pub., 1961. 318 pp. illus. $6.00; Avon, paper 95¢. 61–12019. (C)
Firsthand accounts by famous archaeologists of their work in the Near East, with sections on Egypt; Mesopotamia; Syria and Palestine; and Anatolia, Crete and Greece. The reports are fairly difficult, but of high literary and documentary quality, and of absorbing interest.

Yadin, Yigael. *The Story of Masada.* (Retold for young readers by Gerald Gottlieb.) Random, 1969. 155 pp. illus. 25.5 cm. $3.95. 68–31209. (JH–SH)
Atop a mountain high above the Dead Sea, Herod the Great (37–4 B.C.)

built Masada, a well-defended fortress and a complex of luxurious palaces. There a group of Jewish Zealots (66–73 A.D.) made a last stand against the Roman invaders. The Roman troops then occupied Masada (ca. 73–111 A.D.). There were Christian monks there from ca. 500 to 700 A.D. and then the area was forgotten, visited mainly by nomads and vandals. Yadin's major report on the Masada expedition and excavations, *Masada, Herod's Fortress and Zealot's Last Stand* (Random, 1966; $10.95) has been abridged and rewritten by Gerald Gottlieb for young readers. It will be read with great interest by many nonspecialist adults. Yadin's abilities as a military general were evident in the able organization and direction of a large army of volunteers from many parts of the world. There the successfully completed and formidable task of excavation, study, reconstruction, and interpretation verified the oft-doubted account.

913.37 ARCHEOLOGY—ITALY

Deiss, Joseph Jay. *Herculaneum: Italy's Buried Treasure.* Crowell, 1966. xv+174 pp. illus. $6.95. 66–26638. (SH–C)
Relatively little archeological work has been carried out at Herculaneum, and this book is as much a plea for further excavation as it is a description of this ancient town. Contains brief histories of the resort and the archeological research carried out there since the Eighteenth Century. Well written, numerous photographs and maps, good bibliography and an excellent index. Recommended as reference or collateral reading for students interested in the Roman period of the first century of the Christian Era.

913.38 ARCHEOLOGY—GREECE

Baumann, Hans. *Lion Gate and Labyrinth.* (Tr. by Stella Humphries.) Pantheon, 1967. 182 pp. illus. $4.95. 67–20216. (JH–SH–C)
Biographical, historical, and archeological details are interwoven to provide a readable and informative account of the lives and work of two great pioneers of Mediterranean archeology: Heinrich Schliemann and Arthur Evans. Drawings, maps, diagrams, and 24 pages of magnificent color supplement the text.

Cottrell, Leonard. *Crete: Island of Mystery.* Prentice-Hall, 1965. 71 pp. illus. $3.50. 64–7573. (JH)
Unravels in simple language the complexities of Crete's archeology. It focuses on the myths and legends that led Heinrich Schliemann to Troy, Mycenae, Tiryns, and Crete, and on the contributions made by Sir Arthur Evans at Knossos. It probes the character, lives, and accomplishments of both men. Also concentrates on the work, theory, and method of the archeologist.

Poole, Lynn and Gray Poole. *One Passion, Two Loves: The Story of Heinrich and Sophia Schliemann, Discoverers of Troy.* Crowell, 1966. xv+299 pp. illus. $6.95. 66–25434. (SH–C)
The Schliemanns, are portrayed in this excellent biography as living individuals who were truly interested in uncovering the past history of man. Written with warmth and feeling. It is not only an important contribution

to the historical literature of archeology, but deserves a place among good
literature in English. Numerous illustrations, archeological and personal.

913.42 ARCHEOLOGY—BRITAIN

Birley, Anthony. *Life in Roman Britain.* Putnam's, 1964. xv+176 pp. illus.
$3.50. 65–507. (SH–C)
Describes the native Celts, their hill-forts, timber-built farmsteads, and
patchwork fields, analyzes their trade relations; and makes an excellent case
for the fact that the Roman legions did not walk in upon savages in Britain.
The book literally breathes the life of the Roman period. The feelings, the
tastes, and the smells are all there. Beautifully illustrated.

Hawkins, Gerald A. *Stonehenge Decoded.* (Collaborator, John B. White.)
Doubleday, 1965. xiv+202 pp. illus. $5.95; Dell, 1966, paper $1.95. 65–19933.
(JH–SH–C)
A very readable and stimulating book which contains a history of the studies
of Stonehenge and a review of the information that can be inferred about
the builders of this and other Stone Age monuments. It provides a description
of ways in which this remarkable structure was built during the approximate
period 2000–1500 B. C. Concludes that the structure was a sophisticated observatory designed to mark significant directions of the rising and setting
positions of the sun and the moon.

913.60 ARCHEOLOGY—AFRICA

Cole, Sonia. *The Prehistory of East Africa.* Macmillan, 1963. 382 pp. illus. $7.95;
also NAL (MQ 612) paper 95¢. 63–11790. (C)
A difficult but absorbing and well-organized account of the descent of man
in Africa. Geological background, the relation of early hominids to the apes,
and cultural stages are presented as they have been brought to light by recent
findings. Some dramatic discoveries indicate East Africa as a possible
birthplace for Man.

Fagan, Brian. *Southern Africa.* Praeger, 1965. 222 pp. illus. $7.50. 65–27031.
(SH–C)
Concerns the ironworking peoples in Southern Africa from their earliest appearance in archeological research to the 1900's. One chapter sets the background of the Stone Age cultures of hunters and gatherers. Detailed descriptions of pottery, iron tools, other artifacts, culture as a whole, social
structure, people, etc., of each culture up into the historical record of living
groups (the Bantu). Suggested readings.

Lhote, Henri. *The Search for the Tassili Frescoes* (tr. by Alan H. Broderick).
Dutton, 1959. 236 pp. illus. $7.50. 59–4743. (SH)
A specialized but readable report of the discovery of a large group of pre-
historic paintings in the Tassili-n-Ajjer wasteland of the Central Sahara.
Beautiful reproductions of these rock drawings show the sensitivity of the
peoples that produced them as long as 8,000 years ago.

Pfeiffer, John (Carleton Coon, Consultant). *The Search for Early Man.* Harper
1963. 153 pp. illus. $4.95. LB $4.79. (Horizon Caravel) 63–16371. (JH)

An enthusiastic, accurate, well-illustrated account of recent discoveries in the evolution of prehistoric man. Covers the famous fossils of Olduvai Gorge in East Africa and describes two major digs in France that have shed light on the Neanderthals and other early men.

913.7 ARCHEOLOGY—THE AMERICAS

Baldwin, Gordon C. *The Ancient Ones.* Norton, 1963. 224 pp. illus. $3.75; LB $3.48. 63–11831.
The Navajo called them Anasazi, "the ancient ones," and they lived for at least 13 centuries in the Southwestern United States, beginning about 2,000 years ago. The author describes the archaeological detective work that unearthed their life, how they began as basketweavers and seed gatherers, and moved on to become artisans, farmers, traders and engineers.

Baldwin, Gordon C. *Race Against Time: The Story of Salvage Archeology.* Putnam's, 1966. 191 pp. illus. LB $3.29. 66–14319. (JH–SH)
The emphasis is on salvage in the United States, with short surveys of the situation around the world where attempts are made to obtain prehistoric data before construction, dams, and other projects destroy the sites. A good bibliography of nontechnical publications. The illustrations are excellent, the text is simple and well written.

Benson, Elizabeth P. *The Maya World.* Crowell, 1967. ix+172 pp. illus. $6.95. 67–18523. (JH–SH)
A simplified and well-illustrated introduction to Maya culture. Chapters cover "The place and the people," "The cities," "Agriculture and trade," "Artisans and artifacts," "Science," "Religion," "The end of Maya splendor," and the post-contact period. A brief bibliography lists the reliable semipopular books on the Maya. Illustrations are beautifully reproduced and give an excellent representation of Maya art and architecture.

Coe, Michael D. *America's First Civilization.* (Smithsonian Library). American Heritage (dist. by Van Nostrand), 1968. 159 pp. illus. $4.95. 68–55791. (SH–C)
The origin, development, and eventual demise of the first civilization of Mesoamerica, labeled Olmec by archeologists, is chronicled using data from excavations. A view of the Olmec area as it looks now and how it is used by modern people helps the reader better understand the Olmec. An appendix, index and a list of further readings for those interested in primary and more technical sources, concludes the book.

Jennings, Jesse D., and Edward Norbeck (eds.). *Prehistoric Man in the New World.* U. of Chicago, 1964. x+633 pp. illus. $10.00. 63–18852. (C)
A sound reference work that brings together archaeological and anthropological knowledge about early man in North, Central and South America. Five main geographical areas are dealt with. It includes special studies in geoanthropology, transpacific contacts, connections between North and South American cultures, and a linguistic overview. Though technical, the articles are well written and the book is well organized.

La Farge, Oliver. *A Pictorial History of the American Indian.* Bonanza, 1956. 272 pp. illus. $3.95. 56–11375. American Indian; 1960; $4.95; LB $3.99. Golden Press. (SH)

The forceful writing and profuse illustrations of this oversize book make it exciting reading for all levels. The author, a recognized authority on his subject, explains Indian life from the earliest settlements to the problems of Indian reservations today, with knowledge and sensitivity.

Myron, Robert. *Shadow of the Hawk: Saga of the Mound Builders.* Putnam's, 1964. 189 pp. illus. $3.75. LB $3.68. 64–25762. (JH–SH)
Focuses on the story of two cultures which flourished in the Ohio River Valley—the Hopewell (approx. 100 B.C.–1000 A.D.) and the Adena (approx. 50 B.C.–500 A.D.). Describes what is known of the life of the Mound Builders and speculates on other aspects of their life. Materials used, techniques, and possible sources are presented. The author separates evidence from speculation throughout and gives great vitality to these cultures.

Pendergast, David M. (ed.). *Palenque: The Walker-Caddy Expedition to the Ancient Maya City, 1839–1840.* Univ. of Oklahoma, 1967. xvi+213 pp. illus. $6.95. 66–22722. (C)
Pendergast has done Mayanists a great service by publishing under one cover a summary of the history of the expedition, pertinent portions of Caddy's diary, many of Caddy's excellent drawings, and Walker's official report on the expedition. Caddy's accurate description of the ruins coupled with his drawings constitute an important and valuable early source document on Palenque.

Robbins, Roland Wells, and Evan Jones. *Hidden America.* Knopf, 1959. 264+viii pp. illus. $5.95. 59–9257. (SH)
A practical, first-person account of the author's pick-and-shovel experience around America. He has excavated Thoreau's cabin at Walden Pond, Jefferson's birthplace, and the Saugus ironworks in Massachusetts, among other sites, and he writes about Indian mounds and Viking encampments as well.

Sutton, Ann, and Myron Sutton. *Among the Maya Ruins: The Adventures of John Lloyd Stephens and Frederick Catherwood.* Rand-McNally, 1967. 222 pp. illus. $4.50. 67–11581. (JH–SH)
Extensive quotations from the journals of John Stephens and examples of drawings by Frederick Catherwood make even more vivid a well-written account of the first scientific study of Maya ruins. Recent findings and interpretations of Maya culture are skillfully interwoven with Stephen's appraisals. Brief bibliography and index.

Thompson, John Eric. *The Rise and Fall of Maya Civilization.* (2nd ed.). U. of Oklahoma, 1966. xiv+288 pp. illus. $5.95. 54–10054. (SH)
A comprehensive description of the Mayan civilization that flourished in Central America from as early as 500 B.C., and had almost completely declined before the first Spanish arrived in the New World.

Thompson, John Eric. *Maya Archaeologist.* U. of Oklahoma, 1963. xvii+284 pp. illus. $5.00. 63–8994. (SH)
A leading authority on the Maya civilization of Mexico and Central America tells of the years he spent searching out and excavating Mayan sites. This warm and often humorous book is especially interesting if read with the author's *The Rise and Fall of Maya Civilization.*

Wedel, Waldo R. *Prehistoric Man on the Great Plains.* U. of Oklahoma, 1961. xviii+355 pp. illus. $5.95. 61–9002. (C)
A region-wide, nontechnical survey of archaeological knowledge of the Great Plains area of the United States. Briefly discusses the tools and methods of archaeology and the relation of the Plains environment to man, then gives a section-by-section description of the archaeology of the area.

Wolf, Eric R. *Sons of the Shaking Earth.* U. of Chicago, 1959. vii+303 pp. illus. $5.50. Phoenix P90 paper $1.50. 59–12290. (SH)
The story of the peoples and cultures of Middle America—Mexico and Guatemala—from prehistoric times through Mayan, Aztec, and Spanish civilizations to revolutionary and modern times. This detailed study covers geography, biology and evolution, language, agriculture and other cultural institutions and the succession of the cultures themselves. Extensive bibliographic notes are at the end.

913.98 ARCHEOLOGY—ARCTIC

Giddings, J. Louis. *Ancient Men of the Arctic.* Knopf, 1967. xxxi+391+xv pp. illus. $10.00. 65–11122. (SH–C)
This is the most comprehensive and detailed book for specialists and laymen on Arctic archeology, covering a summary reconstruction of Arctic prehistory. The text is supplemented with illustrations, photographs, and an extensive bibliography.

Author Index

Subject And Title Index

(Subjects are in capital letters)

DIRECTORY OF PUBLISHERS

AAAS—American Association for the Advancement of Science, 1515 Massachusetts Avenue, N.W., Washington DC 20005

Abelard—Abelard-Schuman Limited, 257 Park Avenue South, New York NY 10010

Academic—Academic Press, Inc., 111 Fifth Avenue, New York NY 10003

Addison-Wesley—Addison-Wesley Publishing Co., Inc., Reading MA 01867

Aldine—Aldine Publishing Co., 529 S. Wabash Avenue, Chicago IL 60605

Allyn—Allyn and Bacon, Inc., 470 Atlantic Avenue, Boston MA 02210

AGI—American Geological Institute, 2201 M Street, N.W., Washington DC 20037

American Heritage—American Heritage Publishing Co., Inc., 551 Fifth Avenue, New York NY 10007

ALA—American Library Association, 50 East Huron Street, Chicago IL 60611

Amer. Met. Lab.—American Meteorite Laboratory, P.O. Box 2098, Denver CO 80201

Amer. Public Health Assn.—American Public Health Association, 1740 Broadway, New York NY 10019

Amer. Tech. Soc.—American Technical Society, 848 East 58 Street, Chicago IL 60637

Americana—Americana Corporation, 575 Lexington Avenue, New York NY 10022

Anchor—(see Doubleday & Co.)

Aquarium—(dist. by E. P. Dutton & Co., Inc.)

Appleton—Appleton-Century-Crofts, 440 Park Avenue South, New York NY 10016

Arco—Arco Publishing Co., Inc., 219 Park Avenue South, New York NY 10003

Edward Arnold (London)—Agent: Howard Moorespark, 444 East 82nd Street, New York NY 10028

Atheneum—Atheneum Publishers, 122 East 42 Street, New York NY 10017

Athlone—dist. by Oxford University Press, Inc., 200 Madison Avenue, New York NY 10016

Audel—Theodore Audel & Co., 4300 West 62nd Street, Indianapolis IN 46268

Avon—Avon Books, 959 Eighth Avenue, New York NY 10019

Barnes and Noble—Barnes and Noble, Inc., 105 Fifth Avenue, New York NY 10003

Barrows—M. Barrows & Co. (dist. by Morrow)

Basic—Basic Books, Inc., Publishers, 404 Park Avenue South, New York NY 10016

Beacon—Beacon Press, 25 Beacon Street, Boston MA 02108

Benefic—Benefic Press (pub. div. of Beckly-Cardy Co.), 10300 West Roosevelt Road, Westchester IL 60153

Benjamin—The Benjamin Company, 485 Madison Avenue, New York NY 10022

Blaisdell—Blaisdell Publishing Company, 275 Wyman Street, Waltham MA 02154

Blakiston—(see McGraw-Hill Book Co.)

Bobbs-Merrill—Bobbs-Merrill Co., Inc., 4300 West 62nd Street, Indianapolis IN 46268

Bowker—R. R. Bowker Company, 1180 Avenue of the Americas, New York NY 10036

Boxwood—Boxwood Press, P.O. Box 7171, Pittsburgh PA 15213

Branford—Charles T. Branford Company, 28 Union Street, Newton Centre MA 02159

Braziller—George Braziller, Inc., 1 Park Avenue, New York NY 10016

Britannica—Encyclopaedia Britannica Press, 425 North Michigan Avenue, Chicago IL 60611

Brooks/Cole—Brooks/Cole Publishing Co., Belmont CA 94022

Wm. C. Brown—William C. Brown Company, Publishers, 135 South Locust Street, Dubuque IA 52001

Burgess—Burgess Publishing Company, 426 South Sixth Street, Minneapolis MN 55415

Butterworth—Butterworth & Company, Publishers, 14 Curity Avenue, Toronto 16, Ontario

Cambridge—Cambridge Book Company, Inc. (subs. of Cowles Communications), 488 Madison Avenue, New York NY 10022

Cambridge U.—Cambridge University Press, 32 East 57th Street, New York NY 10022

Jacques Cattell—Jacques Cattell Press, Box 5001, Tempe AZ 85281

Chapman—(Chapman and Hall, London)—Distributed by Barnes and Noble, Inc., 105 Fifth Avenue, New York NY 10003

Chelsea—Chelsea Publishing Company, 159 East Tremont Avenue, Bronx NY 10453

Chemical Pub.—Chemical Publishing Co., Inc., 200 Park Avenue South, New York NY 10003

Chemical Rubber—The Chemical Rubber Co., 2310 Superior Avenue, Cleveland OH 44114

Chilton—Chilton Book Company, 401 Walnut Street, Philadelphia PA 19106

Collins—William Collins Sons & Company, Ltd., 215 Park Avenue South, New York NY 10003

Columbia—Columbia University Press, 440 West 110 Street, New York NY 10025

Compton—F. E. Compton and Company, 1000 North Dearborn Street, Chicago IL 60610

Comstock—(see Cornell University Press)

Cornell U.—Cornell University Press, 124 Roberts Place, Ithaca NY 14850

Coward—Coward-McCann Inc., 200 Madison Avenue, New York NY 10016

Cowles—Cowles Book Company, Inc., 488 Madison Avenue, New York NY 10022

Creative—Creative Educational Society, Inc., 515 North Front Street, Mankato MN 56001

Crest—Fawcett World Library (Crest, Gold Medal, and Premier Books), 67 West 44 Street, New York NY 10036

Criterion—Criterion Books (div. of Intext Educational Publishers) 257 Park Avenue South, New York NY 10010

Crowell—Thomas Y. Crowell Company, 201 Park Avenue South, New York NY 10003

Crown—Crown Publishers, Inc., 419 Park Avenue South, New York NY 10016

Davis—F. A. Davis Co., 1915 Arch Street, Philadelphia PA 19103

Dental Items of Interest—Dental Items of Interest Publishing Company, Inc., 2911 Atlantic Avenue, Brooklyn NY 11207

Dell—Dell Publishing Co., Inc., 750 Third Avenue, New York NY 10017

Devin-Adair—The Devin-Adair Company, 23 East 26 Street, New York NY 10010

Dial—The Dial Press (subs. of Dell Pub. Co.), 750 Third Avenue, New York NY 10017

Dodd, Mead—Dodd, Mead & Company, 79 Madison Avenue, New York NY 10016

Doubleday—Doubleday & Company, Inc., Garden City NY 11530

Dover—Dover Publications, Inc., 180 Varick Street, New York NY 10014

Duell, Sloan & Pearce—Meredith Corporation, 1716 Locust Street, Des Moines IA 50303

Dufour Editions—Dufour Editions, Inc., Chester Springs PA 19425

Dutton—E. P. Dutton & Company, Inc., 201 Park Avenue South, New York NY 10003

Elsevier—American Elsevier Publishing Company, Inc., 52 Vanderbilt Avenue, New York NY 10017

Eriksson—Paul S. Eriksson, Inc., 119 West 57 Street, New York NY 10019

Evans—M. Evans & Company, Inc., 216 East 49th Street, New York NY 10017

Evergreen—Grove Press, Inc., 64 University Place, New York NY 10003

Farrar—Farrar, Straus & Giroux, Inc., 19 Union Square West, New York NY 10003

Fawcett—Fawcett World Library: Crest, Gold Medal, and Premier Books, 67 West 44th Street, New York NY 10036

Ferguson—J. G. Ferguson Publishing Co., 6 North Michigan Avenue, Chicago IL 60602

Field Enterprises—Field Enterprises Educational Corporation, 510 Merchandise Mart Plaza, Chicago IL 60654

Follet—Follet Publishing Company, 201 North Wells Street, Chicago IL 60606

Freeman, Cooper—Freeman, Cooper and Company, 1736 Stockton Street, San Francisco CA 94133

Freeman—W. H. Freeman and Company, Publishers, 660 Market Street, San Francisco CA 94104

Free Press—The Free Press, 866 Third Avenue, New York NY 10022

Funk & Wagnalls—Funk & Wagnalls, 380 Madison Avenue, New York NY 10017

Golden Press—Golden Press, Inc. (div. of Western Pub. Company, Inc.), 1220 Mound Avenue, Racine WN 53404

Grolier—Grolier Incorporated, 575 Lexington Avenue, New York NY 10022

Grosset & Dunlap—Grosset & Dunlap, Inc., 51 Madison Avenue, New York NY 10010

Grossman—Grossman Publishers, Inc., 125 A East 19th Street, New York NY 10003

Hafner—Hafner Publishing Company, Inc., 260 Heights Road, Darien CT 06820

Hale—E. M. Hale and Company, 1202 South Hasting Way, Eau Claire WI 54701

Hammond—Hammond Inc., Maplewood NJ 07040
Harcourt, Brace—Harcourt Brace Jovanovich, Inc., 757 Third Avenue, New York NY 10017
Harper—Harper and Row, Publishers, 49 East 33 Street, New York NY 10016
Hart—Hart Publishing Company, Inc., 510 Avenue of the Americas, New York NY 10010
Harvard—Harvard University Press, 79 Garden Street, Cambridge MA 02138
Harvey—Harvey House, Inc., Publishers, Irvington-on-Hudson NY 10533
Hawthorn—Hawthorn Books, Inc., 70 Fifth Avenue, New York NY 10010
Hayden—Hayden Book Company, Inc., 116 West 14 Street, New York NY 10010
Heath—D. C. Heath & Company, 125 Spring Street, Lexington MA 02173
Hill & Wang—Hill & Wang, Inc., 72 Fifth Avenue, New York NY 10011
Holden-Day—Holden-Day, Inc., 500 Sansome Street, San Francisco CA 94111
Holiday House—Holiday House, Inc., 18 East 56 Street, New York NY 10022
Holt—Holt, Rinehart & Winston, Inc., 383 Madison Avenue, New York NY 10017
Horizon—Horizon Press, 156 Fifth Avenue, New York NY 10010
Houghton Mifflin—Houghton Mifflin Company, 2 Park Street, Boston MA 02107
Howard—Howard University Press, 2400 Sixth Street, N.W., Washington DC 20001
Humanities—Humanities Press, Inc., 303 Park Avenue South, New York NY 10010
Indiana U.—Indiana University Press, Tenth and Morton Streets, Bloomington IN 47401
Industrial—Industrial Press, 200 Madison Avenue, New York NY 10016
International Textbook Company—College Division (div. of Intext Educational Publishers) Scranton PA 18515
International—International Universities Press, 239 Park Avenue South, New York NY 10003
Interscience—John Wiley & Sons, Inc., 605 Third Avenue, New York NY 10016
Iowa State U.—Iowa State University Press, Press Building, Ames IA 50010
John Day—The John Day Company, Inc., 257 Park Avenue South, New York NY 10010
John Hopkins—Johns Hopkins Press, Baltimore MD 21218
Journal of Chem. Ed.—Journal of Chemical Education, 500 Fifth Avenue, New York NY 10036
Knopf—Alfred A. Knopf, Inc., 201 East 50 Street, New York NY 10022
Lantern—Lantern Press, Inc., 257 Park Avenue South, New York NY 10010
Lea & Febiger—Lea & Febiger, 600 South Washington Square, Philadelphia PA 19105
Lippincott—J. B. Lippincott Co., East Washington Square, Philadelphia PA 19105
Little, Brown—Little, Brown & Company, 34 Beacon Street, Boston MA 02106
Lothrop—Lothrop, Lee & Shepard Co., Inc., (div. of Wm Morrow & Co.), 105 Madison Avenue, New York NY 10016
Macfadden—Macfadden-Bartell Corporation, 205 East 42nd Street, New York NY 10017
McGraw-Hill—McGraw-Hill Book Company, 205 West 42nd Street, New York NY 10036
McKay—David McKay Company, Inc., 750 Third Avenue, New York NY 10017
McKnight & McKnight—McKnight & McKnight, U. S. Rt. 66 at Towanda Ave., Bloomington IL 61701
Macmillan—The Macmillan Company, 866 Third Avenue, New York NY 10022
Macrae—Macrae Smith Company, 225 South 15 Street, Philadelphia PA 19102
Mentor—New American Library of World Literature, Inc., 1301 Avenue of the Americas, New York NY 10019
Merck—Merck and Company, Inc., Rahway NJ 07065
Meredith—Meredith Corporation, 1716 Locust Street, Des Moines IA 50303
Merriam—G. and C. Merriam Company, 47 Federal Street, Springfield MA 01101
Merrill—Charles E. Merrill Publishing Company, 1300 Alum Creek Drive, Columbus OH 43216
Messner—Julian Messner, Inc., 1 West 39 Street, New York NY 10018
Methuen—Methuen & Company, 36 Essex Street, Strand, London, W. C. 2 England
M.I.T.—The Massachusetts Institute of Technology Press, 50 Ames Street, Cambridge MA 02142

Morrow—William Morrow & Company, Inc., 105 Madison Avenue, New York NY 10016

Mosby—The C. V. Mosby Company, 3207 Washington Blvd., St. Louis MO 63103

John Murray—John Murray, 50 Albemarle Street, London W. 1 England

National Academy—National Academy of Sciences–National Research Council, 2101 Constitution Avenue, N.W. Washington DC 20036

NCTM—National Council of Teachers of Mathematics, 1201 16th Street, N.W. Washington DC 20036

NEA—National Education Association, 1201 16th Street, N.W. Washington DC 20036

National Geographic—National Geographic Society, 17th & M Streets, N.W. Washington DC 20036

NSTA—National Science Teachers Association, 1201 16th Street, Washington DC 20036

Nat. Hist. Press—Doubleday & Company, Inc., Garden City NY 11530

New American Library—The New American Library, Inc., 1301 Avenue of the Americas, New York NY 10019

New York Graphic—New York Graphic Society, Ltd., 140 Greenwich Avenue, Greenwich, CT 06830

New York U.—New York University Press, Washington Square, New York NY 10003

North-Hollands (dist. by John Wiley & Sons)

Norton—W. W. Norton & Company, Inc., 55 Fifth Avenue, New York NY 10003

Oceana—Oceana Publications, Inc., Dobbs Ferry, NY 10522

Odyssey—Western Publishing Company, Inc., 55 Fifth Avenue, New York NY 10003

Oxford—Oxford Book Company, Inc., 387 Park Avenue South, New York NY 10016

Oxford U.—Oxford University Press, Inc., 200 Madison Avenue, New York NY 10016

Pantheon—Pantheon Books, Inc. (div. of Random House), 201 East 50th Street, New York NY 10022

Penguin—Penguin Books, Inc., 7110 Ambassador Road, Baltimore MD 21207

Pergamon—Pergamon Press, Inc., Maxwell House, Fairview Park, Elmsford NY 10523

Phoenix—University of Chicago Press, 5750 Ellis Avenue, Chicago IL 60637

Pitman—Pitman Publishing Corporation, 6 East 43rd Street, New York NY 10017

Plenum—Plenum Publishing Corporation, 114 Fifth Avenue, New York NY 10011

Praeger—Frederick A. Praeger, Inc., 111 Fourth Avenue, New York NY 10003

Prentice-Hall—Prentice-Hall, Inc., Englewood Cliffs NJ 07632

Princeton—Princeton University Press, Princeton NJ 08540

Prindle—Prindle, Weber & Schmidt, Inc., 53 State Street, Boston MA 02109

Prometheus—(see G. P. Putnam's Sons)

Putnam's—G. P. Putnam's Sons, 200 Madison Avenue, New York NY 10016

Rand McNally—Rand McNally & Company, 8255 Central Park Avenue, Skokie IL (address mail to Box 7600, Chicago IL 60680)

Random House—Random House, Inc., 201 East 50 Street, New York NY 10022

Regnery—Henry Regnery Company, 114 West Illinois Street, Chicago IL 60610

Rider—(see Hayden Book Co., Inc.)

Rockefeller U.—Rockefeller University Press, York Avenue and East 66th Street, New York NY 10021

Ronald—The Ronald Press Company, 79 Madison Avenue, New York NY 10016

Rosen—Richards Rosen Press Inc., 29 East 21 Street, New York NY 10010

Roy—Roy Publishers, Inc., 30 East 74th Street, New York NY 10021

Rutgers Univ.—Rutgers University Press, 30 College Avenue, New Brunswick NJ 08903

Sams—Howard W. Sams & Company, Inc., Publishers, 4300 West 62nd Street, Indianapolis IN 46268

Sampson Low, Marston—(dist. by Ginn and Company)

Saunders—W. B. Saunders Company, (subsidiary of Columbia Broadcasting System) West Washington Square, Philadelphia PA 19105

Scott—Scott, Foresman and Company, 1900 West Lake Avenue, Glenview IL 60025

Scribner's—Charles Scribner's Sons, 597 Fifth Avenue, New York NY 10017

Sheridan—Sheridan House, Inc., 257 Park Avenue South, New York NY 10010

Shorewood—Shorewood Publishers, Inc., 724 Fifth Avenue, New York NY 10019

Signet—New American Library of World Literature, Inc., 1301 Avenue of Americas, New York NY 10019

Silver Burdett—Silver Burdett Company, 250 James Street, Morristown NJ 07960

Simon & Schuster—Simon & Schuster, Inc., 630 Fifth Avenue, New York NY 10020

Singer—The L. W. Singer Company, Inc., 201 East 50th Street, New York NY 10022

Sloane—Distributed by William Morrow & Company, Inc., 105 Madison Avenue, New York NY 10016

Peter Smith—Peter Smith, 6 Lexington Avenue, Gloucester MA 01930

Smithsonian—Smithsonian Institution Press, Washington DC 20560

Southern Illinois U.—Southern Illinois University Press, Carbondale IL 62901

Stackpole—Stackpole Books, Cameron and Keller Streets, Harrisburg PA 17105

Sterling—Sterling Publishing Company, Inc., 419 Park Avenue South, New York NY 10016

St. Martin's—St. Martin's Press, Inc., 175 Fifth Avenue, New York NY 10010

Taplinger—Taplinger Publishing Company, Inc., 29 East Tenth Street, New York NY 10003

Thomas—Charles C. Thomas, Publisher, 301–27 East Lawrence Avenue, Springfield IL 62703

Time, Inc.—Time-Life Books, A Division of Time Inc., Time & Life Building, Rockefeller Center, New York NY 10020 (trade eds. dist. by Little, Brown & Co., library eds. dist. by Silver Burdett Co.)

Universe—Universe Books, 381 Park Avenue South, New York NY 10016

U. of Chicago—University of Chicago Press, 5750 Ellis Avenue, Chicago IL 60637

U. of Kentucky—University Press of Kentucky, Lexington KY 40506

U. of Michigan—University of Michigan Press, Ann Arbor MI 48106

U. of Minnesota—University of Minnesota Press, 2037 University Avenue, S.E., Minneapolis MN 55455

U. of Missouri—University of Missouri Press, Columbia MO 65201

U. of Oklahoma—University of Oklahoma Press, 1005 Asp Avenue, Norman OK 73069

U. of Texas—University of Texas Press, Box 7819, University Station, Austin TX 78712

U. of Toronto—University of Toronto Press, St. George Campus, Toronto 181, Ont.

U. of Washington—University of Washington Press, Seattle WA 98015

U. S. Bureau of Mines—(Order from Supt. of Documents, U. S. Government Printing Office, Washington DC 20402)

U. S. G.P.O.—U. S. Government Printing Office, Superintendent of Documents, Washington DC 20402

Vanguard—Vanguard Press, Inc., 424 Madison Avenue, New York NY 10017

Van Nostrand—Van Nostrand Reinhold Company, 450 West 33rd Street, New York NY 10001

Viking—The Viking Press, Inc., 625 Madison Avenue, New York NY 10022

Vintage—Random House, Inc., 457 Madison Avenue, New York NY 10022

Wadsworth—Wadsworth Publishing Company, Belmont CA 94022

Walck—Henry Z. Walck, Inc., 19 Union Square West, New York NY 10003

Walker—Walker and Company, 720 Fifth Avenue, New York NY 10019

Warne—Frederick Warne & Company, Inc., 101 Fifth Avenue, New York NY 10003

Washburn—Ives Washburn, Inc., 750 Third Avenue, New York NY 10017

Watts—Franklin Watts, Inc., 575 Lexington Avenue, New York NY 10022

Wiley—John Wiley and Sons, Inc., 605 Third Avenue, New York NY 10016

Williams & Wilkins—The Williams & Wilkins Company, 428 East Preston Street, Baltimore MD 21202

Wilson—H. W. Wilson Company, 950 University Avenue, Bronx NY 10452

World—The World Publishing Company, 110 East 59th Street, New York NY 10022

Worth—Worth Publishers, Inc., 70 Fifth Avenue, New York NY 10011

Yale—Yale University Press, 149 York Street, New Haven CT 06511